博文视点云原生精品丛书

Kubernetes 权威指南

第4版

从Docker到Kubernetes实践全接触

龚 正　吴治辉
崔秀龙　闫健勇　编著

电子工业出版社
Publishing House of Electronics Industry
北京·BEIJING

内 容 简 介

Kubernetes 是由谷歌开源的 Docker 容器集群管理系统，为容器化的应用提供了资源调度、部署运行、服务发现、扩容及缩容等一整套功能。本书从架构师、开发人员和运维人员的角度，阐述了 Kubernetes 的基本概念、实践指南、核心原理、开发指导、运维指南、新特性演进等内容，图文并茂、内容丰富、由浅入深、讲解全面；并围绕在生产环境中可能出现的问题，给出了大量的典型案例，比如安全配置方案、网络方案、共享存储方案、高可用方案及 Trouble Shooting 技巧等，有很强的实战指导意义。本书内容随着 Kubernetes 的版本更新不断完善，目前涵盖了 Kubernetes 从 1.0 到 1.14 版本的主要特性，努力为 Kubernetes 用户提供全方位的 Kubernetes 技术指南。本书源码已上传至 GitHub 的 kubeguide/K8sDefinitiveGuide-V4-Sourcecode 目录，可自行下载本书源码进行练习。

无论是对于软件工程师、测试工程师、运维工程师、软件架构师、技术经理，还是对于资深 IT 人士，本书都极具参考价值。

未经许可，不得以任何方式复制或抄袭本书之部分或全部内容。
版权所有，侵权必究。

图书在版编目（CIP）数据

Kubernetes 权威指南：从 Docker 到 Kubernetes 实践全接触 / 龚正等编著. —4 版. —北京：电子工业出版社，2019.6
（博文视点云原生精品丛书）
ISBN 978-7-121-36235-4

Ⅰ.①K… Ⅱ.①龚… Ⅲ.①Linux 操作系统－程序设计－指南 Ⅳ.①TP316.85-62

中国版本图书馆 CIP 数据核字（2019）第 060814 号

责任编辑：张国霞
印　　刷：三河市良远印务有限公司
装　　订：三河市良远印务有限公司
出版发行：电子工业出版社
　　　　　北京市海淀区万寿路 173 信箱　　邮编 100036
开　　本：787×980　1/16　印张：51.5　字数：1072 千字
版　　次：2016 年 1 月第 1 版
　　　　　2019 年 6 月第 4 版
印　　次：2021 年 4 月第 10 次印刷
印　　数：31501～34500 册　　定价：168.00 元

凡所购买电子工业出版社图书有缺损问题，请向购买书店调换。若书店售缺，请与本社发行部联系，联系及邮购电话：（010）88254888，88258888。
质量投诉请发邮件至 zlts@phei.com.cn，盗版侵权举报请发邮件至 dbqq@phei.com.cn。
本书咨询联系方式：010-51260888-819，faq@phei.com.cn。

推荐序

经过作者们多年的实践经验积累及近一年的精心准备，本书终于与我们见面了。我有幸作为首批读者，提前见证和学习了在云时代引领业界技术方向的 Kubernetes 和 Docker 的最新动态。

从内容上讲，本书从一个开发者的角度去理解、分析和解决问题：从基础入门到架构原理，从运行机制到开发源码，再从系统运维到应用实践，讲解全面。本书图文并茂，内容丰富，由浅入深，对基本原理阐述清晰，对系统架构分析透彻，对实践经验讲解深刻。

我认为本书值得推荐的原因有以下几点。

首先，作者的所有观点和经验，均是在多年建设、维护大型应用系统的过程中积累形成的。例如，读者通过学习书中的 Kubernetes 开发指南、集群管理等章节的内容，不仅可以直接提高开发技能，还可以解决在实践过程中经常遇到的各种关键问题。书中的这些内容具有很高的借鉴和推广意义。

其次，通过大量的实例操作和详尽的源码解析，本书可以帮助读者深刻理解 Kubernetes 的各种概念。例如，书中介绍了使用 Java 程序访问 Kubernetes API 的几种方法，读者参照其中的案例，只要稍做修改，再结合实际的应用需求，就可以将这些方法用于正在开发的项目中，达到事半功倍的效果，对有一定 Java 基础的专业人士快速学习 Kubernetes 的各种细节和实践操作十分有利。

再次，为了让初学者快速入门，本书配备了即时在线交流工具和专业后台技术支持团队。如果你在开发和应用的过程中遇到各类相关问题，均可直接联系该团队的开发支持专家。

最后，我们可以看到，容器化技术已经成为计算模型演化的一个开端，Kubernetes 作为谷歌开源的 Docker 容器集群管理技术，在这场新的技术革命中扮演着重要的角色。

Kubernetes正在被众多知名公司及企业采用，例如Google、VMware、CoreOS、腾讯、京东等，因此，Kubernetes站在了容器新技术变革的浪潮之巅，将具有不可预估的发展前景和商业价值。

无论你是架构师、开发者、运维人员，还是对容器技术比较好奇的读者，本书都是一本不可多得的带你从入门到进阶的Kubernetes精品书，值得阅读！

初瑞

中国移动业务支撑中心高级经理

自 序

本书第 1 版出版于 2016 年，几年过去，Kubernetes 已从一个新生事物发展为一个影响全球 IT 技术的基础设施平台，也推动了云原生应用、微服务架构、Service Mesh 等热门技术的普及和落地。现在，Kubernetes 已经成为明星项目，其开源项目拥有超过两万名贡献者，成为开源历史上发展速度超快的项目之一。

在这几年里：

Kubernetes 背后的重要开源公司 RedHat 被 IBM 大手笔收购，使 RedHat 基于 Kubernetes 架构的先进 PaaS 平台——OpenShift 成为 IBM 在云计算基础设施中的重要筹码；

Kubernetes 的两位核心创始人 Joe Beda 和 Craig McLuckie 所创立的提供 Kubernetes 咨询和技术支持的初创公司 Heptio 也被虚拟化领域的巨头 VMware 收购；

Oracle 收购了丹麦的一家初创公司 Wercker，然后开发了 Click2Kube，这是面向 Oracle 裸机云（Oracle Bare Metal Cloud）的一键式 Kubernetes 集群安装工具；

世界 500 强中的一些大型企业也决定以 Kubernetes 为基础重构内部 IT 平台架构，大数据系统的一些用户也在努力将其生产系统从庞大的大数据专有技术栈中剥离出来靠拢 Kubernetes。

Kubernetes 是将"一切以服务（Service）为中心，一切围绕服务运转"作为指导思想的创新型产品，这是它的一个亮点。它的功能和架构设计自始至终地遵循了这一指导思想，构建在 Kubernetes 上的系统不仅可以独立运行在物理机、虚拟机集群或者企业私有云上，也可以被托管在公有云上。

Kubernetes 的另一个亮点是自动化。在 Kubernetes 的解决方案中，一个服务可以自我扩展、自我诊断，并且容易升级，在收到服务扩容的请求后，Kubernetes 会触发调度流程，最终在选定的目标节点上启动相应数量的服务实例副本，这些服务实例副本在启动成功后会自动加入负载均衡器中并生效，整个过程无须额外的人工操作。另外，Kubernetes 会定时巡查每个服务的所有实例的可用性，确保服务实例的数量始终保持为预期的数量，当它发现某个实例不可用时，会自动重启该实例或者在其他节点上重新调度、运行一个新实例，这样，一个复杂的过程无须人工干预即可全部自动完成。试想一下，如果一个包括几十个节点且运行着几万个容器的复杂系统，其负载均衡、故障检测和故障修复等都需要人工介入进行处理，其工作量将多大。

通常，我们会把 Kubernetes 看作 Docker 的上层架构，就好像 Java 与 J2EE 的关系一样：J2EE 是以 Java 为基础的企业级软件架构，Kubernetes 则以 Docker 为基础打造了一个云计算时代的全新分布式系统架构。但 Kubernetes 与 Docker 之间还存在着更为复杂的关系，从表面上看，似乎 Kubernetes 离不开 Docker，但实际上在 Kubernetes 的架构里，Docker 只是其目前支持的两种底层容器技术之一，另一种容器技术则是 Rocket，Rocket 为 CoreOS 推出的竞争产品。

Kubernetes 之所以同时支持 Docker 和 Rocket 这两种互相竞争的容器技术，是有深刻的历史原因的。快速发展的 Docker 打败了谷歌名噪一时的开源容器技术 lmctfy，并迅速风靡世界。但是，作为一个已经对全球 IT 公司产生重要影响的技术，Docker 容器标准的制定不可能被任何一个公司主导。于是，CoreOS 推出了与 Docker 抗衡的开源容器项目 Rocket，动员一些知名 IT 公司一起主导容器技术的标准化，并与谷歌共同发起基于 CoreOS+ Rocket+Kubernetes 的新项目 Tectonic，使容器技术分裂态势加剧。最后，Linux 基金会于 2015 年 6 月宣布成立开放容器技术项目（Open Container Project），谷歌、CoreOS 及 Docker 都加入了该项目。OCP 项目成立后，Docker 公司放弃了自己的独家控制权，Docker 容器格式也被 OCP 采纳为新标准的基础，Docker 负责起草 OCP 草案规范的初稿文档，并提交自己的容器执行引擎的源码作为 OCP 项目的启动资源。

2015 年 7 月，谷歌正式宣布加入 OpenStack 阵营，其目标是确保 Linux 容器及其关联的容器管理技术 Kubernetes 能够被 OpenStack 生态圈所接纳，这也意味着对数据中心控制平面的争夺已经结束，以容器为代表的应用形态与以虚拟化为代表的系统形态将会完美融合于 OpenStack 之上，并与软件定义网络和软件定义存储一起主导下一代数据中心。

谷歌凭借着几十年大规模容器使用的丰富经验，步步为营，先是祭出 Kubernetes 这个神器，然后掌控了容器技术的制定标准，最后入驻 OpenStack 阵营全力支持 Kubernetes 的发展。可以预测，Kubernetes 的影响力可能超过十年，所以，我们每个 IT 人都有理由重视这门新技术。

谁能比别人领先一步掌握新技术，谁就能在竞争中赢得先机。慧与中国通信和媒体解决方案领域的资深专家团一起分工协作、并行研究，并废寝忘食地合力撰写，才促成了这部巨著的出版。经过这些年的高速发展，Kubernetes 先后发布了十几个大版本，每个版本都带来了大量的新特性，能够处理的应用场景也越来越丰富。本书遵循从入门到精通的学习路线，涵盖了入门、安装指南、实践指南、核心原理、开发指南、运维指南、新特性演进等内容，内容翔实、图文并茂，几乎囊括了 Kubernetes 当前主流版本的方方面面，无论是对于软件工程师、测试工程师、运维工程师、软件架构师、技术经理，还是对于资深 IT 人士，本书都极具参考价值。

吴治辉
HPE 资深架构师

读者服务

轻松注册成为博文视点社区用户（www.broadview.com.cn），扫码直达本书页面。

- **提交勘误**：您对书中内容的修改意见可在 提交勘误 处提交，若被采纳，将获赠博文视点社区积分（在您购买电子书时，积分可用来抵扣相应金额）。
- **交流互动**：在页面下方 读者评论 处留下您的疑问或观点，与我们和其他读者一同学习交流。

页面入口：http://www.broadview.com.cn/36235

目　录

第 1 章　Kubernetes 入门 1
1.1　Kubernetes 是什么 2
1.2　为什么要用 Kubernetes 5
1.3　从一个简单的例子开始 6
1.3.1　环境准备 7
1.3.2　启动 MySQL 服务 7
1.3.3　启动 Tomcat 应用 10
1.3.4　通过浏览器访问网页 12
1.4　Kubernetes 的基本概念和术语 13
1.4.1　Master 16
1.4.2　Node 16
1.4.3　Pod 19
1.4.4　Label 24
1.4.5　Replication Controller 28
1.4.6　Deployment 31
1.4.7　Horizontal Pod Autoscaler 34
1.4.8　StatefulSet 36
1.4.9　Service 37
1.4.10　Job 45
1.4.11　Volume 45
1.4.12　Persistent Volume 49
1.4.13　Namespace 51
1.4.14　Annotation 52

| | 1.4.15 | ConfigMap | 53 |
| | 1.4.16 | 小结 | 54 |

第 2 章 Kubernetes 安装配置指南 ... 55

- 2.1 系统要求 ... 56
- 2.2 使用 kubeadm 工具快速安装 Kubernetes 集群 ... 57
  - 2.2.1 安装 kubeadm 和相关工具 ... 57
  - 2.2.2 kubeadm config ... 58
  - 2.2.3 下载 Kubernetes 的相关镜像 ... 59
  - 2.2.4 运行 kubeadm init 命令安装 Master ... 59
  - 2.2.5 安装 Node，加入集群 ... 61
  - 2.2.6 安装网络插件 ... 62
  - 2.2.7 验证 Kubernetes 集群是否安装完成 ... 63
- 2.3 以二进制文件方式安装 Kubernetes 集群 ... 64
  - 2.3.1 Master 上的 etcd、kube-apiserver、kube-controller-manager、kube-scheduler 服务 ... 66
  - 2.3.2 Node 上的 kubelet、kube-proxy 服务 ... 71
- 2.4 Kubernetes 集群的安全设置 ... 73
  - 2.4.1 基于 CA 签名的双向数字证书认证方式 ... 73
  - 2.4.2 基于 HTTP Base 或 Token 的简单认证方式 ... 78
- 2.5 Kubernetes 集群的网络配置 ... 80
- 2.6 内网中的 Kubernetes 相关配置 ... 80
  - 2.6.1 Docker Private Registry（私有 Docker 镜像库） ... 80
  - 2.6.2 kubelet 配置 ... 81
- 2.7 Kubernetes 的版本升级 ... 81
  - 2.7.1 二进制升级 ... 81
  - 2.7.2 使用 kubeadm 进行集群升级 ... 82
- 2.8 Kubernetes 核心服务配置详解 ... 84
  - 2.8.1 公共配置参数 ... 84
  - 2.8.2 kube-apiserver 启动参数 ... 85
  - 2.8.3 kube-controller-manager 启动参数 ... 97
  - 2.8.4 kube-scheduler 启动参数 ... 107

  2.8.5 kubelet 启动参数 113
  2.8.6 kube-proxy 启动参数 128
 2.9 CRI（容器运行时接口）详解 132
  2.9.1 CRI 概述 132
  2.9.2 CRI 的主要组件 133
  2.9.3 Pod 和容器的生命周期管理 133
  2.9.4 面向容器级别的设计思路 135
  2.9.5 尝试使用新的 Docker-CRI 来创建容器 136
  2.9.6 CRI 的进展 137
 2.10 kubectl 命令行工具用法详解 137
  2.10.1 kubectl 用法概述 137
  2.10.2 kubectl 子命令详解 139
  2.10.3 kubectl 参数列表 142
  2.10.4 kubectl 输出格式 143
  2.10.5 kubectl 操作示例 145

第 3 章 深入掌握 Pod 149
 3.1 Pod 定义详解 150
 3.2 Pod 的基本用法 156
 3.3 静态 Pod 161
 3.4 Pod 容器共享 Volume 162
 3.5 Pod 的配置管理 165
  3.5.1 ConfigMap 概述 165
  3.5.2 创建 ConfigMap 资源对象 165
  3.5.3 在 Pod 中使用 ConfigMap 173
  3.5.4 使用 ConfigMap 的限制条件 179
 3.6 在容器内获取 Pod 信息（Downward API） 180
  3.6.1 环境变量方式：将 Pod 信息注入为环境变量 180
  3.6.2 环境变量方式：将容器资源信息注入为环境变量 182
  3.6.3 Volume 挂载方式 184
 3.7 Pod 生命周期和重启策略 186
 3.8 Pod 健康检查和服务可用性检查 187

- 3.9 玩转 Pod 调度 190
  - 3.9.1 Deployment 或 RC：全自动调度 193
  - 3.9.2 NodeSelector：定向调度 194
  - 3.9.3 NodeAffinity：Node 亲和性调度 197
  - 3.9.4 PodAffinity：Pod 亲和与互斥调度策略 198
  - 3.9.5 Taints 和 Tolerations（污点和容忍）.. 202
  - 3.9.6 Pod Priority Preemption：Pod 优先级调度 206
  - 3.9.7 DaemonSet：在每个 Node 上都调度一个 Pod 209
  - 3.9.8 Job：批处理调度 211
  - 3.9.9 Cronjob：定时任务 215
  - 3.9.10 自定义调度器 219
- 3.10 Init Container（初始化容器）... 220
- 3.11 Pod 的升级和回滚 224
  - 3.11.1 Deployment 的升级 225
  - 3.11.2 Deployment 的回滚 231
  - 3.11.3 暂停和恢复 Deployment 的部署操作，以完成复杂的修改 234
  - 3.11.4 使用 kubectl rolling-update 命令完成 RC 的滚动升级 236
  - 3.11.5 其他管理对象的更新策略 239
- 3.12 Pod 的扩缩容 240
  - 3.12.1 手动扩缩容机制 240
  - 3.12.2 自动扩缩容机制 241
- 3.13 使用 StatefulSet 搭建 MongoDB 集群 264
  - 3.13.1 前提条件 264
  - 3.13.2 创建 StatefulSet 265
  - 3.13.3 查看 MongoDB 集群的状态 269
  - 3.13.4 StatefulSet 的常见应用场景 271

第 4 章 深入掌握 Service 276
- 4.1 Service 定义详解 277
- 4.2 Service 的基本用法 279
  - 4.2.1 多端口 Service 282
  - 4.2.2 外部服务 Service 283

4.3 Headless Service 284
    4.3.1 自定义 SeedProvider 285
    4.3.2 通过 Service 动态查找 Pod 286
    4.3.3 Cassandra 集群中新节点的自动添加 289
4.4 从集群外部访问 Pod 或 Service 291
    4.4.1 将容器应用的端口号映射到物理机 291
    4.4.2 将 Service 的端口号映射到物理机 292
4.5 DNS 服务搭建和配置指南 294
    4.5.1 在创建 DNS 服务之前修改每个 Node 上 kubelet 的启动参数 296
    4.5.2 创建 CoreDNS 应用 297
    4.5.3 服务名的 DNS 解析 301
    4.5.4 CoreDNS 的配置说明 302
    4.5.5 Pod 级别的 DNS 配置说明 304
4.6 Ingress：HTTP 7 层路由机制 306
    4.6.1 创建 Ingress Controller 和默认的 backend 服务 307
    4.6.2 定义 Ingress 策略 311
    4.6.3 客户端访问 http://mywebsite.com/demo 313
    4.6.4 Ingress 的策略配置技巧 316
    4.6.5 Ingress 的 TLS 安全设置 320

第 5 章 核心组件运行机制 326

5.1 Kubernetes API Server 原理解析 327
    5.1.1 Kubernetes API Server 概述 327
    5.1.2 API Server 架构解析 330
    5.1.3 独特的 Kubernetes Proxy API 接口 334
    5.1.4 集群功能模块之间的通信 336
5.2 Controller Manager 原理解析 337
    5.2.1 Replication Controller 338
    5.2.2 Node Controller 339
    5.2.3 ResourceQuota Controller 341
    5.2.4 Namespace Controller 343
    5.2.5 Service Controller 与 Endpoints Controller 343

5.3		Scheduler 原理解析	344
5.4		kubelet 运行机制解析	348
	5.4.1	节点管理	349
	5.4.2	Pod 管理	349
	5.4.3	容器健康检查	351
	5.4.4	cAdvisor 资源监控	352
5.5		kube-proxy 运行机制解析	354

第 6 章 深入分析集群安全机制 358

6.1		API Server 认证管理	359
6.2		API Server 授权管理	361
	6.2.1	ABAC 授权模式详解	362
	6.2.2	Webhook 授权模式详解	365
	6.2.3	RBAC 授权模式详解	368
6.3		Admission Control	384
6.4		Service Account	388
6.5		Secret 私密凭据	393
6.6		Pod 的安全策略配置	396
	6.6.1	PodSecurityPolicy 的工作机制	397
	6.6.2	PodSecurityPolicy 配置详解	399
	6.6.3	Pod 的安全设置详解	406

第 7 章 网络原理 410

7.1		Kubernetes 网络模型	411
7.2		Docker 网络基础	413
	7.2.1	网络命名空间	413
	7.2.2	Veth 设备对	416
	7.2.3	网桥	419
	7.2.4	iptables 和 Netfilter	421
	7.2.5	路由	424
7.3		Docker 的网络实现	426
7.4		Kubernetes 的网络实现	435
	7.4.1	容器到容器的通信	435

   7.4.2 Pod 之间的通信 436
 7.5 Pod 和 Service 网络实战 439
 7.6 CNI 网络模型 454
   7.6.1 CNM 模型 454
   7.6.2 CNI 模型 455
   7.6.3 在 Kubernetes 中使用网络插件 467
 7.7 Kubernetes 网络策略 467
   7.7.1 网络策略配置说明 468
   7.7.2 在 Namespace 级别设置默认的网络策略 470
   7.7.3 NetworkPolicy 的发展 472
 7.8 开源的网络组件 472
   7.8.1 Flannel 472
   7.8.2 Open vSwitch 477
   7.8.3 直接路由 483
   7.8.4 Calico 容器网络和网络策略实战 486

第 8 章 共享存储原理 508

 8.1 共享存储机制概述 509
 8.2 PV 详解 510
   8.2.1 PV 的关键配置参数 511
   8.2.2 PV 生命周期的各个阶段 515
 8.3 PVC 详解 516
 8.4 PV 和 PVC 的生命周期 518
   8.4.1 资源供应 518
   8.4.2 资源绑定 519
   8.4.3 资源使用 519
   8.4.4 资源释放 519
   8.4.5 资源回收 519
 8.5 StorageClass 详解 521
   8.5.1 StorageClass 的关键配置参数 521
   8.5.2 设置默认的 StorageClass 524

8.6 动态存储管理实战：GlusterFS 524
8.6.1 准备工作 525
8.6.2 创建 GlusterFS 管理服务容器集群 525
8.6.3 创建 Heketi 服务 528
8.6.4 为 Heketi 设置 GlusterFS 集群 530
8.6.5 定义 StorageClass 533
8.6.6 定义 PVC 534
8.6.7 Pod 使用 PVC 的存储资源 536
8.7 CSI 存储机制详解 537
8.7.1 CSI 的设计背景 538
8.7.2 CSI 存储插件的关键组件和部署架构 539
8.7.3 CSI 存储插件的使用示例 540
8.7.4 CSI 的发展 556

第 9 章 Kubernetes 开发指南 560
9.1 REST 简述 561
9.2 Kubernetes API 详解 563
9.2.1 Kubernetes API 概述 563
9.2.2 Kubernetes API 版本的演进策略 570
9.2.3 API Groups（API 组） 571
9.2.4 API REST 的方法说明 573
9.2.5 API Server 响应说明 575
9.3 使用 Java 程序访问 Kubernetes API 577
9.3.1 Jersey 577
9.3.2 Fabric8 590
9.3.3 使用说明 591
9.3.4 其他客户端库 615
9.4 Kubernetes API 的扩展 616
9.4.1 使用 CRD 扩展 API 资源 617
9.4.2 使用 API 聚合机制扩展 API 资源 626

第 10 章 Kubernetes 集群管理 635

10.1 Node 的管理 636
10.1.1 Node 的隔离与恢复 636
10.1.2 Node 的扩容 637

10.2 更新资源对象的 Label 638

10.3 Namespace：集群环境共享与隔离 639
10.3.1 创建 Namespace 639
10.3.2 定义 Context（运行环境） 640
10.3.3 设置工作组在特定 Context 环境下工作 641

10.4 Kubernetes 资源管理 643
10.4.1 计算资源管理 645
10.4.2 资源配置范围管理（LimitRange） 655
10.4.3 资源服务质量管理（Resource QoS） 662
10.4.4 资源配额管理（Resource Quotas） 670
10.4.5 ResourceQuota 和 LimitRange 实践 676
10.4.6 资源管理总结 685

10.5 资源紧缺时的 Pod 驱逐机制 686
10.5.1 驱逐策略 686
10.5.2 驱逐信号 686
10.5.3 驱逐阈值 688
10.5.4 驱逐监控频率 689
10.5.5 节点的状况 689
10.5.6 节点状况的抖动 690
10.5.7 回收 Node 级别的资源 690
10.5.8 驱逐用户的 Pod 691
10.5.9 资源最少回收量 692
10.5.10 节点资源紧缺情况下的系统行为 692
10.5.11 可调度的资源和驱逐策略实践 694
10.5.12 现阶段的问题 694

10.6 Pod Disruption Budget（主动驱逐保护） 695

10.7 Kubernetes 集群的高可用部署方案 697
10.7.1 手工方式的高可用部署方案 698

- 10.7.2 使用kubeadm的高可用部署方案 ... 709
- 10.8 Kubernetes集群监控 ... 717
  - 10.8.1 通过Metrics Server监控Pod和Node的CPU和内存资源使用数据 ... 717
  - 10.8.2 Prometheus+Grafana集群性能监控平台搭建 ... 720
- 10.9 集群统一日志管理 ... 732
  - 10.9.1 系统部署架构 ... 733
  - 10.9.2 创建Elasticsearch RC和Service ... 733
  - 10.9.3 在每个Node上启动Fluentd ... 736
  - 10.9.4 运行Kibana ... 738
- 10.10 Kubernetes的审计机制 ... 742
- 10.11 使用Web UI（Dashboard）管理集群 ... 746
- 10.12 Helm：Kubernetes应用包管理工具 ... 750
  - 10.12.1 Helm概述 ... 750
  - 10.12.2 Helm的主要概念 ... 751
  - 10.12.3 安装Helm ... 751
  - 10.12.4 Helm的常见用法 ... 752
  - 10.12.5 --set的格式和限制 ... 756
  - 10.12.6 更多的安装方法 ... 757
  - 10.12.7 helm upgrade和helm rollback：应用的更新或回滚 ... 757
  - 10.12.8 helm install/upgrade/rollback命令的常用参数 ... 758
  - 10.12.9 helm delete：删除一个Release ... 759
  - 10.12.10 helm repo：仓库的使用 ... 759
  - 10.12.11 自定义Chart ... 759
  - 10.12.12 对Chart目录结构和配置文件的说明 ... 759
  - 10.12.13 对Chart.yaml文件的说明 ... 760
  - 10.12.14 快速制作自定义的Chart ... 761
  - 10.12.15 搭建私有Repository ... 761

第11章 Trouble Shooting指导 ... 763

- 11.1 查看系统Event ... 764
- 11.2 查看容器日志 ... 766
- 11.3 查看Kubernetes服务日志 ... 767

11.4 常见问题 769
　　11.4.1 由于无法下载 pause 镜像导致 Pod 一直处于 Pending 状态 769
　　11.4.2 Pod 创建成功，但 RESTARTS 数量持续增加 771
　　11.4.3 通过服务名无法访问服务 772
11.5 寻求帮助 773

第 12 章　Kubernetes 开发中的新功能 777

12.1 对 Windows 容器的支持 778
　　12.1.1 Windows Node 部署 778
　　12.1.2 Windows 容器支持的 Kubernetes 特性和发展趋势 790
12.2 对 GPU 的支持 791
　　12.2.1 环境准备 792
　　12.2.2 在容器中使用 GPU 资源 795
　　12.2.3 发展趋势 797
12.3 Pod 的垂直扩缩容 797
　　12.3.1 前提条件 798
　　12.3.2 安装 Vertical Pod Autoscaler 798
　　12.3.3 为 Pod 设置垂直扩缩容 798
　　12.3.4 注意事项 800
12.4 Kubernetes 的演进路线和开发模式 801

第 1 章
Kubernetes 入门

1.1　Kubernetes 是什么

Kubernetes 是什么？

首先，它是一个全新的基于容器技术的分布式架构领先方案。这个方案虽然还很新，但它是谷歌十几年以来大规模应用容器技术的经验积累和升华的重要成果。确切地说，Kubernetes 是谷歌严格保密十几年的秘密武器——Borg 的一个开源版本。Borg 是谷歌的一个久负盛名的内部使用的大规模集群管理系统，它基于容器技术，目的是实现资源管理的自动化，以及跨多个数据中心的资源利用率的最大化。十几年以来，谷歌一直通过 Borg 系统管理着数量庞大的应用程序集群。由于谷歌员工都签署了保密协议，即便离职也不能泄露 Borg 的内部设计，所以外界一直无法了解关于它的更多信息。直到 2015 年 4 月，传闻许久的 Borg 论文伴随 Kubernetes 的高调宣传被谷歌首次公开，大家才得以了解它的更多内幕。正是由于站在 Borg 这个前辈的肩膀上，汲取了 Borg 过去十年间的经验与教训，所以 Kubernetes 一经开源就一鸣惊人，并迅速称霸容器领域。

其次，如果我们的系统设计遵循了 Kubernetes 的设计思想，那么传统系统架构中那些和业务没有多大关系的底层代码或功能模块，都可以立刻从我们的视线中消失，我们不必再费心于负载均衡器的选型和部署实施问题，不必再考虑引入或自己开发一个复杂的服务治理框架，不必再头疼于服务监控和故障处理模块的开发。总之，使用 Kubernetes 提供的解决方案，我们不仅节省了不少于 30%的开发成本，还可以将精力更加集中于业务本身，而且由于 Kubernetes 提供了强大的自动化机制，所以系统后期的运维难度和运维成本大幅度降低。

然后，Kubernetes 是一个开放的开发平台。与 J2EE 不同，它不局限于任何一种语言，没有限定任何编程接口，所以不论是用 Java、Go、C++还是用 Python 编写的服务，都可以被映射为 Kubernetes 的 Service（服务），并通过标准的 TCP 通信协议进行交互。此外，Kubernetes 平台对现有的编程语言、编程框架、中间件没有任何侵入性，因此现有的系统也很容易改造升级并迁移到 Kubernetes 平台上。

最后，Kubernetes 是一个完备的分布式系统支撑平台。Kubernetes 具有完备的集群管理能力，包括多层次的安全防护和准入机制、多租户应用支撑能力、透明的服务注册和服务发现机制、内建的智能负载均衡器、强大的故障发现和自我修复能力、服务滚动升级和在线扩容能力、可扩展的资源自动调度机制，以及多粒度的资源配额管理能力。同时，

Kubernetes 提供了完善的管理工具，这些工具涵盖了包括开发、部署测试、运维监控在内的各个环节。因此，Kubernetes 是一个全新的基于容器技术的分布式架构解决方案，并且是一个一站式的完备的分布式系统开发和支撑平台。

在正式开始本章的 Hello World 之旅之前，我们首先要学习 Kubernetes 的一些基本知识，这样才能理解 Kubernetes 提供的解决方案。

在 Kubernetes 中，Service 是分布式集群架构的核心，一个 Service 对象拥有如下关键特征。

◎ 拥有唯一指定的名称（比如 mysql-server）。
◎ 拥有一个虚拟 IP（Cluster IP、Service IP 或 VIP）和端口号。
◎ 能够提供某种远程服务能力。
◎ 被映射到提供这种服务能力的一组容器应用上。

Service 的服务进程目前都基于 Socket 通信方式对外提供服务，比如 Redis、Memcache、MySQL、Web Server，或者是实现了某个具体业务的特定 TCP Server 进程。虽然一个 Service 通常由多个相关的服务进程提供服务，每个服务进程都有一个独立的 Endpoint（IP+Port）访问点，但 Kubernetes 能够让我们通过 Service（虚拟 Cluster IP +Service Port）连接到指定的 Service。有了 Kubernetes 内建的透明负载均衡和故障恢复机制，不管后端有多少服务进程，也不管某个服务进程是否由于发生故障而被重新部署到其他机器，都不会影响对服务的正常调用。更重要的是，这个 Service 本身一旦创建就不再变化，这意味着我们再也不用为 Kubernetes 集群中服务的 IP 地址变来变去的问题而头疼了。

容器提供了强大的隔离功能，所以有必要把为 Service 提供服务的这组进程放入容器中进行隔离。为此，Kubernetes 设计了 Pod 对象，将每个服务进程都包装到相应的 Pod 中，使其成为在 Pod 中运行的一个容器（Container）。为了建立 Service 和 Pod 间的关联关系，Kubernetes 首先给每个 Pod 都贴上一个标签（Label），给运行 MySQL 的 Pod 贴上 name=mysql 标签，给运行 PHP 的 Pod 贴上 name=php 标签，然后给相应的 Service 定义标签选择器（Label Selector），比如 MySQL Service 的标签选择器的选择条件为 name=mysql，意为该 Service 要作用于所有包含 name=mysql Label 的 Pod。这样一来，就巧妙解决了 Service 与 Pod 的关联问题。

这里先简单介绍 Pod 的概念。首先，Pod 运行在一个被称为节点（Node）的环境中，这个节点既可以是物理机，也可以是私有云或者公有云中的一个虚拟机，通常在一个节点上运行几百个 Pod；其次，在每个 Pod 中都运行着一个特殊的被称为 Pause 的容器，其他容器则为业务容器,这些业务容器共享 Pause 容器的网络栈和 Volume 挂载卷,因此它们之间的通信和数据交换更为高效,在设计时我们可以充分利用这一特性将一组密切相关的服务进程放入同一个 Pod 中;最后,需要注意的是,并不是每个 Pod 和它里面运行的容器都能被映射到一个 Service 上,只有提供服务(无论是对内还是对外)的那组 Pod 才会被映射为一个服务。

在集群管理方面,Kubernetes 将集群中的机器划分为一个 Master 和一些 Node。在 Master 上运行着集群管理相关的一组进程 kube-apiserver、kube-controller-manager 和 kube-scheduler,这些进程实现了整个集群的资源管理、Pod 调度、弹性伸缩、安全控制、系统监控和纠错等管理功能,并且都是自动完成的。Node 作为集群中的工作节点,运行真正的应用程序,在 Node 上 Kubernetes 管理的最小运行单元是 Pod。在 Node 上运行着 Kubernetes 的 kubelet、kube-proxy 服务进程,这些服务进程负责 Pod 的创建、启动、监控、重启、销毁,以及实现软件模式的负载均衡器。

最后,看看传统的 IT 系统中服务扩容和服务升级这两个难题,以及 Kubernetes 所提供的全新解决思路。服务的扩容涉及资源分配(选择哪个节点进行扩容)、实例部署和启动等环节,在一个复杂的业务系统中,这两个难题基本上靠人工一步步操作才得以解决,费时费力又难以保证实施质量。

在 Kubernetes 集群中,只需为需要扩容的 Service 关联的 Pod 创建一个 RC(Replication Controller),服务扩容以至服务升级等令人头疼的问题都迎刃而解。在一个 RC 定义文件中包括以下 3 个关键信息。

◎ 目标 Pod 的定义。
◎ 目标 Pod 需要运行的副本数量(Replicas)。
◎ 要监控的目标 Pod 的标签。

在创建好 RC(系统将自动创建好 Pod)后,Kubernetes 会通过在 RC 中定义的 Label 筛选出对应的 Pod 实例并实时监控其状态和数量,如果实例数量少于定义的副本数量,则会根据在 RC 中定义的 Pod 模板创建一个新的 Pod,然后将此 Pod 调度到合适的 Node 上启动运行,直到 Pod 实例的数量达到预定目标。这个过程完全是自动化的,无须人工干预。有了 RC,服务扩容就变成一个纯粹的简单数字游戏了,只需修改 RC 中的副本数量即可。后续的服务升级也将通过修改 RC 来自动完成。

以将在第 2 章中介绍的 PHP+Redis 留言板应用为例,只要为 PHP 留言板程序(frontend)创建一个有 3 个副本的 RC+Service,为 Redis 读写分离集群创建两个 RC:写节点

（redis-master）创建一个单副本的 RC+Service，读节点（redis-slaver）创建一个有两个副本的 RC+Service，就可以快速完成整个集群的搭建过程，是不是很简单？

1.2 为什么要用 Kubernetes

使用 Kubernetes 的理由很多，最重要的理由是，IT 行业从来都是由新技术驱动的。

当前，Docker 这门容器化技术已经被很多公司采用，从单机走向集群已成为必然，云计算的蓬勃发展正在加速这一进程。Kubernetes 作为当前被业界广泛认可和看好的基于 Docker 的大规模容器化分布式系统解决方案，得到了以谷歌为首的 IT 巨头们的大力宣传和持续推进。

2015 年，谷歌联合 20 多家公司一起建立了 CNCF(Cloud Native Computing Foundation，云原生计算基金会）开源组织来推广 Kubernetes，并由此开创了云原生应用（Cloud Native Application）的新时代。作为 CNCF "钦定"的官方云原生平台，Kubernetes 正在颠覆应用程序的开发方式。

如今，数百家厂商和技术社区共同构建了非常强大的云原生生态，市面上几乎所有提供云基础设施的公司都以原生形式将 Kubernetes 作为底层平台，可以预见，会有大量的新系统选择 Kubernetes，不论这些新系统是运行在企业的本地服务器上，还是被托管到公有云上。

使用 Kubernetes 会收获哪些好处呢？

首先，可以"轻装上阵"地开发复杂系统。以前需要很多人（其中不乏技术达人）一起分工协作才能设计、实现和运维的分布式系统，在采用 Kubernetes 解决方案之后，只需一个精悍的小团队就能轻松应对。在这个团队里，只需一名架构师负责系统中服务组件的架构设计，几名开发工程师负责业务代码的开发，一名系统兼运维工程师负责 Kubernetes 的部署和运维，因为 Kubernetes 已经帮我们做了很多。

其次，可以全面拥抱微服务架构。微服务架构的核心是将一个巨大的单体应用分解为很多小的互相连接的微服务，一个微服务可能由多个实例副本支撑，副本的数量可以随着系统的负荷变化进行调整。微服务架构使得每个服务都可以独立开发、升级和扩展，因此系统具备很高的稳定性和快速迭代能力，开发者也可以自由选择开发技术。谷歌、亚马逊、eBay、Netflix 等大型互联网公司都采用了微服务架构，谷歌更是将微服务架构的基础设施直接打包到Kubernetes解决方案中，让我们可以直接应用微服务架构解决复杂业务系统的架构问题。

再次，可以随时随地将系统整体"搬迁"到公有云上。Kubernetes最初的设计目标就是让用户的应用运行在谷歌自家的公有云GCE中，华为云（CCE）、阿里云（ACK）和腾讯云（TKE）先后宣布支持Kubernetes集群，未来会有更多的公有云及私有云支持Kubernetes。同时，在Kubernetes的架构方案中完全屏蔽了底层网络的细节，基于Service的虚拟IP地址（Cluster IP）的设计思路让架构与底层的硬件拓扑无关，我们无须改变运行期的配置文件，就能将系统从现有的物理机环境无缝迁移到公有云上。

然后，Kubernetes内在的服务弹性扩容机制可以让我们轻松应对突发流量。在服务高峰期，我们可以选择在公有云中快速扩容某些Service的实例副本以提升系统的吞吐量，这样不仅节省了公司的硬件投入，还大大改善了用户体验。中国铁路总公司的12306购票系统，在客流高峰期（如节假日）就租用了阿里云进行分流。

最后，Kubernetes系统架构超强的横向扩容能力可以让我们的竞争力大大提升。对于互联网公司来说，用户规模等价于资产，因此横向扩容能力是衡量互联网业务系统竞争力的关键指标。我们利用Kubernetes提供的工具，不用修改代码，就能将一个Kubernetes集群从只包含几个Node的小集群平滑扩展到拥有上百个Node的大集群，甚至可以在线完成集群扩容。只要微服务架构设计得合理，能够在多个云环境中进行弹性伸缩，系统就能够承受大量用户并发访问带来的巨大压力。

1.3 从一个简单的例子开始

考虑到Kubernetes提供的PHP+Redis留言板的Hello World例子对于绝大多数刚接触Kubernetes的人来说比较复杂，难以顺利上手和实践，所以在此将其替换成一个简单得多的Java Web应用例子，可以让新手快速上手和实践。

此Java Web应用的结构比较简单，是一个运行在Tomcat里的Web App，如图1.1所示，JSP页面通过JDBC直接访问MySQL数据库并展示数据。出于演示和简化的目的，只要程序正确连接到了数据库，就会自动完成对应的Table的创建与初始化数据的准备工作。所以，当我们通过浏览器访问此应用时，就会显示一个表格的页面，数据则来自数据库。

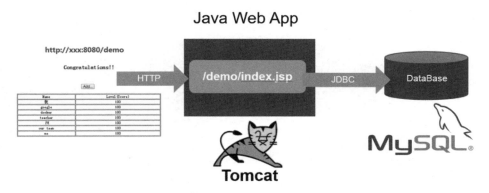

图 1.1 Java Web 应用的结构

此应用需要启动两个容器：Web App 容器和 MySQL 容器，并且 Web App 容器需要访问 MySQL 容器。在 Docker 时代，假设我们在一个宿主机上启动了这两个容器，就需要把 MySQL 容器的 IP 地址通过环境变量注入 Web App 容器里；同时，需要将 Web App 容器的 8080 端口映射到宿主机的 8080 端口，以便在外部访问。在本章的这个例子里，我们介绍在 Kubernetes 时代是如何达到这个目标的。

1.3.1 环境准备

首先，安装 Kubernetes 和下载相关镜像，本书建议采用 VirtualBox 或者 VMware Workstation 在本机虚拟一个 64 位的 CentOS 7 虚拟机作为学习环境。虚拟机采用 NAT 的网络模式以便连接外网，然后使用 kubeadm 快速安装一个 Kubernetes 集群（安装步骤详见 2.2 节的说明）。之后就可以在这个 Kubernetes 集群中进行练习了。

注：本书示例中的 Docker 镜像下载地址为 https://hub.docker.com/u/kubeguide/。

1.3.2 启动 MySQL 服务

首先，为 MySQL 服务创建一个 RC 定义文件 mysql-rc.yaml，下面给出了该文件的完整内容和解释：

```
apiVersion: v1
kind: ReplicationController        # 副本控制器 RC
metadata:
  name: mysql                      # RC 的名称，全局唯一
```

```yaml
spec:
  replicas: 1                              # Pod副本的期待数量
  selector:
    app: mysql                             # 符合目标的Pod拥有此标签
  template:                                # 根据此模板创建Pod的副本（实例）
    metadata:
      labels:
        app: mysql                         # Pod副本拥有的标签，对应RC的Selector
    spec:
      containers:                          # Pod内容器的定义部分
      - name: mysql                        # 容器的名称
        image: mysql                       # 容器对应的Docker Image
        ports:
        - containerPort: 3306              # 容器应用监听的端口号
        env:                               # 注入容器内的环境变量
        - name: MYSQL_ROOT_PASSWORD
          value: "123456"
```

以上 YAML 定义文件中的 kind 属性用来表明此资源对象的类型，比如这里的值为 ReplicationController，表示这是一个 RC；在 spec 一节中是 RC 的相关属性定义，比如 spec.selector 是 RC 的 Pod 标签选择器，即监控和管理拥有这些标签的 Pod 实例，确保在当前集群中始终有且仅有 replicas 个 Pod 实例在运行，这里设置 replicas=1，表示只能运行一个 MySQL Pod 实例。当在集群中运行的 Pod 数量少于 replicas 时，RC 会根据在 spec.template 一节中定义的 Pod 模板来生成一个新的 Pod 实例，spec.template.metadata. labels 指定了该 Pod 的标签，需要特别注意的是：这里的 labels 必须匹配之前的 spec.selector，否则此 RC 每创建一个无法匹配 Label 的 Pod，就会不停地尝试创建新的 Pod，陷入恶性循环中。

在创建好 mysql-rc.yaml 文件后，为了将它发布到 Kubernetes 集群中，我们在 Master 上执行命令：

```
# kubectl create -f mysql-rc.yaml
replicationcontroller "mysql" created
```

接下来，用 kubectl 命令查看刚刚创建的 RC：

```
# kubectl get rc
NAME      DESIRED   CURRENT   AGE
mysql     1         1         1m
```

查看 Pod 的创建情况时，可以运行下面的命令：

```
# kubectl get pods
NAME              READY       STATUS          RESTARTS       AGE
mysql-c95jc       1/1         Running         0              2m
```

我们看到一个名为 mysql-xxxxx 的 Pod 实例，这是 Kubernetes 根据 mysql 这个 RC 的定义自动创建的 Pod。由于 Pod 的调度和创建需要花费一定的时间，比如需要一定的时间来确定调度到哪个节点上，以及下载 Pod 里的容器镜像需要一段时间，所以我们一开始看到 Pod 的状态显示为 Pending。在 Pod 成功创建完成以后，状态最终会被更新为 Running。

我们通过 docker ps 指令查看正在运行的容器，发现提供 MySQL 服务的 Pod 容器已经创建并正常运行了，此外会发现 MySQL Pod 对应的容器还多创建了一个来自谷歌的 pause 容器，这就是 Pod 的 "根容器"，详见 1.4.3 节的说明。

```
# docker ps | grep mysql
   72ca992535b4     mysql
"docker-entrypoint.sh"       12 minutes ago      Up 12 minutes
k8s_mysql.86dc506e_mysql-c95jc_default_511d6705-5051-11e6-a9d8-000c29ed42c1_9f89
d0b4
   76c1790aad27             gcr.io/google_containers/pause-amd64:3.0
"/pause"                     12 minutes ago      Up 12 minutes
k8s_POD.16b20365_mysql-c95jc_default_511d6705-5051-11e6-a9d8-000c29ed42c1_28520a
ba
```

最后，创建一个与之关联的 Kubernetes Service——MySQL 的定义文件（文件名为 mysql-svc.yaml），完整的内容和解释如下：

```
apiVersion: v1
kind: Service                          # 表明是 Kubernetes Service
metadata:
  name: mysql                          # Service 的全局唯一名称
spec:
  ports:
    - port: 3306                       # Service 提供服务的端口号
  selector:                            # Service 对应的 Pod 拥有这里定义的标签
    app: mysql
```

其中，metadata.name 是 Service 的服务名（ServiceName）；port 属性则定义了 Service 的虚端口；spec.selector 确定了哪些 Pod 副本（实例）对应本服务。类似地，我们通过 kubectl create 命令创建 Service 对象。

运行 kubectl 命令，创建 Service：

```
# kubectl create -f mysql-svc.yaml
service "mysql" created
```

运行 kubectl 命令查看刚刚创建的 Service：

```
# kubectl get svc
NAME         CLUSTER-IP        EXTERNAL-IP    PORT(S)      AGE
mysql        169.169.253.143   <none>         3306/TCP     48s
```

可以发现，MySQL 服务被分配了一个值为 169.169.253.143 的 Cluster IP 地址。随后，Kubernetes 集群中其他新创建的 Pod 就可以通过 Service 的 Cluster IP+端口号 3306 来连接和访问它了。

通常，Cluster IP 是在 Service 创建后由 Kubernetes 系统自动分配的，其他 Pod 无法预先知道某个 Service 的 Cluster IP 地址，因此需要一个服务发现机制来找到这个服务。为此，最初时，Kubernetes 巧妙地使用了 Linux 环境变量（Environment Variable）来解决这个问题，后面会详细说明其机制。现在只需知道，根据 Service 的唯一名称，容器可以从环境变量中获取 Service 对应的 Cluster IP 地址和端口，从而发起 TCP/IP 连接请求。

1.3.3 启动 Tomcat 应用

上面定义和启动了 MySQL 服务，接下来采用同样的步骤完成 Tomcat 应用的启动过程。首先，创建对应的 RC 文件 myweb-rc.yaml，内容如下：

```
apiVersion: v1
kind: ReplicationController
metadata:
  name: myweb
spec:
  replicas: 2
  selector:
    app: myweb
  template:
    metadata:
      labels:
        app: myweb
    spec:
      containers:
        - name: myweb
          image: kubeguide/tomcat-app:v1
```

```
        ports:
        - containerPort: 8080
        env:
        - name: MYSQL_SERVICE_HOST
          value: 169.169.253.143
```

注意：在 Tomcat 容器内，应用将使用环境变量 MYSQL_SERVICE_HOST 的值连接 MySQL 服务。更安全可靠的用法是使用服务的名称 mysql，详见本章 Service 的概念和第 4 章的说明。运行下面的命令，完成 RC 的创建和验证工作：

```
#kubectl create -f myweb-rc.yaml
replicationcontroller "myweb" created

# kubectl get pods
NAME              READY    STATUS    RESTARTS   AGE
mysql-c95jc       1/1      Running   0          2h
myweb-g9pmm       1/1      Running   0          3s
```

最后，创建对应的 Service。以下是完整的 YAML 定义文件（myweb-svc.yaml）：

```
apiVersion: v1
kind: Service
metadata:
  name: myweb
spec:
  type: NodePort
  ports:
    - port: 8080
      nodePort: 30001
  selector:
    app: myweb
```

type=NodePort 和 nodePort=30001 的两个属性表明此 Service 开启了 NodePort 方式的外网访问模式。在 Kubernetes 集群之外，比如在本机的浏览器里，可以通过 30001 这个端口访问 myweb（对应到 8080 的虚端口上）。运行 kubectl create 命令进行创建：

```
# kubectl create -f myweb-svc.yaml
You have exposed your service on an external port on all nodes in your
cluster.  If you want to expose this service to the external internet, you may
need to set up firewall rules for the service port(s) (tcp:30001) to serve traffic.
See http://releases.k8s.io/release-1.3/docs/user-guide/services-firewalls.md
for more details.
service "myweb" created
```

我们看到上面有提示信息，意思是需要把 30001 这个端口在防火墙上打开，以便外部的访问能穿过防火墙。

运行 kubectl 命令，查看创建的 Service：

```
# kubectl get services
NAME              CLUSTER-IP          EXTERNAL-IP      PORT(S)      AGE
mysql             169.169.253.143     <none>           3306/TCP     2m
myweb             169.169.149.215     <nodes>          8080/TCP     1m
kubernetes        169.169.0.1         <none>           443/TCP      10m
```

至此，我们的第 1 个 Kubernetes 例子便搭建完成了，我们将在下一节中验证结果。

1.3.4 通过浏览器访问网页

经过上面的几个步骤，我们终于成功实现了 Kubernetes 上第 1 个例子的部署搭建工作。现在一起来见证成果吧！在你的笔记本上打开浏览器，输入 http://虚拟机 IP:30001/demo/。

比如虚拟机 IP 为 192.168.18.131（可以通过#ip a 命令进行查询），在浏览器里输入地址 http:// 192.168.18.131:30001/demo/后，可以看到如图 1.2 所示的网页界面。

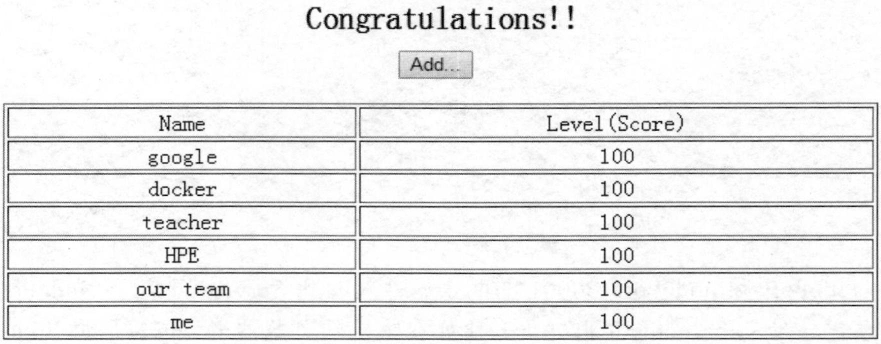

Congratulations!!

Add...

Name	Level(Score)
google	100
docker	100
teacher	100
HPE	100
our team	100
me	100

图 1.2　通过浏览器访问 Tomcat 应用

如果看不到这个网页界面，那么可能有几个原因，比如因为防火墙的问题无法访问 30001 端口，或者因为是通过代理上网的，浏览器错把虚拟机的 IP 地址当作远程地址了。可以在虚拟机上直接运行 curl 192.168.18.131:30001 来验证此端口能否被访问，如果还是不能访问，就肯定不是机器的问题了。

接下来，可以尝试单击"Add..."按钮添加一条记录并提交，如图 1.3 所示，在提交以后，数据就被写入 MySQL 数据库中了。

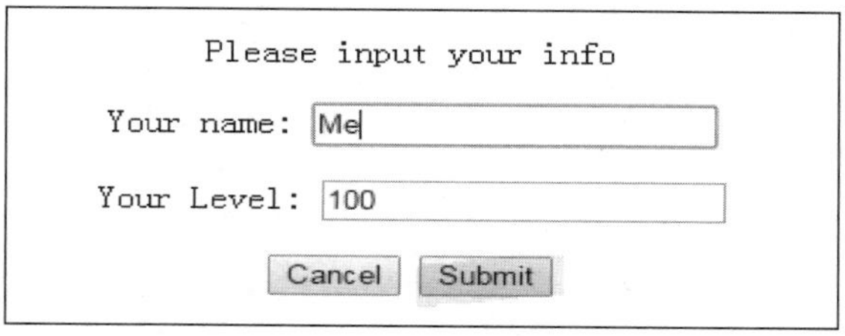

图 1.3　在留言板网页添加新的留言

至此，我们终于完成了 Kubernetes 上的 Tomcat 例子，这个例子并不是很复杂。我们也看到，相对于传统的分布式应用的部署方式，在 Kubernetes 之上我们仅仅通过一些很容易理解的配置文件和相关的简单命令就完成了对整个集群的部署，这让我们惊诧于 Kubernetes 的创新和强大。

下一节，我们将对 Kubernetes 中的基本概念和术语进行全面学习，在这之前，读者可以继续研究这个例子里的一些拓展内容，如下所述。

◎ 研究 RC、Service 等配置文件的格式。
◎ 熟悉 kubectl 的子命令。
◎ 手工停止某个 Service 对应的容器进程，然后观察有什么现象发生。
◎ 修改 RC 文件，改变副本数量，重新发布，观察结果。

1.4　Kubernetes 的基本概念和术语

Kubernetes 中的大部分概念如 Node、Pod、Replication Controller、Service 等都可以被看作一种资源对象，几乎所有资源对象都可以通过 Kubernetes 提供的 kubectl 工具（或者 API 编程调用）执行增、删、改、查等操作并将其保存在 etcd 中持久化存储。从这个角度来看，Kubernetes 其实是一个高度自动化的资源控制系统，它通过跟踪对比 etcd 库里保存的"资源期望状态"与当前环境中的"实际资源状态"的差异来实现自动控制和自动纠错的高级功能。

在声明一个 Kubernetes 资源对象的时候，需要注意一个关键属性：apiVersion。以下面的 Pod 声明为例，可以看到 Pod 这种资源对象归属于 v1 这个核心 API。

```yaml
apiVersion: v1
kind: Pod
metadata:
  name: myweb
  labels:
    name: myweb
spec:
  containers:
  - name: myweb
    image: kubeguide/tomcat-app:v1
    ports:
    - containerPort: 8080
```

Kubernetes 平台采用了"核心+外围扩展"的设计思路，在保持平台核心稳定的同时具备持续演进升级的优势。Kubernetes 大部分常见的核心资源对象都归属于 v1 这个核心 API，比如 Node、Pod、Service、Endpoints、Namespace、RC、PersistentVolume 等。在版本迭代过程中，Kubernetes 先后扩展了 extensions/v1beta1、apps/v1beta1、apps/v1beta2 等 API 组，而在 1.9 版本之后引入了 apps/v1 这个正式的扩展 API 组，正式淘汰（deprecated）了 extensions/v1beta1、apps/v1beta1、apps/v1beta2 这三个 API 组。

我们可以采用 YAML 或 JSON 格式声明（定义或创建）一个 Kubernetes 资源对象，每个资源对象都有自己的特定语法格式（可以理解为数据库中一个特定的表），但随着 Kubernetes 版本的持续升级，一些资源对象会不断引入新的属性。为了在不影响当前功能的情况下引入对新特性的支持，我们通常会采用下面两种典型方法。

◎ 方法 1，在设计数据库表的时候，在每个表中都增加一个很长的备注字段，之后扩展的数据以某种格式（如 XML、JSON、简单字符串拼接等）放入备注字段。因为数据库表的结构没有发生变化，所以此时程序的改动范围是最小的，风险也更小，但看起来不太美观。

◎ 方法 2，直接修改数据库表，增加一个或多个新的列，此时程序的改动范围较大，风险更大，但看起来比较美观。

显然，两种方法都不完美。更加优雅的做法是，先采用方法 1 实现这个新特性，经过几个版本的迭代，等新特性变得稳定成熟了以后，可以在后续版本中采用方法 2 升级到正式版。为此，Kubernetes 为每个资源对象都增加了类似数据库表里备注字段的通用属性 Annotations，以实现方法 1 的升级。以 Kubernetes 1.3 版本引入的 Pod 的 Init Container 新特性为例，一开始，Init Container 的定义是在 Annotations 中声明的，如下面代码中粗体部分所示，是不是很不美观？

```
apiVersion: v1
kind: Pod
metadata:
  name: myapp-pod
  labels:
    app: myapp
  annotations:
    pod.beta.kubernetes.io/init-containers: '[
      {
        "name": "init-mydb",
        "image": "busybox",
        "command": [......]
      }
    ]'
spec:
  containers:
......
```

在 Kubernetes 1.8 版本以后，Init container 特性完全成熟，其定义被放入 Pod 的 spec.initContainers 一节，看起来优雅了很多：

```
apiVersion: v1
kind: Pod
metadata:
  name: myapp-pod
  labels:
    app: myapp
spec:
  # These containers are run during pod initialization
  initContainers:
  - name: init-mydb
    image: busybox
    command:
    - xxx
```

在 Kubernetes 1.8 中，资源对象中的很多 Alpha、Beta 版本的 Annotations 被取消，升级成了常规定义方式，在学习 Kubernetes 的过程中需要特别注意。

下面介绍 Kubernetes 中重要的资源对象。

1.4.1 Master

Kubernetes 里的 Master 指的是集群控制节点,在每个 Kubernetes 集群里都需要有一个 Master 来负责整个集群的管理和控制,基本上 Kubernetes 的所有控制命令都发给它,它负责具体的执行过程,我们后面执行的所有命令基本都是在 Master 上运行的。Master 通常会占据一个独立的服务器(高可用部署建议用 3 台服务器),主要原因是它太重要了,是整个集群的"首脑",如果它宕机或者不可用,那么对集群内容器应用的管理都将失效。

在 Master 上运行着以下关键进程。

◎ Kubernetes API Server(kube-apiserver):提供了 HTTP Rest 接口的关键服务进程,是 Kubernetes 里所有资源的增、删、改、查等操作的唯一入口,也是集群控制的入口进程。
◎ Kubernetes Controller Manager(kube-controller-manager):Kubernetes 里所有资源对象的自动化控制中心,可以将其理解为资源对象的"大总管"。
◎ Kubernetes Scheduler(kube-scheduler):负责资源调度(Pod 调度)的进程,相当于公交公司的"调度室"。

另外,在 Master 上通常还需要部署 etcd 服务,因为 Kubernetes 里的所有资源对象的数据都被保存在 etcd 中。

1.4.2 Node

除了 Master,Kubernetes 集群中的其他机器被称为 Node,在较早的版本中也被称为 Minion。与 Master 一样,Node 可以是一台物理主机,也可以是一台虚拟机。Node 是 Kubernetes 集群中的工作负载节点,每个 Node 都会被 Master 分配一些工作负载(Docker 容器),当某个 Node 宕机时,其上的工作负载会被 Master 自动转移到其他节点上。

在每个 Node 上都运行着以下关键进程。

◎ kubelet:负责 Pod 对应的容器的创建、启停等任务,同时与 Master 密切协作,实现集群管理的基本功能。
◎ kube-proxy:实现 Kubernetes Service 的通信与负载均衡机制的重要组件。
◎ Docker Engine(docker):Docker 引擎,负责本机的容器创建和管理工作。

Node 可以在运行期间动态增加到 Kubernetes 集群中,前提是在这个节点上已经正确安装、配置和启动了上述关键进程,在默认情况下 kubelet 会向 Master 注册自己,这也是 Kubernetes 推荐的 Node 管理方式。一旦 Node 被纳入集群管理范围,kubelet 进程就会定时向 Master 汇报自身的情报,例如操作系统、Docker 版本、机器的 CPU 和内存情况,以及当前有哪些 Pod 在运行等,这样 Master 就可以获知每个 Node 的资源使用情况,并实现高效均衡的资源调度策略。而某个 Node 在超过指定时间不上报信息时,会被 Master 判定为"失联",Node 的状态被标记为不可用(Not Ready),随后 Master 会触发"工作负载大转移"的自动流程。

我们可以执行下述命令查看在集群中有多少个 Node:

```
# kubectl get nodes
NAME           STATUS    ROLES     AGE    VERSION
k8s-node-1     Ready     <none>    350d   v1.14.0
```

然后,通过 kubectl describe node <node_name> 查看某个 Node 的详细信息:

```
$ kubectl describe node k8s-node-1
Name:               k8s-node-1
Roles:              <none>
Labels:             beta.kubernetes.io/arch=amd64
                    beta.kubernetes.io/os=linux
                    kubernetes.io/arch=amd64
                    kubernetes.io/hostname=k8s-node-1
                    kubernetes.io/os=linux
Annotations:        node.alpha.kubernetes.io/ttl: 0
CreationTimestamp:  Tue, 17 Apr 2018 02:29:11 +0800
Taints:             <none>
Unschedulable:      false
Conditions:
  Type               Status  LastHeartbeatTime                 LastTransitionTime                Reason                        Message
  ----               ------  -----------------                 ------------------                ------                        -------
  NetworkUnavailable False   Fri, 29 Mar 2019 22:48:49 +0800   Fri, 29 Mar 2019 22:48:49 +0800   CalicoIsUp                    Calico is running on this node
  MemoryPressure     False   Wed, 03 Apr 2019 01:04:25 +0800   Wed, 05 Dec 2018 00:58:41 +0800   KubeletHasSufficientMemory    kubelet has sufficient memory available
  DiskPressure       False   Wed, 03 Apr 2019 01:04:25 +0800   Wed, 05 Dec 2018 00:58:41 +0800   KubeletHasNoDiskPressure      kubelet has no disk pressure
  PIDPressure        False   Wed, 03 Apr 2019 01:04:25 +0800   Tue, 17 Apr 2018
```

```
02:29:08 +0800    KubeletHasSufficientPID    kubelet has sufficient PID available
      Ready              True       Wed, 03 Apr 2019 01:04:25 +0800    Sat, 30 Mar 2019
09:30:48 +0800    KubeletReady               kubelet is posting ready status
    Addresses:
      InternalIP:  192.168.18.3
      Hostname:    k8s-node-1
    Capacity:
     cpu:                  4
     ephemeral-storage:    49250820Ki
     hugepages-1Gi:        0
     hugepages-2Mi:        0
     memory:               1867048Ki
     pods:                 110
    Allocatable:
     cpu:                  4
     ephemeral-storage:    45389555637
     hugepages-1Gi:        0
     hugepages-2Mi:        0
     memory:               1764648Ki
     pods:                 110
    System Info:
     Machine ID:                 83ae24d58f78439e8197b00e29a99093
     System UUID:                D0E74D56-035F-21F5-7340-DB6F90069202
     Boot ID:                    23c1c362-f9f0-49fe-a4d9-a5db919ca93d
     Kernel Version:             3.10.0-693.el7.x86_64
     OS Image:                   CentOS Linux 7 (Core)
     Operating System:           linux
     Architecture:               amd64
     Container Runtime Version:  docker://18.9.2
     Kubelet Version:            v1.14.0
     Kube-Proxy Version:         v1.14.0
    PodCIDR:                     10.244.0.0/24
    Non-terminated Pods:         (2 in total)
      Namespace              Name                     CPU Requests  CPU Limits
Memory Requests  Memory Limits  AGE
      ---------              ----                     ------------  ----------
---------------  -------------  ---
      kube-system            coredns-767997f5b5-sqkqw  100m (2%)    0 (0%)
70Mi (4%)        170Mi (9%)     3h16m
      kube-system            kube-flannel-ds-bvph5     100m (2%)    500m (12%)
50Mi (2%)        150Mi (8%)     4d1h
```

```
Allocated resources:
  (Total limits may be over 100 percent, i.e., overcommitted.)
  Resource              Requests        Limits
  --------              --------        ------
  cpu                   200m (5%)       500m (12%)
  memory                120Mi (6%)      320Mi (18%)
  ephemeral-storage     0 (0%)          0 (0%)
Events:                 <none>
```

上述命令展示了 Node 的如下关键信息。

- Node 的基本信息：名称、标签、创建时间等。
- Node 当前的运行状态：Node 启动后会做一系列的自检工作，比如磁盘空间是否不足（DiskPressure）、内存是否不足（MemoryPressure）、网络是否正常（NetworkUnavailable）、PID 资源是否充足（PIDPressure）。在一切正常时设置 Node 为 Ready 状态（Ready=True），该状态表示 Node 处于健康状态，Master 将可以在其上调度新的任务了（如启动 Pod）。
- Node 的主机地址与主机名。
- Node 上的资源数量：描述 Node 可用的系统资源，包括 CPU、内存数量、最大可调度 Pod 数量等。
- Node 可分配的资源量：描述 Node 当前可用于分配的资源量。
- 主机系统信息：包括主机 ID、系统 UUID、Linux kernel 版本号、操作系统类型与版本、Docker 版本号、kubelet 与 kube-proxy 的版本号等。
- 当前运行的 Pod 列表概要信息。
- 已分配的资源使用概要信息，例如资源申请的最低、最大允许使用量占系统总量的百分比。
- Node 相关的 Event 信息。

1.4.3　Pod

Pod 是 Kubernetes 最重要的基本概念，如图 1.4 所示是 Pod 的组成示意图，我们看到每个 Pod 都有一个特殊的被称为"根容器"的 Pause 容器。Pause 容器对应的镜像属于 Kubernetes 平台的一部分，除了 Pause 容器，每个 Pod 还包含一个或多个紧密相关的用户业务容器。

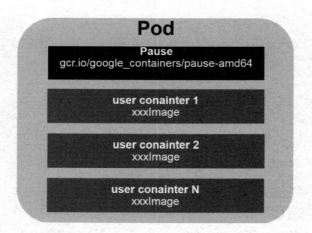

图 1.4　Pod 的组成示意图

为什么 Kubernetes 会设计出一个全新的 Pod 的概念并且 Pod 有这样特殊的组成结构？

原因之一：在一组容器作为一个单元的情况下，我们难以简单地对"整体"进行判断及有效地行动。比如，一个容器死亡了，此时算是整体死亡么？是 N/M 的死亡率么？引入业务无关并且不易死亡的 Pause 容器作为 Pod 的根容器，以它的状态代表整个容器组的状态，就简单、巧妙地解决了这个难题。

原因之二：Pod 里的多个业务容器共享 Pause 容器的 IP，共享 Pause 容器挂接的 Volume，这样既简化了密切关联的业务容器之间的通信问题，也很好地解决了它们之间的文件共享问题。

Kubernetes 为每个 Pod 都分配了唯一的 IP 地址，称之为 Pod IP，一个 Pod 里的多个容器共享 Pod IP 地址。Kubernetes 要求底层网络支持集群内任意两个 Pod 之间的 TCP/IP 直接通信，这通常采用虚拟二层网络技术来实现，例如 Flannel、Open vSwitch 等，因此我们需要牢记一点：在 Kubernetes 里，一个 Pod 里的容器与另外主机上的 Pod 容器能够直接通信。

Pod 其实有两种类型：普通的 Pod 及静态 Pod（Static Pod）。后者比较特殊，它并没被存放在 Kubernetes 的 etcd 存储里，而是被存放在某个具体的 Node 上的一个具体文件中，并且只在此 Node 上启动、运行。而普通的 Pod 一旦被创建，就会被放入 etcd 中存储，随后会被 Kubernetes Master 调度到某个具体的 Node 上并进行绑定（Binding），随后该 Pod 被对应的 Node 上的 kubelet 进程实例化成一组相关的 Docker 容器并启动。在默认情况下，当 Pod 里的某个容器停止时，Kubernetes 会自动检测到这个问题并且重新启动这个 Pod（重启 Pod 里的所有容器），如果 Pod 所在的 Node 宕机，就会将这个 Node 上的所有 Pod 重新调度到其他节点上。Pod、容器与 Node 的关系如图 1.5 所示。

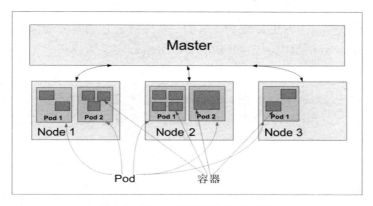

图 1.5　Pod、容器与 Node 的关系

Kubernetes 里的所有资源对象都可以采用 YAML 或者 JSON 格式的文件来定义或描述，下面是我们在之前的 Hello World 例子里用到的 myweb 这个 Pod 的资源定义文件：

```
apiVersion: v1
kind: Pod
metadata:
  name: myweb
  labels:
    name: myweb
spec:
containers:
- name: myweb
  image: kubeguide/tomcat-app:v1
  ports:
  - containerPort: 8080
  env:
  - name: MYSQL_SERVICE_HOST
    value: 'mysql'
  - name: MYSQL_SERVICE_PORT
    value: '3306'
```

Kind 为 Pod 表明这是一个 Pod 的定义，metadata 里的 name 属性为 Pod 的名称，在 metadata 里还能定义资源对象的标签，这里声明 myweb 拥有一个 name=myweb 的标签。在 Pod 里所包含的容器组的定义则在 spec 一节中声明，这里定义了一个名为 myweb、对应镜像为 kubeguide/tomcat-app:v1 的容器，该容器注入了名为 MYSQL_SERVICE_HOST='mysql'和 MYSQL_SERVICE_PORT='3306'的环境变量（env 关键字），并且在 8080 端口（containerPort）启动容器进程。Pod 的 IP 加上这里的容器端口（containerPort），组成了一个新的概念——Endpoint，它代表此 Pod 里的一个服务进程的对外通信地址。一个 Pod 也存在具有多个 Endpoint 的情况，比如当我们把 Tomcat 定义为一个 Pod 时，可以对外暴露管理端口与服务端口这两个 Endpoint。

我们所熟悉的 Docker Volume 在 Kubernetes 里也有对应的概念——Pod Volume，后者有一些扩展，比如可以用分布式文件系统 GlusterFS 实现后端存储功能；Pod Volume 是被定义在 Pod 上，然后被各个容器挂载到自己的文件系统中的。

这里顺便提一下 Kubernetes 的 Event 概念。Event 是一个事件的记录，记录了事件的最早产生时间、最后重现时间、重复次数、发起者、类型，以及导致此事件的原因等众多信息。Event 通常会被关联到某个具体的资源对象上，是排查故障的重要参考信息，之前我们看到 Node 的描述信息包括了 Event，而 Pod 同样有 Event 记录，当我们发现某个 Pod 迟迟无法创建时，可以用 kubectl describe pod xxxx 来查看它的描述信息，以定位问题的成因，比如下面这个 Event 记录信息表明 Pod 里的一个容器被探针检测为失败一次：

```
Events:
  FirstSeen    LastSeen   Count   From               SubobjectPath              Type      Reason    Message
  ---------    --------   -----   ----               -------------              ----      ------
  10h          12m        32      {kubelet k8s-node-1}   spec.containers{kube2sky}   Warning   Unhealthy   Liveness probe failed: Get http://172.17.1.2:8080/healthz: net/http: request canceled (Client.Timeout exceeded while awaiting headers)
```

每个 Pod 都可以对其能使用的服务器上的计算资源设置限额，当前可以设置限额的计算资源有 CPU 与 Memory 两种，其中 CPU 的资源单位为 CPU（Core）的数量，是一个绝对值而非相对值。

对于绝大多数容器来说，一个 CPU 的资源配额相当大，所以在 Kubernetes 里通常以千分之一的 CPU 配额为最小单位，用 m 来表示。通常一个容器的 CPU 配额被定义为 100～300m，即占用 0.1～0.3 个 CPU。由于 CPU 配额是一个绝对值，所以无论在拥有一个 Core 的机器上，还是在拥有 48 个 Core 的机器上，100m 这个配额所代表的 CPU 的使用量都是一样的。与 CPU 配额类似，Memory 配额也是一个绝对值，它的单位是内存字节数。

在 Kubernetes 里，一个计算资源进行配额限定时需要设定以下两个参数。

◎ Requests：该资源的最小申请量，系统必须满足要求。
◎ Limits：该资源最大允许使用的量，不能被突破，当容器试图使用超过这个量的资源时，可能会被 Kubernetes "杀掉" 并重启。

通常，我们会把 Requests 设置为一个较小的数值，符合容器平时的工作负载情况下的资源需求，而把 Limit 设置为峰值负载情况下资源占用的最大量。下面这段定义表明 MySQL 容器申请最少 0.25 个 CPU 及 64MiB 内存，在运行过程中 MySQL 容器所能使用的资源配额为 0.5 个 CPU 及 128MiB 内存：

```
spec:
  containers:
  - name: db
    image: mysql
    resources:
      requests:
        memory: "64Mi"
        cpu: "250m"
      limits:
        memory: "128Mi"
        cpu: "500m"
```

本节最后给出 Pod 及 Pod 周边对象的示意图作为总结，如图 1.6 所示，后面部分还会涉及这张图里的对象和概念，以进一步加强理解。

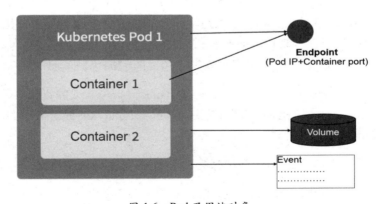

图 1.6　Pod 及周边对象

1.4.4 Label

Label（标签）是 Kubernetes 系统中另外一个核心概念。一个 Label 是一个 key=value 的键值对，其中 key 与 value 由用户自己指定。Label 可以被附加到各种资源对象上，例如 Node、Pod、Service、RC 等，一个资源对象可以定义任意数量的 Label，同一个 Label 也可以被添加到任意数量的资源对象上。Label 通常在资源对象定义时确定，也可以在对象创建后动态添加或者删除。

我们可以通过给指定的资源对象捆绑一个或多个不同的 Label 来实现多维度的资源分组管理功能，以便灵活、方便地进行资源分配、调度、配置、部署等管理工作。例如，部署不同版本的应用到不同的环境中；监控和分析应用（日志记录、监控、告警）等。一些常用的 Label 示例如下。

- 版本标签："release" : "stable"、"release" : "canary"。
- 环境标签："environment":"dev"、"environment":"qa"、"environment":"production"。
- 架构标签："tier" : "frontend"、"tier" : "backend"、"tier" : "middleware"。
- 分区标签："partition" : "customerA"、"partition" : "customerB"。
- 质量管控标签："track" : "daily"、"track" : "weekly"。

Label 相当于我们熟悉的"标签"。给某个资源对象定义一个 Label，就相当于给它打了一个标签，随后可以通过 Label Selector（标签选择器）查询和筛选拥有某些 Label 的资源对象，Kubernetes 通过这种方式实现了类似 SQL 的简单又通用的对象查询机制。

Label Selector 可以被类比为 SQL 语句中的 where 查询条件，例如，name=redis-slave 这个 Label Selector 作用于 Pod 时，可以被类比为 select * from pod where pod's name = 'redis-slave'这样的语句。当前有两种 Label Selector 表达式：基于等式的（Equality-based）和基于集合的（Set-based），前者采用等式类表达式匹配标签，下面是一些具体的例子。

- name = redis-slave：匹配所有具有标签 name=redis-slave 的资源对象。
- env != production：匹配所有不具有标签 env=production 的资源对象，比如 env=test 就是满足此条件的标签之一。

后者则使用集合操作类表达式匹配标签，下面是一些具体的例子。

- name in (redis-master, redis-slave)：匹配所有具有标签 name=redis-master 或者 name= redis-slave 的资源对象。
- name not in (php-frontend)：匹配所有不具有标签 name=php-frontend 的资源对象。

可以通过多个 Label Selector 表达式的组合实现复杂的条件选择，多个表达式之间用 ","进行分隔即可，几个条件之间是 "AND" 的关系，即同时满足多个条件，比如下面的例子：

```
name=redis-slave,env!=production
name notin (php-frontend),env!=production
```

以 myweb Pod 为例，Label 被定义在其 metadata 中：

```
apiVersion: v1
kind: Pod
metadata:
  name: myweb
  labels:
    app: myweb
```

管理对象 RC 和 Service 则通过 Selector 字段设置需要关联 Pod 的 Label：

```
apiVersion: v1
kind: ReplicationController
metadata:
  name: myweb
spec:
  replicas: 1
  selector:
    app: myweb
  template:
    ……

apiVersion: v1
kind: Service
metadata:
  name: myweb
spec:
  selector:
    app: myweb
  ports:
  - port: 8080
```

其他管理对象如 Deployment、ReplicaSet、DaemonSet 和 Job 则可以在 Selector 中使用基于集合的筛选条件定义，例如：

```
selector:
 matchLabels:
    app: myweb
 matchExpressions:
    - {key: tier, operator: In, values: [frontend]}
    - {key: environment, operator: NotIn, values: [dev]}
```

matchLabels用于定义一组Label,与直接写在Selector中的作用相同;matchExpressions用于定义一组基于集合的筛选条件,可用的条件运算符包括 In、NotIn、Exists 和 DoesNotExist。

如果同时设置了matchLabels和matchExpressions,则两组条件为AND关系,即需要同时满足所有条件才能完成Selector的筛选。

Label Selector在Kubernetes中的重要使用场景如下。

◎ kube-controller进程通过在资源对象RC上定义的Label Selector来筛选要监控的Pod副本数量,使Pod副本数量始终符合预期设定的全自动控制流程。
◎ kube-proxy进程通过Service的Label Selector来选择对应的Pod,自动建立每个Service到对应Pod的请求转发路由表,从而实现Service的智能负载均衡机制。
◎ 通过对某些Node定义特定的Label,并且在Pod定义文件中使用NodeSelector这种标签调度策略,kube-scheduler进程可以实现Pod定向调度的特性。

在前面的留言板例子中,我们只使用了一个name=XXX的Label Selector。看一个更复杂的例子:假设为Pod定义了3个Label:release、env和role,不同的Pod定义了不同的Label值,如图1.7所示,如果设置"role=frontend"的Label Selector,则会选取到Node 1和Node 2上的Pod。

如果设置"release=beta"的Label Selector,则会选取到Node 2和Node 3上的Pod,如图1.8所示。

总之,使用Label可以给对象创建多组标签,Label和Label Selector共同构成了Kubernetes系统中核心的应用模型,使得被管理对象能够被精细地分组管理,同时实现了整个集群的高可用性。

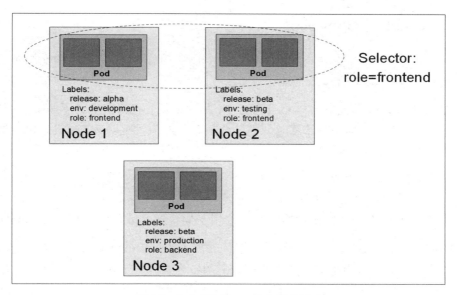

图 1.7　Label Selector 的作用范围 1

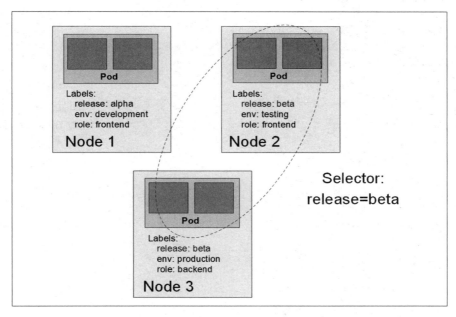

图 1.8　Label Selector 的作用范围 2

1.4.5 Replication Controller

在上一节的例子中已经对 Replication Controller（简称 RC）的定义和作用做了一些说明，本节对 RC 的概念进行深入描述。

RC 是 Kubernetes 系统中的核心概念之一，简单来说，它其实定义了一个期望的场景，即声明某种 Pod 的副本数量在任意时刻都符合某个预期值，所以 RC 的定义包括如下几个部分。

◎ Pod 期待的副本数量。
◎ 用于筛选目标 Pod 的 Label Selector。
◎ 当 Pod 的副本数量小于预期数量时，用于创建新 Pod 的 Pod 模板（template）。

下面是一个完整的 RC 定义的例子，即确保拥有 tier=frontend 标签的这个 Pod（运行 Tomcat 容器）在整个 Kubernetes 集群中始终只有一个副本：

```
apiVersion: v1
kind: ReplicationController
metadata:
  name: frontend
spec:
  replicas: 1
  selector:
    tier: frontend
  template:
    metadata:
      labels:
        app: app-demo
        tier: frontend
    spec:
      containers:
      - name: tomcat-demo
        image: tomcat
        imagePullPolicy: IfNotPresent
        env:
        - name: GET_HOSTS_FROM
          value: dns
        ports:
        - containerPort: 80
```

在我们定义了一个 RC 并将其提交到 Kubernetes 集群中后，Master 上的 Controller

Manager 组件就得到通知,定期巡检系统中当前存活的目标 Pod,并确保目标 Pod 实例的数量刚好等于此 RC 的期望值,如果有过多的 Pod 副本在运行,系统就会停掉一些 Pod,否则系统会再自动创建一些 Pod。可以说,通过 RC,Kubernetes 实现了用户应用集群的高可用性,并且大大减少了系统管理员在传统 IT 环境中需要完成的许多手工运维工作(如主机监控脚本、应用监控脚本、故障恢复脚本等)。

下面以有 3 个 Node 的集群为例,说明 Kubernetes 如何通过 RC 来实现 Pod 副本数量自动控制的机制。假如在我们的 RC 里定义 redis-slave 这个 Pod 需要保持两个副本,系统将可能在其中的两个 Node 上创建 Pod。图 1.9 描述了在两个 Node 上创建 redis-slave Pod 的情形。

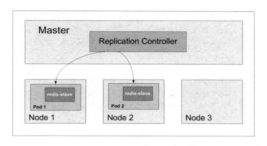

图 1.9　在两个 Node 上创建 redis-slave Pod

假设 Node 2 上的 Pod 2 意外终止,则根据 RC 定义的 replicas 数量 2,Kubernetes 将会自动创建并启动一个新的 Pod,以保证在整个集群中始终有两个 redis-slave Pod 运行。

如图 1.10 所示,系统可能选择 Node 3 或者 Node 1 来创建一个新的 Pod。

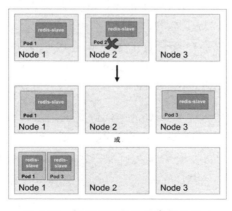

图 1.10　根据 RC 定义创建新的 Pod

此外，在运行时，我们可以通过修改 RC 的副本数量，来实现 Pod 的动态缩放（Scaling），这可以通过执行 kubectl scale 命令来一键完成：

```
$ kubectl scale rc redis-slave --replicas=3
scaled
```

Scaling 的执行结果如图 1.11 所示。

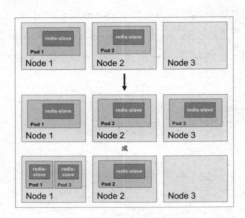

图 1.11 Scaling 的执行结果

需要注意的是，删除 RC 并不会影响通过该 RC 已创建好的 Pod。为了删除所有 Pod，可以设置 replicas 的值为 0，然后更新该 RC。另外，kubectl 提供了 stop 和 delete 命令来一次性删除 RC 和 RC 控制的全部 Pod。

应用升级时，通常会使用一个新的容器镜像版本替代旧版本。我们希望系统平滑升级，比如在当前系统中有 10 个对应的旧版本的 Pod，则最佳的系统升级方式是旧版本的 Pod 每停止一个，就同时创建一个新版本的 Pod，在整个升级过程中此消彼长，而运行中的 Pod 数量始终是 10 个，几分钟以后，当所有的 Pod 都已经是新版本时，系统升级完成。通过 RC 机制，Kubernetes 很容易就实现了这种高级实用的特性，被称为"滚动升级"（Rolling Update），具体的操作方法详见 3.11 节的说明。

Replication Controller 由于与 Kubernetes 代码中的模块 Replication Controller 同名，同时"Replication Controller"无法准确表达它的本意，所以在 Kubernetes 1.2 中，升级为另外一个新概念——Replica Set，官方解释其为"下一代的 RC"。Replica Set 与 RC 当前的唯一区别是，Replica Sets 支持基于集合的 Label selector（Set-based selector），而 RC 只支持基于等式的 Label Selector（equality-based selector），这使得 Replica Set 的功能更强。下面是

等价于之前 RC 例子的 Replica Set 的定义（省去了 Pod 模板部分的内容）：

```
apiVersion: extensions/v1beta1
kind: ReplicaSet
metadata:
  name: frontend
spec:
  selector:
    matchLabels:
      tier: frontend
    matchExpressions:
      - {key: tier, operator: In, values: [frontend]}
  template:
  ......
```

kubectl 命令行工具适用于 RC 的绝大部分命令同样适用于 Replica Set。此外，我们当前很少单独使用 Replica Set，它主要被 Deployment 这个更高层的资源对象所使用，从而形成一整套 Pod 创建、删除、更新的编排机制。我们在使用 Deployment 时，无须关心它是如何创建和维护 Replica Set 的，这一切都是自动发生的。

Replica Set 与 Deployment 这两个重要的资源对象逐步替代了之前 RC 的作用，是 Kubernetes 1.3 里 Pod 自动扩容（伸缩）这个告警功能实现的基础，也将继续在 Kubernetes 未来的版本中发挥重要的作用。

最后总结一下 RC（Replica Set）的一些特性与作用。

- 在大多数情况下，我们通过定义一个 RC 实现 Pod 的创建及副本数量的自动控制。
- 在 RC 里包括完整的 Pod 定义模板。
- RC 通过 Label Selector 机制实现对 Pod 副本的自动控制。
- 通过改变 RC 里的 Pod 副本数量，可以实现 Pod 的扩容或缩容。
- 通过改变 RC 里 Pod 模板中的镜像版本，可以实现 Pod 的滚动升级。

1.4.6 Deployment

Deployment 是 Kubernetes 在 1.2 版本中引入的新概念，用于更好地解决 Pod 的编排问题。为此，Deployment 在内部使用了 Replica Set 来实现目的，无论从 Deployment 的作用与目的、YAML 定义，还是从它的具体命令行操作来看，我们都可以把它看作 RC 的一次升级，两者的相似度超过 90%。

Deployment 相对于 RC 的一个最大升级是我们可以随时知道当前 Pod "部署" 的进度。实际上由于一个 Pod 的创建、调度、绑定节点及在目标 Node 上启动对应的容器这一完整过程需要一定的时间，所以我们期待系统启动 N 个 Pod 副本的目标状态，实际上是一个连续变化的"部署过程"导致的最终状态。

Deployment 的典型使用场景有以下几个。

◎ 创建一个 Deployment 对象来生成对应的 Replica Set 并完成 Pod 副本的创建。
◎ 检查 Deployment 的状态来看部署动作是否完成（Pod 副本数量是否达到预期的值）。
◎ 更新 Deployment 以创建新的 Pod（比如镜像升级）。
◎ 如果当前 Deployment 不稳定，则回滚到一个早先的 Deployment 版本。
◎ 暂停 Deployment 以便于一次性修改多个 PodTemplateSpec 的配置项，之后再恢复 Deployment，进行新的发布。
◎ 扩展 Deployment 以应对高负载。
◎ 查看 Deployment 的状态，以此作为发布是否成功的指标。
◎ 清理不再需要的旧版本 ReplicaSets。

除了 API 声明与 Kind 类型等有所区别，Deployment 的定义与 Replica Set 的定义很类似：

```
apiVersion: extensions/v1beta1          apiVersion: v1
kind: Deployment                        kind: ReplicaSet
metadata:                               metadata:
  name: nginx-deployment                  name: nginx-repset
```

下面通过运行一些例子来直观地感受 Deployment 的概念。创建一个名为 tomcat-deployment.yaml 的 Deployment 描述文件，内容如下：

```
apiVersion: extensions/v1beta1
kind: Deployment
metadata:
  name: frontend
spec:
  replicas: 1
  selector:
    matchLabels:
      tier: frontend
    matchExpressions:
```

```
      - {key: tier, operator: In, values: [frontend]}
  template:
    metadata:
      labels:
        app: app-demo
        tier: frontend
    spec:
      containers:
      - name: tomcat-demo
        image: tomcat
        imagePullPolicy: IfNotPresent
        ports:
        - containerPort: 8080
```

运行下述命令创建 Deployment：

```
# kubectl create -f tomcat-deployment.yaml
deployment "tomcat-deploy" created
```

运行下述命令查看 Deployment 的信息：

```
# kubectl get deployments
NAME            DESIRED   CURRENT   UP-TO-DATE   AVAILABLE   AGE
tomcat-deploy   1         1         1            1           4m
```

对上述输出中涉及的数量解释如下。

- DESIRED：Pod 副本数量的期望值，即在 Deployment 里定义的 Replica。
- CURRENT：当前 Replica 的值，实际上是 Deployment 创建的 Replica Set 里的 Replica 值，这个值不断增加，直到达到 DESIRED 为止，表明整个部署过程完成。
- UP-TO-DATE：最新版本的 Pod 的副本数量，用于指示在滚动升级的过程中，有多少个 Pod 副本已经成功升级。
- AVAILABLE：当前集群中可用的 Pod 副本数量，即集群中当前存活的 Pod 数量。

运行下述命令查看对应的 Replica Set，我们看到它的命名与 Deployment 的名称有关系：

```
# kubectl get rs
NAME                       DESIRED   CURRENT   AGE
tomcat-deploy-1640611518   1         1         1m
```

运行下述命令查看创建的 Pod，我们发现 Pod 的命名以 Deployment 对应的 Replica Set

的名称为前缀,这种命名很清晰地表明了一个 Replica Set 创建了哪些 Pod,对于 Pod 滚动升级这种复杂的过程来说,很容易排查错误:

```
# kubectl get pods
NAME                               READY   STATUS    RESTARTS   AGE
tomcat-deploy-1640611518-zhrsc     1/1     Running   0          3m
```

运行 kubectl describe deployments,可以清楚地看到 Deployment 控制的 Pod 的水平扩展过程,具体内容可参见第 3 章的说明,这里不再赘述。

Pod 的管理对象,除了 RC 和 Deployment,还包括 ReplicaSet、DaemonSet、StatefulSet、Job 等,分别用于不同的应用场景中,将在第 3 章进行详细介绍。

1.4.7　Horizontal Pod Autoscaler

通过手工执行 kubectl scale 命令,我们可以实现 Pod 扩容或缩容。如果仅仅到此为止,显然不符合谷歌对 Kubernetes 的定位目标——自动化、智能化。在谷歌看来,分布式系统要能够根据当前负载的变化自动触发水平扩容或缩容,因为这一过程可能是频繁发生的、不可预料的,所以手动控制的方式是不现实的。

因此,在 Kubernetes 的 1.0 版本实现后,就有人在默默研究 Pod 智能扩容的特性了,并在 Kubernetes 1.1 中首次发布重量级新特性——Horizontal Pod Autoscaling(Pod 横向自动扩容,HPA)。在 Kubernetes 1.2 中 HPA 被升级为稳定版本(apiVersion: autoscaling/v1),但仍然保留了旧版本(apiVersion: extensions/v1beta1)。Kubernetes 从 1.6 版本开始,增强了根据应用自定义的指标进行自动扩容和缩容的功能,API 版本为 autoscaling/v2alpha1,并不断演进。

HPA 与之前的 RC、Deployment 一样,也属于一种 Kubernetes 资源对象。通过追踪分析指定 RC 控制的所有目标 Pod 的负载变化情况,来确定是否需要有针对性地调整目标 Pod 的副本数量,这是 HPA 的实现原理。当前,HPA 有以下两种方式作为 Pod 负载的度量指标。

◎ CPUUtilizationPercentage。
◎ 应用程序自定义的度量指标,比如服务在每秒内的相应请求数(TPS 或 QPS)。

CPUUtilizationPercentage 是一个算术平均值,即目标 Pod 所有副本自身的 CPU 利用率的平均值。一个 Pod 自身的 CPU 利用率是该 Pod 当前 CPU 的使用量除以它的 Pod Request 的值,比如定义一个 Pod 的 Pod Request 为 0.4,而当前 Pod 的 CPU 使用量为 0.2,则它的

CPU 使用率为 50%，这样就可以算出一个 RC 控制的所有 Pod 副本的 CPU 利用率的算术平均值了。如果某一时刻 CPUUtilizationPercentage 的值超过 80%，则意味着当前 Pod 副本数量很可能不足以支撑接下来更多的请求，需要进行动态扩容，而在请求高峰时段过去后，Pod 的 CPU 利用率又会降下来，此时对应的 Pod 副本数应该自动减少到一个合理的水平。如果目标 Pod 没有定义 Pod Request 的值，则无法使用 CPUUtilizationPercentage 实现 Pod 横向自动扩容。除了使用 CPUUtilizationPercentage，Kubernetes 从 1.2 版本开始也在尝试支持应用程序自定义的度量指标。

在 CPUUtilizationPercentage 计算过程中使用到的 Pod 的 CPU 使用量通常是 1min 内的平均值，通常通过查询 Heapster 监控子系统来得到这个值，所以需要安装部署 Heapster，这样便增加了系统的复杂度和实施 HPA 特性的复杂度。因此，从 1.7 版本开始，Kubernetes 自身孵化了一个基础性能数据采集监控框架——Kubernetes Monitoring Architecture，从而更好地支持 HPA 和其他需要用到基础性能数据的功能模块。在 Kubernetes Monitoring Architecture 中，Kubernetes 定义了一套标准化的 API 接口 Resource Metrics API，以方便客户端应用程序（如 HPA）从 Metrics Server 中获取目标资源对象的性能数据，例如容器的 CPU 和内存使用数据。到了 Kubernetes 1.8 版本，Resource Metrics API 被升级为 metrics.k8s.io/v1beta1，已经接近生产环境中的可用目标了。

下面是 HPA 定义的一个具体例子：

```
apiVersion: autoscaling/v1
kind: HorizontalPodAutoscaler
metadata:
 name: php-apache
 namespace: default
spec:
 maxReplicas: 10
 minReplicas: 1
 scaleTargetRef:
   kind: Deployment
   name: php-apache
 targetCPUUtilizationPercentage: 90
```

根据上面的定义，我们可以知道这个 HPA 控制的目标对象为一个名为 php-apache 的 Deployment 里的 Pod 副本，当这些 Pod 副本的 CPUUtilizationPercentage 的值超过 90%时会触发自动动态扩容行为，在扩容或缩容时必须满足的一个约束条件是 Pod 的副本数为 1~10。

除了可以通过直接定义 YAML 文件并且调用 kubectl create 的命令来创建一个 HPA 资源对象的方式，还可以通过下面的简单命令行直接创建等价的 HPA 对象：

```
# kubectl autoscale deployment php-apache --cpu-percent=90 --min=1 --max=10
```

第 2 章将会给出一个完整的 HPA 例子来说明其用法和功能。

1.4.8 StatefulSet

在 Kubernetes 系统中，Pod 的管理对象 RC、Deployment、DaemonSet 和 Job 都面向无状态的服务。但现实中有很多服务是有状态的，特别是一些复杂的中间件集群，例如 MySQL 集群、MongoDB 集群、Akka 集群、ZooKeeper 集群等，这些应用集群有 4 个共同点。

（1）每个节点都有固定的身份 ID，通过这个 ID，集群中的成员可以相互发现并通信。

（2）集群的规模是比较固定的，集群规模不能随意变动。

（3）集群中的每个节点都是有状态的，通常会持久化数据到永久存储中。

（4）如果磁盘损坏，则集群里的某个节点无法正常运行，集群功能受损。

如果通过 RC 或 Deployment 控制 Pod 副本数量来实现上述有状态的集群，就会发现第 1 点是无法满足的，因为 Pod 的名称是随机产生的，Pod 的 IP 地址也是在运行期才确定且可能有变动的，我们事先无法为每个 Pod 都确定唯一不变的 ID。另外，为了能够在其他节点上恢复某个失败的节点，这种集群中的 Pod 需要挂接某种共享存储，为了解决这个问题，Kubernetes 从 1.4 版本开始引入了 PetSet 这个新的资源对象，并且在 1.5 版本时更名为 StatefulSet，StatefulSet 从本质上来说，可以看作 Deployment/RC 的一个特殊变种，它有如下特性。

◎ StatefulSet 里的每个 Pod 都有稳定、唯一的网络标识，可以用来发现集群内的其他成员。假设 StatefulSet 的名称为 kafka，那么第 1 个 Pod 叫 kafka-0，第 2 个叫 kafka-1，以此类推。

◎ StatefulSet 控制的 Pod 副本的启停顺序是受控的，操作第 n 个 Pod 时，前 $n-1$ 个 Pod 已经是运行且准备好的状态。

◎ StatefulSet 里的 Pod 采用稳定的持久化存储卷，通过 PV 或 PVC 来实现，删除 Pod 时默认不会删除与 StatefulSet 相关的存储卷（为了保证数据的安全）。

StatefulSet 除了要与 PV 卷捆绑使用以存储 Pod 的状态数据，还要与 Headless Service 配合使用，即在每个 StatefulSet 定义中都要声明它属于哪个 Headless Service。Headless Service 与普通 Service 的关键区别在于，它没有 Cluster IP，如果解析 Headless Service 的 DNS 域名，则返回的是该 Service 对应的全部 Pod 的 Endpoint 列表。StatefulSet 在 Headless Service 的基础上又为 StatefulSet 控制的每个 Pod 实例都创建了一个 DNS 域名，这个域名的格式为：

```
$(podname).$(headless service name)
```

比如一个 3 节点的 Kafka 的 StatefulSet 集群对应的 Headless Service 的名称为 kafka，StatefulSet 的名称为 kafka，则 StatefulSet 里的 3 个 Pod 的 DNS 名称分别为 kafka-0.kafka、kafka-1.kafka、kafka-2.kafka，这些 DNS 名称可以直接在集群的配置文件中固定下来。

1.4.9 Service

1. 概述

Service 服务也是 Kubernetes 里的核心资源对象之一，Kubernetes 里的每个 Service 其实就是我们经常提起的微服务架构中的一个微服务，之前讲解 Pod、RC 等资源对象其实都是为讲解 Kubernetes Service 做铺垫的。图 1.12 显示了 Pod、RC 与 Service 的逻辑关系。

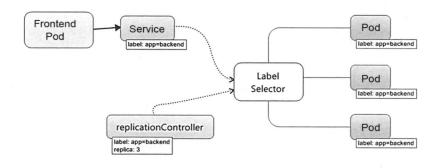

图 1.12 Pod、RC 与 Service 的逻辑关系

从图 1.12 中可以看到，Kubernetes 的 Service 定义了一个服务的访问入口地址，前端的应用（Pod）通过这个入口地址访问其背后的一组由 Pod 副本组成的集群实例，Service 与其后端 Pod 副本集群之间则是通过 Label Selector 来实现无缝对接的。RC 的作用实际上

是保证 Service 的服务能力和服务质量始终符合预期标准。

通过分析、识别并建模系统中的所有服务为微服务——Kubernetes Service，我们的系统最终由多个提供不同业务能力而又彼此独立的微服务单元组成的，服务之间通过 TCP/IP 进行通信，从而形成了强大而又灵活的弹性网格，拥有强大的分布式能力、弹性扩展能力、容错能力，程序架构也变得简单和直观许多，如图 1.13 所示。

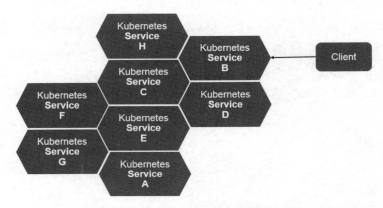

图 1.13　Kubernetes 提供的微服务网格架构

既然每个 Pod 都会被分配一个单独的 IP 地址，而且每个 Pod 都提供了一个独立的 Endpoint（Pod IP+ContainerPort）以被客户端访问，现在多个 Pod 副本组成了一个集群来提供服务，那么客户端如何来访问它们呢？一般的做法是部署一个负载均衡器（软件或硬件），为这组 Pod 开启一个对外的服务端口如 8000 端口，并且将这些 Pod 的 Endpoint 列表加入 8000 端口的转发列表，客户端就可以通过负载均衡器的对外 IP 地址+服务端口来访问此服务。客户端的请求最后会被转发到哪个 Pod，由负载均衡器的算法所决定。

Kubernetes 也遵循上述常规做法，运行在每个 Node 上的 kube-proxy 进程其实就是一个智能的软件负载均衡器，负责把对 Service 的请求转发到后端的某个 Pod 实例上，并在内部实现服务的负载均衡与会话保持机制。但 Kubernetes 发明了一种很巧妙又影响深远的设计：Service 没有共用一个负载均衡器的 IP 地址，每个 Service 都被分配了一个全局唯一的虚拟 IP 地址，这个虚拟 IP 被称为 Cluster IP。这样一来，每个服务就变成了具备唯一 IP 地址的通信节点，服务调用就变成了最基础的 TCP 网络通信问题。

我们知道，Pod 的 Endpoint 地址会随着 Pod 的销毁和重新创建而发生改变，因为新 Pod 的 IP 地址与之前旧 Pod 的不同。而 Service 一旦被创建，Kubernetes 就会自动为它分配一个可用的 Cluster IP，而且在 Service 的整个生命周期内，它的 Cluster IP 不会发生改变。于是,服务发现这个棘手的问题在 Kubernetes 的架构里也得以轻松解决:只要用 Service 的 Name 与 Service 的 Cluster IP 地址做一个 DNS 域名映射即可完美解决问题。现在想想,这真是一个很棒的设计。

说了这么久,下面动手创建一个 Service 来加深对它的理解。创建一个名为 tomcat-service.yaml 的定义文件,内容如下:

```
apiVersion: v1
kind: Service
metadata:
  name: tomcat-service
spec:
  ports:
  - port: 8080
  selector:
    tier: frontend
```

上述内容定义了一个名为 tomcat-service 的 Service,它的服务端口为 8080,拥有 "tier = frontend" 这个 Label 的所有 Pod 实例都属于它,运行下面的命令进行创建:

```
#kubectl create -f tomcat-server.yaml
service "tomcat-service" created
```

我们之前在 tomcat-deployment.yaml 里定义的 Tomcat 的 Pod 刚好拥有这个标签,所以刚才创建的 tomcat-service 已经对应一个 Pod 实例,运行下面的命令可以查看 tomcat-service 的 Endpoint 列表,其中 172.17.1.3 是 Pod 的 IP 地址,端口 8080 是 Container 暴露的端口:

```
# kubectl get endpoints
NAME              ENDPOINTS              AGE
kubernetes        192.168.18.131:6443    15d
tomcat-service    172.17.1.3:8080        1m
```

你可能有疑问:"说好的 Service 的 Cluster IP 呢?怎么没有看到?"运行下面的命令即可看到 tomct-service 被分配的 Cluster IP 及更多的信息:

```
# kubectl get svc tomcat-service -o yaml
apiVersion: v1
kind: Service
metadata:
  creationTimestamp: 2016-07-21T17:05:52Z
  name: tomcat-service
```

```
  namespace: default
  resourceVersion: "23964"
  selfLink: /api/v1/namespaces/default/services/tomcat-service
  uid: 61987d3c-4f65-11e6-a9d8-000c29ed42c1
spec:
  clusterIP: 169.169.65.227
  ports:
  - port: 8080
    protocol: TCP
    targetPort: 8080
  selector:
    tier: frontend
  sessionAffinity: None
  type: ClusterIP
status:
  loadBalancer: {}
```

在 spec.ports 的定义中，targetPort 属性用来确定提供该服务的容器所暴露（EXPOSE）的端口号，即具体业务进程在容器内的 targetPort 上提供 TCP/IP 接入；port 属性则定义了 Service 的虚端口。前面定义 Tomcat 服务时没有指定 targetPort，则默认 targetPort 与 port 相同。

接下来看看 Service 的多端口问题。

很多服务都存在多个端口的问题，通常一个端口提供业务服务，另外一个端口提供管理服务，比如 Mycat、Codis 等常见中间件。Kubernetes Service 支持多个 Endpoint，在存在多个 Endpoint 的情况下，要求每个 Endpoint 都定义一个名称来区分。下面是 Tomcat 多端口的 Service 定义样例：

```
apiVersion: v1
kind: Service
metadata:
  name: tomcat-service
spec:
  ports:
  - port: 8080
    name: service-port
  - port: 8005
    name: shutdown-port
  selector:
    tier: frontend
```

多端口为什么需要给每个端口都命名呢？这就涉及 Kubernetes 的服务发现机制了，接下来进行讲解。

2. Kubernetes 的服务发现机制

任何分布式系统都会涉及"服务发现"这个基础问题，大部分分布式系统都通过提供特定的 API 接口来实现服务发现功能，但这样做会导致平台的侵入性比较强，也增加了开发、测试的难度。Kubernetes 则采用了直观朴素的思路去解决这个棘手的问题。

首先，每个 Kubernetes 中的 Service 都有唯一的 Cluster IP 及唯一的名称，而名称是由开发者自己定义的，部署时也没必要改变，所以完全可以被固定在配置中。接下来的问题就是如何通过 Service 的名称找到对应的 Cluster IP。

最早时 Kubernetes 采用了 Linux 环境变量解决这个问题，即每个 Service 都生成一些对应的 Linux 环境变量（ENV），并在每个 Pod 的容器启动时自动注入这些环境变量。以下是 tomcat-service 产生的环境变量条目：

```
TOMCAT_SERVICE_SERVICE_HOST=169.169.41.218
TOMCAT_SERVICE_SERVICE_PORT_SERVICE_PORT=8080
TOMCAT_SERVICE_SERVICE_PORT_SHUTDOWN_PORT=8005
TOMCAT_SERVICE_SERVICE_PORT=8080
TOMCAT_SERVICE_PORT_8005_TCP_PORT=8005
TOMCAT_SERVICE_PORT=tcp://169.169.41.218:8080
TOMCAT_SERVICE_PORT_8080_TCP_ADDR=169.169.41.218
TOMCAT_SERVICE_PORT_8080_TCP=tcp://169.169.41.218:8080
TOMCAT_SERVICE_PORT_8080_TCP_PROTO=tcp
TOMCAT_SERVICE_PORT_8080_TCP_PORT=8080
TOMCAT_SERVICE_PORT_8005_TCP=tcp://169.169.41.218:8005
TOMCAT_SERVICE_PORT_8005_TCP_ADDR=169.169.41.218
TOMCAT_SERVICE_PORT_8005_TCP_PROTO=tcp
```

在上述环境变量中，比较重要的是前 3 个环境变量。可以看到，每个 Service 的 IP 地址及端口都有标准的命名规范，遵循这个命名规范，就可以通过代码访问系统环境变量来得到所需的信息，实现服务调用。

考虑到通过环境变量获取 Service 地址的方式仍然不太方便、不够直观，后来 Kubernetes 通过 Add-On 增值包引入了 DNS 系统，把服务名作为 DNS 域名，这样程序就可以直接使用服务名来建立通信连接了。目前，Kubernetes 上的大部分应用都已经采用了 DNS 这种新兴的服务发现机制，后面会讲解如何部署 DNS 系统。

3. 外部系统访问 Service 的问题

为了更深入地理解和掌握 Kubernetes，我们需要弄明白 Kubernetes 里的 3 种 IP，这 3 种 IP 分别如下。

◎ Node IP：Node 的 IP 地址。
◎ Pod IP：Pod 的 IP 地址。
◎ Cluster IP：Service 的 IP 地址。

首先，Node IP 是 Kubernetes 集群中每个节点的物理网卡的 IP 地址，是一个真实存在的物理网络，所有属于这个网络的服务器都能通过这个网络直接通信，不管其中是否有部分节点不属于这个 Kubernetes 集群。这也表明在 Kubernetes 集群之外的节点访问 Kubernetes 集群之内的某个节点或者 TCP/IP 服务时，都必须通过 Node IP 通信。

其次，Pod IP 是每个 Pod 的 IP 地址，它是 Docker Engine 根据 docker0 网桥的 IP 地址段进行分配的，通常是一个虚拟的二层网络，前面说过，Kubernetes 要求位于不同 Node 上的 Pod 都能够彼此直接通信，所以 Kubernetes 里一个 Pod 里的容器访问另外一个 Pod 里的容器时，就是通过 Pod IP 所在的虚拟二层网络进行通信的，而真实的 TCP/IP 流量是通过 Node IP 所在的物理网卡流出的。

最后说说 Service 的 Cluster IP，它也是一种虚拟的 IP，但更像一个"伪造"的 IP 网络，原因有以下几点。

◎ Cluster IP 仅仅作用于 Kubernetes Service 这个对象，并由 Kubernetes 管理和分配 IP 地址（来源于 Cluster IP 地址池）。
◎ Cluster IP 无法被 Ping，因为没有一个"实体网络对象"来响应。
◎ Cluster IP 只能结合 Service Port 组成一个具体的通信端口，单独的 Cluster IP 不具备 TCP/IP 通信的基础，并且它们属于 Kubernetes 集群这样一个封闭的空间，集群外的节点如果要访问这个通信端口，则需要做一些额外的工作。
◎ 在 Kubernetes 集群内，Node IP 网、Pod IP 网与 Cluster IP 网之间的通信，采用的是 Kubernetes 自己设计的一种编程方式的特殊路由规则，与我们熟知的 IP 路由有很大的不同。

根据上面的分析和总结，我们基本明白了：Service 的 Cluster IP 属于 Kubernetes 集群内部的地址，无法在集群外部直接使用这个地址。那么矛盾来了：实际上在我们开发的业务系统中肯定多少有一部分服务是要提供给 Kubernetes 集群外部的应用或者用户来使用的，典型的例子就是 Web 端的服务模块，比如上面的 tomcat-service，那么用户怎么访问它？

采用 NodePort 是解决上述问题的最直接、有效的常见做法。以 tomcat-service 为例，在 Service 的定义里做如下扩展即可（见代码中的粗体部分）：

```
apiVersion: v1
kind: Service
metadata:
  name: tomcat-service
spec:
  type: NodePort
  ports:
  - port: 8080
    nodePort: 31002
  selector:
    tier: frontend
```

其中，nodePort:31002 这个属性表明手动指定 tomcat-service 的 NodePort 为 31002，否则 Kubernetes 会自动分配一个可用的端口。接下来在浏览器里访问 http://<nodePort IP>:31002/，就可以看到 Tomcat 的欢迎界面了，如图 1.14 所示。

图 1.14 通过 NodePort 访问 Service

NodePort 的实现方式是在 Kubernetes 集群里的每个 Node 上都为需要外部访问的 Service 开启一个对应的 TCP 监听端口,外部系统只要用任意一个 Node 的 IP 地址+具体的 NodePort 端口号即可访问此服务,在任意 Node 上运行 netstat 命令,就可以看到有 NodePort 端口被监听:

```
# netstat -tlp | grep 31002
tcp6   0   0 [::]:31002        [::]:*          LISTEN      1125/kube-proxy
```

但 NodePort 还没有完全解决外部访问 Service 的所有问题,比如负载均衡问题。假如在我们的集群中有 10 个 Node,则此时最好有一个负载均衡器,外部的请求只需访问此负载均衡器的 IP 地址,由负载均衡器负责转发流量到后面某个 Node 的 NodePort 上,如图 1.15 所示。

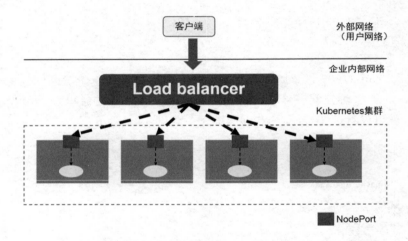

图 1.15 NodePort 与 Load balancer

图 1.15 中的 Load balancer 组件独立于 Kubernetes 集群之外,通常是一个硬件的负载均衡器,或者是以软件方式实现的,例如 HAProxy 或者 Nginx。对于每个 Service,我们通常需要配置一个对应的 Load balancer 实例来转发流量到后端的 Node 上,这的确增加了工作量及出错的概率。于是 Kubernetes 提供了自动化的解决方案,如果我们的集群运行在谷歌的公有云 GCE 上,那么只要把 Service 的 type=NodePort 改为 type=LoadBalancer,Kubernetes 就会自动创建一个对应的 Load balancer 实例并返回它的 IP 地址供外部客户端使用。其他公有云提供商只要实现了支持此特性的驱动,则也可以达到上述目的。此外,裸机上的类似机制(Bare Metal Service Load Balancers)也在被开发。

1.4.10 Job

批处理任务通常并行（或者串行）启动多个计算进程去处理一批工作项（work item），在处理完成后，整个批处理任务结束。从 1.2 版本开始，Kubernetes 支持批处理类型的应用，我们可以通过 Kubernetes Job 这种新的资源对象定义并启动一个批处理任务 Job。与 RC、Deployment、ReplicaSet、DaemonSet 类似，Job 也控制一组 Pod 容器。从这个角度来看，Job 也是一种特殊的 Pod 副本自动控制器，同时 Job 控制 Pod 副本与 RC 等控制器的工作机制有以下重要差别。

（1）Job 所控制的 Pod 副本是短暂运行的，可以将其视为一组 Docker 容器，其中的每个 Docker 容器都仅仅运行一次。当 Job 控制的所有 Pod 副本都运行结束时，对应的 Job 也就结束了。Job 在实现方式上与 RC 等副本控制器不同，Job 生成的 Pod 副本是不能自动重启的，对应 Pod 副本的 RestartPoliy 都被设置为 Never。因此，当对应的 Pod 副本都执行完成时，相应的 Job 也就完成了控制使命，即 Job 生成的 Pod 在 Kubernetes 中是短暂存在的。Kubernetes 在 1.5 版本之后又提供了类似 crontab 的定时任务——CronJob，解决了某些批处理任务需要定时反复执行的问题。

（2）Job 所控制的 Pod 副本的工作模式能够多实例并行计算，以 TensorFlow 框架为例，可以将一个机器学习的计算任务分布到 10 台机器上，在每台机器上都运行一个 worker 执行计算任务，这很适合通过 Job 生成 10 个 Pod 副本同时启动运算。

在第 3 章会继续深入讲解 Job 的实现原理及对应的案例。

1.4.11 Volume

Volume（存储卷）是 Pod 中能够被多个容器访问的共享目录。Kubernetes 的 Volume 概念、用途和目的与 Docker 的 Volume 比较类似，但两者不能等价。首先，Kubernetes 中的 Volume 被定义在 Pod 上，然后被一个 Pod 里的多个容器挂载到具体的文件目录下；其次，Kubernetes 中的 Volume 与 Pod 的生命周期相同，但与容器的生命周期不相关，当容器终止或者重启时，Volume 中的数据也不会丢失。最后，Kubernetes 支持多种类型的 Volume，例如 GlusterFS、Ceph 等先进的分布式文件系统。

Volume 的使用也比较简单，在大多数情况下，我们先在 Pod 上声明一个 Volume，然后在容器里引用该 Volume 并挂载（Mount）到容器里的某个目录上。举例来说，我们要给之前的 Tomcat Pod 增加一个名为 datavol 的 Volume，并且挂载到容器的/mydata-data 目录上，则只要对 Pod 的定义文件做如下修正即可（注意代码中的粗体部分）：

```
template:
  metadata:
    labels:
      app: app-demo
      tier: frontend
  spec:
    volumes:
      - name: datavol
        emptyDir: {}
    containers:
    - name: tomcat-demo
      image: tomcat
      volumeMounts:
        - mountPath: /mydata-data
          name: datavol
      imagePullPolicy: IfNotPresent
```

除了可以让一个 Pod 里的多个容器共享文件、让容器的数据写到宿主机的磁盘上或者写文件到网络存储中，Kubernetes 的 Volume 还扩展出了一种非常有实用价值的功能，即容器配置文件集中化定义与管理，这是通过 ConfigMap 这种新的资源对象来实现的，后面会详细说明。

Kubernetes 提供了非常丰富的 Volume 类型，下面逐一进行说明。

1. emptyDir

一个 emptyDir Volume 是在 Pod 分配到 Node 时创建的。从它的名称就可以看出，它的初始内容为空，并且无须指定宿主机上对应的目录文件，因为这是 Kubernetes 自动分配的一个目录，当 Pod 从 Node 上移除时，emptyDir 中的数据也会被永久删除。emptyDir 的一些用途如下。

◎ 临时空间，例如用于某些应用程序运行时所需的临时目录，且无须永久保留。
◎ 长时间任务的中间过程 CheckPoint 的临时保存目录。
◎ 一个容器需要从另一个容器中获取数据的目录（多容器共享目录）。

目前，用户无法控制 emptyDir 使用的介质种类。如果 kubelet 的配置是使用硬盘，那么所有 emptyDir 都将被创建在该硬盘上。Pod 在将来可以设置 emptyDir 是位于硬盘、固态硬盘上还是基于内存的 tmpfs 上，上面的例子便采用了 emptyDir 类的 Volume。

2. hostPath

hostPath 为在 Pod 上挂载宿主机上的文件或目录，它通常可以用于以下几方面。

- ◎ 容器应用程序生成的日志文件需要永久保存时，可以使用宿主机的高速文件系统进行存储。
- ◎ 需要访问宿主机上 Docker 引擎内部数据结构的容器应用时，可以通过定义 hostPath 为宿主机/var/lib/docker 目录，使容器内部应用可以直接访问 Docker 的文件系统。

在使用这种类型的 Volume 时，需要注意以下几点。

- ◎ 在不同的 Node 上具有相同配置的 Pod，可能会因为宿主机上的目录和文件不同而导致对 Volume 上目录和文件的访问结果不一致。
- ◎ 如果使用了资源配额管理，则 Kubernetes 无法将 hostPath 在宿主机上使用的资源纳入管理。

在下面的例子中使用宿主机的/data 目录定义了一个 hostPath 类型的 Volume：

```
volumes:
- name: "persistent-storage"
  hostPath:
    path: "/data"
```

3. gcePersistentDisk

使用这种类型的 Volume 表示使用谷歌公有云提供的永久磁盘（Persistent Disk，PD）存放 Volume 的数据，它与 emptyDir 不同，PD 上的内容会被永久保存，当 Pod 被删除时，PD 只是被卸载（Unmount），但不会被删除。需要注意的是，你需要先创建一个 PD，才能使用 gcePersistentDisk。

使用 gcePersistentDisk 时有以下一些限制条件。

- ◎ Node（运行 kubelet 的节点）需要是 GCE 虚拟机。
- ◎ 这些虚拟机需要与 PD 存在于相同的 GCE 项目和 Zone 中。

通过 gcloud 命令即可创建一个 PD：

```
gcloud compute disks create --size=500GB --zone=us-central1-a my-data-disk
```

定义 gcePersistentDisk 类型的 Volume 的示例如下：

```
volumes:
- name: test-volume
  # This GCE PD must already exist.
  gcePersistentDisk:
    pdName: my-data-disk
    fsType: ext4
```

4. awsElasticBlockStore

与 GCE 类似，该类型的 Volume 使用亚马逊公有云提供的 EBS Volume 存储数据，需要先创建一个 EBS Volume 才能使用 awsElasticBlockStore。

使用 awsElasticBlockStore 的一些限制条件如下。

◎ Node（运行 kubelet 的节点）需要是 AWS EC2 实例。
◎ 这些 AWS EC2 实例需要与 EBS Volume 存在于相同的 region 和 availability-zone 中。
◎ EBS 只支持单个 EC2 实例挂载一个 Volume。

通过 aws ec2 create-volume 命令可以创建一个 EBS Volume：

```
aws ec2 create-volume --availability-zone eu-west-1a --size 10 --volume-type gp2
```

定义 awsElasticBlockStore 类型的 Volume 的示例如下：

```
volumes:
- name: test-volume
  # This AWS EBS volume must already exist.
  awsElasticBlockStore:
    volumeID: aws://<availability-zone>/<volume-id>
    fsType: ext4
```

5. NFS

使用 NFS 网络文件系统提供的共享目录存储数据时，我们需要在系统中部署一个 NFS Server。定义 NFS 类型的 Volume 的示例如下：

```
volumes:
  - name: nfs
```

```
nfs:
    # 改为你的 NFS 服务器地址
    server: nfs-server.localhost
    path: "/"
```

6. 其他类型的 Volume

◎ iscsi：使用 iSCSI 存储设备上的目录挂载到 Pod 中。
◎ flocker：使用 Flocker 管理存储卷。
◎ glusterfs：使用开源 GlusterFS 网络文件系统的目录挂载到 Pod 中。
◎ rbd：使用 Ceph 块设备共享存储（Rados Block Device）挂载到 Pod 中。
◎ gitRepo：通过挂载一个空目录，并从 Git 库 clone 一个 git repository 以供 Pod 使用。
◎ secret：一个 Secret Volume 用于为 Pod 提供加密的信息，你可以将定义在 Kubernetes 中的 Secret 直接挂载为文件让 Pod 访问。Secret Volume 是通过 TMFS（内存文件系统）实现的，这种类型的 Volume 总是不会被持久化的。

1.4.12　Persistent Volume

之前提到的 Volume 是被定义在 Pod 上的，属于计算资源的一部分，而实际上，网络存储是相对独立于计算资源而存在的一种实体资源。比如在使用虚拟机的情况下，我们通常会先定义一个网络存储，然后从中划出一个"网盘"并挂接到虚拟机上。Persistent Volume（PV）和与之相关联的 Persistent Volume Claim（PVC）也起到了类似的作用。

PV 可以被理解成 Kubernetes 集群中的某个网络存储对应的一块存储，它与 Volume 类似，但有以下区别。

◎ PV 只能是网络存储，不属于任何 Node，但可以在每个 Node 上访问。
◎ PV 并不是被定义在 Pod 上的，而是独立于 Pod 之外定义的。
◎ PV 目前支持的类型包括：gcePersistentDisk、AWSElasticBlockStore、AzureFile、AzureDisk、FC（Fibre Channel）、Flocker、NFS、iSCSI、RBD（Rados Block Device）、CephFS、Cinder、GlusterFS、VsphereVolume、Quobyte Volumes、VMware Photon、Portworx Volumes、ScaleIO Volumes 和 HostPath（仅供单机测试）。

下面给出了 NFS 类型的 PV 的一个 YAML 定义文件，声明了需要 5Gi 的存储空间：

```
apiVersion: v1
kind: PersistentVolume
```

```
metadata:
  name: pv0003
spec:
  capacity:
    storage: 5Gi
  accessModes:
    - ReadWriteOnce
  nfs:
    path: /somepath
    server: 172.17.0.2
```

比较重要的是 PV 的 accessModes 属性，目前有以下类型。

◎ ReadWriteOnce：读写权限，并且只能被单个 Node 挂载。
◎ ReadOnlyMany：只读权限，允许被多个 Node 挂载。
◎ ReadWriteMany：读写权限，允许被多个 Node 挂载。

如果某个 Pod 想申请某种类型的 PV，则首先需要定义一个 PersistentVolumeClaim 对象：

```
kind: PersistentVolumeClaim
apiVersion: v1
metadata:
  name: myclaim
spec:
  accessModes:
    - ReadWriteOnce
  resources:
    requests:
      storage: 8Gi
```

然后，在 Pod 的 Volume 定义中引用上述 PVC 即可：

```
volumes:
  - name: mypd
    persistentVolumeClaim:
      claimName: myclaim
```

最后说说 PV 的状态。PV 是有状态的对象，它的状态有以下几种。

◎ Available：空闲状态。
◎ Bound：已经绑定到某个 PVC 上。

- Released：对应的 PVC 已经被删除，但资源还没有被集群收回。
- Failed：PV 自动回收失败。

共享存储的原理解析和实践指南详见第 8 章的说明。

1.4.13 Namespace

Namespace（命名空间）是 Kubernetes 系统中的另一个非常重要的概念，Namespace 在很多情况下用于实现多租户的资源隔离。Namespace 通过将集群内部的资源对象"分配"到不同的 Namespace 中，形成逻辑上分组的不同项目、小组或用户组，便于不同的分组在共享使用整个集群的资源的同时还能被分别管理。

Kubernetes 集群在启动后会创建一个名为 default 的 Namespace，通过 kubectl 可以查看：

```
$ kubectl get namespaces
NAME            LABELS              STATUS
default         <none>              Active
```

接下来，如果不特别指明 Namespace，则用户创建的 Pod、RC、Service 都将被系统创建到这个默认的名为 default 的 Namespace 中。

Namespace 的定义很简单。如下所示的 YAML 定义了名为 development 的 Namespace。

```
apiVersion: v1
kind: Namespace
metadata:
  name: development
```

一旦创建了 Namespace，我们在创建资源对象时就可以指定这个资源对象属于哪个 Namespace。比如在下面的例子中定义了一个名为 busybox 的 Pod，并将其放入 development 这个 Namespace 里：

```
apiVersion: v1
kind: Pod
metadata:
  name: busybox
  namespace: development
spec:
  containers:
```

```
    - image: busybox
      command:
        - sleep
        - "3600"
      name: busybox
```

此时使用 kubectl get 命令查看,将无法显示:

```
$ kubectl get pods
NAME       READY      STATUS      RESTARTS      AGE
```

这是因为如果不加参数,则 kubectl get 命令将仅显示属于 default 命名空间的资源对象。

可以在 kubectl 命令中加入 --namespace 参数来查看某个命名空间中的对象:

```
# kubectl get pods --namespace=development
NAME        READY      STATUS       RESTARTS      AGE
busybox     1/1        Running      0             1m
```

当给每个租户创建一个 Namespace 来实现多租户的资源隔离时,还能结合 Kubernetes 的资源配额管理,限定不同租户能占用的资源,例如 CPU 使用量、内存使用量等。关于资源配额管理的问题,在后面的章节中会详细介绍。

1.4.14　Annotation

Annotation(注解)与 Label 类似,也使用 key/value 键值对的形式进行定义。不同的是 Label 具有严格的命名规则,它定义的是 Kubernetes 对象的元数据(Metadata),并且用于 Label Selector。Annotation 则是用户任意定义的附加信息,以便于外部工具查找。在很多时候,Kubernetes 的模块自身会通过 Annotation 标记资源对象的一些特殊信息。

通常来说,用 Annotation 来记录的信息如下。

◎ build 信息、release 信息、Docker 镜像信息等,例如时间戳、release id 号、PR 号、镜像 Hash 值、Docker Registry 地址等。
◎ 日志库、监控库、分析库等资源库的地址信息。
◎ 程序调试工具信息,例如工具名称、版本号等。
◎ 团队的联系信息,例如电话号码、负责人名称、网址等。

1.4.15　ConfigMap

为了能够准确和深刻理解 Kubernetes ConfigMap 的功能和价值，我们需要从 Docker 说起。我们知道，Docker 通过将程序、依赖库、数据及配置文件"打包固化"到一个不变的镜像文件中的做法，解决了应用的部署的难题，但这同时带来了棘手的问题，即配置文件中的参数在运行期如何修改的问题。我们不可能在启动 Docker 容器后再修改容器里的配置文件，然后用新的配置文件重启容器里的用户主进程。为了解决这个问题，Docker 提供了两种方式：

◎ 在运行时通过容器的环境变量来传递参数；
◎ 通过 Docker Volume 将容器外的配置文件映射到容器内。

这两种方式都有其优势和缺点，在大多数情况下，后一种方式更合适我们的系统，因为大多数应用通常从一个或多个配置文件中读取参数。但这种方式也有明显的缺陷：我们必须在目标主机上先创建好对应的配置文件，然后才能映射到容器里。

上述缺陷在分布式情况下变得更为严重，因为无论采用哪种方式，写入（修改）多台服务器上的某个指定文件，并确保这些文件保持一致，都是一个很难完成的目标。此外，在大多数情况下，我们都希望能集中管理系统的配置参数，而不是管理一堆配置文件。针对上述问题，Kubernetes 给出了一个很巧妙的设计实现，如下所述。

首先，把所有的配置项都当作 key-value 字符串，当然 value 可以来自某个文本文件，比如配置项 password=123456、user=root、host=192.168.8.4 用于表示连接 FTP 服务器的配置参数。这些配置项可以作为 Map 表中的一个项，整个 Map 的数据可以被持久化存储在 Kubernetes 的 Etcd 数据库中，然后提供 API 以方便 Kubernetes 相关组件或客户应用 CRUD 操作这些数据，上述专门用来保存配置参数的 Map 就是 Kubernetes ConfigMap 资源对象。

接下来，Kubernetes 提供了一种内建机制，将存储在 etcd 中的 ConfigMap 通过 Volume 映射的方式变成目标 Pod 内的配置文件，不管目标 Pod 被调度到哪台服务器上，都会完成自动映射。进一步地，如果 ConfigMap 中的 key-value 数据被修改，则映射到 Pod 中的"配置文件"也会随之自动更新。于是，Kubernetes ConfigMap 就成了分布式系统中最为简单（使用方法简单，但背后实现比较复杂）且对应用无侵入的配置中心。ConfigMap 配置集中化的一种简单方案如图 1.16 所示。

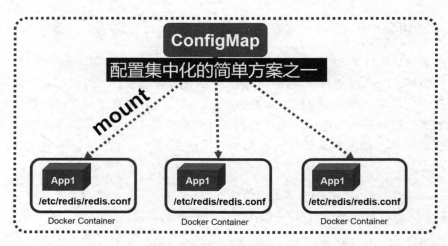

图 1.16 ConfigMap 配置集中化的一种简单方案

1.4.16 小结

上述这些组件是 Kubernetes 系统的核心组件,它们共同构成了 Kubernetes 系统的框架和计算模型。通过对它们进行灵活组合,用户就可以快速、方便地对容器集群进行配置、创建和管理。除了本章所介绍的核心组件,在 Kubernetes 系统中还有许多辅助配置的资源对象,例如 LimitRange、ResourceQuota。另外,一些系统内部使用的对象 Binding、Event 等请参考 Kubernetes 的 API 文档。

在第 2 章中,我们将深入实践并全面掌握 Kubernetes 的各种使用技巧。

第 2 章
Kubernetes 安装配置指南

2.1 系统要求

Kubernetes 系统由一组可执行程序组成，用户可以通过 GitHub 上的 Kubernetes 项目页下载编译好的二进制包，或者下载源代码并编译后进行安装。

安装 Kubernetes 对软件和硬件的系统要求如表 2.1 所示。

表 2.1 安装 Kubernetes 对软件和硬件的系统要求

软　硬　件	最　低　配　置	推　荐　配　置
CPU 和内存	Master：至少 2 core 和 4GB 内存 Node：至少 4 core 和 16GB 内存	Master：4 core 和 16GB 内存。 Node：应根据需要运行的容器数量进行配置
Linux 操作系统	基于 x86_64 架构的各种 Linux 发行版本，包括 Red Hat Linux、CentOS、Fedora、Ubuntu 等，Kernel 版本要求在 3.10 及以上。kubeadm 推荐关闭交换空间的使用，简单说来，执行 swapoff -a 命令，然后在/etc/fstab 中删除对 swap 的加载，并重新启动服务器即可。 也可以在谷歌的 GCE（Google Compute Engine）或者 Amazon 的 AWS（Amazon Web Service）云平台上进行安装	Red Hat Linux 7 CentOS 7
etcd	3.0 版本及以上 下载和安装说明见 https://github.com/coreos/etcd/releases	3.3 版本
Docker（注）	18.03 版本及以上 下载和安装说明见 https://www.docker.com	18.09 版本

Kubernetes 需要容器运行时（Container Runtime Interface，CRI）的支持，目前官方支持的容器运行时包括：Docker、Containerd、CRI-O 和 frakti。容器运行时 CRI 的原理详见 3.9 节的说明。本节以 Docker 作为容器运行环境，推荐版本为 Docker CE 18.09。

宿主机操作系统以 CentOS Linux 7 为例，使用 Systemd 系统完成对 Kubernetes 服务的配置。其他 Linux 发行版的服务配置请参考相关的系统管理手册。为了便于管理，常见的做法是将 Kubernetes 服务程序配置为 Linux 系统开机自启动的服务。

需要注意的是，CentOS Linux 7 默认启动了防火墙服务（firewalld），而 Kubernetes 的 Master 与工作 Node 之间会有大量的网络通信，安全的做法是在防火墙上配置各组件需要相互通信的端口号，具体要配置的端口号详见 2.8 节各服务启动参数中监听的端口号。在安全的内部网络环境中可以关闭防火墙服务：

```
# systemctl disable firewalld
# systemctl stop firewalld
```

另外，建议在主机上禁用 SELinux，让容器可以读取主机文件系统：

```
# setenforce 0
```

或修改系统文件/etc/sysconfig/selinux，将 SELINUX=enforcing 修改成 SELINUX=disabled，然后重启 Linux。

2.2 使用 kubeadm 工具快速安装 Kubernetes 集群

最简单的方法是使用 yum install kubernetes 命令安装 Kubernetes 集群，但仍需修改各组件的启动参数，才能完成对 Kubernetes 集群的配置，整个过程比较复杂，也容易出错。Kubernetes 从 1.4 版本开始引入了命令行工具 kubeadm，致力于简化集群的安装过程，并解决 Kubernetes 集群的高可用问题。在 Kubernetes 1.13 版本中，kubeadm 工具进入 GA 阶段，宣称已经为生产环境应用准备就绪。本节先讲解基于 kubeadm 的安装过程（以 CentOS 7 为例）。

2.2.1 安装 kubeadm 和相关工具

首先配置 yum 源，官方 yum 源的地址为 https://packages.cloud.google.com/yum/repos/kubernetes-el7-x86_64。如果无法访问官方 yum 源的地址，则也可以使用国内的一个 yum 源，地址为 http://mirrors.aliyun.com/kubernetes/yum/repos/kubernetes-el7-x86_64/，yum 源的配置文件/etc/yum.repos.d/kubernetes.repo 的内容如下：

```
[kubernetes]
name=Kubernetes Repository
baseurl=http://mirrors.aliyun.com/kubernetes/yum/repos/kubernetes-el7-x86_64/
enabled=1
gpgcheck=0
```

然后运行 yum install 命令安装 kubeadm 和相关工具：

```
# yum install -y kubelet kubeadm kubectl --disableexcludes=kubernetes
```

运行下面的命令，启动 Docker 服务（如果已安装 Docker，则无须再次启动）和 kubelet

服务，并设置为开机自动启动：

```
# systemctl enable docker && systemctl start docker
# systemctl enable kubelet && systemctl start kubelet
```

2.2.2 kubeadm config

kubeadm 已经进入 GA 阶段，其控制面初始化和加入节点步骤都支持大量的可定制内容，因此 kubeadm 还提供了配置文件功能用于复杂定制。同时，kubeadm 将配置文件以 ConfigMap 的形式保存到集群之中，便于后续的查询和升级工作。kubeadm config 子命令提供了对这一组功能的支持：

◎ kubeadm config upload from-file：由配置文件上传到集群中生成 ConfigMap。
◎ kubeadm config upload from-flags：由配置参数生成 ConfigMap。
◎ kubeadm config view：查看当前集群中的配置值。
◎ kubeadm config print init-defaults：输出 kubeadm init 默认参数文件的内容。
◎ kubeadm config print join-defaults：输出 kubeadm join 默认参数文件的内容。
◎ kubeadm config migrate：在新旧版本之间进行配置转换。
◎ kubeadm config images list：列出所需的镜像列表。
◎ kubeadm config images pull：拉取镜像到本地。

例如，执行 kubeadm config print init-defaults，可以取得默认的初始化参数文件：

```
# kubeadm config print init-defaults > init.default.yaml
```

对生成的文件进行编辑，可以按需生成合适的配置。例如，若需要定制镜像仓库的地址，以及 Pod 的地址范围，则可以使用如下配置：

```
apiVersion: kubeadm.k8s.io/v1beta1
kind: ClusterConfiguration
imageRepository: docker.io/dustise
kubernetesVersion: v1.14.0
networking:
  podSubnet: "192.168.0.0/16"
```

将上面的内容保存为 init-config.yaml 备用。

2.2.3　下载 Kubernetes 的相关镜像

为了从国内的镜像托管站点获得镜像加速支持，建议修改 Docker 的配置文件，增加 Registry Mirror 参数，将镜像配置写入配置参数中，例如 echo '{"registry-mirrors": ["https://registry.docker-cn.com"]}' > /etc/docker/daemon.json，然后重启 Docker 服务。

使用 config images pull 子命令下载所需镜像，例如：

```
# kubeadm config images pull --config=init-config.yaml
[config/images] Pulled docker.io/dustise/kube-apiserver:v1.14.0
[config/images] Pulled docker.io/dustise/kube-controller-manager:v1.14.0
[config/images] Pulled docker.io/dustise/kube-scheduler:v1.14.0
[config/images] Pulled docker.io/dustise/kube-proxy:v1.14.0
[config/images] Pulled docker.io/dustise/pause:3.1
[config/images] Pulled docker.io/dustise/etcd:3.3.10
[config/images] Pulled docker.io/dustise/coredns:1.3.1
```

在镜像下载完成之后，就可以进行安装了。

2.2.4　运行 kubeadm init 命令安装 Master

至此，准备工作已就绪，执行 kubeadm init 命令即可一键安装 Kubernetes 的 Master。

在开始之前需要注意：kubeadm 的安装过程不涉及网络插件（CNI）的初始化，因此 kubeadm 初步安装完成的集群不具备网络功能，任何 Pod 包括自带的 CoreDNS 都无法正常工作。而网络插件的安装往往对 kubeadm init 命令的参数有一定的要求。例如，安装 Calico 插件时需要指定--pod-network-cidr=192.168.0.0/16，详情可参考 https://kubernetes.io/docs/setup/independent/create-cluster-kubeadm/#pod-network。

接下来使用 kubeadm init 命令，使用前面创建的配置文件进行集群控制面的初始化：

```
# kubeadm init --config=init-config.yaml
```

运行后，控制台将输出如下内容：

```
[init] Using Kubernetes version: v1.14.0
......
[preflight] Pulling images required for setting up a Kubernetes cluster
......
[kubelet-start] Activating the kubelet service
......
```

```
[certs] Using certificateDir folder "/etc/kubernetes/pki"
[certs] Generating "ca" certificate and key
[certs] Generating "apiserver" certificate and key
......
[kubeconfig] Using kubeconfig folder "/etc/kubernetes"
[kubeconfig] Writing "admin.conf" kubeconfig file
[kubeconfig] Writing "kubelet.conf" kubeconfig file
......
[kubeconfig] Writing "scheduler.conf" kubeconfig file
[control-plane] Using manifest folder "/etc/kubernetes/manifests"
[control-plane] Creating static Pod manifest for "kube-apiserver"
```

等待一段时间后,Kubernetes 的 Master 安装成功,显示如下信息:

```
[addons] Applied essential addon: CoreDNS
[addons] Applied essential addon: kube-proxy

Your Kubernetes control-plane has initialized successfully!

To start using your cluster, you need to run the following as a regular user:

  mkdir -p $HOME/.kube
  sudo cp -i /etc/kubernetes/admin.conf $HOME/.kube/config
  sudo chown $(id -u):$(id -g) $HOME/.kube/config

You should now deploy a pod network to the cluster.
Run "kubectl apply -f [podnetwork].yaml" with one of the options listed at:
  https://kubernetes.io/docs/concepts/cluster-administration/addons/

Then you can join any number of worker nodes by running the following on each as root:

kubeadm join 10.211.55.30:6443 --token ah9koe.nvuvz2v60iam0e0d \
    --discovery-token-ca-cert-hash
sha256:9ded80601bc7f5568a9a7ece7ee13fd73be193777641054420a080f778b330fc
```

按照提示执行下面的命令,复制配置文件到普通用户的 home 目录下:

```
# mkdir -p $HOME/.kube
# sudo cp -i /etc/kubernetes/admin.conf $HOME/.kube/config
# sudo chown $(id -u):$(id -g) $HOME/.kube/config
```

这样就在 Master 上安装了 Kubernetes,但在集群内还是没有可用的工作 Node,并缺乏对容器网络的配置。

这里需要注意 kubeadm init 命令执行完成后的最后几行提示信息,其中包含加入节点的指令(kubeadm join)和所需的 Token。

此时可以用 kubectl 命令验证在 2.2.2 节中提到的 ConfigMap:

```
# kubectl get -n kube-system configmap
kubectl get -n kube-system configmap
NAME                                  DATA   AGE
coredns                               1      48m
extension-apiserver-authentication    6      48m
kube-proxy                            2      48m
kubeadm-config                        2      48m
kubelet-config-1.14                   1      48m
```

可以看到其中生成了名为 kubeadm-config 的 ConfigMap 对象。

2.2.5　安装 Node,加入集群

对于新节点的添加,系统准备和 Kubernetes yum 源的配置过程是一致的,在 Node 主机上执行下面的安装过程。

(1)安装 kubeadm 和相关工具:

```
# yum install kubelet kubeadm --disableexcludes=kubernetes
```

运行下面的命令启动 Docker 服务与 kubelet 服务,并将其设置为开机自启动:

```
# systemctl enable docker && systemctl start docker
# systemctl enable kubelet && systemctl start kubelet
```

(2)为 kubeadm 命令生成配置文件。创建文件 join-config.yaml,内容如下:

```
apiVersion: kubeadm.k8s.io/v1beta1
kind: JoinConfiguration
discovery:
  bootstrapToken:
    apiServerEndpoint: 10.211.55.30:6443
    token: ah9koe.nvuvz2v60iam0e0d
    unsafeSkipCAVerification: true
  tlsBootstrapToken: ah9koe.nvuvz2v60iam0e0d
```

其中，apiServerEndpoint 的值来自 Master 服务器的地址，token 和 tlsBootstrapToken 的值就来自于使用 kubeadm init 安装 Master 的最后一行提示信息。

（3）执行 kubeadm join 命令，将本 Node 加入集群：

```
# kubeadm join --config=join-config.yaml
[preflight] Running pre-flight checks
......
[preflight] Reading configuration from the cluster...
[preflight] FYI: You can look at this config file with 'kubectl -n kube-system get cm kubeadm-config -oyaml'
[kubelet-start] Downloading configuration for the kubelet from the "kubelet-config-1.14" ConfigMap in the kube-system namespace
[kubelet-start] Writing kubelet configuration to file "/var/lib/kubelet/config.yaml"
[kubelet-start] Writing kubelet environment file with flags to file "/var/lib/kubelet/kubeadm-flags.env"
[kubelet-start] Activating the kubelet service
[kubelet-start] Waiting for the kubelet to perform the TLS Bootstrap...

This node has joined the cluster:
* Certificate signing request was sent to apiserver and a response was received.
* The Kubelet was informed of the new secure connection details.

Run 'kubectl get nodes' on the control-plane to see this node join the cluster.
```

kubeadm 在 Master 上也安装了 kubelet，在默认情况下并不参与工作负载。如果希望安装一个单机 All-In-One 的 Kubernetes 环境，则可以执行下面的命令（删除 Node 的 Label "node-role.kubernetes.io/master"），让 Master 成为一个 Node：

```
# kubectl taint nodes --all node-role.kubernetes.io/master-
```

2.2.6 安装网络插件

执行 kubectl get nodes 命令，会发现 Kubernetes 提示 Master 为 NotReady 状态，这是因为还没有安装 CNI 网络插件：

```
# kubectl get nodes
NAME              STATUS     ROLES    AGE    VERSION
node-kubeadm-1    NotReady   master   67m    v1.14.0
node-kubeadm-2    NotReady   <none>   2m9s   v1.14.0
```

下面根据 kubeadm 的提示安装 CNI 网络插件。对于 CNI 网络插件，可以有许多选择，请参考 https://kubernetes.io/docs/setup/independent/create-cluster-kubeadm/#pod-network 的说明。

例如，选择 weave 插件，执行下面的命令即可一键完成安装：

```
# kubectl apply -f "https://cloud.weave.works/k8s/net?k8s-version=$(kubectl version | base64 | tr -d '\n')"
serviceaccount/weave-net created
clusterrole.rbac.authorization.k8s.io/weave-net created
clusterrolebinding.rbac.authorization.k8s.io/weave-net created
role.rbac.authorization.k8s.io/weave-net created
rolebinding.rbac.authorization.k8s.io/weave-net created
daemonset.extensions/weave-net created
```

2.2.7 验证 Kubernetes 集群是否安装完成

执行下面的命令，验证 Kubernetes 集群的相关 Pod 是否都正常创建并运行：

```
# kubectl get pods --all-namespaces
NAMESPACE     NAME                                READY   STATUS    RESTARTS   AGE
kube-system   coredns-7d8f5b649d-2lckr            1/1     Running   0          71m
kube-system   coredns-7d8f5b649d-stssz            1/1     Running   0          71m
kube-system   etcd-node1                          1/1     Running   0          70m
kube-system   kube-apiserver-node1                1/1     Running   0          70m
kube-system   kube-controller-manager-node1       1/1     Running   0          70m
kube-system   kube-proxy-tnm4c                    1/1     Running   0          71m
kube-system   kube-proxy-xjwgw                    1/1     Running   0          10m
kube-system   kube-scheduler-node1                1/1     Running   0          70m
kube-system   weave-net-nsfmn                     2/2     Running   0          61s
kube-system   weave-net-qfk5x                     2/2     Running   0          61s
```

如果发现有状态错误的 Pod，则可以执行 kubectl --namespace=kube-system describe pod <pod_name> 来查看错误原因，常见的错误原因是镜像没有下载完成。

至此，通过 kubeadm 工具就实现了 Kubernetes 集群的快速搭建。如果安装失败，则可以执行 kubeadm reset 命令将主机恢复原状，重新执行 kubeadm init 命令，再次进行安装。

2.3 以二进制文件方式安装 Kubernetes 集群

本节以二进制文件方式安装 Kubernetes 集群,并对每个组件的配置进行详细说明。

从 Kubernetes 发布官网 https://github.com/kubernetes/kubernetes/releases 找到对应的版本号,单击 CHANGELOG,找到已编译好的二进制文件的下载页面,如图 2.1 和图 2.2 所示。本书基于 Kubernetes 1.14 版本进行说明。

在压缩包 kubernetes.tar.gz 内包含了 Kubernetes 的服务程序文件、文档和示例;在压缩包 kubernetes-src.tar.gz 内则包含了全部源代码。也可以直接下载 Server Binaries 中的 kubernetes-server-linux-amd64.tar.gz 文件,其中包含了 Kubernetes 需要运行的全部服务程序文件。主要的服务程序文件列表如表 2.2 所示。

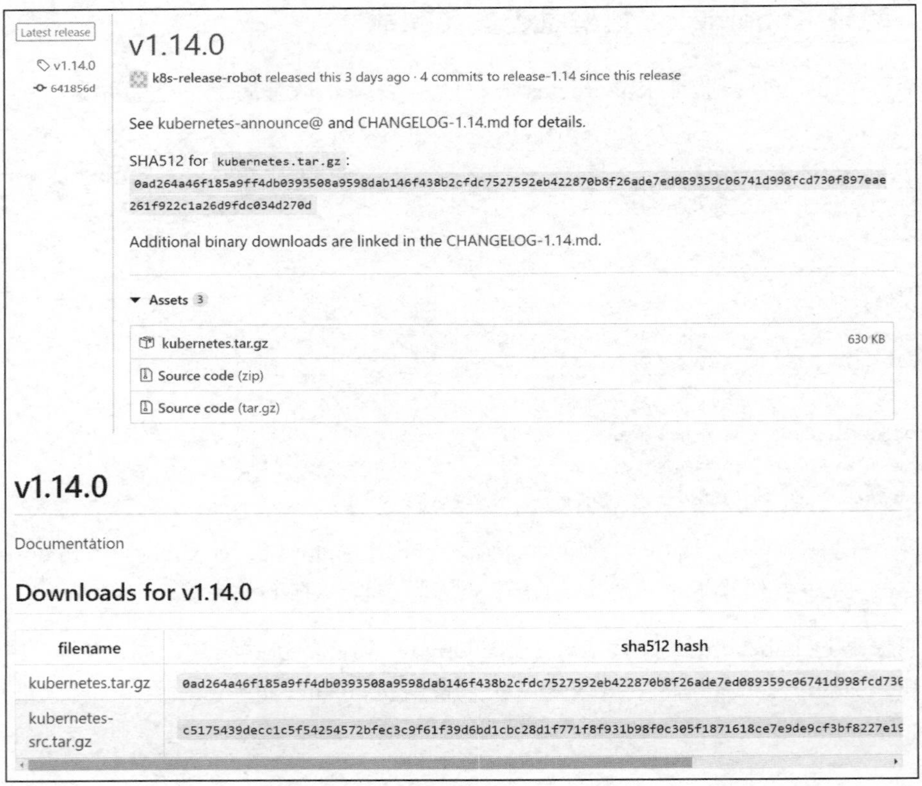

图 2.1 GitHub 上 Kubernetes 的下载页面一

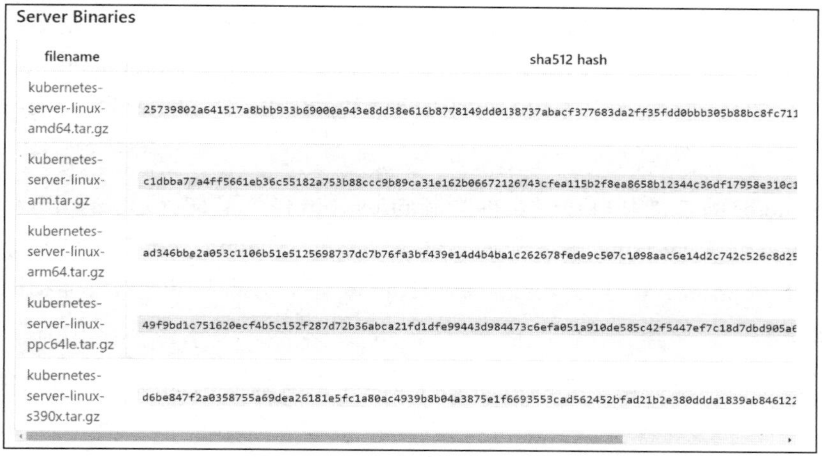

Server Binaries

filename	sha512 hash
kubernetes-server-linux-amd64.tar.gz	25739802a641517a8bbb933b69000a943e8dd38e616b8778149dd0138737abacf377683da2ff35fdd0bbb305b88bc8fc711
kubernetes-server-linux-arm.tar.gz	c1dbba77a4ff5661eb36c55182a753b88ccc9b89ca31e162b06672126743cfea115b2f8ea8658b12344c36df17958e310c1
kubernetes-server-linux-arm64.tar.gz	ad346bbe2a053c1106b51e5125698737dc7b76fa3bf439e14d4b4ba1c262678fede9c507c1098aac6e14d2c742c526c8d25
kubernetes-server-linux-ppc64le.tar.gz	49f9bd1c751620ecf4b5c152f287d72b36abca21fd1dfe99443d984473c6efa051a910de585c42f5447ef7c18d7dbd905a6
kubernetes-server-linux-s390x.tar.gz	d6be847f2a0358755a69dea26181e5fc1a80ac4939b8b04a3875e1f6693553cad562452bfad21b2e380ddda1839ab846122

图 2.2　GitHub 上 Kubernetes 的下载页面二

表 2.2　主要的服务程序文件列表

文　件　名	说　　明
kube-apiserver	kube-apiserver 主程序
kube-apiserver.docker_tag	kube-apiserver docker 镜像的 tag
kube-apiserver.tar	kube-apiserver docker 镜像文件
kube-controller-manager	kube-controller-manager 主程序
kube-controller-manager.docker_tag	kube-controller-manager docker 镜像的 tag
kube-controller-manager.tar	kube-controller-manager docker 镜像文件
kube-scheduler	kube-scheduler 主程序
kube-scheduler.docker_tag	kube-scheduler docker 镜像的 tag
kube-scheduler.tar	kube-scheduler docker 镜像文件
kubelet	kubelet 主程序
kube-proxy	kube-proxy 主程序
kube-proxy.tar	kube-proxy docker 镜像文件
kubectl	客户端命令行工具
kubeadm	Kubernetes 集群安装的命令行工具
hyperkube	包含了所有服务的程序，可以启动任一服务
cloud-controller-manager	提供与云提供商服务对接的各种 Controller
apiextensions-apiserver	提供实现自定义资源对象的扩展 API Server

Kubernetes 的主要服务程序都可以通过直接运行二进制文件加上启动参数完成运行。在 Kubernetes 的 Master 上需要部署 etcd、kube-apiserver、kube-controller-manager、kube-scheduler 服务进程，在工作 Node 上需要部署 docker、kubelet 和 kube-proxy 服务进程。

将 Kubernetes 的二进制可执行文件复制到 /usr/bin 目录下，然后在 /usr/lib/system/system 目录下为各服务创建 systemd 服务配置文件（完整的 systemd 系统知识请参考 Linux 的相关手册），这样就完成了软件的安装。要使 Kubernetes 正常工作，需要详细配置各个服务的启动参数。

下面主要介绍各服务最主要的启动参数，每个服务完整启动的参数说明详见 2.8 节。

2.3.1 Master 上的 etcd、kube-apiserver、kube-controller-manager、kube-scheduler 服务

1. etcd 服务

etcd 服务作为 Kubernetes 集群的主数据库，在安装 Kubernetes 各服务之前需要首先安装和启动。

从 GitHub 官网（https://github.com/coreos/etcd/releases）下载 etcd 二进制文件，将 etcd 和 etcdctl 文件复制到 /usr/bin 目录下。

设置 systemd 服务配置文件 /usr/lib/systemd/system/etcd.service：

```
[Unit]
Description=Etcd Server
After=network.target

[Service]
Type=simple
WorkingDirectory=/var/lib/etcd/
EnvironmentFile=-/etc/etcd/etcd.conf
ExecStart=/usr/bin/etcd

[Install]
WantedBy=multi-user.target
```

其中，WorkingDirectory（/var/lib/etcd/）表示 etcd 数据保存的目录，需要在启动 etcd

服务之前创建。

配置文件/etc/etcd/etcd.conf 通常不需要特别的参数设置（详细的参数配置内容参见官方文档），etcd 默认使用 http://127.0.0.1:2379 地址供客户端连接。

配置完成后，通过 systemctl start 命令启动 etcd 服务。同时，使用 systemctl enable 命令将服务加入开机启动列表中：

```
# systemctl daemon-reload
# systemctl enable etcd.service
# systemctl start etcd.service
```

通过执行 etcdctl cluster-health，可以验证 etcd 是否正确启动：

```
# etcdctl endpoint health
127.0.0.1:2379 is healthy: successfully committed proposal: took = 1.33112ms
```

2. kube-apiserver 服务

将 kube-apiserver、kube-controller-manager 和 kube-scheduler 文件复制到/usr/bin 目录。

设置 systemd 服务配置文件/usr/lib/systemd/system/kube-apiserver.service，内容如下：

```
[Unit]
Description=Kubernetes API Server
Documentation=https://github.com/GoogleCloudPlatform/kubernetes
After=etcd.service
Wants=etcd.service

[Service]
EnvironmentFile=/etc/kubernetes/apiserver
ExecStart=/usr/bin/kube-apiserver $KUBE_API_ARGS
Restart=on-failure
Type=notify
LimitNOFILE=65536

[Install]
WantedBy=multi-user.target
```

配置文件/etc/kubernetes/apiserver 的内容包括了 kube-apiserver 的全部启动参数，主要的配置参数在变量 KUBE_API_ARGS 中指定。

```
# cat /etc/kubernetes/apiserver
```

```
    KUBE_API_ARGS="--etcd-servers=http://127.0.0.1:2379
--insecure-bind-address=0.0.0.0 --insecure-port=8080
--service-cluster-ip-range=169.169.0.0/16 --service-node-port-range=1-65535
--enable-admission-plugins=NamespaceLifecycle,LimitRanger,ServiceAccount,Default
StorageClass,DefaultTolerationSeconds,MutatingAdmissionWebhook,ValidatingAdmissi
onWebhook,ResourceQuota --logtostderr=false --log-dir=/var/log/kubernetes --v=0"
```

对启动参数说明如下。

- --etcd-servers：指定 etcd 服务的 URL。
- --storage-backend：指定 etcd 的版本，从 Kubernetes 1.6 开始，默认为 etcd 3。注意，在 Kubernetes 1.6 之前的版本中没有这个参数，kube-apiserver 默认使用 etcd 2，对于正在运行的 1.5 或旧版本的 Kubernetes 集群，etcd 提供了数据升级方案，详见 etcd 文档（https://coreos.com/etcd/docs/latest/upgrades/upgrade_3_0.html）。
- --insecure-bind-address：API Server 绑定主机的非安全 IP 地址，设置 0.0.0.0 表示绑定所有 IP 地址。
- --insecure-port：API Server 绑定主机的非安全端口号，默认为 8080。
- --service-cluster-ip-range：Kubernetes 集群中 Service 的虚拟 IP 地址范围，以 CIDR 格式表示，例如 169.169.0.0/16，该 IP 范围不能与物理机的 IP 地址有重合。
- --service-node-port-range：Kubernetes 集群中 Service 可使用的物理机端口号范围，默认值为 30000～32767。
- --enable-admission-plugins：Kubernetes 集群的准入控制设置，各控制模块以插件的形式依次生效。
- --logtostderr：设置为 false 表示将日志写入文件，不写入 stderr。
- --log-dir：日志目录。
- --v：日志级别。

3. kube-controller-manager 服务

kube-controller-manager 服务依赖于 kube-apiserver 服务，设置 systemd 服务配置文件 /usr/lib/systemd/system/kube-controller-manager.service，内容如下：

```
[Unit]
Description=Kubernetes Controller Manager
Documentation=https://github.com/GoogleCloudPlatform/kubernetes
After=kube-apiserver.service
Requires=kube-apiserver.service
```

```
[Service]
EnvironmentFile=/etc/kubernetes/controller-manager
ExecStart=/usr/bin/kube-controller-manager $KUBE_CONTROLLER_MANAGER_ARGS
Restart=on-failure
LimitNOFILE=65536

[Install]
WantedBy=multi-user.target
```

配置文件/etc/kubernetes/controller-manager 的内容包含了 kube-controller-manager 的全部启动参数，主要的配置参数在变量 KUBE_CONTROLLER_MANAGER_ARGS 中指定：

```
# cat /etc/kubernetes/controller-manager
KUBE_CONTROLLER_MANAGER_ARGS="--kubeconfig=/etc/kubernetes/kubeconfig
--logtostderr=false --log-dir=/var/log/kubernetes --v=0"
```

对启动参数说明如下。

◎ --kubeconfig：设置与 API Server 连接的相关配置，例如：

```
apiVersion: v1
kind: Config
users:
- name: client
  user:
clusters:
- name: default
  cluster:
    server: http://192.168.18.3:8080
contexts:
- context:
    cluster: default
    user: client
  name: default
current-context: default
```

◎ --logtostderr：设置为 false 表示将日志写入文件，不写入 stderr。
◎ --log-dir：日志目录。
◎ --v：日志级别。

4. kube-scheduler 服务

kube-scheduler 服务也依赖于 kube-apiserver 服务，设置 systemd 服务配置文件 /usr/lib/systemd/system/kube-scheduler.service，内容如下：

```
[Unit]
Description=Kubernetes Controller Manager
Documentation=https://github.com/GoogleCloudPlatform/kubernetes
After=kube-apiserver.service
Requires=kube-apiserver.service

[Service]
EnvironmentFile=/etc/kubernetes/scheduler
ExecStart=/usr/bin/kube-scheduler $KUBE_SCHEDULER_ARGS
Restart=on-failure
LimitNOFILE=65536

[Install]
WantedBy=multi-user.target
```

配置文件/etc/kubernetes/scheduler 的内容包括了 kube-scheduler 的全部启动参数，主要的配置参数在变量 KUBE_SCHEDULER_ARGS 中指定：

```
# cat /etc/kubernetes/scheduler
KUBE_SCHEDULER_ARGS="--kubeconfig=/etc/kubernetes/kubeconfig
--logtostderr=false --log-dir=/var/log/kubernetes --v=0"
```

对启动参数说明如下。

- --kubeconfig：设置与 API Server 连接的相关配置，可以与 kube-controller-manager 使用的 kubeconfig 文件相同。
- --logtostderr：设置为 false 表示将日志写入文件，不写入 stderr。
- --log-dir：日志目录。
- --v：日志级别。

配置完成后，执行 systemctl start 命令按顺序启动这 3 个服务，同时，使用 systemctl enable 命令将服务加入开机启动列表中：

```
# systemctl daemon-reload
# systemctl enable kube-apiserver.service
# systemctl start kube-apiserver.service
# systemctl enable kube-controller-manager
```

```
# systemctl start kube-controller-manager
# systemctl enable kube-scheduler
# systemctl start kube-scheduler
```

通过 systemctl status <service_name> 验证服务的启动状态，running 表示启动成功。

至此，Master 上所需的服务就全部启动完成了。

2.3.2 Node 上的 kubelet、kube-proxy 服务

在工作 Node 上需要预先安装好 Docker Daemon 并且正常启动。Docker 的安装和启动详见 Docker 官网 http://www.docker.com 的说明文档。

1. kubelet 服务

kubelet 服务依赖于 Docker 服务，设置 systemd 服务配置文件/usr/lib/systemd/system/kubelet.service，内容如下：

```
[Unit]
Description=Kubernetes Kubelet Server
Documentation=https://github.com/GoogleCloudPlatform/kubernetes
After=docker.service
Requires=docker.service

[Service]
WorkingDirectory=/var/lib/kubelet
EnvironmentFile=/etc/kubernetes/kubelet
ExecStart=/usr/bin/kubelet $KUBELET_ARGS
Restart=on-failure

[Install]
WantedBy=multi-user.target
```

其中，WorkingDirectory 表示 kubelet 保存数据的目录，需要在启动 kubelet 服务之前创建。

配置文件/etc/kubernetes/kubelet 的内容包括了 kubelet 的全部启动参数，主要的配置参数在变量 KUBELET_ARGS 中指定：

```
# cat /etc/kubernetes/kubelet
KUBELET_ARGS="--kubeconfig=/etc/kubernetes/kubeconfig
```

```
--hostname-override=192.168.18.3 --logtostderr=false
--log-dir=/var/log/kubernetes --v=0"
```

对启动参数说明如下。

◎ --kubeconfig：设置与 API Server 连接的相关配置，可以与 kube-controller-manager 使用的 kubeconfig 文件相同。
◎ --hostname-override：设置本 Node 的名称。
◎ --logtostderr：设置为 false 表示将日志写入文件，不写入 stderr。
◎ --log-dir：日志目录。
◎ --v：日志级别。

2. kube-proxy 服务

kube-proxy 服务依赖于 network 服务，设置 systemd 服务配置文件/usr/lib/systemd/system/kube-proxy.service，内容如下：

```
[Unit]
Description=Kubernetes Kube-Proxy Server
Documentation=https://github.com/GoogleCloudPlatform/kubernetes
After=network.target
Requires=network.service

[Service]
EnvironmentFile=/etc/kubernetes/proxy
ExecStart=/usr/bin/kube-proxy $KUBE_PROXY_ARGS
Restart=on-failure
LimitNOFILE=65536

[Install]
WantedBy=multi-user.target
```

配置文件/etc/kubernetes/proxy 的内容包括了 kube-proxy 的全部启动参数，主要的配置参数在变量 KUBE_PROXY_ARGS 中指定：

```
# cat /etc/kubernetes/proxy
KUBE_PROXY_ARGS="--kubeconfig=/etc/kubernetes/kubeconfig --logtostderr=false --log-dir=/var/log/kubernetes --v=2"
```

对启动参数说明如下。

- ◎ --kubeconfig：设置与 API Server 连接的相关配置，可以与 kube-controller-manager 使用的 kubeconfig 文件相同。
- ◎ --logtostderr：设置为 false 表示将日志写入文件，不写入 stderr。
- ◎ --log-dir：日志目录。
- ◎ --v：日志级别。

配置完成后，通过 systemctl 启动 kubelet 和 kube-proxy 服务：

```
# systemctl daemon-reload
# systemctl enable kubelet.service
# systemctl start kubelet.service
# systemctl enable kube-proxy
# systemctl start kube-proxy
```

kubelet 默认采用向 Master 自动注册本 Node 的机制，在 Master 上查看各 Node 的状态，状态为 Ready 表示 Node 已经成功注册并且状态为可用：

```
# kubectl get nodes
NAME              STATUS      AGE
192.168.18.3      Ready       1m
```

等所有 Node 的状态都为 Ready 之后，一个 Kubernetes 集群就启动完成了。接下来可以创建 Pod、Deployment、Service 等资源对象来部署容器应用了。

2.4 Kubernetes 集群的安全设置

2.4.1 基于 CA 签名的双向数字证书认证方式

在一个安全的内网环境中，Kubernetes 的各个组件与 Master 之间可以通过 kube-apiserver 的非安全端口 http://<kube-apiserver-ip>:8080 进行访问。但如果 API Server 需要对外提供服务，或者集群中的某些容器也需要访问 API Server 以获取集群中的某些信息，则更安全的做法是启用 HTTPS 安全机制。Kubernetes 提供了基于 CA 签名的双向数字证书认证方式和简单的基于 HTTP Base 或 Token 的认证方式，其中 CA 证书方式的安全性最高。本节先介绍如何以 CA 证书的方式配置 Kubernetes 集群，要求 Master 上的 kube-apiserver、kube-controller-manager、kube-scheduler 进程及各 Node 上的 kubelet、 kube-proxy 进程进行 CA 签名双向数字证书安全设置。

基于 CA 签名的双向数字证书的生成过程如下。

（1）为 kube-apiserver 生成一个数字证书，并用 CA 证书签名。

（2）为 kube-apiserver 进程配置证书相关的启动参数，包括 CA 证书（用于验证客户端证书的签名真伪）、自己的经过 CA 签名后的证书及私钥。

（3）为每个访问 Kubernetes API Server 的客户端（如 kube-controller-manager、kube-scheduler、kubelet、kube-proxy 及调用 API Server 的客户端程序 kubectl 等）进程都生成自己的数字证书，也都用 CA 证书签名，在相关程序的启动参数里增加 CA 证书、自己的证书等相关参数。

1. 设置 kube-apiserver 的 CA 证书相关的文件和启动参数

使用 OpenSSL 工具在 Master 服务器上创建 CA 证书和私钥相关的文件：

```
# openssl genrsa -out ca.key 2048
# openssl req -x509 -new -nodes -key ca.key -subj "/CN=k8s-master" -days 5000 -out ca.crt
# openssl genrsa -out server.key 2048
```

注意：在生成 ca.crt 时，-subj 参数中 "/CN" 的值为 Master 主机名。

准备 master_ssl.cnf 文件，该文件用于 x509 v3 版本的证书。在该文件中主要需要设置 Master 服务器的 hostname（k8s-master）、IP 地址（192.168.18.3），以及 Kubernetes Master Service 的虚拟服务名称（kubernetes.default 等）和该虚拟服务的 ClusterIP 地址（169.169.0.1）。

master_ssl.cnf 文件的示例如下：

```
[req]
req_extensions = v3_req
distinguished_name = req_distinguished_name
[req_distinguished_name]
[ v3_req ]
basicConstraints = CA:FALSE
keyUsage = nonRepudiation, digitalSignature, keyEncipherment
subjectAltName = @alt_names
[alt_names]
DNS.1 = kubernetes
```

```
DNS.2 = kubernetes.default
DNS.3 = kubernetes.default.svc
DNS.4 = kubernetes.default.svc.cluster.local
DNS.5 = k8s-master
IP.1 = 169.169.0.1
IP.2 = 192.168.18.3
```

基于 master_ssl.cnf 创建 server.csr 和 server.crt 文件。在生成 server.csr 时，-subj 参数中 "/CN" 的值需为 Master 的主机名：

```
# openssl req -new -key server.key -subj "/CN=k8s-master" -config master_ssl.cnf -out server.csr
# openssl x509 -req -in server.csr -CA ca.crt -CAkey ca.key -CAcreateserial -days 5000 -extensions v3_req -extfile master_ssl.cnf -out server.crt
```

在全部执行完后会生成 6 个文件：ca.crt、ca.key、ca.srl、server.crt、server.csr、server.key。

将这些文件复制到一个目录下（例如/var/run/kubernetes/），然后设置 kube-apiserver 的三个启动参数 "--client-ca-file" "--tls-cert-file" 和 "--tls-private-key-file"，分别代表 CA 根证书文件、服务端证书文件和服务端私钥文件：

```
--client-ca-file=/var/run/kubernetes/ca.crt
--tls-private-key-file=/var/run/kubernetes/server.key
--tls-cert-file=/var/run/kubernetes/server.crt
```

同时，可以关闭非安全端口（设置--insecure-port=0），设置安全端口为 6443（默认值）：

```
--insecure-port=0
--secure-port=6443
```

最后重启 kube-apiserver 服务。

2. 设置 kube-controller-manager 的客户端证书、私钥和启动参数

代码如下：

```
$ openssl genrsa -out cs_client.key 2048
$ openssl req -new -key cs_client.key -subj "/CN=k8s-master" -out cs_client.csr
$ openssl x509 -req -in cs_client.csr -CA ca.crt -CAkey ca.key -CAcreateserial -out cs_client.crt -days 5000
```

其中，在生成 cs_client.crt 时，-CA 参数和-CAkey 参数使用的是 API Server 的 ca.crt 和 ca.key 文件。然后将这些文件复制到一个目录下（例如/var/run/kubernetes/）。

接下来创建/etc/kubernetes/kubeconfig 文件（kube-controller-manager 与 kube-scheduler 共用），配置客户端证书等相关参数，内容如下：

```
apiVersion: v1
kind: Config
users:
- name: controllermanager
  user:
    client-certificate: /var/run/kubernetes/cs_client.crt
    client-key: /var/run/kubernetes/cs_client.key
clusters:
- name: local
  cluster:
    certificate-authority: /var/run/kubernetes/ca.crt
    server: https://192.168.18.3:6443
contexts:
- context:
    cluster: local
    user: controllermanager
  name: my-context
current-context: my-context
```

然后设置 kube-controller-manager 服务的启动参数：

```
--service-account-key-file=/var/run/kubernetes/server.key
--root-ca-file=/var/run/kubernetes/ca.crt
--kubeconfig=/etc/kubernetes/kubeconfig
```

最后重启 kube-controller-manager 服务。

3. 设置 kube-scheduler 启动参数

kube-scheduler 复用上一步 kube-controller-manager 创建的客户端证书，配置启动参数：

```
--kubeconfig=/etc/kubernetes/kubeconfig
```

重启 kube-scheduler 服务。

4. 设置每个 Node 上 kubelet 的客户端证书、私钥和启动参数

首先，复制 kube-apiserver 的 ca.crt 和 ca.key 文件到 Node 上，在生成 kubelet_client.crt 时 -CA 参数和 -CAkey 参数使用的是 API Server 的 ca.crt 和 ca.key 文件；在生成

kubelet_client.csr 时,将 -subj 参数中的"/CN"设置为本 Node 的 IP 地址:

```
$ openssl genrsa -out kubelet_client.key 2048
$ openssl req -new -key kubelet_client.key -subj "/CN=192.168.18.4" -out kubelet_client.csr
$ openssl x509 -req -in kubelet_client.csr -CA ca.crt -CAkey ca.key -CAcreateserial -out kubelet_client.crt -days 5000
```

将这些文件复制到一个目录下(例如/var/run/kubernetes/)。

接下来创建/etc/kubernetes/kubeconfig 文件(kubelet 和 kube-proxy 进程共用),配置客户端证书等相关参数,内容如下:

```
apiVersion: v1
kind: Config
users:
- name: kubelet
  user:
    client-certificate: /etc/kubernetes/ssl_keys/kubelet_client.crt
    client-key: /etc/kubernetes/ssl_keys/kubelet_client.key
clusters:
- name: local
  cluster:
    certificate-authority: /etc/kubernetes/ssl_keys/ca.crt
    server: https://192.168.18.3:6443
contexts:
- context:
    cluster: local
    user: kubelet
  name: my-context
current-context: my-context
```

然后设置 kubelet 服务的启动参数:

```
--kubeconfig=/etc/kubelet/kubeconfig
```

最后重启 kubelet 服务。

5. 设置 kube-proxy 的启动参数

kube-proxy 复用上一步 kubelet 创建的客户端证书,配置启动参数:

```
--kubeconfig=/etc/kubernetes/kubeconfig
```

重启 kube-proxy 服务。

至此，一个基于 CA 的双向数字证书认证的 Kubernetes 集群环境就搭建完成了。

6. 设置 kubectl 客户端使用安全方式访问 API Server

在使用 kubectl 对 Kubernetes 集群进行操作时，默认使用非安全端口 8080 对 API Server 进行访问，也可以设置为安全访问 API Server 的模式，需要设置 3 个证书相关的参数"—certificate -authority""--client-certificate"和"--client-key"，分别表示用于 CA 授权的证书、客户端证书和客户端密钥。

- --certificate-authority：使用为 kube-apiserver 生成的 ca.crt 文件。
- --client-certificate：使用为 kube-controller-manager 生成的 cs_client.crt 文件。
- --client-key：使用为 kube-controller-manager 生成的 cs_client.key 文件。

同时，指定 API Server 的 URL 地址为 HTTPS 安全地址（例如 https://k8s-master:443），最后输入需要执行的子命令，即可对 API Server 进行安全访问了：

```
# kubectl --server=https://192.168.18.3:6443
--certificate-authority=/etc/kubernetes/ssl_keys/ca.crt
--client-certificate=/etc/kubernetes/ssl_keys/cs_client.crt
--client-key=/etc/kubernetes/ssl_keys/cs_client.key get nodes
   NAME            STATUS      AGE
   k8s-node-1      Ready       1h
```

2.4.2 基于 HTTP Base 或 Token 的简单认证方式

除了提供了基于 CA 的双向数字证书认证方式，Kubernetes 也提供了基于 HTTP Base 或 Token 的简单认证方式。各组件与 API Server 之间的通信方式仍然采用 HTTPS，但不使用 CA 数字证书。

采用基于 HTTP Base 或 Token 的简单认证方式时，API Server 对外暴露 HTTPS 端口，客户端提供用户名、密码或 Token 来完成认证过程。需要说明的是，kubectl 命令行工具比较特殊，它同时支持 CA 双向认证和简单认证两种模式与 API Server 通信，其他客户端组件只能配置为双向安全认证或非安全模式与 API Server 通信。

1. 基于 HTTP Base 认证的配置过程

（1）创建包括用户名、密码和 UID 的文件 basic_auth_file，放置在合适的目录下，例如 /etc/kuberntes 目录。需要注意的是，这是一个纯文本文件，用户名、密码都是明文。

```
# vi /etc/kubernetes/basic_auth_file
admin,admin,1
system,system,2
```

（2）设置 kube-apiserver 的启动参数 "--basic-auth-file"，使用上述文件提供安全认证：

```
--secure-port=6443
--basic-auth-file=/etc/kubernetes/basic_auth_file
```

然后，重启 API Server 服务。

（3）使用 kubectl 通过指定的用户名和密码来访问 API Server：

```
# kubectl --server=https://192.168.18.3:6443 --username=admin --password=admin --insecure-skip-tls-verify=true get nodes
NAME            STATUS      AGE
k8s-node-1      Ready       1h
```

2. 基于 Token 认证的配置过程

（1）创建包括用户名、密码和 UID 的文件 token_auth_file，放置在合适的目录下，例如 /etc/kuberntes 目录。需要注意的是，这是一个纯文本文件，用户名、密码都是明文。

```
$ cat /etc/kubernetes/token_auth_file
admin,admin,1
system,system,2
```

（2）设置 kube-apiserver 的启动参数 "--token-auth-file"，使用上述文件提供安全认证：

```
--secure-port=6443
--token-auth-file=/etc/kubernetes/token_auth_file
```

然后，重启 API Server 服务。

（3）用 curl 验证和访问 API Server：

```
$ curl -k --header "Authorization:Bearer admin" https://192.168.18.3:6443/version
{
  "major": "1",
```

```
"minor": "14",
"gitVersion": "v1.14.0",
"gitCommit": "641856db18352033a0d96dbc99153fa3b27298e5",
"gitTreeState": "clean",
"buildDate": "2019-03-25T15:45:25Z",
"goVersion": "go1.12.1",
"compiler": "gc",
"platform": "linux/amd64"
}
```

2.5　Kubernetes集群的网络配置

在多个Node组成的Kubernetes集群内，跨主机的容器间网络互通是Kubernetes集群能够正常工作的前提条件。Kubernetes本身并不会对跨主机的容器网络进行设置，这需要额外的工具来实现。除了谷歌公有云GCE平台提供的网络设置，一些开源的工具包括Flannel、Open vSwitch、Weave、Calico等都能够实现跨主机的容器间网络互通。随着CNI网络模型的逐渐成熟，Kubernetes将优先使用CNI网络插件打通跨主机的容器网络。具体的网络原理及主流开源网络工具的原理和配置详见第7章的说明。

2.6　内网中的Kubernetes相关配置

Kubernetes在能够访问Internet网络的环境中使用起来非常方便：一方面，在docker.io和gcr.io网站中已经存在了大量官方制作的Docker镜像；另一方面，GCE、AWS提供的云平台已经成熟，用户通过租用一定的空间来部署Kubernetes集群也很简便。

但是，许多企业内部由于安全原因无法访问Internet，需要通过创建一个内部的私有Docker Registry，并修改一些Kubernetes的配置，来启动内网中的Kubernetes集群。

2.6.1　Docker Private Registry（私有Docker镜像库）

使用Docker提供的Registry镜像创建一个私有镜像仓库。

详细的安装步骤请参考Docker的官方文档 https://docs.docker.com/registry/deploying/。

2.6.2 kubelet 配置

由于在 Kubernetes 中是以 Pod 而不是以 Docker 容器为管理单元的，在 kubelet 创建 Pod 时，还通过启动一个名为 k8s.gcr.io/pause:3.1 的镜像来实现 Pod 的概念。

该镜像存在于谷歌镜像库 k8s.gcr.io 中，需要通过一台能够连上 Internet 的服务器将其下载，导出文件，再 push 到私有 Docker Registry 中。

之后，可以给每个 Node 的 kubelet 服务都加上启动参数--pod-infra-container-image，指定为私有 Docker Registry 中 pause 镜像的地址。例如：

```
# cat /etc/kubernetes/kubelet
KUBELET_ARGS="--kubeconfig=/etc/kubernetes/kubeconfig
--hostname-override=192.168.18.3 --log-dir=/var/log/kubernetes --v=0
--pod-infra-container-image=myregistry/pause:3.1"
```

然后重启 kubelet 服务：

```
# systemctl restart kubelet
```

通过以上设置就在内网环境中搭建了一个企业内部的私有容器云平台。

2.7 Kubernetes 的版本升级

2.7.1 二进制升级

Kubernetes 的版本升级需要考虑到不要让当前集群中正在运行的容器受到影响。应对集群中的各 Node 逐个进行隔离，然后等待在其上运行的容器全部执行完成，再更新该 Node 上的 kubelet 和 kube-proxy 服务，将全部 Node 都更新完成后，再更新 Master 的服务。

- ◎ 通过官网获取最新版本的二进制包 kubernetes.tar.gz，解压后提取服务的二进制文件。
- ◎ 逐个隔离 Node，等待在其上运行的全部容器工作完成后，更新 kubelet 和 kube-proxy 服务文件，然后重启这两个服务。
- ◎ 更新 Master 的 kube-apiserver、kube-controller-manager、kube-scheduler 服务文件并重启。

2.7.2　使用 kubeadm 进行集群升级

kubeadm 提供了 upgrade 命令用于对 kubeadm 安装的 Kubernetes 集群进行升级。这一功能提供了从 1.10 到 1.11、从 1.11 到 1.12、从 1.12 到 1.13 及从 1.13 到 1.14 的升级能力，这里试用一下从 1.13 到 1.14 的升级。

开始之前需要注意：

◎ 虽然 kubeadm 的升级不会触及工作负载，但还是要在升级之前做好备份。
◎ 升级过程可能会因为 Pod 的变化而造成容器重启。

继续以 CentOS 7 环境为例，首先需要升级的是 kubeadm：

```
# yum install -y kubeadm-1.14.0 --disableexcludes=kubernetes
```

查看 kubeadm 的版本：

```
# kubeadm version
kubeadm version: &version.Info{Major:"1", Minor:"14", GitVersion:"v1.14.0",
GitCommit:"641856db18352033a0d96dbc99153fa3b27298e5", GitTreeState:"clean",
BuildDate:"2019-03-25T15:51:21Z", GoVersion:"go1.12.1", Compiler:"gc",
Platform:"linux/amd64"}
```

接下来查看 kubeadm 的升级计划：

```
# kubeadm upgrade plan
```

会出现预备升级的内容描述：

```
[preflight] Running pre-flight checks.
[upgrade] Making sure the cluster is healthy:
[upgrade/config] Making sure the configuration is correct:
[upgrade/config] Reading configuration from the cluster...
[upgrade/config] FYI: You can look at this config file with 'kubectl -n kube-system get cm kubeadm-config -oyaml'
[upgrade] Fetching available versions to upgrade to
[upgrade/versions] Cluster version: v1.13.2
[upgrade/versions] kubeadm version: v1.14.0

Awesome, you're up-to-date! Enjoy!
```

按照任务指引进行升级：

```
# kubeadm upgrade apply 1.14.0
```

```
[preflight] Running pre-flight checks.
[upgrade] Making sure the cluster is healthy:
[upgrade/config] Making sure the configuration is correct:
[upgrade/config] Reading configuration from the cluster...
[upgrade/config] FYI: You can look at this config file with 'kubectl -n kube-system get cm kubeadm-config -oyaml'
[upgrade/version] You have chosen to change the cluster version to "v1.14.0"
[upgrade/versions] Cluster version: v1.13.2
[upgrade/versions] kubeadm version: v1.14.0
[upgrade/confirm] Are you sure you want to proceed with the upgrade? [y/N]:
```

输入"y",确认之后,开始进行升级。

运行完成之后,再次查询版本:

```
# kubectl version
Client Version: version.Info{Major:"1", Minor:"13", GitVersion:"v1.13.2",
GitCommit:"cff46ab41ff0bb44d8584413b598ad8360ec1def", GitTreeState:"clean",
BuildDate:"2019-01-10T23:35:51Z", GoVersion:"go1.11.4", Compiler:"gc",
Platform:"linux/amd64"}
    Server Version: version.Info{Major:"1", Minor:"14", GitVersion:"v1.14.0",
GitCommit:"641856db18352033a0d96dbc99153fa3b27298e5", GitTreeState:"clean",
BuildDate:"2019-03-25T15:45:25Z", GoVersion:"go1.12.1", Compiler:"gc",
Platform:"linux/amd64"}
```

可以看到,虽然 kubectl 还是 1.13.2,服务端的控制平面已经升级到了 1.14.0,但是查看 Node 版本,会发现 Node 版本还是滞后的:

```
# kubectl get nodes
NAME              STATUS    ROLES     AGE     VERSION
node-kubeadm-1    Ready     master    15m     v1.13.2
node-kubeadm-2    Ready     <none>    13m     v1.13.2
```

然后可以对节点配置进行升级:

```
# kubeadm upgrade node config --kubelet-version 1.14.0
```

接下来,按照和二进制升级一样的步骤将 kubectl 升级,这样就完成了集群的整体升级。

```
# kubectl get nodes
NAME              STATUS    ROLES     AGE     VERSION
node-kubeadm-1    Ready     master    25m     v1.14.0
node-kubeadm-2    Ready     <none>    22m     v1.14.0
```

2.8 Kubernetes 核心服务配置详解

2.1 节对 Kubernetes 各服务启动进程的关键配置参数进行了简要说明,实际上 Kubernetes 的每个服务都提供了许多可配置的参数。这些参数涉及安全性、性能优化及功能扩展(Plugin)等方方面面。全面理解和掌握这些参数的含义和配置,对 Kubernetes 的生产部署及日常运维都有很大的帮助。

每个服务的可用参数都可以通过运行"cmd --help"命令进行查看,其中 cmd 为具体的服务启动命令,例如 kube-apiserver、kube-controller-manager、kube-scheduler、kubelet、kube-proxy 等。另外,可以通过在命令的配置文件(例如/etc/kubernetes/kubelet 等)中添加"--参数名=参数取值"语句来完成对某个参数的配置。

本节将对 Kubernetes 所有服务的参数进行全面介绍,为了方便学习和查阅,对每个服务的参数都用一个小节进行详细说明。

2.8.1 公共配置参数

公共配置参数适用于所有服务,如表 2.3 所示的参数可用于 kube-apiserver、kube-controller-manager、kube-scheduler、kubelet、kube-proxy。本节对这些参数进行统一说明,不再在每个服务的参数列表中列出。

表 2.3 公共配置参数表

参数名和类型	说明
--alsologtostderr	设置为 true 表示将日志输出到文件的同时输出到 stderr
-h, --help	查看参数列表的帮助信息
--log-backtrace-at traceLocation	记录日志每到"file:行号"时打印一次 stack trace,默认值为 0
--log-dir string	日志文件路径
--log-file string	设置日志文件名
--log-flush-frequency duration	设置 flush 日志文件的时间间隔,默认值为 5s
--logtostderr	设置为 true 表示将日志输出到 stderr,不输出到日志文件
--skip-headers	设置为 true 表示在日志信息中不显示 header 信息
--stderrthreshold severity	将该 threshold 级别之上的日志输出到 stderr,默认值为 2

续表

参数名和类型	说明
-v, --v Level	设置日志级别
--version version[=true]	设置为 true 表示显示版本信息然后退出
--vmodule moduleSpec	设置基于 glog 模块的详细日志级别，格式为 pattern=N，以逗号分隔

2.8.2 kube-apiserver 启动参数

对 kube-apiserver 启动参数的详细说明如表 2.4 所示。

表 2.4 对 kube-apiserver 启动参数的详细说明

参数名和类型	说明
[通用参数]	
--advertise-address ip	用于广播自己的 IP 地址给集群的所有成员，在不指定该地址时将使用 --bind-address 定义的 IP 地址
--cloud-provider-gce-lb-src-cidrs cidrs	GCE 防火墙上开放的负载均衡器源 IP CIDR 列表，默认值为 130.211.0.0/22、209.85.152.0/22、209.85.204.0/22、35.191.0.0/16
--cors-allowed-origins strings	CORS（跨域资源共享）设置允许访问的源域列表，以逗号分隔，并可使用正则表达式匹配子网。如果不指定，则表示不启用 CORS
--default-not-ready-toleration-seconds int	等待 notReady:NoExecute 的 toleration 秒数，默认值为 300。默认会给所有未设置 toleration 的 Pod 添加该设置
--default-unreachable-toleration-seconds int	等待 unreachable:NoExecute 的 toleration 秒数，默认值为 300。默认会给所有未设置 toleration 的 Pod 添加该设置
--external-hostname string	用于生成该 Master 的对外 URL 地址，例如用于 swagger api 文档中的 URL 地址
--feature-gates mapStringBool	用于实验性质的特性开关组，每个开关以 key=value 形式表示。当前可用的开关包括： APIListChunking=true|false (BETA - default=true) APIResponseCompression=true|false (ALPHA - default=false) AllAlpha=true|false (ALPHA - default=false) AppArmor=true|false (BETA - default=true) AttachVolumeLimit=true|false (BETA - default=true) BalanceAttachedNodeVolumes=true|false (ALPHA - default=false)

续表

参数名和类型	说 明
	BlockVolume=true|false (BETA - default=true)
	BoundServiceAccountTokenVolume=true|false (ALPHA - default=false)
	CPUManager=true|false (BETA - default=true)
	CRIContainerLogRotation=true|false (BETA - default=true)
	CSIBlockVolume=true|false (BETA - default=true)
	CSIDriverRegistry=true|false (BETA - default=true)
	CSIInlineVolume=true|false (ALPHA - default=false)
	CSIMigration=true|false (ALPHA - default=false)
	CSIMigrationAWS=true|false (ALPHA - default=false)
	CSIMigrationGCE=true|false (ALPHA - default=false)
	CSIMigrationOpenStack=true|false (ALPHA - default=false)
	CSINodeInfo=true|false (BETA - default=true)
	CustomCPUCFSQuotaPeriod=true|false (ALPHA - default=false)
	CustomResourcePublishOpenAPI=true|false (ALPHA - default=false)
	CustomResourceSubresources=true|false (BETA - default=true)
	CustomResourceValidation=true|false (BETA - default=true)
	CustomResourceWebhookConversion=true|false (ALPHA - default=false)
	DebugContainers=true|false (ALPHA - default=false)
	DevicePlugins=true|false (BETA - default=true)
	DryRun=true|false (BETA - default=true)
	DynamicAuditing=true|false (ALPHA - default=false)
	DynamicKubeletConfig=true|false (BETA - default=true)
	ExpandCSIVolumes=true|false (ALPHA - default=false)
	ExpandInUsePersistentVolumes=true|false (ALPHA - default=false)
	ExpandPersistentVolumes=true|false (BETA - default=true)
	ExperimentalCriticalPodAnnotation=true|false (ALPHA - default=false)
	ExperimentalHostUserNamespaceDefaulting=true|false (BETA - default=false)
	HyperVContainer=true|false (ALPHA - default=false)
	KubeletPodResources=true|false (ALPHA - default=false)
	LocalStorageCapacityIsolation=true|false (BETA - default=true)

参数名和类型	说明
	MountContainers=true|false (ALPHA - default=false)
	NodeLease=true|false (BETA - default=true)PodShareProcessNamespace=true|false (BETA - default=true)
	ProcMountType=true|false (ALPHA - default=false)
	QOSReserved=true|false (ALPHA - default=false)
	ResourceLimitsPriorityFunction=true|false (ALPHA - default=false)
	ResourceQuotaScopeSelectors=true|false (BETA - default=true)
	RotateKubeletClientCertificate=true|false (BETA - default=true)
	RotateKubeletServerCertificate=true|false (BETA - default=true)
	RunAsGroup=true|false (BETA - default=true)
	RuntimeClass=true|false (BETA - default=true)
	SCTPSupport=true|false (ALPHA - default=false)
	ScheduleDaemonSetPods=true|false (BETA - default=true)
	ServerSideApply=true|false (ALPHA - default=false)
	ServiceNodeExclusion=true|false (ALPHA - default=false)
	StorageVersionHash=true|false (ALPHA - default=false)
	StreamingProxyRedirects=true|false (BETA - default=true)
	SupportNodePidsLimit=true|false (ALPHA - default=false)
	SupportPodPidsLimit=true|false (BETA - default=true)
	Sysctls=true|false (BETA - default=true)
	TTLAfterFinished=true|false (ALPHA - default=false)
	TaintBasedEvictions=true|false (BETA - default=true)
	TaintNodesByCondition=true|false (BETA - default=true)
	TokenRequest=true|false (BETA - default=true)
	TokenRequestProjection=true|false (BETA - default=true)
	ValidateProxyRedirects=true|false (BETA - default=true)
	VolumeSnapshotDataSource=true|false (ALPHA - default=false)
	VolumeSubpathEnvExpansion=true|false (ALPHA - default=false)
	WinDSR=true|false (ALPHA - default=false)
	WinOverlay=true|false (ALPHA - default=false)
	WindowsGMSA=true|false (ALPHA - default=false)

续表

参数名和类型	说明
--master-service-namespace string	已弃用，设置 Master 服务所在的命名空间，默认值为 default
--max-mutating-requests-inflight int	同时处理的最大突变请求数量，默认值为 200，超过该数量的请求将被拒绝。设置为 0 表示无限制
--max-requests-inflight int	同时处理的最大请求数量，默认值为 400，超过该数量的请求将被拒绝。设置为 0 表示无限制
--min-request-timeout int	最小请求处理超时时间，单位为 s，默认值为 1800s，目前仅用于 watch request handler，其将会在该时间值上加一个随机时间作为请求的超时时间
--request-timeout duration	请求处理超时时间，可以被--min-request-timeout 参数覆盖，默认值为 1m0s
--target-ram-mb int	API Server 的内存限制，单位为 MB，常用于设置缓存大小
[etcd 相关参数]	
--default-watch-cache-size int	设置默认 watch 缓存的大小，设置为 0 表示不缓存，默认值为 100
--delete-collection-workers int	启动 DeleteCollection 的工作线程数，用于提高清理 Namespace 的效率，默认值为 1
--enable-garbage-collector	设置为 true 表示启用垃圾回收器。必须与 kube-controller-manager 的该参数设置为相同的值，默认值为 true
--encryption-provider-config string	在 etcd 中存储机密信息的配置文件
--etcd-cafile string	到 etcd 安全连接使用的 SSL CA 文件
--etcd-certfile string	到 etcd 安全连接使用的 SSL 证书文件
--etcd-compaction-interval duration	压缩请求的时间间隔，设置为 0 表示不压缩，默认值为 5m0s
--etcd-count-metric-poll-period duration	按类型查询 etcd 中资源数量的时间频率，默认值为 1m0s
--etcd-keyfile string	到 etcd 安全连接使用的 SSL key 文件
--etcd-prefix string	在 etcd 中保存 Kubernetes 集群数据的根目录名，默认值为/registry
--etcd-servers strings	以逗号分隔的 etcd 服务 URL 列表，etcd 服务以 http://ip:port 格式表示
--etcd-servers-overrides	按资源覆盖 etcd 服务的设置，以逗号分隔。单个覆盖格式为：group/resource#servers，其中 servers 格式为 http://ip:port，以分号分隔
--storage-backend string	设置持久化存储类型，可选项为 etcd2、etcd3，从 Kubernetes 1.6 版本开始默认值为 etcd3
--storage-media-type	持久化存储后端的介质类型。某些资源类型只能使用特定类型的介质进行保存，将忽略这个参数的设置，默认值为 application/vnd.kubernetes.protobuf

续表

参数名和类型	说　明
--watch-cache	设置为 true 表示缓存 watch，默认值为 true
--watch-cache-sizes strings	设置各资源对象 watch 缓存大小的列表，以逗号分隔，每个资源对象的设置格式为 resource#size，当 watch-cache 被设置为 true 时生效
[安全服务相关参数]	
--bind-address ip	Kubernetes API Server 在本地址的 6443 端口开启安全的 HTTPS 服务，默认值为 0.0.0.0
--cert-dir string	TLS 证书所在的目录，默认值为/var/run/kubernetes。如果设置了--tls-cert-file 和 --tls-private-key-file，则该设置将被忽略，默认值为/var/run/kubernetes
--http2-max-streams-per-connection int	服务器为客户端提供的 HTTP/2 连接中最大流（stream）数量限制，设置为 0 表示使用 Golang 程序的默认值
--secure-port int	设置 API Server 使用的 HTTPS 安全模式端口号，设置为 0 表示不启用 HTTPS，默认值为 6443
--tls-cert-file string	包含 x509 证书的文件路径，用于 HTTPS 认证
--tls-cipher-suites strings	服务器端加密算法列表，以逗号分隔，若不进行设置，则使用 Go cipher suites 的默认列表。可选加密算法包括：TLS_ECDHE_ECDSA_WITH_AES_128_CBC_SHA TLS_ECDHE_ECDSA_WITH_AES_128_CBC_SHA256 TLS_ECDHE_ECDSA_WITH_AES_128_GCM_SHA256 TLS_ECDHE_ECDSA_WITH_AES_256_CBC_SHA TLS_ECDHE_ECDSA_WITH_AES_256_GCM_SHA384 TLS_ECDHE_ECDSA_WITH_CHACHA20_POLY1305 TLS_ECDHE_ECDSA_WITH_RC4_128_SHA TLS_ECDHE_RSA_WITH_3DES_EDE_CBC_SHA TLS_ECDHE_RSA_WITH_AES_128_CBC_SHA TLS_ECDHE_RSA_WITH_AES_128_CBC_SHA256 TLS_ECDHE_RSA_WITH_AES_128_GCM_SHA256 TLS_ECDHE_RSA_WITH_AES_256_CBC_SHA TLS_ECDHE_RSA_WITH_AES_256_GCM_SHA384 TLS_ECDHE_RSA_WITH_CHACHA20_POLY1305 TLS_ECDHE_RSA_WITH_RC4_128_SHA TLS_RSA_WITH_3DES_EDE_CBC_SHA

续表

参数名和类型	说　　明
	TLS_RSA_WITH_AES_128_CBC_SHA
	TLS_RSA_WITH_AES_128_CBC_SHA256
	TLS_RSA_WITH_AES_128_GCM_SHA256
	TLS_RSA_WITH_AES_256_CBC_SHA
	TLS_RSA_WITH_AES_256_GCM_SHA384
	TLS_RSA_WITH_RC4_128_SHA
--tls-min-version string	设置支持的最小 TLS 版本号，可选的版本号包括：VersionTLS10、VersionTLS11、VersionTLS12
--tls-private-key-file string	包含 x509 证书与 tls-cert-file 对应的私钥文件路径
--tls-sni-cert-key namedCertKey	x509 证书与私钥文件路径对，如果有多对设置，则需要指定多次--tls-sni-cert-key 参数，默认值为[]。常用配置示例如"example.key,example.crt"或"foo.crt,foo.key:*.foo.com,foo.com"等
[不安全服务相关参数]	
--address ip	已弃用，设置绑定的不安全 IP 地址，建议使用--bind-address
--insecure-bind-address ip	绑定的不安全 IP 地址，与--insecure-port 共同使用，默认值为 localhost。设置为 0.0.0.0 表示使用全部网络接口，默认值为 127.0.0.1
--insecure-port int	提供非安全认证访问的监听端口，默认值为 8080。应在防火墙中进行配置，以使外部客户端不可以通过非安全端口访问 API Server
--port int	已弃用，设置绑定的不安全端口号，建议使用--secure-port
[审计相关参数]	
--audit-dynamic-configuration	设置启用动态审计配置，要求启用 DynamicAuditing feature 开关
--audit-log-batch-buffer-size int	审计日志持久化 Event 的缓存大小，仅用于批量模式，默认值为 10000
--audit-log-batch-max-size int	审计日志最大批量大小，仅用于批量模式，默认值为 1
--audit-log-batch-max-wait duration	审计日志持久化 Event 的最长等待时间，仅用于批量模式
--audit-log-batch-throttle-burst int	审计日志批量处理允许的并发最大数量，仅当之前没有启用过 ThrottleQPS 时生效，仅用于批量模式
--audit-log-batch-throttle-enable	设置是否启用批处理并发处理，仅用于批量模式
--audit-log-batch-throttle-qps float32	设置每秒处理批次的最大值，仅用于批量模式

续表

参数名和类型	说 明
--audit-log-format string	审计日志的记录格式,设置为 legacy 表示按每行文本方式记录日志;设置为 json 表示使用 JSON 格式进行记录,默认值为 json
--audit-log-maxage int	审计日志文件保留的最长天数
--audit-log-maxbackup int	审计日志文件的个数
--audit-log-maxsize int	审计日志文件的单个大小限制,单位 MB,默认值为 100MB
--audit-log-mode string	审计日志记录模式,包括同步模式 blocking 或 blocking-strict 和异步模式 batch,默认值为 blocking
--audit-log-path string	审计日志文件的全路径
--audit-log-truncate-enabled	设置是否启用记录 Event 分批截断机制
--audit-log-truncate-max-batch-size int	设置每批次最大可保存 Event 的字节数,超过时自动分成新的批次,默认值为 10485760
--audit-log-truncate-max-event-size int	设置可保存 Event 的最大字节数,超过时自动移除第 1 个请求和应答,仍然超限时将丢弃该 Event,默认值为 102400
--audit-log-version string	审计日志的 API 版本号,默认值为 audit.k8s.io/v1
--audit-policy-file string	审计策略配置文件的全路径
--audit-webhook-batch-buffer-size int	当使用 Webhook 保存审计日志时,审计日志持久化 Event 的缓存大小,仅用于批量模式,默认值为 10000
--audit-webhook-batch-max-size int	当使用 Webhook 保存审计日志时,审计日志的最大批量大小,仅用于批量模式,默认值为 400
--audit-webhook-batch-max-wait duration	当使用 Webhook 保存审计日志时,审计日志持久化 Event 的最长等待时间,仅用于批量模式,默认值为 30s
--audit-webhook-batch-throttle-burst int	当使用 Webhook 保存审计日志时,审计日志批量处理允许的并发最大数量,仅当之前没有启用过 ThrottleQPS 时生效,仅用于批量模式,默认值为 15
--audit-webhook-batch-throttle-enable	当使用 Webhook 保存审计日志时,设置是否启用批处理并发处理,仅用于批量模式,默认值为 true
--audit-webhook-batch-throttle-qps float32	当使用 Webhook 保存审计日志时,设置每秒处理批次的最大值,仅用于批量模式,默认值为 10
--audit-webhook-config-file string	当使用 Webhook 保存审计日志时 Webhook 配置文件的全路径,格式为 kubeconfig 格式

续表

参数名和类型	说 明
--audit-webhook-initial-backoff duration	当使用 Webhook 保存审计日志时，对第 1 个失败请求重试的等待时间，默认值为 10s
--audit-webhook-mode string	当使用 Webhook 保存审计日志时审计日志记录模式，包括同步模式 blocking 或 blocking-strict 和异步模式 batch，默认值为 batch
--audit-webhook-truncate-enabled	当使用 Webhook 保存审计日志时，设置是否启用记录 Event 分批截断机制
--audit-webhook-truncate-max-batch-size int	当使用 Webhook 保存审计日志时，设置每批次最大可保存 Event 的字节数，超过时自动分成新的批次，默认值为 10485760
--audit-webhook-truncate-max-event-size int	当使用 Webhook 保存审计日志时，设置可保存 Event 的最大字节数，超过时自动移除第 1 个请求和应答，仍然超限时将丢弃该 Event，默认值为 102400
--audit-webhook-version string	当使用 Webhook 保存审计日志时审计日志的 API 版本号，默认值为 audit.k8s.io/v1
[其他特性参数]	
--contention-profiling	当性能分析功能打开时，设置是否启用锁竞争分析功能
--profiling	设置为 true 表示打开性能分析功能，可以通过<host>:<port>/debug/pprof/地址查看程序栈、线程等系统信息，默认值为 true
[认证相关参数]	
--anonymous-auth	设置为 true 表示 APIServer 的安全端口可以接收匿名请求,不会被任何 authentication 拒绝的请求将被标记为匿名请求。匿名请求的用户名为 system:anonymous，用户组为 system:unauthenticated，默认值为 true
--api-audiences strings	API 标识符列表，服务账户令牌身份验证器将验证针对 API 使用的令牌是否绑定到设置的至少一个 API 标识符。如果设置了--service-account-issuer 标志但未设置此标志，则此字段默认包含颁发者 URL 的单个元素列表
--authentication-token-webhook-cache-ttl duration	将 Webhook Token Authenticator 返回的响应保存在缓存内的时间，默认值为 2m0s
--authentication-token-webhook-config-file string	Webhook 相关的配置文件，将用于 Token Authentication
--basic-auth-file string	设置该文件通过 HTTP 基本认证的方式访问 API Server 的安全端口
--client-ca-file string	如果指定，则该客户端证书将被用于认证
--enable-bootstrap-token-auth	设置在 TLS 认证引导时是否允许使用 kube-system 命名空间中类型为 bootstrap.kubernetes.io/token 的 secret

续表

参数名和类型	说明
--oidc-ca-file string	在该文件内设置鉴权机构，OpenID Server 的证书将被其中一个机构验证。如果不设置，则将使用主机的 root CA 证书
--oidc-client-id string	OpenID Connect 的客户端 ID，在设置 oidc-issuer-url 时必须设置这个 ID
--oidc-groups-claim string	定制的 OpenID Connect 用户组声明的设置，以字符串数组的形式表示，实验用
--oidc-groups-prefix string	设置 OpenID 用户组的前缀
--oidc-issuer-url string	OpenID 发行者的 URL 地址，仅支持 HTTPS scheme，用于验证 OIDC JSON Web Token
--oidc-required-claim mapStringString	用于描述 ID 令牌所需的声明，以 key=value 形式表示，设置之后，该声明必须存在于匹配的 ID 令牌中，系统会对此进行验证。可以重复该参数以设置多个声明
--oidc-signing-algs strings	设置允许的 JOSE 非对称签名算法列表，以逗号分隔。JWT header 中 alg 的值不在该列表中时将被拒绝。值由 RFC7518（https://tools.ietf.org/html/rfc7518#section-3.1）定义，默认值为[RS256]
--oidc-username-claim string	OpenID claim 的用户名，默认值为 sub，实验用
--oidc-username-prefix string	设置 OpenID 用户名的前缀，未指定时使用发行方 URL 作为前缀以避免冲突，设置为'-'表示不使用前缀
--requestheader-allowed-names strings	允许的客户端证书中的 common names 列表，通过 header 中由--requestheader-username-headers 参数指定的字段获取。未设置时表示经过--requestheader-client-ca-file 验证的客户端证书都会被认可
--requestheader-client-ca-file string	用于验证客户端证书的根证书，在信任--requestheader-username-headers 参数中的用户名之前进行验证
--requestheader-extra-headers-prefix strings	待审查请求 header 的前缀列表，建议用 X-Remote-Extra-
--requestheader-group-headers strings	待审查请求 header 的用户组的列表，建议用 X-Remote-Group
--requestheader-username-headers strings	待审查请求 header 的用户名的列表，通常用 X-Remote-User
--service-account-issuer string	设置 Service Account 颁发者的标识符。颁发者将在已颁发令牌的"iss"字段中断言此标识符，以字符串或 ULI 格式表示
--service-account-key-file stringArray	包含 PEM-encoded x509 RSA 公钥和私钥的文件路径，用于验证 Service Account 的 Token。若不指定，则使用--tls-private-key-file 指定的文件

续表

参数名和类型	说明
--service-account-lookup	设置为 true 时,系统会到 etcd 验证 Service Account Token 是否存在
--service-account-max-token-expiration duration	设置 Service Account 令牌颁发者创建的令牌的最长有效期。如果一个合法 TokenRequest 请求申请的有效期更长,则以该最长有效期为准
--token-auth-file string	用于访问 API Server 安全端口的 Token 认证文件路径
[授权相关参数]	
--authorization-mode string	到 API Server 的安全访问的认证模式列表,以逗号分隔,可选值包括:AlwaysAllow、AlwaysDeny、ABAC、Webhook、RBAC,默认值为 AlwaysAllow
--authorization-policy-file string	当--authorization-mode 被设置为 ABAC 时使用的 csv 格式的授权配置文件
--authorization-webhook-cache-authorized-ttl duration	将 Webhook Authorizer 返回的已授权响应保存在缓存内的时间,默认值为 5m0s
--authorization-webhook-cache-unauthorized-ttl duration	将 Webhook Authorizer 返回的未授权响应保存在缓存内的时间,默认值为 30s
--authorization-webhook-config-file string	当--authorization-mode 设置为 Webhook 时使用的授权配置文件
[云服务商相关参数]	
--cloud-config string	云服务商的配置文件路径,若不配置,则表示不使用云服务商的配置文件
--cloud-provider string	云服务商的名称,若不配置,则表示不使用云服务商
[API 相关参数]	
--runtime-config mapStringString	一组 key=value 用于运行时的配置信息。apis/<groupVersion>/<resource>可用于打开或关闭对某个 API 版本的支持。api/all 和 api/legacy 特别用于支持所有版本的 API 或支持旧版本的 API
[准入控制相关参数]	
--admission-control string	已弃用,改用--enable-admission-plugins 或--disable-admission-plugins 参数 对发送给 API Server 的请求进行准入控制,配置为一个准入控制器的列表,多个准入控制器之间以逗号分隔。多个准入控制器将按顺序对发送给 API Server 的请求进行拦截和过滤。可配置的准入控制器包括:AlwaysAdmit、AlwaysDeny、AlwaysPullImages、DefaultStorageClass、DefaultTolerationSeconds、DenyEscalatingExec、DenyExecOnPrivileged、EventRateLimit、ExtendedResourceToleration、ImagePolicyWebhook、LimitPodHardAntiAffinityTopology、LimitRanger、MutatingAdmissionWebhook、NamespaceAutoProvision、NamespaceExists、NamespaceLifecycle、NodeRestriction、OwnerReferencesPermissionEnforcement、PersistentVolumeClaimResize、PersistentVolumeLabel、PodNodeSelector、

续表

参数名和类型	说 明
	PodPreset、PodSecurityPolicy、PodTolerationRestriction、Priority、ResourceQuota、SecurityContextDeny、ServiceAccount、StorageObjectInUseProtection、TaintNodesByCondition、ValidatingAdmissionWebhook
--admission-control-config-file string	控制规则的配置文件
--disable-admission-plugins strings	设置禁用的准入控制插件列表，不论其是否在默认启用的插件列表中，以逗号分隔。 默认启用的插件列表为 NamespaceLifecycle、LimitRanger、ServiceAccount、TaintNodesByCondition、Priority、DefaultTolerationSeconds、DefaultStorageClass、PersistentVolumeClaimResize、MutatingAdmissionWebhook、ValidatingAdmissionWebhook、ResourceQuota。 可选插件包括：AlwaysAdmit、AlwaysDeny、AlwaysPullImages、DefaultStorageClass、DefaultTolerationSeconds、DenyEscalatingExec、DenyExecOnPrivileged、EventRateLimit、ExtendedResourceToleration、ImagePolicyWebhook、LimitPodHardAntiAffinityTopology、LimitRanger、MutatingAdmissionWebhook、NamespaceAutoProvision、NamespaceExists、NamespaceLifecycle、NodeRestriction、OwnerReferencesPermissionEnforcement、PersistentVolumeClaimResize、PersistentVolumeLabel、PodNodeSelector、PodPreset、PodSecurityPolicy、PodTolerationRestriction、Priority、ResourceQuota、SecurityContextDeny、ServiceAccount、StorageObjectInUseProtection、TaintNodesByCondition、ValidatingAdmissionWebhook，先后顺序没有影响
--enable-admission-plugins strings	设置启用的准入控制插件列表，以逗号分隔。 默认启用的插件列表为 NamespaceLifecycle、LimitRanger、ServiceAccount、TaintNodesByCondition、Priority、DefaultTolerationSeconds、DefaultStorageClass、PersistentVolumeClaimResize、MutatingAdmissionWebhook、ValidatingAdmissionWebhook、ResourceQuota。 可选插件包括：AlwaysAdmit、AlwaysDeny、AlwaysPullImages、DefaultStorageClass、DefaultTolerationSeconds、DenyEscalatingExec、DenyExecOnPrivileged、EventRateLimit、ExtendedResourceToleration、ImagePolicyWebhook、LimitPodHardAntiAffinityTopology、LimitRanger、MutatingAdmissionWebhook、NamespaceAutoProvision、NamespaceExists、

续表

参数名和类型	说 明
	NamespaceLifecycle、NodeRestriction、OwnerReferencesPermissionEnforcement、PersistentVolumeClaimResize、PersistentVolumeLabel、PodNodeSelector、PodPreset、PodSecurityPolicy、PodTolerationRestriction、Priority、ResourceQuota、SecurityContextDeny、ServiceAccount、StorageObjectInUseProtection、TaintNodesByCondition、ValidatingAdmissionWebhook，先后顺序没有影响
[其他参数]	
--allow-privileged	设置为true时，Kubernetes将允许在Pod中运行拥有系统特权的容器应用，与docker run --privileged 的效果相同
--apiserver-count int	集群中运行的API Server数量，默认值为1
--enable-aggregator-routing	设置为true表示aggregator将请求路由到Endpoint的IP地址，否则路由到ClusterIP
--enable-logs-handler	设置为true表示为API Server安装一个/logs的处理程序，默认值为true
--endpoint-reconciler-type string	设置Endpoint协调器的类型，可选类型包括master-count、lease、none，默认值为lease
--event-ttl duration	Kubernetes事件的保存时间，默认为1h0m0s
--kubelet-certificate-authority string	用于CA授权的cert文件路径
--kubelet-client-certificate string	用于TLS的客户端证书文件路径
--kubelet-client-key string	用于TLS的客户端key文件路径
--kubelet-https	指定kubelet是否使用HTTPS连接，默认值为true
--kubelet-preferred-address-types strings	连接kubelet时使用的节点地址类型（NodeAddressTypes），默认值为列表[Hostname,InternalDNS,InternalIP,ExternalDNS,ExternalIP,LegacyHostIP]，表示可用其中任一地址类型
--kubelet-read-only-port uint	已弃用，为kubelet端口号，默认值为10255
--kubelet-timeout int	kubelet执行操作的超时时间，默认为5s
--kubernetes-service-node-port int	设置Master服务是否使用NodePort模式，如果设置，则Master服务将映射到物理机的端口号；设置为0表示以ClusterIP的形式启动Master服务
--max-connection-bytes-per-sec int	设置为非0的值表示限制每个客户端连接的带宽为xx字节/s，目前仅用于需要长时间执行的请求
--proxy-client-cert-file string	用于在请求期间验证aggregator或kube-apiserver身份的客户端证书文件路径。将请求代理到用户api-server并调用Webhook准入控制插件时，要求此证书在

续表

参数名和类型	说　明
	--requestheader-client-ca-file 指定的文件中包含来自 CA 的签名。该 CA 发布在 kube-system 命名空间中名为 extension-apiserver-authentication 的 Configmap 中
--proxy-client-key-file string	用于在请求期间验证 aggregator 或 kube-apiserver 身份的客户端私钥文件路径
--service-account-signing-key-file string	设置 Service Account 令牌颁发者的当前私钥的文件路径。颁发者用该私钥对颁发的 ID 令牌进行签名。要求启用 TokenRequest 特性开关
--service-cluster-ip-range ipNet	Service 的 Cluster IP（虚拟 IP）池，例如 169.169.0.0/16，这个 IP 地址池不能与物理机所在的网络重合
--service-node-port-range portRange	Service 的 NodePort 能使用的主机端口号范围，默认值为 30000～32767，包括 30000 和 32767

2.8.3　kube-controller-manager 启动参数

对 kube-controller-manager 启动参数的详细说明如表 2.5 所示。

表 2.5　对 kube-controller-manager 启动参数的详细说明

参数名和类型	说　明
[调试相关参数]	
--contention-profiling	当设置打开性能分析时，设置是否打开锁竞争分析
--profiling	设置为 true 表示打开性能分析，可以通过<host>:<port>/debug/pprof/地址查看程序栈、线程等系统信息，默认值为 true
[通用参数]	
--allocate-node-cidrs	设置为 true 表示使用云服务商为 Pod 分配的 CIDRs，仅用于公有云
--cidr-allocator-type string	CIDR 分配器的类型，默认值为 RangeAllocator
--cloud-config string	云服务商的配置文件路径，仅用于公有云
--cloud-provider string	云服务商的名称，仅用于公有云
--cluster-cidr string	集群中 Pod 的可用 CIDR 范围
--cluster-name string	集群的名称，默认值为 kubernetes
--configure-cloud-routes	设置云服务商是否为--allocate-node-cidrs 分配的 CIDR 设置路由，默认值为 true
--controller-start-interval duration	启动各个 controller manager 的时间间隔，默认值为 0s

续表

参数名和类型	说　　明
--controllers strings	要启用的 controller 列表，默认值为"*"，表示启用所有 controller，foo 表示启用名为 foo 的 controller，-foo 表示不启用名为 foo 的 controller。所有 controller 列表包括：attachdetach、bootstrapsigner、cloud-node-lifecycle、clusterrole-aggregation、cronjob、csrapproving、csrcleaner、csrsigning、daemonset、deployment、disruption、endpoint、garbagecollector、horizontalpodautoscaling、job、namespace、nodeipam、nodelifecycle、persistentvolume-binder、persistentvolume-expander、podgc、pv-protection、pvc-protection、replicaset、replicationcontroller、resourcequota、root-ca-cert-publisher、route、service、serviceaccount、serviceaccount-token、statefulset、tokencleaner、ttl、ttl-after-finished。默认不启用的 controller 有 bootstrapsigner 和 tokencleaner
--external-cloud-volume-plugin string	当设置--cloud-provider 为外部云服务商时使用的 Volume 插件
--feature-gates mapStringBool	用于实验性质的特性开关组，每个开关以 key=value 形式表示。当前可用开关包括： APIListChunking=true|false (BETA - default=true) APIResponseCompression=true|false (ALPHA - default=false) AllAlpha=true|false (ALPHA - default=false) AppArmor=true|false (BETA - default=true) AttachVolumeLimit=true|false (BETA - default=true) BalanceAttachedNodeVolumes=true|false (ALPHA - default=false) BlockVolume=true|false (BETA - default=true) BoundServiceAccountTokenVolume=true|false (ALPHA - default=false) CPUManager=true|false (BETA - default=true) CRIContainerLogRotation=true|false (BETA - default=true) CSIBlockVolume=true|false (BETA - default=true) CSIDriverRegistry=true|false (BETA - default=true) CSIInlineVolume=true|false (ALPHA - default=false) CSIMigration=true|false (ALPHA - default=false) CSIMigrationAWS=true|false (ALPHA - default=false) CSIMigrationGCE=true|false (ALPHA - default=false) CSIMigrationOpenStack=true|false (ALPHA - default=false)

续表

参数名和类型	说　　明
	CSINodeInfo=true|false (BETA - default=true)
	CustomCPUCFSQuotaPeriod=true|false (ALPHA - default=false)
	CustomResourcePublishOpenAPI=true|false (ALPHA - default=false)
	CustomResourceSubresources=true|false (BETA - default=true)
	CustomResourceValidation=true|false (BETA - default=true)
	CustomResourceWebhookConversion=true|false (ALPHA - default=false)
	DebugContainers=true|false (ALPHA - default=false)
	DevicePlugins=true|false (BETA - default=true)
	DryRun=true|false (BETA - default=true)
	DynamicAuditing=true|false (ALPHA - default=false)
	DynamicKubeletConfig=true|false (BETA - default=true)
	ExpandCSIVolumes=true|false (ALPHA - default=false)
	ExpandInUsePersistentVolumes=true|false (ALPHA - default=false)
	ExpandPersistentVolumes=true|false (BETA - default=true)
	ExperimentalCriticalPodAnnotation=true|false (ALPHA - default=false)
	ExperimentalHostUserNamespaceDefaulting=true|false (BETA - default=false)
	HyperVContainer=true|false (ALPHA - default=false)
	KubeletPodResources=true|false (ALPHA - default=false)
	LocalStorageCapacityIsolation=true|false (BETA - default=true)
	MountContainers=true|false (ALPHA - default=false)
	NodeLease=true|false (BETA - default=true)
	PodShareProcessNamespace=true|false (BETA - default=true)
	ProcMountType=true|false (ALPHA - default=false)
	QOSReserved=true|false (ALPHA - default=false)
	ResourceLimitsPriorityFunction=true|false (ALPHA - default=false)
	ResourceQuotaScopeSelectors=true|false (BETA - default=true)
	RotateKubeletClientCertificate=true|false (BETA - default=true)
	RotateKubeletServerCertificate=true|false (BETA - default=true)
	RunAsGroup=true|false (BETA - default=true)
	RuntimeClass=true|false (BETA - default=true)

参数名和类型	说明
	SCTPSupport=true|false (ALPHA - default=false)
	ScheduleDaemonSetPods=true|false (BETA - default=true)
	ServerSideApply=true|false (ALPHA - default=false)
	ServiceNodeExclusion=true|false (ALPHA - default=false)
	StorageVersionHash=true|false (ALPHA - default=false)
	StreamingProxyRedirects=true|false (BETA - default=true)
	SupportNodePidsLimit=true|false (ALPHA - default=false)
	SupportPodPidsLimit=true|false (BETA - default=true)
	Sysctls=true|false (BETA - default=true)
	TTLAfterFinished=true|false (ALPHA - default=false)
	TaintBasedEvictions=true|false (BETA - default=true)
	TaintNodesByCondition=true|false (BETA - default=true)
	TokenRequest=true|false (BETA - default=true)
	TokenRequestProjection=true|false (BETA - default=true)
	ValidateProxyRedirects=true|false (BETA - default=true)
	VolumeSnapshotDataSource=true|false (ALPHA - default=false)
	VolumeSubpathEnvExpansion=true|false (ALPHA - default=false)
	WinDSR=true|false (ALPHA - default=false)
	WinOverlay=true|false (ALPHA - default=false)
	WindowsGMSA=true|false (ALPHA - default=false)
--kube-api-burst int32	发送到 API Server 的每秒请求量,默认值为 30
--kube-api-content-type string	发送到 API Server 的请求内容类型,默认值为 application/vnd.kubernetes.protobuf
--kube-api-qps float32	与 API Server 通信的 QPS 值,默认值为 20
--leader-elect	设置为 true 表示进行 leader 选举,用于多个 Master 组件的高可用部署,默认值为 true
--leader-elect-lease-duration duration	leader 选举过程中非 leader 等待选举的时间间隔,默认值为 15s,当--leader-elect=true 时生效
--leader-elect-renew-deadline duration	leader 选举过程中在停止 leading 角色之前再次 renew 的时间间隔,应小于或等于 leader-elect-lease-duration,默认值为 10s,当--leader-elect=true 时生效

续表

参数名和类型	说明
--leader-elect-resource-lock endpoints	在 leader 选举过程中使用哪种资源对象进行锁定操作，可选值包括 endpoints 或 configmap，默认值为 endpoints
--leader-elect-retry-period duration	在 leader 选举过程中获取 leader 角色和 renew 之间的等待时间，默认值为 2s，当 --leader-elect=true 时生效
--min-resync-period duration	最小重新同步的时间间隔，实际重新同步的时间为 MinResyncPeriod 到 2×MinResyncPeriod 之间的一个随机数，默认值为 12h0m0s
--node-monitor-period duration	同步 NodeStatus 的时间间隔，默认值为 5s
--route-reconciliation-period duration	云服务商创建的路由同步时间，默认值为 10s
--use-service-account-credentials	设置为 true 表示为每个 controller 分别设置 Service Account
[Service 控制器相关参数]	
--concurrent-service-syncs int32	设置允许的并发同步 Service 对象的数量，值越大表示同步操作越快，但将会消耗更多的 CPU 和网络资源，默认值为 1
[安全服务相关参数]	
--bind-address ip	在 HTTPS 安全端口提供服务时监听的 IP 地址，默认值为 0.0.0.0
--cert-dir string	TLS 证书所在的目录，如果设置了 --tls-cert-file 和 --tls-private-key-file，则该设置将被忽略
--http2-max-streams-per-connection int	服务器为客户端提供的 HTTP/2 连接中最大流数量限制，设置为 0 表示使用 Golang 程序的默认值
--secure-port int	设置 HTTPS 安全模式的监听端口号，设置为 0 表示不启用 HTTPS，默认值为 10257
--tls-cert-file string	包含 x509 证书的文件路径，用于 HTTPS 认证
--tls-cipher-suites strings	服务器端加密算法列表，以逗号分隔。若不设置，则使用 Go cipher suites 的默认列表。可选加密算法包括： TLS_ECDHE_ECDSA_WITH_AES_128_CBC_SHA TLS_ECDHE_ECDSA_WITH_AES_128_CBC_SHA256 TLS_ECDHE_ECDSA_WITH_AES_128_GCM_SHA256 TLS_ECDHE_ECDSA_WITH_AES_256_CBC_SHA TLS_ECDHE_ECDSA_WITH_AES_256_GCM_SHA384 TLS_ECDHE_ECDSA_WITH_CHACHA20_POLY1305 TLS_ECDHE_ECDSA_WITH_RC4_128_SHA TLS_ECDHE_RSA_WITH_3DES_EDE_CBC_SHA

续表

参数名和类型	说明
	TLS_ECDHE_RSA_WITH_AES_128_CBC_SHA
	TLS_ECDHE_RSA_WITH_AES_128_CBC_SHA256
	TLS_ECDHE_RSA_WITH_AES_128_GCM_SHA256
	TLS_ECDHE_RSA_WITH_AES_256_CBC_SHA
	TLS_ECDHE_RSA_WITH_AES_256_GCM_SHA384
	TLS_ECDHE_RSA_WITH_CHACHA20_POLY1305
	TLS_ECDHE_RSA_WITH_RC4_128_SHA
	TLS_RSA_WITH_3DES_EDE_CBC_SHA
	TLS_RSA_WITH_AES_128_CBC_SHA
	TLS_RSA_WITH_AES_128_CBC_SHA256
	TLS_RSA_WITH_AES_128_GCM_SHA256
	TLS_RSA_WITH_AES_256_CBC_SHA
	TLS_RSA_WITH_AES_256_GCM_SHA384；TLS_RSA_WITH_RC4_128_SHA
--tls-min-version string	设置支持的最小 TLS 版本号，可选的版本号包括：VersionTLS10、VersionTLS11、VersionTLS12
--tls-private-key-file string	包含 x509 证书与 tls-cert-file 对应的私钥文件路径
--tls-sni-cert-key namedCertKey	x509 证书与私钥文件路径对，如果有多对设置，则需要指定多次--tls-sni-cert-key 参数，默认值为[]。可选配置域名后缀，例如"example.key,example.crt"或"*.foo.com,foo.com:foo.key,foo.crt"
[不安全服务相关参数]	
--address ip	已弃用，设置绑定的不安全 IP 地址，建议使用--bind-address
--port int	已弃用，设置绑定的不安全端口号，建议使用--secure-port
[认证相关参数]	
--authentication-kubeconfig string	设置允许在 Kubernetes 核心服务中创建 tokenaccessreviews.authentication.k8s.io 资源对象的 kubeconfig 配置文件，是可选设置，如果将其设置为空，则所有的 Token 请求都被视为匿名，也不会启用客户端 CA 认证
--authentication-skip-lookup	设置为 true 表示跳过认证，设置为 false 表示使用--authentication-kubeconfig 参数指定的配置文件查询集群内认证缺失的配置
--authentication-token-webhook-cache-ttl duration	对 Webhook 令牌认证服务返回响应进行缓存的时间，默认值为 10s

续表

参数名和类型	说明
--authentication-tolerate-lookup-failure	设置为true表示当查询集群内认证缺失的配置失败时仍然认为合法，注意这样可能导致认证服务将所有请求都视为匿名
--client-ca-file string	如果指定，则该客户端证书将被用于认证
--requestheader-allowed-names strings	允许的客户端证书中的common names列表，通过header中由"--requestheader-username-headers"参数指定的字段获取。未设置时表示由"--requestheader-client-ca-file"中认定的任意客户端证书都被允许
--requestheader-client-ca-file string	用于验证客户端证书的根证书，在信任"--requestheader-username-headers"参数中的用户名之前进行验证
--requestheader-extra-headers-prefix strings	待审查请求header的前缀列表，建议用X-Remote-Extra-
--requestheader-group-headers strings	待审查请求header的用户组的列表，建议用X-Remote-Group
--requestheader-username-headers strings	待审查请求header的用户名的列表，通常用X-Remote-User
[授权相关参数]	
--authorization-always-allow-paths strings	设置无须授权的HTTP路径列表，默认值为[/healthz]
--authorization-kubeconfig string	设置允许在Kubernetes核心服务中创建subjectaccessreviews.authorization.k8s.io资源对象的kubeconfig配置文件，是可选设置，如果将其设置为空，则所有未列入白名单的请求都将被拒绝
--authorization-webhook-cache-authorized-ttl duration	对Webhook授权服务返回的已授权响应进行缓存的时间，默认值为10s
--authorization-webhook-cache-unauthorized-ttl duration	对Webhook授权服务返回的未授权响应进行缓存的时间，默认值为10s
[attach/detach 相关参数]	
--attach-detach-reconcile-sync-period duration	Volume的attach、detach等操作的reconciler同步等待时间，必须大于1s，默认值为1m0s
--disable-attach-detach-reconcile-sync	设置为true表示禁用Volume的attach、detach同步操作
[CSR 签名控制器相关参数]	
--cluster-signing-cert-file string	PEM-encoded X509 CA证书文件，用于集群范围的认证，默认值为/etc/kubernetes/ca/ca.pem

续表

参数名和类型	说明
--cluster-signing-key-file string	PEM-encoded RSA 或 ECDSA 私钥文件,用于集群范围的认证,默认值为 /etc/kubernetes/ca/ca.key
--experimental-cluster-signing-duration duration	集群范围签名证书的有效期,默认值为 8760h0m0s
[Deployment 控制器相关参数]	
--concurrent-deployment-syncs int32	设置允许的并发同步 Deployment 对象的数量,值越大表示同步操作越快,但将会消耗更多的 CPU 和网络资源,默认值为 5
--deployment-controller-sync-period duration	同步 Deployment 的时间间隔,默认值为 30s
[已弃用的参数]	以下参数已被弃用
[Endpoint 控制器相关参数]	
--concurrent-endpoint-syncs int32	设置并发执行 Endpoint 同步操作的数量,值越大表示同步操作越快,但将会消耗更多的 CPU 和网络资源,默认值为 5
[GC 控制器相关参数]	
--concurrent-gc-syncs int32	设置并发执行 GC Worker 的数量,默认值为 20
--enable-garbage-collector	设置为 true 表示启用垃圾回收机制,必须与 kube-apiserver 的该参数设置为相同的值,默认值为 true
[HPA 控制器相关参数]	
--horizontal-pod-autoscaler-cpu-initialization-period duration	Pod 启动之后应跳过的初始 CPU 使用率采样时间,默认值为 5m0s
--horizontal-pod-autoscaler-downscale-stabilization duration	Pod 自动扩容器在进行缩容操作之前的等待时间,默认值为 5m0s
--horizontal-pod-autoscaler-initial-readiness-delay duration	Pod 启动之后应跳过的 readiness 检查时间,默认值为 30s
--horizontal-pod-autoscaler-sync-period duration	Pod 自动扩容器的 Pod 数量的同步时间间隔,默认值为 30s
--horizontal-pod-autoscaler-tolerance float	Pod 自动扩容器判断是否需要执行扩缩容操作时"期望值/实际值"的最小比值,默认值为 0.1
[Namespace 控制器相关参数]	

续表

参数名和类型	说明
--concurrent-namespace-syncs int32	设置并发同步Namespace资源对象的数量,值越大表示同步操作越快,但将会消耗更多的CPU和网络资源,默认值为2
--namespace-sync-period duration	更新Namespace状态的同步时间间隔,默认值为5m0s
[Node IPAM控制器相关参数]	
--node-cidr-mask-size int32	Node CIDR的子网掩码设置,默认值为24
--service-cluster-ip-range string	Service的IP范围
[Node Lifecycle控制器相关参数]	
--enable-taint-manager	测试用,设置为true表示启用NoExecute Taints,并将在设置了该taint的Node上驱逐(Evict)所有not-tolerating的Pod,默认值为true
--large-cluster-size-threshold int32	设置Node的数量,用于NodeController根据集群规模是否需要进行Pod Eviction的逻辑判断。设置该值后--secondary-node-eviction-rate将会被隐式重置为0。默认值为50
--node-eviction-rate float32	在zone仍为healthy状态(参考--unhealthy-zone-threshold参数定义的健康状态,zone指整个集群)且该zone中Node失效的情况下,驱逐Pod时每秒处理的Node数量,默认值为0.1
--node-monitor-grace-period duration	监控Node状态的时间间隔,默认值为40s,超过该设置时间后,controller-manager会把Node标记为不可用状态。此值的设置有如下要求: 它应该被设置为kubelet汇报的Node状态时间间隔(参数--node-status-update-frequency=10s)的N倍,N为kubelet状态汇报的重试次数
--node-startup-grace-period duration	Node启动的最大允许时间,若超过此时间无响应,则会标记Node为不可用状态(启动失败),默认值为1m0s
--pod-eviction-timeout duration	在失效Node上删除Pod的超时时间,默认值为5m0s
--secondary-node-eviction-rate float32	在zone为unhealthy状态(参考--unhealthy-zone-threshold参数定义的健康状态,zone指整个集群),且该zone中出现Node失效的情况下,驱逐Pod时每秒处理的Node数量,默认值为0.01。当设置了--large-cluster-size-threshold参数并且集群Node数量少于--large-cluster-size-threshold的值时,该参数被隐式重置为0
--unhealthy-zone-threshold float32	设置在一个zone中有多少比例的Node失效时将被判断为unhealthy,至少有3个Node失效才能进行判断,默认值为0.55
[PV-binder控制器相关参数]	

续表

参数名和类型	说　明
--enable-dynamic-provisioning	设置为 true 表示启用动态 provisioning（需底层存储驱动支持），默认值为 true
--enable-hostpath-provisioner	设置为 true 表示启用 hostPath PV provisioning 机制，仅用于测试，不可用于多 Node 的集群环境
--flex-volume-plugin-dir string	设置 Flex Volume 插件应搜索其他第三方 Volume 插件的全路径，默认值为 /usr/libexec/kubernetes/kubelet-plugins/volume/exec/
--pv-recycler-increment-timeout-nfs int32	使用 nfs scrubber 的 Pod 每增加 1Gi 空间在 ActiveDeadlineSeconds 上增加的时间，默认值为 30s
--pv-recycler-minimum-timeout-hostpath int32	使用 hostPath recycler 的 Pod 的最小 ActiveDeadlineSeconds 秒数，默认值为 60s。实验用
--pv-recycler-minimum-timeout-nfs int32	使用 nfs recycler 的 Pod 的最小 ActiveDeadlineSeconds 秒数，默认值为 300s
--pv-recycler-pod-template-filepath-hostpath string	使用 hostPath recycler 的 Pod 的模板文件全路径
--pv-recycler-pod-template-filepath-nfs string	使用 nfs recycler 的 Pod 的模板文件全路径
--pv-recycler-timeout-increment-hostpath int32	使用 hostPath scrubber 的 Pod 每增加 1Gi 空间在 ActiveDeadlineSeconds 上增加的时间，默认值为 30s
--pvclaimbinder-sync-period duration	同步 PV 和 PVC（容器声明的 PV）的时间间隔，默认值为 15s
[Pod GC 控制器相关参数]	
--terminated-pod-gc-threshold int32	设置可保存的终止 Pod 的数量，超过该数量时，垃圾回收器将进行删除操作。设置为不大于 0 的值表示不启用该功能，默认值为 12500
[ReplicaSet 控制器相关参数]	
--concurrent-replicaset-syncs int32	设置允许的并发同步 Replica Set 对象的数量，值越大表示同步操作越快，但将会消耗更多的 CPU 和网络资源，默认值为 5
[RC 控制器相关参数]	
--concurrent-rc-syncs int32	并发执行 RC 同步操作的协程数，值越大表示同步操作越快，但会消耗更多的 CPU 和网络资源，默认值为 5
[ResourceQuota 控制器相关参数]	
--concurrent-resource-quota-syncs int32	设置允许的并发同步 Replication Controller 对象的数量，值越大表示同步操作越快，但会消耗更多的 CPU 和网络资源

续表

参数名和类型	说 明
--resource-quota-sync-period duration	Resource Quota 使用信息同步的时间间隔，默认值为 5m0s
[ServiceAccount 控制器相关参数]	
--concurrent-serviceaccount-token-syncs int32	设置允许的并发同步 Service Account Token 对象的数量，值越大表示同步操作越快，但会消耗更多的 CPU 和网络资源，默认值为 1
--root-ca-file string	根 CA 证书文件路径，被用于 Service Account 的 Token Secret 中
--service-account-private-key-file string	用于为 Service Account Token 签名的 PEM-encoded RSA 私钥文件路径
[TTL-after-finished 控制器相关参数]	
--concurrent-ttl-after-finished-syncs int32	设置允许的并发同步 TTL-after-finished 控制器 worker 的数量，默认值为 5
[其他参数]	
--kubeconfig string	kubeconfig 配置文件路径，在配置文件中包括 Master 地址信息及必要的认证信息
--master string	API Server 的 URL 地址，设置后不再使用在 kubeconfig 中设置的值

2.8.4 kube-scheduler 启动参数

对 kube-scheduler 启动参数的详细说明如表 2.6 所示。

表 2.6 对 kube-scheduler 启动参数的详细说明

参数名和类型	说 明
[配置相关参数]	
--config string	配置文件的路径
--master string	API Server 的 URL 地址，设置后不再使用在 kubeconfig 中设置的值
--write-config-to string	设置时表示将配置参数写入配置文件，然后退出
[安全服务相关参数]	
--bind-address ip	在 HTTPS 安全端口提供服务时监听的 IP 地址，默认值为 0.0.0.0
--cert-dir string	TLS 证书所在的目录，如果设置了 --tls-cert-file 和 --tls-private-key-file，则该设置将被忽略
--http2-max-streams-per-connection int	服务器为客户端提供的 HTTP/2 连接中的最大流数量限制，设置为 0 表示使用 Golang 的默认值
--secure-port int	设置 HTTPS 安全模式的监听端口号，设置为 0 表示不启用 HTTPS，默认值为 10259

续表

参数名和类型	说明
--tls-cert-file string	包含 x509 证书的文件路径，用于 HTTPS 认证
--tls-cipher-suites strings	服务器端加密算法列表，以逗号分隔，若不设置，则使用 Go cipher suites 的默认列表。可选加密算法包括： TLS_ECDHE_ECDSA_WITH_AES_128_CBC_SHA TLS_ECDHE_ECDSA_WITH_AES_128_CBC_SHA256 TLS_ECDHE_ECDSA_WITH_AES_128_GCM_SHA256 TLS_ECDHE_ECDSA_WITH_AES_256_CBC_SHA TLS_ECDHE_ECDSA_WITH_AES_256_GCM_SHA384 TLS_ECDHE_ECDSA_WITH_CHACHA20_POLY1305 TLS_ECDHE_ECDSA_WITH_RC4_128_SHA TLS_ECDHE_RSA_WITH_3DES_EDE_CBC_SHA TLS_ECDHE_RSA_WITH_AES_128_CBC_SHA TLS_ECDHE_RSA_WITH_AES_128_CBC_SHA256 TLS_ECDHE_RSA_WITH_AES_128_GCM_SHA256 TLS_ECDHE_RSA_WITH_AES_256_CBC_SHA TLS_ECDHE_RSA_WITH_AES_256_GCM_SHA384 TLS_ECDHE_RSA_WITH_CHACHA20_POLY1305 TLS_ECDHE_RSA_WITH_RC4_128_SHA TLS_RSA_WITH_3DES_EDE_CBC_SHA TLS_RSA_WITH_AES_128_CBC_SHA TLS_RSA_WITH_AES_128_CBC_SHA256 TLS_RSA_WITH_AES_128_GCM_SHA256 TLS_RSA_WITH_AES_256_CBC_SHA TLS_RSA_WITH_AES_256_GCM_SHA384 TLS_RSA_WITH_RC4_128_SHA
--tls-min-version string	设置支持的最小 TLS 版本号，可选的版本号包括：VersionTLS10、VersionTLS11、VersionTLS12
--tls-private-key-file string	包含 x509 证书与 tls-cert-file 对应的私钥文件路径
--tls-sni-cert-key namedCertKey	x509 证书与私钥文件路径对，如果有多对设置，则需要指定多次--tls-sni-cert-key 参数，默认值为[]。可选配置域名后缀，例如 "example.key,example.crt" 或 "*.foo.com,foo.com:foo.key,foo.crt"

续表

参数名和类型	说 明
[不安全服务相关参数]	
--address ip	已弃用，设置绑定的不安全 IP 地址，建议使用--bind-address
--port int	已弃用，设置绑定的不安全端口号，建议使用--secure-port
[认证相关参数]	
--authentication-kubeconfig string	设置允许在 Kubernetes 核心服务中创建 tokenaccessreviews.authentication.k8s.io 资源对象的 kubeconfig 配置文件，是可选设置，如果将其设置为空，则所有 Token 请求都被视为匿名，也不会启用客户端 CA 认证
--authentication-skip-lookup	设置为 true 表示跳过认证，设置为 false 表示使用--authentication-kubeconfig 参数指定的配置文件查询集群内认证缺失的配置
--authentication-token-webhook-cache-ttl duration	对 Webhook 令牌认证服务返回的响应进行缓存的时间，默认值为 10s
--authentication-tolerate-lookup-failure	设置为 true 表示当查询集群内认证缺失的配置失败时仍然认为合法，注意这样可能导致认证服务将所有请求都视为匿名
--client-ca-file string	如果指定，则该客户端证书将被用于认证
--requestheader-allowed-names strings	允许的客户端证书中的 common names 列表，通过 header 中由--requestheader-username-headers 参数指定的字段获取。未设置时表示由--requestheader-client-ca-file 中认定的任意客户端证书都被允许
--requestheader-client-ca-file string	用于验证客户端证书的根证书，在信任--requestheader-username-headers 参数中的用户名之前进行验证
--requestheader-extra-headers-prefix strings	待审查请求 header 的前缀列表，建议用 X-Remote-Extra-
--requestheader-group-headers strings	待审查请求 header 的用户组的列表，建议用 X-Remote-Group
--requestheader-username-headers strings	待审查请求 header 的用户名的列表，通常用 X-Remote-User
[授权相关参数]	
--authorization-always-allow-paths strings	设置无须授权的 HTTP 路径列表，默认值为[/healthz]
--authorization-kubeconfig string	设置允许在 Kubernetes 核心服务中创建 subjectaccessreviews.authorization.k8s.io 资源对象的 kubeconfig 配置文件，是可选设置，如果将其设置为空，则所有未列入白名单的请求都将被拒绝
--authorization-webhook-cache-authorized-ttl duration	对 Webhook 授权服务返回的已授权响应进行缓存的时间，默认值为 10s

续表

参数名和类型	说明
--authorization-webhook-cache-unauthorized-ttl duration	对 Webhook 授权服务返回的未授权响应进行缓存的时间，默认值为 10s
[已弃用的参数]	
--algorithm-provider string	设置调度算法，可选项为 ClusterAutoscalerProvider 或 DefaultProvider，默认值为 DefaultProvider
--contention-profiling	设置为 true 表示启用锁竞争性能数据采集，当 profiling 被设置为 true 时生效
--kube-api-burst int32	发送到 API Server 的每秒请求数量，默认值为 100
--kube-api-content-type string	发送到 API Server 的请求内容类型，默认值为 application/vnd.kubernetes.protobuf
--kube-api-qps float32	与 API Server 通信的 QPS 值，默认值为 50
--kubeconfig string	kubeconfig 配置文件路径，在配置文件中包括 Master 的地址信息及必要的认证信息
--lock-object-name string	已弃用，设置 lock 对象的名称，默认值为 kube-scheduler
--lock-object-namespace string	已弃用，设置 lock 对象所在的 Namespace，默认值为 kube-system
--policy-config-file string	调度策略（scheduler policy）配置文件的路径
--policy-configmap string	已弃用，设置包含调度策略配置信息的 ConfigMap 名称，需要先在 Data 元素中定义 key='policy.cfg'，再设置配置内容
--policy-configmap-namespace string	已弃用，设置包含调度策略配置信息的 ConfigMap 所在的 Namespace，默认值为 kube-system
--profiling	打开性能分析，可以通过<host>:<port>/debug/pprof/地址查看栈、线程等系统运行信息，默认值为 true
--scheduler-name string	调度器名称，用于选择哪些 Pod 将被该调度器处理，选择的依据是 Pod 的 annotation 设置，包含 key='scheduler.alpha.kubernetes.io/name' 的 annotation，默认值为 default-scheduler
--use-legacy-policy-config	已弃用，设置调度策略配置文件的路径，将忽略调度策略 ConfigMap 的设置
[Leader 选举相关参数]	
--leader-elect	设置为 true 表示进行 leader 选举，用于多个 Master 组件的高可用部署，默认值为 true
--leader-elect-lease-duration duration	leader 选举过程中非 leader 等待选举的时间间隔，默认值为 15s，当 leader-elect=true 时生效
--leader-elect-renew-deadline duration	leader 选举过程中在停止 leading 角色之前再次 renew 的时间间隔，应小于或等于 leader-elect-lease-duration，默认值为 10s，当 leader-elect=true 时生效

续表

参数名和类型	说明
--leader-elect-resource-lock endpoints	leader 选举过程中将哪种资源对象用于锁定操作，可选值包括 endpoints 或 configmap，默认值为 endpoints
--leader-elect-retry-period duration	leader 选举过程中获取 leader 角色和 renew 之间的等待时间，默认值为 2s，当 leader-elect=true 时生效
[特性开关相关参数]	
--feature-gates mapStringBool	用于实验性质的特性开关组，每个开关以 key=value 形式表示。当前可用开关包括： APIListChunking=true|false (BETA - default=true) APIResponseCompression=true|false (ALPHA - default=false) AllAlpha=true|false (ALPHA - default=false) AppArmor=true|false (BETA - default=true) AttachVolumeLimit=true|false (BETA - default=true) BalanceAttachedNodeVolumes=true|false (ALPHA - default=false) BlockVolume=true|false (BETA - default=true) BoundServiceAccountTokenVolume=true|false (ALPHA - default=false) CPUManager=true|false (BETA - default=true) CRIContainerLogRotation=true|false (BETA - default=true) CSIBlockVolume=true|false (BETA - default=true) CSIDriverRegistry=true|false (BETA - default=true) CSIInlineVolume=true|false (ALPHA - default=false) CSIMigration=true|false (ALPHA - default=false) CSIMigrationAWS=true|false (ALPHA - default=false) CSIMigrationGCE=true|false (ALPHA - default=false) CSIMigrationOpenStack=true|false (ALPHA - default=false) CSINodeInfo=true|false (BETA - default=true) CustomCPUCFSQuotaPeriod=true|false (ALPHA - default=false) CustomResourcePublishOpenAPI=true|false (ALPHA - default=false) CustomResourceSubresources=true|false (BETA - default=true) CustomResourceValidation=true|false (BETA - default=true) CustomResourceWebhookConversion=true|false (ALPHA - default=false) DebugContainers=true|false (ALPHA - default=false)

续表

参数名和类型	说　　明
	DevicePlugins=true|false (BETA - default=true)
	DryRun=true|false (BETA - default=true)
	DynamicAuditing=true|false (ALPHA - default=false)
	DynamicKubeletConfig=true|false (BETA - default=true)
	ExpandCSIVolumes=true|false (ALPHA - default=false)
	ExpandInUsePersistentVolumes=true|false (ALPHA - default=false)
	ExpandPersistentVolumes=true|false (BETA - default=true)
	ExperimentalCriticalPodAnnotation=true|false (ALPHA - default=false)
	ExperimentalHostUserNamespaceDefaulting=true|false (BETA - default=false)
	HyperVContainer=true|false (ALPHA - default=false)
	KubeletPodResources=true|false (ALPHA - default=false)
	LocalStorageCapacityIsolation=true|false (BETA - default=true)
	MountContainers=true|false (ALPHA - default=false)
	NodeLease=true|false (BETA - default=true)
	PodShareProcessNamespace=true|false (BETA - default=true)
	ProcMountType=true|false (ALPHA - default=false)
	QOSReserved=true|false (ALPHA - default=false)
	ResourceLimitsPriorityFunction=true|false (ALPHA - default=false)
	ResourceQuotaScopeSelectors=true|false (BETA - default=true)
	RotateKubeletClientCertificate=true|false (BETA - default=true)
	RotateKubeletServerCertificate=true|false (BETA - default=true)
	RunAsGroup=true|false (BETA - default=true)
	RuntimeClass=true|false (BETA - default=true)
	SCTPSupport=true|false (ALPHA - default=false)
	ScheduleDaemonSetPods=true|false (BETA - default=true)
	ServerSideApply=true|false (ALPHA - default=false)
	ServiceNodeExclusion=true|false (ALPHA - default=false)
	StorageVersionHash=true|false (ALPHA - default=false)
	StreamingProxyRedirects=true|false (BETA - default=true)
	SupportNodePidsLimit=true|false (ALPHA - default=false)

续表

参数名和类型	说明
	SupportPodPidsLimit=true|false (BETA - default=true)
	Sysctls=true|false (BETA - default=true)
	TTLAfterFinished=true|false (ALPHA - default=false)
	TaintBasedEvictions=true|false (BETA - default=true)
	TaintNodesByCondition=true|false (BETA - default=true)
	TokenRequest=true|false (BETA - default=true)
	TokenRequestProjection=true|false (BETA - default=true)
	ValidateProxyRedirects=true|false (BETA - default=true)
	VolumeSnapshotDataSource=true|false (ALPHA - default=false)
	VolumeSubpathEnvExpansion=true|false (ALPHA - default=false)
	WinDSR=true|false (ALPHA - default=false)
	WinOverlay=true|false (ALPHA - default=false)
	WindowsGMSA=true|false (ALPHA - default=false)

2.8.5 kubelet 启动参数

对 kubelet 启动参数的详细说明如表 2.7 所示。

表 2.7 对 kubelet 启动参数的详细说明

参数名和类型	说明
--address ip	已弃用，在--config 指定的配置文件中进行设置。 绑定主机 IP 地址，默认值为 0.0.0.0，表示使用全部网络接口
--allow-privileged	已弃用，设置是否允许以特权模式启动容器，默认值为 true
--allowed-unsafe-sysctls strings	设置允许的非安全 sysctls 或 sysctl 模式白名单，由于操作的是操作系统，所以需小心控制
--anonymous-auth	已弃用，在--config 指定的配置文件中进行设置。 设置为 true 表示 kubelet server 可以接收匿名请求。不会被任何 authentication 拒绝的请求将被标记为匿名请求。匿名请求的用户名为 system:anonymous，用户组为 system:unauthenticated。默认值为 true
--application-metrics-count-limit int	已弃用，为每个容器保存的性能指标的最大数量，默认值为 100

续表

参数名和类型	说 明
--authentication-token-webhook	已弃用,在--config 指定的配置文件中进行设置。 使用 TokenReview API 授权客户端 Token
--authentication-token-webhook-cache-ttl duration	已弃用,在--config 指定的配置文件中进行设置。 将 Webhook Token Authenticator 返回的响应保存在缓存内的时间,默认值为 2m0s
--authorization-mode string	已弃用,在--config 指定的配置文件中进行设置。 到 kubelet server 的安全访问的认证模式,可选值包括:AlwaysAllow、Webhook(使用 SubjectAccessReview API 进行授权),默认值为 AlwaysAllow
--authorization-webhook-cache-authorized-ttl duration	已弃用,在--config 指定的配置文件中进行设置。 Webhook Authorizer 返回"已授权"的应答缓存时间,默认值为 5m0s
--authorization-webhook-cache-unauthorized-ttl duration	已弃用,在--config 指定的配置文件中进行设置。 Webhook Authorizer 返回未授权的应答缓存时间,默认值为 30s
--azure-container-registry-config string	Azure 公有云上镜像库的配置文件路径
--boot-id-file string	已弃用,以逗号分隔的文件列表,使用第 1 个存在 book-id 的文件,默认值为 /proc/sys/kernel/random/boot_id
--bootstrap-checkpoint-path string	[Alpha 版特性] 保存 checkpoint 的目录
--bootstrap-kubeconfig string	将用于获取 kubelet 客户端证书的 kubeconfig 配置文件的路径。如果--kubeconfig 指定的文件不存在,则从 API Server 获取客户端证书。成功时,将在--kubeconfig 指定的路径下生成一个引用客户端证书和密钥的 kubeconfig 文件。客户端证书和密钥文件将被存储在--cert-dir 指向的目录下
--cert-dir string	TLS 证书所在的目录,默认值为/var/run/kubernetes。如果设置了--tls-cert-file 和 --tls-private-key-file,则该设置将被忽略
--cgroup-driver string	已弃用,在--config 指定的配置文件中进行设置。 用于操作本机 cgroup 的驱动模式,支持的选项包括 groupfs 或 systemd,默认值为 cgroupfs
--cgroup-root string	已弃用,在--config 指定的配置文件中进行设置。 为 pods 设置的 root cgroup,如果不设置,则将使用容器运行时的默认设置,默认值为空字符串(表示为两个单引号'')
--cgroups-per-qos	已弃用,在--config 指定的配置文件中进行设置。 设置为 true 表示启用创建 QoS cgroup hierarchy,默认值为 true
--chaos-chance float	随机产生客户端错误的概率,用于测试,默认值为 0.0,即不产生

续表

参数名和类型	说 明
--client-ca-file	已弃用，在--config 指定的配置文件中进行设置。 设置客户端 CA 证书文件，一旦设置该文件，则将对所有客户端请求进行鉴权，验证客户端证书的 CommonName 信息
--cloud-config string	云服务商的配置文件路径
--cloud-provider string	云服务商的名称，默认将自动检测，设置为空表示无云服务商，默认值为 auto-detect
--cluster-dns strings	已弃用，在--config 指定的配置文件中进行设置。 集群内 DNS 服务的 IP 地址，以逗号分隔。仅当 Pod 设置了"dnsPolicy=ClusterFirst"属性时可用。注意，所有 DNS 服务器都必须包含相同的记录组，否则名字解析可能出错
--cluster-domain string	已弃用，在--config 指定的配置文件中进行设置。 集群内 DNS 服务所用的域名
--cni-bin-dir string	[Alpha 版特性] CNI 插件二进制文件所在的目录，默认值为/opt/cni/bin
--cni-conf-dir string	[Alpha 版特性] CNI 插件配置文件所在的目录，默认值为/etc/cni/net.d
--config string	kubelet 主配置文件
--container-hints	已弃用，容器 hints 文件所在的全路径，默认值为/etc/cadvisor/container_hints.json
--container-log-max-files int32	[Beta 版特性] 设置容器日志文件的最大数量，必须不少于 2，默认值为 5。此参数只能与--container-runtime=remote 参数一起使用，应在--config 指定的配置文件中进行设置
--container-log-max-size string	[Beta 版特性] 设置容器日志文件的单文件最大大小，写满时将滚动生成新的文件，默认值为 10MiB。此参数只能与--container-runtime=remote 参数一起使用，应在--config 指定的配置文件中进行设置
--container-runtime string	容器类型，目前支持 docker、remote，默认值为 docker
--container-runtime-endpoint string	[实验性特性] 容器运行时的远程服务 endpoint，在 Linux 系统上支持的类型包括 unix socket 和 tcp endpoint，在 Windows 系统上支持的类型包括 npipe 和 tcp endpoint，例如 unix:///var/run/dockershim.sock 和 npipe:////./pipe/dockershim，默认值为 unix:///var/run/dockershim.sock
--containerd string	已弃用，设置 containerd 的 Endpoint，默认值为 unix:///var/run/containerd.sock
--containerized	将 kubelet 运行在容器中，仅供测试使用，默认值为 false

续表

参数名和类型	说 明
--contention-profiling	已弃用,在--config 指定的配置文件中进行设置。 当设置打开性能分析时,设置是否打开锁竞争分析
--cpu-cfs-quota	已弃用,在--config 指定的配置文件中进行设置。 设置为 true 表示启用 CPU CFS quota,用于设置容器的 CPU 限制,默认值为 true
--cpu-cfs-quota-period duration	已弃用,在--config 指定的配置文件中进行设置。 设置 CPU CFS quota 时间 cpu.cfs_period_us,默认使用 Linux Kernel 的系统默认值 100ms
--cpu-manager-policy string	已弃用,在--config 指定的配置文件中进行设置。 设置 CPU Manager 策略,可选值包括 none、static,默认值为 none
--cpu-manager-reconcile-period NodeStatusUpdateFrequency	已弃用,在--config 指定的配置文件中进行设置。 [Alpha 版特性] 设置 CPU Manager 的调和时间,例如 10s 或 1min,默认值为 NodeStatusUpdateFrequency 的值 10s
--docker string	已弃用,Docker 服务的 Endpoint 地址,默认值为 unix:///var/run/docker.sock
--docker-endpoint string	已弃用,Docker 服务的 Endpoint 地址,默认值为 unix:///var/run/docker.sock
--docker-env-metadata-whitelist string	已弃用,Docker 容器需要使用的环境变量 key 列表,以逗号分隔
--docker-only	已弃用,设置为 true 表示仅报告 Docker 容器的统计信息而不再报告其他统计信息
--docker-root string	已弃用,Docker 根目录的全路径,默认值为/var/lib/docker
--docker-tls	已弃用,连接 Docker 的 TLS 设置
--docker-tls-ca string	已弃用,TLS CA 路径,默认值为 ca.pem
--docker-tls-cert string	已弃用,客户端证书路径,默认值为 cert.pem
--docker-tls-key string	已弃用,私钥文件路径,默认值为 key.pem
--dynamic-config-dir string	设置 kubelet 使用动态配置文件的路径,需要启用 DynamicKubeletConfig 特性开关
--enable-controller-attach-detach	已弃用,在--config 指定的配置文件中进行设置。 设置为 true 表示启用 Attach/Detach Controller 进行调度到该 Node 的 Volume 的 attach 与 detach 操作,同时禁用 kubelet 执行 attach、detach 操作,默认值为 true
--enable-debugging-handlers	已弃用,在--config 指定的配置文件中进行设置。 设置为 true 表示提供远程访问本节点容器的日志、进入容器执行命令等相关 Rest 服务,默认值为 true
--enable-load-reader	已弃用,设置为 true 表示启用 CPU 负载的 reader

续表

参数名和类型	说 明
--enable-server	启动 kubelet 上的 HTTP Rest Server，此 Server 提供了获取在本节点上运行的 Pod 列表、Pod 状态和其他管理监控相关的 Rest 接口，默认值为 true
--enforce-node-allocatable strings	已弃用，在--config 指定的配置文件中进行设置。 本 Node 上 kubelet 资源的分配设置，以逗号分隔，可选配置为'pods'、'system-reserved'和'kube-reserved'。在设置'system-reserved'和'kube-reserved'这两个值时，要求同时设置'--system-reserved-cgroup'和'--kube-reserved-cgroup'这两个参数。参考 https://github.com/kubernetes/community/blob/master/contributors/design-proposals/node-allocatable.md。默认值为 pods
--event-burst int32	已弃用，在--config 指定的配置文件中进行设置。 临时允许的 Event 记录突发的最大数量，默认值为 10，当设置--event-qps>0 时生效
--event-qps int32	已弃用，在--config 指定的配置文件中进行设置。 设置大于 0 的值表示限制每秒能创建的 Event 数量，设置为 0 表示不限制，默认值为 5
--event-storage-age-limit string	已弃用，保存 Event 的最大时间。按事件类型以 key=value 的格式表示，以逗号分隔，事件类型包括 creation、oom 等，"default"表示所有事件的类型，默认 default=0
--event-storage-event-limit string	已弃用，保存 Event 的最大数量。按事件类型以 key=value 格式表示，以逗号分隔，事件类型包括 creation、oom 等，"default"表示所有事件的类型，默认 default=0
--eviction-hard string	已弃用，在--config 指定的配置文件中进行设置。 触发 Pod Eviction 操作的一组硬门限设置，默认 memory.available<100Mi
--eviction-max-pod-grace-period int32	已弃用，在--config 指定的配置文件中进行设置。 终止 Pod 操作为 Pod 自行停止预留的时间，单位为 s。时间到达时，将触发 Pod Eviction 操作。默认值为 0，设置为负数表示使用在 Pod 中指定的值
--eviction-minimum-reclaim string	已弃用，在--config 指定的配置文件中进行设置。 当本节点压力过大时，kubelet 进行 Pod Eviction 操作，进而需要完成资源回收的最小数量的一组设置，例如 imagefs.available=2Gi
--eviction-pressure-transition-period duration	已弃用，在--config 指定的配置文件中进行设置。 kubelet 在触发 Pod Eviction 操作之前等待的最长时间，默认值为 5m0s

续表

参数名和类型	说　明
--eviction-soft string	已弃用，在--config 指定的配置文件中进行设置。 触发 Pod Eviction 操作的一组软门限设置，例如可用内存少于 1.5Gi，与 grace-period 一起生效，在 Pod 的响应时间超过 grace-period 后进行触发
--eviction-soft-grace-period string	已弃用，在--config 指定的配置文件中进行设置。 触发 Pod Eviction 操作的一组软门限等待时间设置，例如 memory.available=1m30s
--exit-on-lock-contention	设置为 true 表示当有文件锁存在时 kubelet 也可以退出
--experimental-allocatable-ignore-eviction	设置为 true 表示计算 Node Allocatable 时忽略硬门限设置。参考 https://github.com/kubernetes/community/blob/master/contributors/design-proposals/node-allocatable.md。默认值为 false
--experimental-bootstrap-kubeconfig string	已弃用，使用--bootstrap-kubeconfig 参数
--experimental-check-node-capabilities-before-mount	[实验性特性] 设置为 true 表示 kubelet 在进行 mount 操作之前对本 Node 上所需组件（二进制文件等）进行检查
--experimental-kernel-memcg-notification	[实验性特性] 设置为 true 表示 kubelet 将会集成 kernel 的 memcg 通知机制，以判断是否达到了内存 Eviction 门限
--experimental-mounter-path string	[实验性特性] mounter 二进制文件的路径。设置为空表示使用默认 mount
--fail-swap-on	已弃用，在--config 指定的配置文件中进行设置。 设置为 true 表示如果主机启用了 swap，kubelet 将无法启动，默认值为 true
--feature-gates string	已弃用，在--config 指定的配置文件中进行设置。 用于实验性质的特性开关组，每个开关以 key=value 形式表示。当前可用开关包括： APIListChunking=true|false (BETA - default=true) APIResponseCompression=true|false (ALPHA - default=false) AllAlpha=true|false (ALPHA - default=false) AppArmor=true|false (BETA - default=true) AttachVolumeLimit=true|false (BETA - default=true) BalanceAttachedNodeVolumes=true|false (ALPHA - default=false) BlockVolume=true|false (BETA - default=true) BoundServiceAccountTokenVolume=true|false (ALPHA - default=false) CPUManager=true|false (BETA - default=true) CRIContainerLogRotation=true|false (BETA - default=true)

续表

参数名和类型	说明
	CSIBlockVolume=true|false (BETA - default=true)
	CSIDriverRegistry=true|false (BETA - default=true)
	CSIInlineVolume=true|false (ALPHA - default=false)
	CSIMigration=true|false (ALPHA - default=false)
	CSIMigrationAWS=true|false (ALPHA - default=false)
	CSIMigrationGCE=true|false (ALPHA - default=false)
	CSIMigrationOpenStack=true|false (ALPHA - default=false)
	CSINodeInfo=true|false (BETA - default=true)
	CustomCPUCFSQuotaPeriod=true|false (ALPHA - default=false)
	CustomResourcePublishOpenAPI=true|false (ALPHA - default=false)
	CustomResourceSubresources=true|false (BETA - default=true)
	CustomResourceValidation=true|false (BETA - default=true)
	CustomResourceWebhookConversion=true|false (ALPHA - default=false)
	DebugContainers=true|false (ALPHA - default=false)
	DevicePlugins=true|false (BETA - default=true)
	DryRun=true|false (BETA - default=true)
	DynamicAuditing=true|false (ALPHA - default=false)
	DynamicKubeletConfig=true|false (BETA - default=true)
	ExpandCSIVolumes=true|false (ALPHA - default=false)
	ExpandInUsePersistentVolumes=true|false (ALPHA - default=false)
	ExpandPersistentVolumes=true|false (BETA - default=true)
	ExperimentalCriticalPodAnnotation=true|false (ALPHA - default=false)
	ExperimentalHostUserNamespaceDefaulting=true|false (BETA - default=false)
	HyperVContainer=true|false (ALPHA - default=false)
	KubeletPodResources=true|false (ALPHA - default=false)
	LocalStorageCapacityIsolation=true|false (BETA - default=true)
	MountContainers=true|false (ALPHA - default=false)
	NodeLease=true|false (BETA - default=true)
	PodShareProcessNamespace=true|false (BETA - default=true)
	ProcMountType=true|false (ALPHA - default=false)

续表

参数名和类型	说明
	QOSReserved=true|false (ALPHA - default=false)
	ResourceLimitsPriorityFunction=true|false (ALPHA - default=false)
	ResourceQuotaScopeSelectors=true|false (BETA - default=true)
	RotateKubeletClientCertificate=true|false (BETA - default=true)
	RotateKubeletServerCertificate=true|false (BETA - default=true)
	RunAsGroup=true|false (BETA - default=true)
	RuntimeClass=true|false (BETA - default=true)
	SCTPSupport=true|false (ALPHA - default=false)
	ScheduleDaemonSetPods=true|false (BETA - default=true)
	ServerSideApply=true|false (ALPHA - default=false)
	ServiceNodeExclusion=true|false (ALPHA - default=false)
	StorageVersionHash=true|false (ALPHA - default=false)
	StreamingProxyRedirects=true|false (BETA - default=true)
	SupportNodePidsLimit=true|false (ALPHA - default=false)
	SupportPodPidsLimit=true|false (BETA - default=true)
	Sysctls=true|false (BETA - default=true)
	TTLAfterFinished=true|false (ALPHA - default=false)
	TaintBasedEvictions=true|false (BETA - default=true)
	TaintNodesByCondition=true|false (BETA - default=true)
	TokenRequest=true|false (BETA - default=true)
	TokenRequestProjection=true|false (BETA - default=true)
	ValidateProxyRedirects=true|false (BETA - default=true)
	VolumeSnapshotDataSource=true|false (ALPHA - default=false)
	VolumeSubpathEnvExpansion=true|false (ALPHA - default=false)
	WinDSR=true|false (ALPHA - default=false)
	WinOverlay=true|false (ALPHA - default=false)
	WindowsGMSA=true|false (ALPHA - default=false)
--file-check-frequency duration	已弃用，在--config指定的配置文件中进行设置。 在File Source作为Pod源的情况下，kubelet定期重新检查文件变化的时间间隔，文件发生变化后，kubelet重新加载更新的文件内容，默认值为20s

续表

参数名和类型	说明
--global-housekeeping-interval duration	已弃用，全局 housekeeping 的时间间隔，默认值为 1m0s
--hairpin-mode string	已弃用，在--config 指定的配置文件中进行设置。 设置 hairpin 模式，表示 kubelet 设置 hairpin NAT 的方式。该模式允许后端 Endpoint 在访问其本身 Service 时能够再次 loadbalance 回自身。可选项包括 promiscuous-bridge、hairpin-veth 和 none，默认值为 promiscuous-bridge
--healthz-bind-address ip	已弃用，在--config 指定的配置文件中进行设置。 healthz 服务监听的 IP 地址，默认值为 127.0.0.1，设置为 0.0.0.0 表示监听全部 IP 地址
--healthz-port int32	已弃用，在--config 指定的配置文件中进行设置。 本地 healthz 服务监听的端口号，默认值为 10248
--host-ipc-sources strings	已弃用，kubelet 允许 Pod 使用宿主机 ipc namespace 的列表，以逗号分隔，默认值为[*]
--host-network-sources strings	已弃用，kubelet 允许 Pod 使用宿主机 network 的列表，以逗号分隔，默认值为[*]
--host-pid-sources strings	已弃用，kubelet 允许 Pod 使用宿主机 pid namespace 的列表，以逗号分隔，默认值为[*]
--hostname-override string	设置本 Node 在集群中的主机名，不设置将使用本机 hostname
--housekeeping-interval duration	对容器进行 housekeeping 操作的时间间隔，默认值为 10s
--http-check-frequency duration	已弃用，在--config 指定的配置文件中进行设置。 在 HTTP URL Source 作为 Pod 源的情况下，kubelet 定期检查 URL 返回的内容是否发生变化的时间周期，作用同 file-check-frequency 参数，默认值为 20s
--image-gc-high-threshold int32	已弃用，在--config 指定的配置文件中进行设置。 镜像垃圾回收上限，磁盘使用空间达到该百分比时，镜像垃圾回收将持续工作，默认值为 90
--image-gc-low-threshold int32	已弃用，在--config 指定的配置文件中进行设置。 镜像垃圾回收下限，磁盘使用空间在达到该百分比之前，镜像垃圾回收将不启用，默认值为 80
--image-pull-progress-deadline duration	如果在该参数值之前还没能开始 pull 镜像的过程，pull 镜像操作将被取消，默认值为 1m0s

续表

参数名和类型	说 明
--image-service-endpoint string	[实验性特性] 远程镜像服务的 Endpoint。未设定时使用--container-runtime-endpoint 的值，在 Linux 系统上支持的类型包括 unix socket 和 tcp endpoint，在 Windows 系统上支持的类型包括 npipe 和 tcp endpoint，例如 unix:///var/run/dockershim.sock 和 npipe:////./pipe/dockershim
--iptables-drop-bit int32	已弃用，在--config 指定的配置文件中进行设置。 标记数据包将被丢弃（Drop）的 fwmark 位设置，有效范围为[0, 31]，默认值为 15
--iptables-masquerade-bit int32	已弃用，在--config 指定的配置文件中进行设置。 标记数据包将进行 SNAT 的 fwmark 位设置,有效范围为[0, 31],必须与 kube-proxy 的相关参数设置一致，默认值为 14
--keep-terminated-pod-volumes	已弃用，设置为 true 表示在 Pod 被删除后仍然保留之前 mount 过的 Volume，常用于 Volume 相关问题的查错
--kube-api-burst int32	已弃用，在--config 指定的配置文件中进行设置。 发送到 API Server 的每秒请求数量，默认值为 10
--kube-api-content-type string	已弃用，在--config 指定的配置文件中进行设置。 发送到 API Server 的请求内容类型，默认值为 application/vnd.kubernetes.protobuf
--kube-api-qps int32	已弃用，在--config 指定的配置文件中进行设置。 与 API Server 通信的 QPS 值，默认值为 5
--kube-reserved mapStringString	已弃用，在--config 指定的配置文件中进行设置。 Kubernetes 系统预留的资源配置，以一组 ResourceName=ResourceQuantity 格式表示，例如 cpu=200m,memory=150G。目前仅支持 CPU 和内存的设置，详见 http://releases.k8s.io/HEAD/docs/user-guide/compute-resources.md，默认值为 none
--kube-reserved-cgroup string	已弃用，在--config 指定的配置文件中进行设置。 用于管理 Kubernetes 的带--kube-reserved 标签组件的计算资源，设置顶层 cgroup 全路径名，例如/kube-reserved，默认值为""
--kubeconfig string	kubeconfig 配置文件路径，在配置文件中包括 Master 地址信息及必要的认证信息，默认值为/var/lib/kubelet/kubeconfig
--kubelet-cgroups string	已弃用，在--config 指定的配置文件中进行设置。 kubelet 运行所需的 cgroups 名称
--lock-file string	[ALPHA 版特性] kubelet 使用的 lock 文件
--log-cadvisor-usage	已弃用，设置为 true 表示将 cAdvisor 容器的使用情况进行日志记录

续表

参数名和类型	说明
--machine-id-file string	已弃用，用于查找 machine-id 的文件列表，使用找到的第 1 个值，默认值为 /etc/machine-id,/var/lib/dbus/machine-id
--make-iptables-util-chains	已弃用，在--config 指定的配置文件中进行设置。 设置为 true 表示 kubelet 将确保 iptables 规则在 Node 上存在，默认值为 true
--manifest-url string	已弃用，在--config 指定的配置文件中进行设置。 为 HTTP URL Source 源类型时，kubelet 用来获取 Pod 定义的 URL 地址，此 URL 返回一组 Pod 定义
--manifest-url-header string	已弃用，在--config 指定的配置文件中进行设置。 访问 menifest URL 地址时使用的 HTTP 头信息，以 key:value 格式表示，例如 a:hello,b:again,c:world
--master-service-namespace string	已弃用，Master 服务的命名空间，默认值为 default
--max-open-files int	已弃用，在--config 指定的配置文件中进行设置。 kubelet 打开文件的最大数量，默认值为 1 000 000
--max-pods int32	已弃用，在--config 指定的配置文件中进行设置。 kubelet 能运行的最大 Pod 数量，默认值为 110
--maximum-dead-containers int32	已弃用，使用--eviction-hard 或--eviction-soft 参数。 可以保留的已停止容器的最大数量
--maximum-dead-containers-per-container int32	已弃用，使用--eviction-hard 或--eviction-soft 参数。 可以保留的每个已停止容器的最大实例数量
--minimum-container-ttl-duration duration	已弃用，使用--eviction-hard 或--eviction-soft 参数。 不再使用的容器被清理之前的最少存活时间，例如 300ms、10s 或 2h45m
--minimum-image-ttl-duration duration	已弃用，使用--eviction-hard 或--eviction-soft 参数。 不再使用的镜像在被清理之前的最少存活时间，例如 300ms、10s 或 2h45m，超过此存活时间的镜像被标记为可被 GC 清理，默认值为 2m0s
--network-plugin string	[Alpha 版特性] 自定义的网络插件的名字，在 Pod 的生命周期中，相关的一些事件会调用此网络插件进行处理
--network-plugin-mtu int32	[Alpha 版特性] 传递给网络插件的 MTU 值，设置为 0 表示使用默认 1460 MTU
--node-ip string	设置本 Node 的 IP 地址
--node-labels mapStringString	[Alpha 版特性] kubelet 注册本 Node 时设置的 Label，Label 以 key=value 的格式表示，多个 Label 以逗号分隔。命名空间 ubernetes.io 中的 Label 必须以

续表

参数名和类型	说明
	kubelet.kubernetes.io 或 node.kubernetes.io 为前缀，或者在以下允许的范围内：beta.kubernetes.io/arch、beta.kubernetes.io/instance-type、beta.kubernetes.io/os、failure-domain.beta.kubernetes.io/region、failure-domain.beta.kubernetes.io/zone、failure-domain.kubernetes.io/region、failure-domain.kubernetes.io/zone、kubernetes.io/arch、kubernetes.io/hostname、kubernetes.io/instance-type 和 kubernetes.io/os
--node-status-max-images int32	[Alpha 版特性] 可以报告的最大镜像数量，默认值为 50，设置为-1 表示没有上限
--node-status-update-frequency duration	已弃用，在--config 指定的配置文件中进行设置。kubelet 向 Master 汇报 Node 状态的时间间隔，默认值为 10s。与 controller-manager 的--node-monitor-grace-period 参数共同起作用
--non-masquerade-cidr string	已弃用，kubelet 向该 IP 段之外的 IP 地址发送的流量将使用 IP Masquerade 技术，默认值为 10.0.0.0/8
--oom-score-adj int32	已弃用，在--config 指定的配置文件中进行设置。kubelet 进程的 oom_score_adj 参数值，有效范围为[-1000, 1000]，默认值为-999
--pod-cidr string	已弃用，在--config 指定的配置文件中进行设置。用于给 Pod 分配 IP 地址的 CIDR 地址池，仅在单机模式中使用。在一个集群中，kubelet 会从 API Server 中获取 CIDR 设置
--pod-infra-container-image string	用于 Pod 内网络命名空间共享的基础 pause 镜像，默认值为 k8s.gcr.io/pause:3.1
--pod-manifest-path string	已弃用，在--config 指定的配置文件中进行设置。Pod Manifest 文件路径，不扫描以"."开头的隐藏文件
--pod-max-pids int	已弃用，在--config 指定的配置文件中进行设置。[Alpha 版特性] 设置一个 Pod 内最大的进程数量，默认值为-1
--pods-per-core int32	已弃用，在--config 指定的配置文件中进行设置。该 kubelet 上每个 core 可运行的 Pod 数量。最大值将被 max-pods 参数限制。默认值为 0 表示不做限制
--port int32	已弃用，在--config 指定的配置文件中进行设置。kubelet 服务监听的本机端口号，默认值为 10250
--protect-kernel-defaults	已弃用，在--config 指定的配置文件中进行设置。设置 kernel tuning 的默认 kubelet 行为。如果 kernel tunables 与 kubelet 默认值不同，kubelet 将报错

续表

参数名和类型	说明
--provider-id string	设置主机数据库中标识 Node 的唯一 ID, 例如 cloudprovider
--qos-reserved mapStringString	已弃用，在--config 指定的配置文件中进行设置。 [Alpha 版特性] 设置在指定的 QoS 级别预留的 Pod 资源请求，以"资源名=百分比"的形式进行设置，例如 memory=50%，可以设置多个。要求启用 QOSReserved 特性开关
--read-only-port int32	已弃用，在--config 指定的配置文件中进行设置。 kubelet 服务监听的只读端口号，默认值为 10255，设置为 0 表示不启用
--really-crash-for-testing	设置为 true 表示发生 panics 情况时崩溃，仅用于测试
--redirect-container-streaming	启用容器流数据并重定向给 API Server。设置为 false 表示 kubelet 将代理 API Server 和容器运行时之间的容器流数据。设置为 true 表示 kubelet 将容器运行时重定向给 API Server，之后 API Server 可以直接访问容器运行时。代理模式更安全，但是会浪费一些性能；重定向模式性能更好，但是安全性较低，因为 API Server 和容器运行时之间的连接可能无法进行身份验证
--register-node	将本 Node 注册到 API Server，默认值为 true
--register-schedulable	已弃用，注册本 Node 为可以被调度的，--register-node 为 false 则无效，默认值为 true
--register-with-taints []api.Taint	设置本 Node 的 taints，格式为 <key>=<value>:<effect>，以逗号分隔。当--register-node=false 时不生效
--registry-burst int32	已弃用，在--config 指定的配置文件中进行设置。 最多同时拉取镜像的数量，默认值为 10
--registry-qps int32	已弃用，在--config 指定的配置文件中进行设置。 在 Pod 创建过程中容器的镜像可能需要从 Registry 中拉取，由于在拉取镜像的过程中会消耗大量带宽，因此可能需要限速，此参数与 registry-burst 一起用来限制每秒拉取多少个镜像，默认值为 5
--resolv-conf string	已弃用，在--config 指定的配置文件中进行设置。 命名服务配置文件，用于容器内应用的 DNS 解析，默认值为/etc/resolv.conf
--root-dir string	kubelet 运行根目录，将保持 Pod 和 Volume 的相关文件，默认值为/var/lib/kubelet
--rotate-certificates	已弃用，在--config 指定的配置文件中进行设置。 [Beta 版特性] 设置当客户端证书过期时 kubelet 自动从 kube-apiserver 请求或滚动新的证书

续表

参数名和类型	说明
--rotate-server-certificates	已弃用，在--config 指定的配置文件中进行设置。 当证书过期时自动从 kube-apiserver 请求或者滚动新的证书，要求启用 RotateKubeletServerCertificate 特性开关，以及对提交的 CertificateSigningRequest 对象进行 approve 操作
--runonce	设置为 true 表示创建完 Pod 之后立即退出 kubelet 进程，与--enable-server 参数互斥
--runtime-cgroups string	为容器 runtime 设置的 cgroup
--runtime-request-timeout duration	已弃用，在--config 指定的配置文件中进行设置。 除了长时间运行的 request，对其他 request 的超时时间设置，包括 pull、logs、exec、attach 等操作。当超时时间到达时，请求会被杀掉，抛出一个错误并会重试。默认值为 2m0s
--seccomp-profile-root string	[Alpha 版特性] seccomp 配置文件目录，默认值为/var/lib/kubelet/seccomp
--serialize-image-pulls	已弃用，在--config 指定的配置文件中进行设置。 按顺序挨个 pull 镜像。建议 Docker 低于 1.9 版本或使用 Aufs storage backend 时将其设置为 true，详见 issue #10959，默认值为 true
--storage-driver-buffer-duration duration	已弃用，将缓存数据写入后端存储的时间间隔，默认值为 1m0s
--storage-driver-db string	已弃用，后端存储的数据库名称，默认值为 cadvisor
--storage-driver-host string	已弃用，后端存储的数据库连接 URL 地址，默认值为 localhost:8086
--storage-driver-password string	已弃用，后端存储的数据库密码，默认值为 root
--storage-driver-secure	已弃用，后端存储的数据库是否用安全连接，默认值为 false
--storage-driver-table string	已弃用，后端存储的数据库表名，默认值为 stats
--storage-driver-user string	已弃用，后端存储的数据库用户名，默认值为 root
--streaming-connection-idle-timeout duration	已弃用，在--config 指定的配置文件中进行设置。 在容器中执行命令或者进行端口转发的过程中会产生输入、输出流，这个参数用来控制连接空闲超时而关闭的时间，如果设置为 5m，则表示在连接超过 5min 没有输入、输出的情况下就被认为是空闲的，而会被自动关闭。默认值为 4h0m0s
--sync-frequency duration	已弃用，在--config 指定的配置文件中进行设置。 同步运行中容器的配置的频率，默认值为 1m0s

续表

参数名和类型	说　　明
--system-cgroups /	已弃用，在--config 指定的配置文件中进行设置。 kubelet 为运行非 kernel 进程设置的 cgroups 名称，默认值为""
--system-reserved mapStringString	已弃用，在--config 指定的配置文件中进行设置。 系统预留的资源配置，以一组 ResourceName=ResourceQuantity 格式表示，例如 cpu=200m,memory=500M。目前仅支持 CPU 和内存的设置，详见 http://releases.k8s.io/HEAD/docs/user-guide/compute-resources.md，默认值为 none
--system-reserved-cgroup string	已弃用，在--config 指定的配置文件中进行设置。 用于管理非 kubernetes 的带--system-reserved 标签组件的计算资源，设置顶层 cgroup 全路径名，例如/system-reserved，默认值为空字符串
--tls-cert-file string	已弃用，在--config 指定的配置文件中进行设置。 包含 x509 证书的文件路径，用于 HTTPS 认证
--tls-cipher-suites strings	已弃用，在--config 指定的配置文件中进行设置。 服务器端加密算法列表，以逗号分隔，如果不设置，则使用 Go cipher suites 的默认列表。可选加密算法包括： TLS_ECDHE_ECDSA_WITH_AES_128_CBC_SHA TLS_ECDHE_ECDSA_WITH_AES_128_CBC_SHA256 TLS_ECDHE_ECDSA_WITH_AES_128_GCM_SHA256 TLS_ECDHE_ECDSA_WITH_AES_256_CBC_SHA TLS_ECDHE_ECDSA_WITH_AES_256_GCM_SHA384 TLS_ECDHE_ECDSA_WITH_CHACHA20_POLY1305 TLS_ECDHE_ECDSA_WITH_RC4_128_SHA TLS_ECDHE_RSA_WITH_3DES_EDE_CBC_SHA TLS_ECDHE_RSA_WITH_AES_128_CBC_SHA TLS_ECDHE_RSA_WITH_AES_128_CBC_SHA256 TLS_ECDHE_RSA_WITH_AES_128_GCM_SHA256 TLS_ECDHE_RSA_WITH_AES_256_CBC_SHA TLS_ECDHE_RSA_WITH_AES_256_GCM_SHA384 TLS_ECDHE_RSA_WITH_CHACHA20_POLY1305 TLS_ECDHE_RSA_WITH_RC4_128_SHA TLS_RSA_WITH_3DES_EDE_CBC_SHA

续表

参数名和类型	说明
	TLS_RSA_WITH_AES_128_CBC_SHA
	TLS_RSA_WITH_AES_128_CBC_SHA256
	TLS_RSA_WITH_AES_128_GCM_SHA256
	TLS_RSA_WITH_AES_256_CBC_SHA
	TLS_RSA_WITH_AES_256_GCM_SHA384；TLS_RSA_WITH_RC4_128_SHA
--tls-min-version string	已弃用，在--config 指定的配置文件中进行设置。 设置支持的最小 TLS 版本号，可选的版本号包括：VersionTLS10、VersionTLS11 和 VersionTLS12
--tls-private-key-file string	已弃用，在--config 指定的配置文件中进行设置。 包含 x509 与 tls-cert-file 对应的私钥文件路径
--volume-plugin-dir string	搜索第三方 Volume 插件的目录，默认值为 /usr/libexec/kubernetes/kubelet-plugins/volume/exec/
--volume-stats-agg-period duration	已弃用，在--config 指定的配置文件中进行设置。 kubelet 计算所有 Pod 和 Volume 的磁盘使用情况聚合值的时间间隔，默认值为 1m0s。设置为 0 表示不启用该计算功能

2.8.6　kube-proxy 启动参数

对 kube-proxy 启动参数的详细说明见表 2.8。

表 2.8　对 kube-proxy 启动参数的详细说明

参数名和类型	说明
--bind-address ip	kube-proxy 绑定主机的 IP 地址，默认值为 0.0.0.0，表示绑定所有 IP 地址
--cleanup	设置为 true 表示在清除 iptables 规则和 IPVS 规则后退出
--cleanup-ipvs	设置为 true 表示在清除 IPVS 规则后再启动 kube-proxy，默认值为 true
--cluster-cidr string	集群中 Pod 的 CIDR 地址范围，用于桥接集群外部流量到内部。用于公有云环境
--config string	kube-proxy 的主配置文件
--config-sync-period duration	从 API Server 更新配置的时间间隔，必须大于 0，默认值为 15m0s
--conntrack-max-per-core int32	跟踪每个 CPU core 的 NAT 连接的最大数量（设置为 0 表示无限制，并忽略 conntrack-min 的值），默认值为 32768

续表

参数名和类型	说明
--conntrack-min int32	最小 conntrack 条目的分配数量，默认值为 131072
--conntrack-tcp-timeout-close-wait duration	当 TCP 连接处于 CLOSE_WAIT 状态时的 NAT 超时时间，默认值为 1h0m0s
--conntrack-tcp-timeout-established	建立 TCP 连接的超时时间，设置为 0 表示无限制，默认值为 24h0m0s
--feature-gates mapStringBool	用于实验性质的特性开关组，每个开关都以 key=value 形式表示。当前可用开关包括： APIListChunking=true|false (BETA - default=true) APIResponseCompression=true|false (ALPHA - default=false) AllAlpha=true|false (ALPHA - default=false) AppArmor=true|false (BETA - default=true) AttachVolumeLimit=true|false (BETA - default=true) BalanceAttachedNodeVolumes=true|false (ALPHA - default=false) BlockVolume=true|false (BETA - default=true) BoundServiceAccountTokenVolume=true|false (ALPHA - default=false) CPUManager=true|false (BETA - default=true) CRIContainerLogRotation=true|false (BETA - default=true) CSIBlockVolume=true|false (BETA - default=true) CSIDriverRegistry=true|false (BETA - default=true) CSIInlineVolume=true|false (ALPHA - default=false) CSIMigration=true|false (ALPHA - default=false) CSIMigrationAWS=true|false (ALPHA - default=false) CSIMigrationGCE=true|false (ALPHA - default=false) CSIMigrationOpenStack=true|false (ALPHA - default=false) CSINodeInfo=true|false (BETA - default=true) CustomCPUCFSQuotaPeriod=true|false (ALPHA - default=false) CustomResourcePublishOpenAPI=true|false (ALPHA - default=false) CustomResourceSubresources=true|false (BETA - default=true) CustomResourceValidation=true|false (BETA - default=true) CustomResourceWebhookConversion=true|false (ALPHA - default=false) DebugContainers=true|false (ALPHA - default=false) DevicePlugins=true|false (BETA - default=true)

续表

参数名和类型	说明
	DryRun=true|false (BETA - default=true)
	DynamicAuditing=true|false (ALPHA - default=false)
	DynamicKubeletConfig=true|false (BETA - default=true)
	ExpandCSIVolumes=true|false (ALPHA - default=false)
	ExpandInUsePersistentVolumes=true|false (ALPHA - default=false)
	ExpandPersistentVolumes=true|false (BETA - default=true)
	ExperimentalCriticalPodAnnotation=true|false (ALPHA - default=false)
	ExperimentalHostUserNamespaceDefaulting=true|false (BETA - default=false)
	HyperVContainer=true|false (ALPHA - default=false)
	KubeletPodResources=true|false (ALPHA - default=false)
	LocalStorageCapacityIsolation=true|false (BETA - default=true)
	MountContainers=true|false (ALPHA - default=false)
	NodeLease=true|false (BETA - default=true)
	PodShareProcessNamespace=true|false (BETA - default=true)
	ProcMountType=true|false (ALPHA - default=false)
	QOSReserved=true|false (ALPHA - default=false)
	ResourceLimitsPriorityFunction=true|false (ALPHA - default=false)
	ResourceQuotaScopeSelectors=true|false (BETA - default=true)
	RotateKubeletClientCertificate=true|false (BETA - default=true)
	RotateKubeletServerCertificate=true|false (BETA - default=true)
	RunAsGroup=true|false (BETA - default=true)
	RuntimeClass=true|false (BETA - default=true)
	SCTPSupport=true|false (ALPHA - default=false)
	ScheduleDaemonSetPods=true|false (BETA - default=true)
	ServerSideApply=true|false (ALPHA - default=false)
	ServiceNodeExclusion=true|false (ALPHA - default=false)
	StorageVersionHash=true|false (ALPHA - default=false)
	StreamingProxyRedirects=true|false (BETA - default=true)
	SupportNodePidsLimit=true|false (ALPHA - default=false)
	SupportPodPidsLimit=true|false (BETA - default=true)

续表

参数名和类型	说明
	Sysctls=true|false (BETA - default=true)
	TTLAfterFinished=true|false (ALPHA - default=false)
	TaintBasedEvictions=true|false (BETA - default=true)
	TaintNodesByCondition=true|false (BETA - default=true)
	TokenRequest=true|false (BETA - default=true)
	TokenRequestProjection=true|false (BETA - default=true)
	ValidateProxyRedirects=true|false (BETA - default=true)
	VolumeSnapshotDataSource=true|false (ALPHA - default=false)
	VolumeSubpathEnvExpansion=true|false (ALPHA - default=false)
	WinDSR=true|false (ALPHA - default=false)
	WinOverlay=true|false (ALPHA - default=false)
	WindowsGMSA=true|false (ALPHA - default=false)
--healthz-bind-address ip	healthz 服务绑定主机 IP 地址，设置为 0.0.0.0 表示使用所有 IP 地址，默认值为 0.0.0.0:10256
--healthz-port int32	healthz 服务监听的主机端口号，设置为 0 表示不启用，默认值为 10256
--hostname-override string	设置本 Node 在集群中的主机名，不设置时将使用本机 hostname
--iptables-masquerade-bit int32	标记数据包将进行 SNAT 的 fwmark 位设置，有效范围为[0, 31]，默认值为 14
--iptables-min-sync-period duration	刷新 iptables 规则的最小时间间隔，例如 5s、1m、2h22m
--iptables-sync-period duration	刷新 iptables 规则的最大时间间隔，例如 5s、1m、2h22m，必须大于 0，默认值为 30s
--ipvs-exclude-cidrs strings	设置在清除 IPVS 规则时过滤掉的 CIDR 列表，以逗号分隔
--ipvs-min-sync-period duration	刷新 IPVS 规则的最小时间间隔，例如 5s、1m、2h22m
--ipvs-scheduler string	设置 IPVS 调度器的类型
--ipvs-sync-period duration	刷新 IPVS 规则的最大时间间隔，例如 5s、1m、2h22m，必须大于 0，默认值为 30s
--kube-api-burst int32	每秒发送到 API Server 的请求的数量，默认值为 10
--kube-api-content-type string	发送到 API Server 的请求内容类型，默认值为 application/vnd.kubernetes.protobuf
--kube-api-qps float32	与 API Server 通信的 QPS 值，默认值为 5
--kubeconfig string	kubeconfig 配置文件路径，在配置文件中包括 Master 地址信息及必要的认证信息

续表

参数名和类型	说 明
--masquerade-all	设置为 true 表示使用纯 iptables 代理，所有网络包都将进行 SNAT 转换
--master string	API Server 的地址
--metrics-bind-address 0.0.0.0	Metrics Server 的监听地址，设置为 0.0.0.0 表示使用所有 IP 地址，默认值为 127.0.0.1:10249
--metrics-port int32	Metrics Server 的监听端口号，设置为 0 表示禁用，默认值为 10249
--nodeport-addresses strings	设置 NodePort 可用的 IP 地址范围，例如 1.2.3.0/24, 1.2.3.4/32，默认值为[]，表示使用所有本地 IP 地址
--oom-score-adj int32	kube-proxy 进程的 oom_score_adj 参数值，有效范围为[-1000,1000]，默认值为 -999
--profiling	设置为 true 表示打开性能分析，可以通过 <host>:<port>/debug/pprof/地址查看程序栈、线程等系统信息，默认值为 true
--proxy-mode ProxyMode	代理模式，可选项为 userspace、iptables、ipvs，默认值为 iptables，当操作系统 kernel 版本或 iptables 版本不够新时，将自动降级为 userspace 模式
--proxy-port-range port-range	进行 Service 代理的本地端口号范围，格式为 begin-end，含两端，未指定时采用随机选择的系统可用的端口号
--udp-timeout duration	保持空闲 UDP 连接的时间，例如 250ms、2s，必须大于 0，仅当 proxy-mode=userspace 时生效，默认值为 250ms

2.9 CRI（容器运行时接口）详解

归根结底，Kubernetes Node（kubelet）的主要功能就是启动和停止容器的组件，我们称之为容器运行时（Container Runtime），其中最知名的就是 Docker 了。为了更具扩展性，Kubernetes 从 1.5 版本开始就加入了容器运行时插件 API，即 Container Runtime Interface，简称 CRI。

2.9.1 CRI 概述

每个容器运行时都有特点，因此不少用户希望 Kubernetes 能够支持更多的容器运行时。Kubernetes 从 1.5 版本开始引入了 CRI 接口规范，通过插件接口模式，Kubernetes 无须重新编译就可以使用更多的容器运行时。CRI 包含 Protocol Buffers、gRPC API、运行库支持及开发中的标准规范和工具。Docker 的 CRI 实现在 Kubernetes 1.6 中被更新为 Beta 版本，并在 kubelet 启动时默认启动。

可替代的容器运行时支持是 Kubernetes 中的新概念。在 Kubernetes 1.3 发布时，rktnetes 项目同时发布，让 rkt 容器引擎成为除 Docker 外的又一选择。然而，不管是 Docker 还是 rkt，都用到了 kubelet 的内部接口，同 kubelet 源码纠缠不清。这种程度的集成需要对 kubelet 的内部机制有非常深入的了解，还会给社区带来管理压力，这就给新生代容器运行时造成了难于跨越的集成壁垒。CRI 接口规范试图用定义清晰的抽象层清除这一壁垒，让开发者能够专注于容器运行时本身。在通向插件式容器支持及建设健康生态环境的路上，这是一小步，也是很重要的一步。

2.9.2　CRI 的主要组件

kubelet 使用 gRPC 框架通过 UNIX Socket 与容器运行时（或 CRI 代理）进行通信。在这个过程中 kubelet 是客户端，CRI 代理（shim）是服务端，如图 2.3 所示。

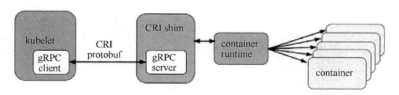

图 2.3　CRI 的主要组件

Protocol Buffers API 包含两个 gRPC 服务：ImageService 和 RuntimeService。

ImageService 提供了从仓库拉取镜像、查看和移除镜像的功能。

RuntimeService 负责 Pod 和容器的生命周期管理，以及与容器的交互（exec/attach/port-forward）。rkt 和 Docker 这样的容器运行时可以使用一个 Socket 同时提供两个服务，在 kubelet 中可以用--container-runtime-endpoint 和--image-service-endpoint 参数设置这个 Socket。

2.9.3　Pod 和容器的生命周期管理

Pod 由一组应用容器组成，其中包含共有的环境和资源约束。在 CRI 里，这个环境被称为 PodSandbox。Kubernetes 有意为容器运行时留下一些发挥空间，它们可以根据自己的内部实现来解释 PodSandbox。对于 Hypervisor 类的运行时，PodSandbox 会具体化为一个虚拟机。其他例如 Docker，会是一个 Linux 命名空间。在 v1alpha1 API 中，kubelet 会创建 Pod 级别的 cgroup 传递给容器运行时，并以此运行所有进程来满足 PodSandbox 对 Pod 的资源保障。

在启动 Pod 之前，kubelet 调用 RuntimeService.RunPodSandbox 来创建环境。这一过程包括为 Pod 设置网络资源（分配 IP 等操作）。PodSandbox 被激活之后，就可以独立地创建、启动、停止和删除不同的容器了。kubelet 会在停止和删除 PodSandbox 之前首先停止和删除其中的容器。

kubelet 的职责在于通过 RPC 管理容器的生命周期，实现容器生命周期的钩子，存活和健康监测，以及执行 Pod 的重启策略等。

RuntimeService 服务包括对 Sandbox 和 Container 操作的方法，下面的伪代码展示了主要的 RPC 方法：

```
service RuntimeService {
    // 沙箱操作
    rpc RunPodSandbox(RunPodSandboxRequest) returns (RunPodSandboxResponse) {}
    rpc StopPodSandbox(StopPodSandboxRequest) returns (StopPodSandboxResponse) {}
    rpc RemovePodSandbox(RemovePodSandboxRequest) returns (RemovePodSandboxResponse) {}
    rpc PodSandboxStatus(PodSandboxStatusRequest) returns (PodSandboxStatusResponse) {}
    rpc ListPodSandbox(ListPodSandboxRequest) returns (ListPodSandboxResponse) {}
    // 容器操作
    rpc CreateContainer(CreateContainerRequest) returns (CreateContainerResponse) {}
    rpc StartContainer(StartContainerRequest) returns (StartContainerResponse) {}
    rpc StopContainer(StopContainerRequest) returns (StopContainerResponse) {}
    rpc RemoveContainer(RemoveContainerRequest) returns (RemoveContainerResponse) {}
    rpc ListContainers(ListContainersRequest) returns (ListContainersResponse) {}
    rpc ContainerStatus(ContainerStatusRequest) returns (ContainerStatusResponse) {}
    ......
}
```

2.9.4 面向容器级别的设计思路

众所周知，Kubernetes 的最小调度单元是 Pod，它曾经可能采用的一个 CRI 设计就是复用 Pod 对象，使得容器运行时可以自行实现控制逻辑和状态转换，这样一来，就能极大地简化 API，让 CRI 能够更广泛地适用于多种容器运行时。但是经过深入讨论之后，Kubernetes 放弃了这一想法。

首先，kubelet 有很多 Pod 级别的功能和机制（例如 crash-loop backoff 机制），如果交给容器运行时去实现，则会造成很重的负担；其次且更重要的是，Pod 标准还在快速演进中。很多新功能（如初始化容器）都是由 kubelet 完成管理的，无须交给容器运行时实现。

CRI 选择了在容器级别进行实现，使得容器运行时能够共享这些通用特性，以获得更快的开发速度。这并不意味着设计哲学的改变——kubelet 要负责、保证容器应用的实际状态和声明状态的一致性。

Kubernetes 为用户提供了与 Pod 及其中的容器进行交互的功能（kubectl exec/attach/port-forward）。kubelet 目前提供了两种方式来支持这些功能。

（1）调用容器的本地方法。

（2）使用 Node 上的工具（例如 nsenter 及 socat）。

因为多数工具都假设 Pod 用 Linux namespace 做了隔离，因此使用 Node 上的工具并不是一种容易移植的方案。在 CRI 中显式定义了这些调用方法，让容器运行时进行具体实现。下面的伪代码显示了 Exec、Attach、PortForward 这几个调用需要实现的 RuntimeService 方法：

```
service RuntimeService {
    ......
    // ExecSync 在容器中同步执行一个命令。
    rpc ExecSync(ExecSyncRequest) returns (ExecSyncResponse) {}
    // Exec 在容器中执行命令
    rpc Exec(ExecRequest) returns (ExecResponse) {}
    // Attach 附着在容器上
    rpc Attach(AttachRequest) returns (AttachResponse) {}
    // PortForward 从 Pod 沙箱中进行端口转发
    rpc PortForward(PortForwardRequest) returns (PortForwardResponse) {}
    ......
}
```

目前还有一个潜在问题是，kubelet 处理所有的请求连接，使其有成为 Node 通信瓶颈的可能。在设计 CRI 时，要让容器运行时能够跳过中间过程。容器运行时可以启动一个单独的流式服务来处理请求（还能对 Pod 的资源使用情况进行记录），并将服务地址返回给 kubelet。这样 kubelet 就能反馈信息给 API Server，使之可以直接连接到容器运行时提供的服务，并连接到客户端。

2.9.5 尝试使用新的 Docker-CRI 来创建容器

要尝试新的 Kubelet-CRI-Docker 集成，只需为 kubelet 启动参数加上 --enable-cri=true 开关来启动 CRI。这个选项从 Kubernetes 1.6 开始已经作为 kubelet 的默认选项了。如果不希望使用 CRI，则可以设置 --enable-cri=false 来关闭这个功能。

查看 kubelet 的日志，可以看到启用 CRI 和创建 gRPC Server 的日志：

```
I0603 15:08:28.953332    3442 container_manager_linux.go:250] Creating
Container Manager object based on Node Config: {RuntimeCgroupsName: SystemCgroupsName:
KubeletCgroupsName: ContainerRuntime:docker CgroupsPerQOS:true CgroupRoot:/
CgroupDriver:cgroupfs ProtectKernelDefaults:false EnableCRI:true
NodeAllocatableConfig:{KubeReservedCgroupName: SystemReservedCgroupName:
EnforceNodeAllocatable:map[pods:{}] KubeReserved:map[] SystemReserved:map[]
HardEvictionThresholds:[{Signal:memory.available Operator:LessThan
Value:{Quantity:100Mi Percentage:0} GracePeriod:0s MinReclaim:<nil>}]}
ExperimentalQOSReserved:map[]}
......
I0603 15:08:29.060283    3442 kubelet.go:573] Starting the GRPC server for the
docker CRI shim.
```

创建一个 Deployment：

```
$ kubectl run nginx --image=nginx
deployment "nginx " created
```

查看 Pod 的详细信息，可以看到将会创建沙箱（Sandbox）的 Event：

```
$ kubectl describe pod nginx
......
Events:
...From                      Type      Reason           Message
..-------------------        -----     ---------------  ------------------------
```

```
    ...default-scheduler    Normal   Scheduled           Successfully assigned nginx
to k8s-node-1
    ...kubelet, k8s-node-1  Normal   SandboxReceived     Pod sandbox received, it will
be created.
    ......
```

这表明 kubelet 使用了 CRI 接口来创建容器。

2.9.6　CRI 的进展

目前已经有多款开源 CRI 项目可用于 Kubernetes：Docker、CRI-O、Containerd、frakti（基于 Hypervisor 的容器运行时），各 CRI 运行时的安装手册可参考官网 https://kubernetes.io/docs/setup/cri/ 的说明。

2.10　kubectl 命令行工具用法详解

kubectl 作为客户端 CLI 工具，可以让用户通过命令行对 Kubernetes 集群进行操作。本节对 kubectl 的子命令和用法进行详细说明。

2.10.1　kubectl 用法概述

kubectl 命令行的语法如下：

```
$ kubectl [command] [TYPE] [NAME] [flags]
```

其中，command、TYPE、NAME、flags 的含义如下。

（1）command：子命令，用于操作 Kubernetes 集群资源对象的命令，例如 create、delete、describe、get、apply 等。

（2）TYPE：资源对象的类型，区分大小写，能以单数、复数或者简写形式表示。例如以下 3 种 TYPE 是等价的。

```
$ kubectl get pod pod1
$ kubectl get pods pod1
$ kubectl get po pod1
```

（3）NAME：资源对象的名称，区分大小写。如果不指定名称，系统则将返回属于TYPE的全部对象的列表，例如$ kubectl get pods 将返回所有 Pod 的列表。

（4）flags：kubectl 子命令的可选参数，例如使用"-s"指定 API Server 的 URL 地址而不用默认值。

kubectl 可操作的资源对象类型及其缩写如表 2.9 所示。

表 2.9　kubectl 可操作的资源对象类型及其缩写

资源对象类型	缩　写
clusters	
componentstatuses	cs
configmaps	cm
daemonsets	ds
deployments	deploy
endpoints	ep
events	ev
horizontalpodautoscalers	hpa
ingresses	ing
Jobs	
limitranges	limits
nodes	no
namespaces	ns
networkpolicies	
statefulsets	
persistentvolumeclaims	pvc
persistentvolumes	pv
pods	po
podsecuritypolicies	psp
podtemplates	
replicasets	rs

续表

资源对象类型	缩　写
replicationcontrollers	rc
resourcequotas	quota
cronjob	
secrets	
serviceaccounts	
services	svc
storageclasses	sc
thirdpartyresources	

在一个命令行中也可以同时对多个资源对象进行操作，以多个 TYPE 和 NAME 的组合表示，示例如下。

◎ 获取多个 Pod 的信息：

```
$ kubectl get pods pod1 pod2
```

◎ 获取多种对象的信息：

```
$ kubectl get pod/pod1 rc/rc1
```

◎ 同时应用多个 YAML 文件，以多个 -f file 参数表示：

```
$ kubectl get pod -f pod1.yaml -f pod2.yaml
$ kubectl create -f pod1.yaml -f rc1.yaml -f service1.yaml
```

2.10.2　kubectl 子命令详解

kubectl 的子命令非常丰富，涵盖了对 Kubernetes 集群的主要操作，包括资源对象的创建、删除、查看、修改、配置、运行等。详细的子命令如表 2.10 所示。

表 2.10　kubectl 子命令详解

子命令	语　法	说　明
annotate	kubectl annotate (-f FILENAME | TYPE NAME | TYPE/NAME) KEY_1=VAL_1 ... KEY_N=VAL_N [--overwrite] [--all] [--resource-version=version] [flags]	添加或更新资源对象的 annotation 信息

续表

子命令	语 法	说 明
api-versions	kubectl api-versions [flags]	列出当前系统支持的 API 版本列表，格式为"group/version"
apply	kubectl apply -f FILENAME [flags]	从配置文件或 stdin 中对资源对象进行配置更新
attach	kubectl attach POD -c CONTAINER [flags]	附着到一个正在运行的容器上
auth	kubectl auth [flags] [options]	检测 RBAC 权限设置
autoscale	kubectl autoscale (-f FILENAME | TYPE NAME | TYPE/NAME) [--min=MINPODS] --max=MAXPODS [--cpu-percent=CPU] [flags]	对 Deployment、ReplicaSet 或 ReplicationController 进行水平自动扩容和缩容的设置
cluster-info	kubectl cluster-info [flags]	显示集群 Master 和内置服务的信息
completion	kubectl completion SHELL [flags]	输出 shell 命令的执行结果码（bash 或 zsh）
config	kubectl config SUBCOMMAND [flags]	修改 kubeconfig 文件
convert	kubectl convert –f FILENAME [flags]	转换配置文件为不同的 API 版本
cordon	kubectl cordon NODE [flags]	将 Node 标记为 unschedulable，即"隔离"出集群调度范围
create	kubectl create –f FILENAME [flags]	从配置文件或 stdin 中创建资源对象
delete	kubectl delete (-f FILENAME | TYPE [NAME | /NAME | -l label | --all]) [flags]	根据配置文件、stdin、资源名称或 label selector 删除资源对象
describe	kubectl describe (-f FILENAME | TYPE [NAME_PREFIX | /NAME | -l label]) [flags]	描述一个或多个资源对象的详细信息
diff	kubectl diff -f FILENAME [options]	查看配置文件与当前系统中正在运行的资源对象的差异
drain	kubectl drain NODE [flags]	首先将 Node 设置为 unschedulable，然后删除在该 Node 上运行的所有 Pod，但不会删除不由 API Server 管理的 Pod
edit	kubectl edit (-f FILENAME | TYPE NAME | TYPE/NAME) [flags]	编辑资源对象的属性，在线更新
exec	kubectl exec POD [-c CONTAINER] [-i] [-t] [flags] [-- COMMAND [args...]]	执行一个容器中的命令

续表

子命令	语　法	说　明
explain	kubectl explain [--include-extended-apis=true] [--recursive=false] [flags]	对资源对象属性的详细说明
expose	kubectl expose (-f FILENAME | TYPE NAME | TYPE/NAME) [--port=port] [--protocol=TCP|UDP] [--target-port=number-or-name] [--name=name] [----external-ip=external-ip-of-service] [--type=type] [flags]	将已经存在的一个 RC、Service、Deployment 或 Pod 暴露为一个新的 Service
get	kubectl get (-f FILENAME | TYPE [NAME | /NAME | -l label]) [--watch] [--sort-by=FIELD] [[-o | --output]=OUTPUT_FORMAT] [flags]	显示一个或多个资源对象的概要信息
label	kubectl label (-f FILENAME | TYPE NAME | TYPE/NAME) KEY_1=VAL_1 ... KEY_N=VAL_N [--overwrite] [--all] [--resource-version=version] [flags]	设置或更新资源对象的 labels
logs	kubectl logs POD [-c CONTAINER] [--follow] [flags]	在屏幕上打印一个容器的日志
patch	kubectl patch (-f FILENAME | TYPE NAME | TYPE/NAME) --patch PATCH [flags] kubectl patch (-f FILENAME | TYPE NAME | TYPE/NAME) --patch PATCH [flags]	以 merge 形式对资源对象的部分字段的值进行修改
plugin	kubectl plugin [flags] [options]	在 kubectl 命令行使用用户自定义的插件
port-forward	kubectl port-forward POD [LOCAL_PORT:]REMOTE_PORT [...[LOCAL_PORT_N:]REMOTE_PORT_N] [flags]	将本机的某个端口号映射到 Pod 的端口号，通常用于测试
proxy	kubectl proxy [--port=PORT] [--www=static-dir] [--www-prefix=prefix] [--api-prefix=prefix] [flags]	将本机某个端口号映射到 API Server
replace	kubectl replace -f FILENAME [flags]	从配置文件或 stdin 替换资源对象
rolling-update	kubectl rolling-update OLD_CONTROLLER_NAME ([NEW_CONTROLLER_NAME] --image=NEW_CONTAINER_IMAGE | -f NEW_CONTROLLER_SPEC) [flags]	对 RC 进行滚动升级
rollout	kubectl rollout SUBCOMMAND [flags]	对 Deployment 进行管理,可用操作包括: history、pause、resume、undo、status

续表

子命令	语 法	说 明
run	kubectl run NAME --image=image [--env="key=value"] [--port=port] [--replicas=replicas] [--dry-run=bool] [--overrides=inline-json] [flags]	基于一个镜像在 Kubernetes 集群上启动一个 Deployment
scale	kubectl scale (-f FILENAME | TYPE NAME | TYPE/NAME) --replicas=COUNT [--resource-version=version] [--current-replicas=count] [flags]	扩容、缩容一个 Deployment、ReplicaSet、RC 或 Job 中 Pod 的数量
set	kubectl set SUBCOMMAND [flags]	设置资源对象的某个特定信息，目前仅支持修改容器的镜像
taint	kubectl taint NODE NAME KEY_1=VAL_1:TAINT_EFFECT_1 ... KEY_N=VAL_N:TAINT_EFFECT_N [flags]	设置 Node 的 taint 信息，用于将特定的 Pod 调度到特定的 Node 的操作，为 Alpha 版本的功能
top	kubectl top node kubectl top pod	查看 Node 或 Pod 的资源使用情况，需要在集群中运行 Metrics Server
uncordon	kubectl uncordon NODE [flags]	将 Node 设置为 schedulable
version	kubectl version [--client] [flags]	打印系统的版本信息

2.10.3 kubectl 参数列表

kubectl 命令行的公共启动参数如表 2.11 所示。

表 2.11 kubectl 命令行的公共启动参数

参数名和取值示例	说 明
--alsologtostderr=false	设置为 true 表示将日志输出到文件的同时输出到 stderr
--as=''	设置本次操作的用户名
--certificate-authority=''	用于 CA 授权的 cert 文件路径
--client-certificate=''	用于 TLS 的客户端证书文件路径
--client-key=''	用于 TLS 的客户端 key 文件路径
--cluster=''	设置要使用的 kubeconfig 中的 cluster 名
--context=''	设置要使用的 kubeconfig 中的 context 名
--insecure-skip-tls-verify=false	设置为 true 表示跳过 TLS 安全验证模式，将使得 HTTPS 连接不安全

续表

参数名和取值示例	说　　明
--kubeconfig=""	kubeconfig 配置文件路径，在配置文件中包括 Master 的地址信息及必要的认证信息
--log-backtrace-at=:0	记录日志每到"file:行号"时打印一次 stack trace
--log-dir=""	日志文件路径
--log-flush-frequency=5s	设置 flush 日志文件的时间间隔
--logtostderr=true	设置为 true 表示将日志输出到 stderr，不输出到日志文件
--match-server-version=false	设置为 true 表示客户端版本号需要与服务端一致
--namespace=""	设置本次操作所在的 Namespace
--password=""	设置 API Server 的 basic authentication 的密码
-s, --server=""	设置 API Server 的 URL 地址，默认值为 localhost:8080
--stderrthreshold=2	在该 threshold 级别之上的日志将输出到 stderr
--token=""	设置访问 API Server 的安全 Token
--user=""	指定用户名（应在 kubeconfig 配置文件中设置过）
--username=""	设置 API Server 的 basic authentication 的用户名
--v=0	glog 日志级别
--vmodule=	glog 基于模块的详细日志级别

每个子命令（如 create、delete、get 等）还有特定的 flags 参数，可以通过$ kubectl [command] --help 命令进行查看。

2.10.4　kubectl 输出格式

kubectl 命令可以用多种格式对结果进行显示，输出的格式通过-o 参数指定：

```
$ kubectl [command] [TYPE] [NAME] -o=<output_format>
```

根据不同子命令的输出结果，可选的输出格式如表 2.12 所示。

表 2.12　kubectl 命令的可选输出格式列表

输　出　格　式	说　　明
-o=custom-columns=<spec>	根据自定义列名进行输出，以逗号分隔
-o=custom-columns-file=<filename>	从文件中获取自定义列名进行输出
-o=json	以 JSON 格式显示结果

续表

输出格式	说明
-o=jsonpath=<template>	输出 jsonpath 表达式定义的字段信息
-o=jsonpath-file=<filename>	输出 jsonpath 表达式定义的字段信息,来源于文件
-o=name	仅输出资源对象的名称
-o=wide	输出额外信息。对于 Pod,将输出 Pod 所在的 Node 名称
-o=yaml	以 YAML 格式显示结果

常用的输出格式示例如下。

(1) 显示 Pod 的更多信息:

```
$ kubectl get pod <pod-name> -o wide
```

(2) 以 YAML 格式显示 Pod 的详细信息:

```
$ kubectl get pod <pod-name> -o yaml
```

(3) 以自定义列名显示 Pod 的信息:

```
$ kubectl get pod <pod-name>
-o=custom-columns=NAME:.metadata.name,RSRC:.metadata.resourceVersion
```

(4) 基于文件的自定义列名输出:

```
$ kubectl get pods <pod-name> -o=custom-columns-file=template.txt
```

template.txt 文件的内容为:

```
NAME                    RSRC
metadata.name           metadata.resourceVersion
```

输出结果为:

```
NAME        RSRC
pod-name    52305
```

另外,可以将输出结果按某个字段排序,通过--sort-by 参数以 jsonpath 表达式进行指定:

```
$ kubectl [command] [TYPE] [NAME] --sort-by=<jsonpath_exp>
```

例如,按照名称进行排序:

```
$ kubectl get pods --sort-by=.metadata.name
```

2.10.5 kubectl 操作示例

本节将一些常用的 kubectl 操作作为示例进行说明。

1. 创建资源对象

根据 YAML 配置文件一次性创建 Service 和 RC：

```
$ kubectl create -f my-service.yaml -f my-rc.yaml
```

根据 <directory> 目录下所有 .yaml、.yml、.json 文件的定义进行创建：

```
$ kubectl create -f <directory>
```

2. 查看资源对象

查看所有 Pod 列表：

```
$ kubectl get pods
```

查看 RC 和 Service 列表：

```
$ kubectl get rc,service
```

3. 描述资源对象

显示 Node 的详细信息：

```
$ kubectl describe nodes <node-name>
```

显示 Pod 的详细信息：

```
$ kubectl describe pods/<pod-name>
```

显示由 RC 管理的 Pod 的信息：

```
$ kubectl describe pods <rc-name>
```

4. 删除资源对象

基于 pod.yaml 定义的名称删除 Pod：

```
$ kubectl delete -f pod.yaml
```

删除所有包含某个 Label 的 Pod 和 Service：

```
$ kubectl delete pods,services -l name=<label-name>
```

删除所有 Pod：

```
$ kubectl delete pods --all
```

5. 执行容器的命令

执行 Pod 的 date 命令，默认使用 Pod 中的第 1 个容器执行：

```
$ kubectl exec <pod-name> date
```

指定 Pod 中的某个容器执行 date 命令：

```
$ kubectl exec <pod-name> -c <container-name> date
```

通过 bash 获得 Pod 中某个容器的 TTY，相当于登录容器：

```
$ kubectl exec -ti <pod-name> -c <container-name> /bin/bash
```

6. 查看容器的日志

查看容器输出到 stdout 的日志：

```
$ kubectl logs <pod-name>
```

跟踪查看容器的日志，相当于 tail -f 命令的结果：

```
$ kubectl logs -f <pod-name> -c <container-name>
```

7. 创建或更新资源对象

用法和 kubectl create 类似，逻辑稍有差异：如果目标资源对象不存在，则进行创建；否则进行更新，例如：

```
$ kubectl apply -f app.yaml
```

8. 在线编辑运行中的资源对象

可以使用 kubectl edit 命令编辑运行中的资源对象，例如使用下面的命令编辑运行中的一个 Deployment：

```
$ kubectl edit deploy nginx
```

在命令执行之后，会通过 YAML 格式展示该对象的定义和状态，用户可以对代码进行编辑和保存，从而完成对在线资源的直接修改。

9. 将 Pod 的开放端口映射到本地

将集群上 Pod 的 80 端口映射到本地的 8888 端口，在浏览器 http://127.0.0.1:8888 中就能够访问到容器提供的服务了：

```
# kubectl port-forward --address 0.0.0.0 \
pod/nginx-6ddbbc47fb-sfdcv 8888:80
```

10. 在 Pod 和本地之间复制文件

把 Pod 上的 /etc/fstab 复制到本地的 /tmp 目录：

```
# kubectl cp nginx-6ddbbc47fb-sfdcv:/etc/fstab /tmp
```

11. 资源对象的标签设置

为 default namespace 设置 testing=true 标签：

```
# kubectl label namespaces default testing=true
```

12. 检查可用的 API 资源类型列表

该命令经常用于检查特定类型的资源是否已经定义，列出所有资源对象类型：

```
# kubectl api-resources
```

13. 使用命令行插件

为了扩展 kubectl 的功能，Kubernetes 从 1.8 版本开始引入插件机制，在 1.14 版本时达到稳定版。

用户自定义插件的可执行文件名需要以"kubectl-"开头，复制到 $PATH 中的某个目录（如 /usr/local/bin），然后就可以通过 kubectl <plugin-name> 运行自定义插件了。例如，实现一个名为 hello 的插件，其功能为在屏幕上输出字符串"hello world"：

新建名为 kubectl-hello 的可执行脚本文件，其内容为

```
echo "hello world"
```

复制 kubectl-hello 文件到 /usr/local/bin/ 目录下，就完成了安装插件的工作。

然后在 kubectl 命令后带上插件名称就能使用这个插件了：

```
# kubectl hello
hello world
```

使用 kubectl plugin list 命令可以查看当前系统中已安装的插件列表：

```
# kubectl plugin list
The following kubectl-compatible plugins are available:

/usr/local/bin/kubectl-hello
/usr/local/bin/kubectl-foo
/usr/local/bin/kubectl-bar
```

更完整的插件开发示例可以从 https://github.com/kubernetes/sample-cli-plugin 找到。

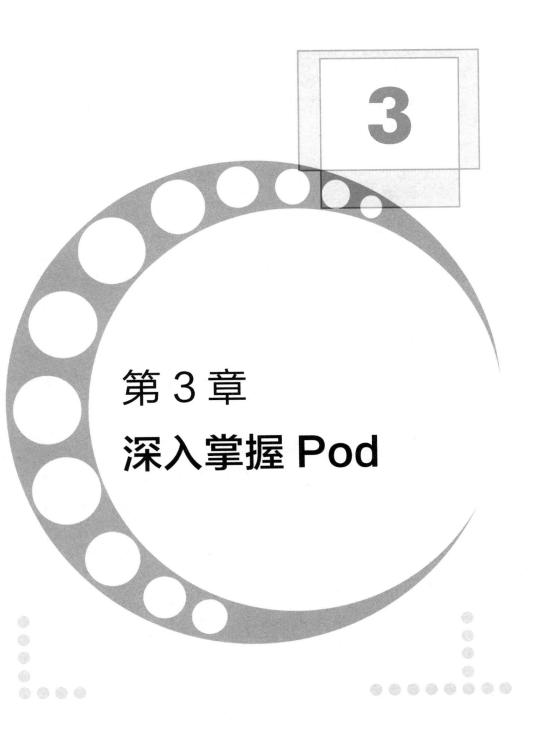

第 3 章
深入掌握 Pod

接下来，让我们深入探索 Pod 的应用、配置、调度、升级及扩缩容，开始 Kubernetes 容器编排之旅。

本章将对 Kubernetes 如何发布与管理容器应用进行详细说明和示例，主要包括 Pod 和容器的使用、应用配置管理、Pod 的控制和调度管理、Pod 的升级和回滚，以及 Pod 的扩缩容机制等内容。

3.1　Pod 定义详解

YAML 格式的 Pod 定义文件的完整内容如下：

```
apiVersion: v1
kind: Pod
metadata:
  name: string
  namespace: string
  labels:
    - name: string
  annotations:
    - name: string
spec:
  containers:
  - name: string
    image: string
    imagePullPolicy: [Always | Never | IfNotPresent]
    command: [string]
    args: [string]
    workingDir: string
    volumeMounts:
    - name: string
      mountPath: string
      readOnly: boolean
    ports:
    - name: string
      containerPort: int
      hostPort: int
      protocol: string
    env:
    - name: string
```

```
      value: string
    resources:
      limits:
        cpu: string
        memory: string
      requests:
        cpu: string
        memory: string
    livenessProbe:
      exec:
        command: [string]
      httpGet:
        path: string
        port: number
        host: string
        scheme: string
        httpHeaders:
        - name: string
          value: string
      tcpSocket:
        port: number
      initialDelaySeconds: 0
      timeoutSeconds: 0
      periodSeconds: 0
      successThreshold: 0
      failureThreshold: 0
    securityContext:
      privileged: false
  restartPolicy: [Always | Never | OnFailure]
  nodeSelector: object
  imagePullSecrets:
  - name: string
  hostNetwork: false
  volumes:
  - name: string
    emptyDir: {}
    hostPath:
      path: string
    secret:
      secretName: string
      items:
```

```
        - key: string
          path: string
    configMap:
      name: string
      items:
        - key: string
          path: string
```

对各属性的详细说明如表 3.1 所示。

表 3.1 对 Pod 定义文件模板中各属性的详细说明

属性名称	取值类型	是否必选	取值说明
version	String	Required	版本号，例如 v1
kind	String	Required	Pod
metadata	Object	Required	元数据
metadata.name	String	Required	Pod 的名称，命名规范需符合 RFC 1035 规范
metadata.namespace	String	Required	Pod 所属的命名空间，默认值为 default
metadata.labels[]	List		自定义标签列表
metadata.annotation[]	List		自定义注解列表
Spec	Object	Required	Pod 中容器的详细定义
spec.containers[]	List	Required	Pod 中的容器列表
spec.containers[].name	String	Required	容器的名称，需符合 RFC 1035 规范
spec.containers[].image	String	Required	容器的镜像名称
spec.containers[].imagePullPolicy	String		镜像拉取策略，可选值包括：Always、Never、IfNotPresent，默认值为 Always。 （1）Always：表示每次都尝试重新拉取镜像。 （2）IfNotPresent：表示如果本地有该镜像，则使用本地的镜像，本地不存在时拉取镜像。 （3）Never：表示仅使用本地镜像。 另外，如果包含如下设置，系统将默认设置 imagePullPolicy=Always，如下所述： （1）不设置 imagePullPolicy，也未指定镜像的 tag； （2）不设置 imagePullPolicy，镜像 tag 为 latest； （3）启用名为 AlwaysPullImages 的准入控制器（Admission Controller）

续表

属性名称	取值类型	是否必选	取值说明
spec.containers[].command[]	List		容器的启动命令列表，如果不指定，则使用镜像打包时使用的启动命令
spec.containers[].args[]	List		容器的启动命令参数列表
spec.containers[].workingDir	String		容器的工作目录
spec.containers[].volumeMounts[]	List		挂载到容器内部的存储卷配置
spec.containers[].volumeMounts[].name	String		引用 Pod 定义的共享存储卷的名称，需使用 volumes[]部分定义的共享存储卷名称
spec.containers[].volumeMounts[].mountPath	String		存储卷在容器内 Mount 的绝对路径，应少于 512 个字符
spec.containers[].volumeMounts[].readOnly	Boolean		是否为只读模式，默认为读写模式
spec.containers[].ports[]	List		容器需要暴露的端口号列表
spec.containers[].ports[].name	String		端口的名称
spec.containers[].ports[].containerPort	Int		容器需要监听的端口号
spec.containers[].ports[].hostPort	Int		容器所在主机需要监听的端口号，默认与 containerPort 相同。设置 hostPort 时，同一台宿主机将无法启动该容器的第 2 份副本
spec.containers[].ports[].protocol	String		端口协议，支持 TCP 和 UDP，默认值为 TCP
spec.containers[].env[]	List		容器运行前需设置的环境变量列表
spec.containers[].env[].name	String		环境变量的名称
spec.containers[].env[].value	String		环境变量的值
spec.containers[].resources	Object		资源限制和资源请求的设置，详见第 10 章的说明
spec.containers[].resources.limits	Object		资源限制的设置
spec.containers[].resources.limits.cpu	String		CPU 限制，单位为 core 数，将用于 docker run --cpu-shares 参数
spec.containers[].resources.limits.memory	String		内存限制，单位可以为 MiB、GiB 等，将用于 docker run --memory 参数
spec.containers[].resources.requests	Object		资源限制的设置
spec.containers[].resources.requests.cpu	String		CPU 请求，单位为 core 数，容器启动的初始可用数量

续表

属性名称	取值类型	是否必选	取值说明
spec.containers[].resources.requests.memory	String		内存请求，单位可以为 MiB、GiB 等，容器启动的初始可用数量
spec.volumes[]	List		在该 Pod 上定义的共享存储卷列表
spec.volumes[].name	String		共享存储卷的名称，在一个 Pod 中每个存储卷定义一个名称，应符合 RFC 1035 规范。容器定义部分的 containers[].volumeMounts[].name 将引用该共享存储卷的名称。 Volume 的类型包括：emptyDir、hostPath、gcePersistentDisk、awsElasticBlockStore、gitRepo、secret、nfs、iscsi、glusterfs、persistentVolumeClaim、rbd、flexVolume、cinder、cephfs、flocker、downwardAPI、fc、azureFile、configMap、vsphereVolume，可以定义多个 Volume，每个 Volume 的 name 保持唯一。本节讲解 emptyDir、hostPath、secret、configMap 这 4 种 Volume，其他类型 Volume 的设置方式详见第 1 章的说明
spec.volumes[].emptyDir	Object		类型为 emptyDir 的存储卷，表示与 Pod 同生命周期的一个临时目录，其值为一个空对象：emptyDir: {}
spec.volumes[].hostPath	Object		类型为 hostPath 的存储卷，表示挂载 Pod 所在宿主机的目录，通过 volumes[].hostPath.path 指定
spec.volumes[].hostPath.path	String		Pod 所在主机的目录，将被用于容器中 mount 的目录
spec.volumes[].secret	Object		类型为 secret 的存储卷，表示挂载集群预定义的 secret 对象到容器内部
spec.volumes[].configMap	Object		类型为 configMap 的存储卷，表示挂载集群预定义的 configMap 对象到容器内部
spec.volumes[].livenessProbe	Object		对 Pod 内各容器健康检查的设置，当探测无响应几次之后，系统将自动重启该容器。可以设置的方法包括：exec、httpGet 和 tcpSocket。对一个容器仅需设置一种健康检查方法

续表

属 性 名 称	取值类型	是否必选	取 值 说 明
spec.volumes[].livenessProbe.exec	Object		对 Pod 内各容器健康检查的设置，exec 方式
spec.volumes[].livenessProbe.exec.command[]	String		exec 方式需要指定的命令或者脚本
spec.volumes[].livenessProbe.httpGet	Object		对 Pod 内各容器健康检查的设置，HTTPGet 方式。需指定 path、port
spec.volumes[].livenessProbe.tcpSocket	Object		对 Pod 内各容器健康检查的设置，tcpSocket 方式
spec.volumes[].livenessProbe.initialDelaySeconds	Number		容器启动完成后首次探测的时间，单位为 s
spec.volumes[].livenessProbe.timeoutSeconds	Number		对容器健康检查的探测等待响应的超时时间设置，单位为 s，默认值为 1s。若超过该超时时间设置，则将认为该容器不健康，会重启该容器
spec.volumes[].livenessProbe.periodSeconds	Number		对容器健康检查的定期探测时间设置，单位为 s，默认 10s 探测一次
spec.restartPolicy	String		Pod 的重启策略，可选值为 Always、OnFailure，默认值为 Always。 (1) Always: Pod 一旦终止运行，则无论容器是如何终止的，kubelet 都将重启它。 (2) OnFailure: 只有 Pod 以非零退出码终止时，kubelet 才会重启该容器。如果容器正常结束（退出码为 0），则 kubelet 将不会重启它。 (3) Never: Pod 终止后，kubelet 将退出码报告给 Master，不会再重启该 Pod
spec.nodeSelector	Object		设置 Node 的 Label，以 key:value 格式指定，Pod 将被调度到具有这些 Label 的 Node 上
spec.imagePullSecrets	Object		pull 镜像时使用的 Secret 名称，以 name:secretkey 格式指定
spec.hostNetwork	Boolean		是否使用主机网络模式，默认值为 false。设置为 true 表示容器使用宿主机网络，不再使用 Docker 网桥，该 Pod 将无法在同一台宿主机上启动第 2 个副本

3.2 Pod 的基本用法

在对 Pod 的用法进行说明之前，有必要先对 Docker 容器中应用的运行要求进行说明。

在使用 Docker 时，可以使用 docker run 命令创建并启动一个容器。而在 Kubernetes 系统中对长时间运行容器的要求是：其主程序需要一直在前台执行。如果我们创建的 Docker 镜像的启动命令是后台执行程序，例如 Linux 脚本：

```
nohup ./start.sh &
```

则在 kubelet 创建包含这个容器的 Pod 之后运行完该命令，即认为 Pod 执行结束，将立刻销毁该 Pod。如果为该 Pod 定义了 ReplicationController，则系统会监控到该 Pod 已经终止，之后根据 RC 定义中 Pod 的 replicas 副本数量生成一个新的 Pod。一旦创建新的 Pod，就在执行完启动命令后陷入无限循环的过程中。这就是 Kubernetes 需要我们自己创建的 Docker 镜像并以一个前台命令作为启动命令的原因。

对于无法改造为前台执行的应用，也可以使用开源工具 Supervisor 辅助进行前台运行的功能。Supervisor 提供了一种可以同时启动多个后台应用，并保持 Supervisor 自身在前台执行的机制，可以满足 Kubernetes 对容器的启动要求。关于 Supervisor 的安装和使用，请参考官网 http://supervisord.org 的文档说明。

接下来对 Pod 对容器的封装和应用进行说明。

Pod 可以由 1 个或多个容器组合而成。在上一节 Guestbook 的例子中，名为 frontend 的 Pod 只由一个容器组成：

```
apiVersion: v1
kind: Pod
metadata:
  name: frontend
  labels:
    name: frontend
spec:
  containers:
  - name: frontend
    image: kubeguide/guestbook-php-frontend
    env:
    - name: GET_HOSTS_FROM
```

```
        value: env
      ports:
      - containerPort: 80
```

这个 frontend Pod 在成功启动之后,将启动 1 个 Docker 容器。

另一种场景是,当 frontend 和 redis 两个容器应用为紧耦合的关系,并组合成一个整体对外提供服务时,应将这两个容器打包为一个 Pod,如图 3.1 所示。

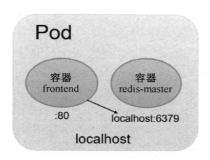

图 3.1 包含两个容器的 Pod

配置文件 frontend-localredis-pod.yaml 的内容如下:

```
apiVersion: v1
kind: Pod
metadata:
  name: redis-php
  labels:
    name: redis-php
spec:
  containers:
  - name: frontend
    image: kubeguide/guestbook-php-frontend:localredis
    ports:
    - containerPort: 80
  - name: redis
    image: kubeguide/redis-master
    ports:
    - containerPort: 6379
```

属于同一个 Pod 的多个容器应用之间相互访问时仅需要通过 localhost 就可以通信,使得这一组容器被"绑定"在了一个环境中。

在 Docker 容器 kubeguide/guestbook-php-frontend:localredis 的 PHP 网页中，直接通过 URL 地址 "localhost:6379" 对同属于一个 Pod 的 redis-master 进行访问。guestbook.php 的内容如下：

```php
<?
set_include_path('.:/usr/local/lib/php');
error_reporting(E_ALL);
ini_set('display_errors', 1);
require 'Predis/Autoloader.php';
Predis\Autoloader::register();

if (isset($_GET['cmd']) === true) {
  $host = 'localhost';
  if (getenv('REDIS_HOST') && strlen(getenv('REDIS_HOST')) > 0 ) {
    $host = getenv('REDIS_HOST');
  }
  header('Content-Type: application/json');
  if ($_GET['cmd'] == 'set') {
    $client = new Predis\Client([
      'scheme' => 'tcp',
      'host'   => $host,
      'port'   => 6379,
    ]);

    $client->set($_GET['key'], $_GET['value']);
    print('{"message": "Updated"}');
  } else {
    $host = 'localhost';
    if (getenv('REDIS_HOST') && strlen(getenv('REDIS_HOST')) > 0 ) {
      $host = getenv('REDIS_HOST');
    }
    $client = new Predis\Client([
      'scheme' => 'tcp',
      'host'   => $host,
      'port'   => 6379,
    ]);

    $value = $client->get($_GET['key']);
    print('{"data": "' . $value . '"}');
  }
} else {
```

```
    phpinfo();
} ?>
```

运行 kubectl create 命令创建该 Pod:

```
$ kubectl create -f frontend-localredis-pod.yaml
pod "redis-php" created
```

查看已经创建的 Pod:

```
# kubectl get pods
NAME        READY   STATUS    RESTARTS   AGE
redis-php   2/2     Running   0          10m
```

可以看到 READY 信息为 2/2, 表示 Pod 中的两个容器都成功运行了。

查看这个 Pod 的详细信息,可以看到两个容器的定义及创建的过程(Event 事件信息):

```
# kubectl describe pod redis-php
Name:           redis-php
Namespace:      default
Node:           k8s/192.168.18.3
Start Time:     Thu, 28 Jul 2016 12:28:21 +0800
Labels:         name=redis-php
Status:         Running
IP:             172.17.1.4
Controllers:    <none>
Containers:
  frontend:
    Container ID:   docker://ccc8616f8df1fb19abbd0ab189a36e6f6628b78ba7b97b1077d86e7fc224ee08
    Image:          kubeguide/guestbook-php-frontend:localredis
    Image ID:       docker://sha256:d014f67384a11186e135b95a7ed0d794674f7ce258f0dce47267c3052a0d0fa9
    Port:           80/TCP
    State:          Running
      Started:      Thu, 28 Jul 2016 12:28:22 +0800
    Ready:          True
    Restart Count:  0
    Environment Variables:  <none>
  redis:
    Container ID:   docker://c0b19362097cda6dd5b8ed7d8eaaaf43aeeb969ee023ef255604bde089808075
    Image:          kubeguide/redis-master
```

```
      Image ID:      docker://sha256:405a0b586f7ebeb545ec65be0e914311159d1baedccd3a93e9d3e3b249ec5cbd
      Port:                    6379/TCP
      State:                   Running
        Started:               Thu, 28 Jul 2016 12:28:23 +0800
      Ready:                   True
      Restart Count:           0
      Environment Variables:   <none>
  Conditions:
    Type           Status
    Initialized    True
    Ready          True
    PodScheduled   True
  Volumes:
    default-token-97j21:
      Type:        Secret (a volume populated by a Secret)
      SecretName:  default-token-97j21
  QoS Tier:        BestEffort
  Events:
    FirstSeen    LastSeen    Count    From              SubobjectPath    Type      Reason     Message
    ---------    --------    -----    ----              -------------    -----     ------     -------
    18m          18m         1        {default-scheduler }                         Normal     Scheduled      Successfully assigned redis-php to k8s-node-1
    18m          18m         1        {kubelet k8s-node-1}    spec.containers{frontend}    Normal    Pulled       Container image "kubeguide/guestbook-php-frontend:localredis" already present on machine
    18m          18m         1        {kubelet k8s-node-1}    spec.containers{frontend}    Normal    Created      Created container with docker id ccc8616f8df1
    18m          18m         1        {kubelet k8s-node-1}    spec.containers{frontend}    Normal    Started      Started container with docker id ccc8616f8df1
    18m          18m         1        {kubelet k8s-node-1}    spec.containers{redis}       Normal    Pulled       Container image "kubeguide/redis-master" already present on machine
    18m          18m         1        {kubelet k8s-node-1}    spec.containers{redis}       Normal    Created      Created container with docker id c0b19362097c
    18m          18m         1        {kubelet k8s-node-1}    spec.containers{redis}       Normal    Started      Started container with docker id c0b19362097c
```

3.3 静态 Pod

静态 Pod 是由 kubelet 进行管理的仅存在于特定 Node 上的 Pod。它们不能通过 API Server 进行管理，无法与 ReplicationController、Deployment 或者 DaemonSet 进行关联，并且 kubelet 无法对它们进行健康检查。静态 Pod 总是由 kubelet 创建的，并且总在 kubelet 所在的 Node 上运行。

创建静态 Pod 有两种方式：配置文件方式和 HTTP 方式。

1. 配置文件方式

首先，需要设置 kubelet 的启动参数 "--pod-manifest-path"（或者在 kubelet 配置文件中设置 staticPodPath，这也是新版本推荐的设置方式，--pod-manifest-path 参数将被逐渐弃用），指定 kubelet 需要监控的配置文件所在的目录，kubelet 会定期扫描该目录，并根据该目录下的.yaml 或.json 文件进行创建操作。

假设配置目录为/etc/kubelet.d/，配置启动参数为--pod-manifest-path=/etc/kubelet.d/，然后重启 kubelet 服务。

在目录/etc/kubelet.d 中放入 static-web.yaml 文件，内容如下：

```
apiVersion: v1
kind: Pod
metadata:
  name: static-web
  labels:
    name: static-web
spec:
  containers:
  - name: static-web
    image: nginx
    ports:
    - name: web
      containerPort: 80
```

等待一会儿，查看本机中已经启动的容器：

```
# docker ps
CONTAINER ID    IMAGE    COMMAND    CREATED    STATUS    PORTS    NAMES
```

```
2292ea231ab1    nginx    "nginx -g 'daemon off"    1 minute ago    1m
k8s_static-web.68ee0075_static-web-k8s-node-1_default_78c7efddebf191c949cbb7aa22
a927c8_401b96d0
```

可以看到一个 Nginx 容器已经被 kubelet 成功创建了出来。

到 Master 上查看 Pod 列表，可以看到这个 static pod：

```
# kubectl get pods
NAME                READY    STATUS     RESTARTS    AGE
static-web-node1    1/1      Running    0           5m
```

由于静态 Pod 无法通过 API Server 直接管理，所以在 Master 上尝试删除这个 Pod 时，会使其变成 Pending 状态，且不会被删除。

```
# kubectl delete pod static-web-node1
pod "static-web-node1" deleted

# kubectl get pods
NAME                READY    STATUS     RESTARTS    AGE
static-web-node1    0/1      Pending    0           1s
```

删除该 Pod 的操作只能是到其所在 Node 上将其定义文件 static-web.yaml 从 /etc/kubelet.d 目录下删除。

```
# rm /etc/kubelet.d/static-web.yaml
# docker ps
// 无容器运行
```

2. HTTP 方式

通过设置 kubelet 的启动参数 "--manifest-url"，kubelet 将会定期从该 URL 地址下载 Pod 的定义文件，并以 .yaml 或 .json 文件的格式进行解析，然后创建 Pod。其实现方式与配置文件方式是一致的。

3.4 Pod 容器共享 Volume

同一个 Pod 中的多个容器能够共享 Pod 级别的存储卷 Volume。Volume 可以被定义为各种类型，多个容器各自进行挂载操作，将一个 Volume 挂载为容器内部需要的目录，如图 3.2 所示。

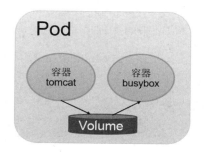

图 3.2　Pod 中多个容器共享 Volume

在下面的例子中，在 Pod 内包含两个容器：tomcat 和 busybox，在 Pod 级别设置 Volume "app-logs"，用于 tomcat 向其中写日志文件，busybox 读日志文件。

配置文件 pod-volume-applogs.yaml 的内容如下：

```yaml
apiVersion: v1
kind: Pod
metadata:
  name: volume-pod
spec:
  containers:
  - name: tomcat
    image: tomcat
    ports:
    - containerPort: 8080
    volumeMounts:
    - name: app-logs
      mountPath: /usr/local/tomcat/logs
  - name: busybox
    image: busybox
    command: ["sh", "-c", "tail -f /logs/catalina*.log"]
    volumeMounts:
    - name: app-logs
      mountPath: /logs
  volumes:
  - name: app-logs
    emptyDir: {}
```

这里设置的 Volume 名为 app-logs，类型为 emptyDir（也可以设置为其他类型，详见第 1 章对 Volume 概念的说明），挂载到 tomcat 容器内的 /usr/local/tomcat/logs 目录，同时挂载到 logreader 容器内的/logs 目录。tomcat 容器在启动后会向/usr/local/tomcat/logs 目录写文件，logreader 容器就可以读取其中的文件了。

logreader 容器的启动命令为 tail -f /logs/catalina*.log，我们可以通过 kubectl logs 命令查看 logreader 容器的输出内容：

```
# kubectl logs volume-pod -c busybox
...........
29-Jul-2016 12:55:59.626 INFO [localhost-startStop-1] org.apache.catalina.startup.HostConfig.deployDirectory Deploying web application directory /usr/local/tomcat/webapps/manager
    29-Jul-2016 12:55:59.722 INFO [localhost-startStop-1] org.apache.catalina.startup.HostConfig.deployDirectory Deployment of web application directory /usr/local/tomcat/webapps/manager has finished in 96 ms
    29-Jul-2016 12:55:59.740 INFO [main] org.apache.coyote.AbstractProtocol.start Starting ProtocolHandler ["http-apr-8080"]
    29-Jul-2016 12:55:59.794 INFO [main] org.apache.coyote.AbstractProtocol.start Starting ProtocolHandler ["ajp-apr-8009"]
    29-Jul-2016 12:56:00.604 INFO [main] org.apache.catalina.startup.Catalina.start Server startup in 4052 ms
```

这个文件为 tomcat 生成的日志文件/usr/local/tomcat/logs/catalina.<date>.log 的内容。登录 tomcat 容器进行查看：

```
# kubectl exec -ti volume-pod -c tomcat -- ls /usr/local/tomcat/logs
catalina.2016-07-29.log      localhost_access_log.2016-07-29.txt
host-manager.2016-07-29.log  manager.2016-07-29.log


# kubectl exec -ti volume-pod -c tomcat -- tail /usr/local/tomcat/logs/catalina.2016-07-29.log
...........
    29-Jul-2016 12:55:59.722 INFO [localhost-startStop-1] org.apache.catalina.startup.HostConfig.deployDirectory Deployment of web application directory /usr/local/tomcat/webapps/manager has finished in 96 ms
    29-Jul-2016 12:55:59.740 INFO [main] org.apache.coyote.AbstractProtocol.start Starting ProtocolHandler ["http-apr-8080"]
    29-Jul-2016 12:55:59.794 INFO [main] org.apache.coyote.AbstractProtocol.start Starting ProtocolHandler ["ajp-apr-8009"]
    29-Jul-2016 12:56:00.604 INFO [main] org.apache.catalina.startup.Catalina.start Server startup in 4052 ms
```

3.5 Pod 的配置管理

应用部署的一个最佳实践是将应用所需的配置信息与程序进行分离,这样可以使应用程序被更好地复用,通过不同的配置也能实现更灵活的功能。将应用打包为容器镜像后,可以通过环境变量或者外挂文件的方式在创建容器时进行配置注入,但在大规模容器集群的环境中,对多个容器进行不同的配置将变得非常复杂。从 Kubernetes 1.2 开始提供了一种统一的应用配置管理方案——ConfigMap。本节对 ConfigMap 的概念和用法进行详细描述。

3.5.1 ConfigMap 概述

ConfigMap 供容器使用的典型用法如下。

(1)生成为容器内的环境变量。

(2)设置容器启动命令的启动参数(需设置为环境变量)。

(3)以 Volume 的形式挂载为容器内部的文件或目录。

ConfigMap 以一个或多个 key:value 的形式保存在 Kubernetes 系统中供应用使用,既可以用于表示一个变量的值(例如 apploglevel=info),也可以用于表示一个完整配置文件的内容(例如 server.xml=<?xml...>...)。

可以通过 YAML 配置文件或者直接使用 kubectl create configmap 命令行的方式来创建 ConfigMap。

3.5.2 创建 ConfigMap 资源对象

1. 通过 YAML 配置文件方式创建

下面的例子 cm-appvars.yaml 描述了将几个应用所需的变量定义为 ConfigMap 的用法:

```
cm-appvars.yaml
apiVersion: v1
kind: ConfigMap
```

```
metadata:
  name: cm-appvars
data:
  apploglevel: info
  appdatadir: /var/data
```

执行 kubectl create 命令创建该 ConfigMap：

```
$kubectl create -f cm-appvars.yaml
configmap "cm-appvars" created
```

查看创建好的 ConfigMap：

```
# kubectl get configmap
NAME          DATA      AGE
cm-appvars    2         3s

# kubectl describe configmap cm-appvars
Name:           cm-appvars
Namespace:      default
Labels:         <none>
Annotations:    <none>

Data
====
appdatadir:     9 bytes
apploglevel:    4 bytes

# kubectl get configmap cm-appvars -o yaml
apiVersion: v1
data:
  appdatadir: /var/data
  apploglevel: info
kind: ConfigMap
metadata:
  creationTimestamp: 2016-07-28T19:57:16Z
  name: cm-appvars
  namespace: default
  resourceVersion: "78709"
  selfLink: /api/v1/namespaces/default/configmaps/cm-appvars
  uid: 7bb2e9c0-54fd-11e6-9dcd-000c29dc2102
```

下面的例子 cm-appconfigfiles.yaml 描述了将两个配置文件 server.xml 和

logging.properties 定义为 ConfigMap 的用法，设置 key 为配置文件的别名，value 则是配置文件的全部文本内容：

```yaml
cm-appconfigfiles.yaml
apiVersion: v1
kind: ConfigMap
metadata:
  name: cm-appconfigfiles
data:
  key-serverxml: |
    <?xml version='1.0' encoding='utf-8'?>
    <Server port="8005" shutdown="SHUTDOWN">
      <Listener className="org.apache.catalina.startup.VersionLoggerListener" />
      <Listener className="org.apache.catalina.core.AprLifecycleListener" SSLEngine="on" />
      <Listener className="org.apache.catalina.core.JreMemoryLeakPreventionListener" />
      <Listener className="org.apache.catalina.mbeans.GlobalResourcesLifecycleListener" />
      <Listener className="org.apache.catalina.core.ThreadLocalLeakPreventionListener" />
      <GlobalNamingResources>
        <Resource name="UserDatabase" auth="Container"
                  type="org.apache.catalina.UserDatabase"
                  description="User database that can be updated and saved"
                  factory="org.apache.catalina.users.MemoryUserDatabaseFactory"
                  pathname="conf/tomcat-users.xml" />
      </GlobalNamingResources>

      <Service name="Catalina">
        <Connector port="8080" protocol="HTTP/1.1"
                   connectionTimeout="20000"
                   redirectPort="8443" />
        <Connector port="8009" protocol="AJP/1.3" redirectPort="8443" />
        <Engine name="Catalina" defaultHost="localhost">
          <Realm className="org.apache.catalina.realm.LockOutRealm">
            <Realm className="org.apache.catalina.realm.UserDatabaseRealm"
                   resourceName="UserDatabase"/>
          </Realm>
          <Host name="localhost"  appBase="webapps"
                unpackWARs="true" autoDeploy="true">
```

```
            <Valve className="org.apache.catalina.valves.AccessLogValve"
directory="logs"
              prefix="localhost_access_log" suffix=".txt"
              pattern="%h %l %u %t &quot;%r&quot; %s %b" />

        </Host>
      </Engine>
    </Service>
  </Server>
  key-loggingproperties: "handlers
    =1catalina.org.apache.juli.FileHandler, 2localhost.org.apache.juli.
FileHandler,
    3manager.org.apache.juli.FileHandler, 4host-manager.org.apache.juli.
FileHandler,
    java.util.logging.ConsoleHandler\r\n\r\n.handlers= 1catalina.org.apache.
juli.FileHandler,

java.util.logging.ConsoleHandler\r\n\r\n1catalina.org.apache.juli.FileHandler.level
      = FINE\r\n1catalina.org.apache.juli.FileHandler.directory
      = ${catalina.base}/logs\r\n1catalina.org.apache.juli.FileHandler.prefix
      = catalina.\r\n\r\n2localhost.org.apache.juli.FileHandler.level
      = FINE\r\n2localhost.org.apache.juli.FileHandler.directory
      = ${catalina.base}/logs\r\n2localhost.org.apache.juli.FileHandler.prefix
      = localhost.\r\n\r\n3manager.org.apache.juli.FileHandler.level
      = FINE\r\n3manager.org.apache.juli.FileHandler.directory
      = ${catalina.base}/logs\r\n3manager.org.apache.juli.FileHandler.prefix
      = manager.\r\n\r\n4host-manager.org.apache.juli.FileHandler.level
      = FINE\r\n4host-manager.org.apache.juli.FileHandler.directory
      = ${catalina.base}/logs\r\n4host-manager.org.apache.juli.FileHandler.
prefix = host-manager.\r\n\r\n\rjava.util.logging.ConsoleHandler.level
      = FINE\r\ njava.util.logging.ConsoleHandler.formatter
      = java.util.logging.SimpleFormatter\r\n\r\n\r\norg.apache.catalina.core.
ContainerBase.[Catalina].[localhost].level
      = INFO\r\norg.apache.catalina.core.ContainerBase.[Catalina].[localhost].
handlers
      = 2localhost.org.apache.juli.FileHandler\r\n\r\norg.apache.catalina.core.
ContainerBase.[Catalina].[localhost].[/manager].level
      = INFO\r\norg.apache.catalina.core.ContainerBase.[Catalina].[localhost].
[/manager].handlers
      = 3manager.org.apache.juli.FileHandler\r\n\r\norg.apache.catalina.core.
ContainerBase.[Catalina].[localhost].[/host-manager].level
```

```
        = INFO\r\norg.apache.catalina.core.ContainerBase.[Catalina].[localhost].
[/host-manager].handlers
        = 4host-manager.org.apache.juli.FileHandler\r\n\r\n"
```

执行 kubectl create 命令创建该 ConfigMap：

```
$kubectl create -f cm-appconfigfiles.yaml
configmap "cm-appconfigfiles" created
```

查看创建好的 ConfigMap：

```
# kubectl get configmap cm-appconfigfiles
NAME                DATA      AGE
cm-appconfigfiles   2         14s

# kubectl describe configmap cm-appconfigfiles
Name:           cm-appconfigfiles
Namespace:      default
Labels:         <none>
Annotations:    <none>

Data
====
key-loggingproperties:   1809 bytes
key-serverxml:           1686 bytes
```

查看已创建的 ConfigMap 的详细内容，可以看到两个配置文件的全文：

```
# kubectl get configmap cm-appconfigfiles -o yaml
apiVersion: v1
data:
  key-loggingproperties: "handlers = 1catalina.org.apache.juli.FileHandler,
2localhost.org.apache.juli.FileHandler,
    3manager.org.apache.juli.FileHandler, 4host-manager.org.apache.juli.
FileHandler,
    java.util.logging.ConsoleHandler\r\n\r\n.handlers = 1catalina.org.apache.
juli.FileHandler,
    java.util.logging.ConsoleHandler\r\n\r\n1catalina.org.apache.juli.
FileHandler.level
    = FINE\r\n1catalina.org.apache.juli.FileHandler.directory
    = ${catalina.base}/logs\r\n1catalina.org.apache.juli.FileHandler.prefix
    = catalina.\r\n\r\n2localhost.org.apache.juli.FileHandler.level
    = FINE\r\n2localhost.org.apache.juli.FileHandler.directory
```

```
          = ${catalina.base}/logs\r\n2localhost.org.apache.juli.FileHandler.prefix
          = localhost.\r\n\r\n3manager.org.apache.juli.FileHandler.level
          = FINE\r\n3manager.org.apache.juli.FileHandler.directory
          = ${catalina.base}/logs\r\n3manager.org.apache.juli.FileHandler.prefix
          = manager.\r\n\r\n4host-manager.org.apache.juli.FileHandler.level
          = FINE\r\n4host-manager.org.apache.juli.FileHandler.directory
          = ${catalina.base}/logs\r\n4host-manager.org.apache.juli.FileHandler.
prefix =
          host-manager.\r\n\r\njava.util.logging.ConsoleHandler.level = FINE\r\njava.
util.logging.ConsoleHandler.formatter
          = java.util.logging.SimpleFormatter\r\n\r\n\r\norg.apache.catalina.core.
ContainerBase.[Catalina].[localhost].level
          = INFO\r\norg.apache.catalina.core.ContainerBase.[Catalina].[localhost].
handlers
          = 2localhost.org.apache.juli.FileHandler\r\n\r\norg.apache.catalina.core.
ContainerBase.[Catalina].[localhost].[/manager].level
          = INFO\r\norg.apache.catalina.core.ContainerBase.[Catalina].[localhost].
[/manager].handlers
          = 3manager.org.apache.juli.FileHandler\r\n\r\norg.apache.catalina.core.
ContainerBase.[Catalina].[localhost].[/host-manager].level
          = INFO\r\norg.apache.catalina.core.ContainerBase.[Catalina].[localhost].
[/host-manager].handlers
          = 4host-manager.org.apache.juli.FileHandler\r\n\r\n"
    key-serverxml: |
        <?xml version='1.0' encoding='utf-8'?>
        <Server port="8005" shutdown="SHUTDOWN">
          <Listener className="org.apache.catalina.startup.VersionLoggerListener" />
          <Listener className="org.apache.catalina.core.AprLifecycleListener"
SSLEngine="on" />
          <Listener className="org.apache.catalina.core.
JreMemoryLeakPreventionListener" />
          <Listener className="org.apache.catalina.mbeans.
GlobalResourcesLifecycleListener" />
          <Listener className="org.apache.catalina.core.
ThreadLocalLeakPreventionListener" />
          <GlobalNamingResources>
            <Resource name="UserDatabase" auth="Container"
                type="org.apache.catalina.UserDatabase"
                description="User database that can be updated and saved"
                factory="org.apache.catalina.users.MemoryUserDatabaseFactory"
                pathname="conf/tomcat-users.xml" />
```

```
        </GlobalNamingResources>

        <Service name="Catalina">
          <Connector port="8080" protocol="HTTP/1.1"
                  connectionTimeout="20000"
                  redirectPort="8443" />
          <Connector port="8009" protocol="AJP/1.3" redirectPort="8443" />
          <Engine name="Catalina" defaultHost="localhost">
            <Realm className="org.apache.catalina.realm.LockOutRealm">
              <Realm className="org.apache.catalina.realm.UserDatabaseRealm"
                  resourceName="UserDatabase"/>
            </Realm>
            <Host name="localhost"  appBase="webapps"
                unpackWARs="true" autoDeploy="true">
              <Valve className="org.apache.catalina.valves.AccessLogValve" directory="logs"
                  prefix="localhost_access_log" suffix=".txt"
                  pattern="%h %l %u %t &quot;%r&quot; %s %b" />

            </Host>
          </Engine>
        </Service>
      </Server>
kind: ConfigMap
metadata:
  creationTimestamp: 2016-07-29T00:52:18Z
  name: cm-appconfigfiles
  namespace: default
  resourceVersion: "85054"
  selfLink: /api/v1/namespaces/default/configmaps/cm-appconfigfiles
  uid: b30d5019-5526-11e6-9dcd-000c29dc2102
```

2. 通过 kubectl 命令行方式创建

不使用 YAML 文件，直接通过 kubectl create configmap 也可以创建 ConfigMap，可以使用参数--from-file 或--from-literal 指定内容，并且可以在一行命令中指定多个参数。

（1）通过--from-file 参数从文件中进行创建，可以指定 key 的名称，也可以在一个命令行中创建包含多个 key 的 ConfigMap，语法为：

```
# kubectl create configmap NAME --from-file=[key=]source
```

```
--from-file=[key=]source
```

（2）通过--from-file 参数从目录中进行创建，该目录下的每个配置文件名都被设置为 key，文件的内容被设置为 value，语法为：

```
# kubectl create configmap NAME --from-file=config-files-dir
```

（3）使用--from-literal 时会从文本中进行创建，直接将指定的 key#=value#创建为 ConfigMap 的内容，语法为：

```
# kubectl create configmap NAME --from-literal=key1=value1 --from-literal=key2=value2
```

下面对这几种用法举例说明。

例如，在当前目录下含有配置文件 server.xml，可以创建一个包含该文件内容的 ConfigMap：

```
# kubectl create configmap cm-server.xml --from-file=server.xml
configmap "cm-server.xml" created

# kubectl describe configmap cm-server.xml
Name:           cm-server.xml
Namespace:      default
Labels:         <none>
Annotations:    <none>

Data
====
server.xml:     6458 bytes
```

假设在 configfiles 目录下包含两个配置文件 server.xml 和 logging.properties，创建一个包含这两个文件内容的 ConfigMap：

```
# kubectl create configmap cm-appconf --from-file=configfiles
configmap "cm-appconf" created

# kubectl describe configmap cm-appconf
Name:           cm-appconf
Namespace:      default
Labels:         <none>
Annotations:    <none>
```

```
Data
====
logging.properties:      3354 bytes
server.xml:              6458 bytes
```

使用--from-literal 参数进行创建的示例如下：

```
# kubectl create configmap cm-appenv --from-literal=loglevel=info
--from-literal =appdatadir=/var/data
configmap "cm-appenv" created

# kubectl describe configmap cm-appenv
Name:           cm-appenv
Namespace:      default
Labels:         <none>
Annotations:    <none>

Data
====
appdatadir:     9 bytes
loglevel:       4 bytes
```

容器应用对 ConfigMap 的使用有以下两种方法。

（1）通过环境变量获取 ConfigMap 中的内容。

（2）通过 Volume 挂载的方式将 ConfigMap 中的内容挂载为容器内部的文件或目录。

3.5.3 在 Pod 中使用 ConfigMap

1. 通过环境变量方式使用 ConfigMap

以前面创建的 ConfigMap "cm-appvars" 为例：

```
apiVersion: v1
kind: ConfigMap
metadata:
  name: cm-appvars
data:
  apploglevel: info
  appdatadir: /var/data
```

在 Pod "cm-test-pod" 的定义中,将 ConfigMap "cm-appvars" 中的内容以环境变量(APPLOGLEVEL 和 APPDATADIR)方式设置为容器内部的环境变量,容器的启动命令将显示这两个环境变量的值("env | grep APP"):

```yaml
apiVersion: v1
kind: Pod
metadata:
  name: cm-test-pod
spec:
  containers:
  - name: cm-test
    image: busybox
    command: [ "/bin/sh", "-c", "env | grep APP" ]
    env:
    - name: APPLOGLEVEL              # 定义环境变量的名称
      valueFrom:                     # key "apploglevel" 对应的值
        configMapKeyRef:
          name: cm-appvars           # 环境变量的值取自 cm-appvars:
          key: apploglevel           # key 为 apploglevel
    - name: APPDATADIR               # 定义环境变量的名称
      valueFrom:                     # key "appdatadir" 对应的值
        configMapKeyRef:
          name: cm-appvars           # 环境变量的值取自 cm-appvars
          key: appdatadir            # key 为 appdatadir
  restartPolicy: Never
```

使用 kubectl create -f 命令创建该 Pod,由于是测试 Pod,所以该 Pod 在执行完启动命令后将会退出,并且不会被系统自动重启(restartPolicy=Never):

```
# kubectl create -f cm-test-pod.yaml
pod "cm-test-pod" created
```

使用 kubectl get pods --show-all 查看已经停止的 Pod:

```
# kubectl get pods --show-all
NAME            READY    STATUS       RESTARTS    AGE
cm-test-pod     0/1      Completed    0           8s
```

查看该 Pod 的日志,可以看到启动命令 "env | grep APP" 的执行结果如下:

```
# kubectl logs cm-test-pod
APPDATADIR=/var/data
APPLOGLEVEL=info
```

说明容器内部的环境变量使用ConfigMap cm-appvars中的值进行了正确设置。

Kubernetes从1.6版本开始，引入了一个新的字段envFrom，实现了在Pod环境中将ConfigMap（也可用于Secret资源对象）中所有定义的key=value自动生成为环境变量：

```yaml
apiVersion: v1
kind: Pod
metadata:
  name: cm-test-pod
spec:
  containers:
  - name: cm-test
    image: busybox
    command: [ "/bin/sh", "-c", "env" ]
    envFrom:
    - configMapRef
      name: cm-appvars         # 根据cm-appvars中的key=value自动生成环境变量
  restartPolicy: Never
```

通过这个定义，在容器内部将会生成如下环境变量：

```
apploglevel=info
appdatadir=/var/data
```

需要说明的是，环境变量的名称受POSIX命名规范（[a-zA-Z_][a-zA-Z0-9_]*）约束，不能以数字开头。如果包含非法字符，则系统将跳过该条环境变量的创建，并记录一个Event来提示环境变量无法生成，但并不阻止Pod的启动。

2. 通过volumeMount使用ConfigMap

在如下所示的cm-appconfigfiles.yaml例子中包含两个配置文件的定义：server.xml和logging.properties。

```yaml
cm-appconfigfiles.yaml
apiVersion: v1
kind: ConfigMap
metadata:
  name: cm-appconfigfiles
data:
  key-serverxml: |
    <?xml version='1.0' encoding='utf-8'?>
    <Server port="8005" shutdown="SHUTDOWN">
```

```xml
        <Listener className="org.apache.catalina.startup.VersionLoggerListener" />
        <Listener className="org.apache.catalina.core.AprLifecycleListener" SSLEngine="on" />
        <Listener className="org.apache.catalina.core.JreMemoryLeakPreventionListener" />
        <Listener className="org.apache.catalina.mbeans.GlobalResourcesLifecycleListener" />
        <Listener className="org.apache.catalina.core.ThreadLocalLeakPreventionListener" />
        <GlobalNamingResources>
          <Resource name="UserDatabase" auth="Container"
                    type="org.apache.catalina.UserDatabase"
                    description="User database that can be updated and saved"
                    factory="org.apache.catalina.users.MemoryUserDatabaseFactory"
                    pathname="conf/tomcat-users.xml" />
        </GlobalNamingResources>

        <Service name="Catalina">
          <Connector port="8080" protocol="HTTP/1.1"
                     connectionTimeout="20000"
                     redirectPort="8443" />
          <Connector port="8009" protocol="AJP/1.3" redirectPort="8443" />
          <Engine name="Catalina" defaultHost="localhost">
            <Realm className="org.apache.catalina.realm.LockOutRealm">
              <Realm className="org.apache.catalina.realm.UserDatabaseRealm"
                     resourceName="UserDatabase"/>
            </Realm>
            <Host name="localhost" appBase="webapps"
                  unpackWARs="true" autoDeploy="true">
              <Valve className="org.apache.catalina.valves.AccessLogValve" directory="logs"
                     prefix="localhost_access_log" suffix=".txt"
                     pattern="%h %l %u %t &quot;%r&quot; %s %b" />

            </Host>
          </Engine>
        </Service>
      </Server>
  key-loggingproperties: "handlers
    = 1catalina.org.apache.juli.FileHandler,
2localhost.org.apache.juli.FileHandler,
```

```
                3manager.org.apache.juli.FileHandler,
4host-manager.org.apache.juli.FileHandler,
            java.util.logging.ConsoleHandler\r\n\r\n.handlers =
1catalina.org.apache.juli.FileHandler,
java.util.logging.ConsoleHandler\r\n\r\n1catalina.org.apache.juli.FileHandler.level
    = FINE\r\n1catalina.org.apache.juli.FileHandler.directory =
${catalina.base}/logs\r\n1catalina.org.apache.juli.FileHandler.prefix
    = catalina.\r\n\r\n2localhost.org.apache.juli.FileHandler.level =
FINE\r\n2localhost.org.apache.juli.FileHandler.directory
    = ${catalina.base}/logs\r\n2localhost.org.apache.juli.FileHandler.prefix =
localhost.\r\n\r\n3manager.org.apache.juli.FileHandler.level
    = FINE\r\n3manager.org.apache.juli.FileHandler.directory =
${catalina.base}/logs\r\n3manager.org.apache.juli.FileHandler.prefix
    = manager.\r\n\r\n4host-manager.org.apache.juli.FileHandler.level =
FINE\r\n4host-manager.org.apache.juli.FileHandler.directory
    = ${catalina.base}/logs\r\n4host-manager.org.apache.juli.FileHandler.
prefix =
            host-manager.\r\n\r\njava.util.logging.ConsoleHandler.level =
FINE\r\njava.util.logging.ConsoleHandler.formatter
    = java.util.logging.SimpleFormatter\r\n\r\n\r\norg.apache.catalina.core.
ContainerBase.[Catalina].[localhost].level
    = INFO\r\norg.apache.catalina.core.ContainerBase.[Catalina].[localhost].
handlers
    = 2localhost.org.apache.juli.FileHandler\r\n\r\norg.apache.catalina.core.
ContainerBase.[Catalina].[localhost].[/manager].level
    = INFO\r\norg.apache.catalina.core.ContainerBase.[Catalina].[localhost].
[/manager].handlers
    = 3manager.org.apache.juli.FileHandler\r\n\r\norg.apache.catalina.core.
ContainerBase.[Catalina].[localhost].[/host-manager].level
    = INFO\r\norg.apache.catalina.core.ContainerBase.[Catalina].[localhost].
[/host-manager].handlers
    = 4host-manager.org.apache.juli.FileHandler\r\n\r\n"
```

在Pod "cm-test-app" 的定义中,将ConfigMap "cm-appconfigfiles" 中的内容以文件的形式mount到容器内部的/configfiles目录下。Pod配置文件cm-test-app.yaml的内容如下:

```
apiVersion: v1
kind: Pod
metadata:
  name: cm-test-app
spec:
```

```yaml
  containers:
  - name: cm-test-app
    image: kubeguide/tomcat-app:v1
    ports:
    - containerPort: 8080
    volumeMounts:
    - name: serverxml                          # 引用Volume的名称
      mountPath: /configfiles                  # 挂载到容器内的目录
  volumes:
  - name: serverxml                            # 定义Volume的名称
    configMap:
      name: cm-appconfigfiles                  # 使用ConfigMap "cm-appconfigfiles"
      items:
      - key: key-serverxml                     # key=key-serverxml
        path: server.xml                       # value将server.xml文件名进行挂载
      - key: key-loggingproperties             # key=key-loggingproperties
        path: logging.properties               # value将logging.properties文件名进行挂载
```

创建该Pod：

```
# kubectl create -f cm-test-app.yaml
pod "cm-test-app" created
```

登录容器，查看到在/configfiles目录下存在server.xml和logging.properties文件，它们的内容就是ConfigMap "cm-appconfigfiles"中两个key定义的内容：

```
# kubectl exec -ti cm-test-app -- bash
root@cm-test-app:/# cat /configfiles/server.xml
<?xml version='1.0' encoding='utf-8'?>
<Server port="8005" shutdown="SHUTDOWN">
......

root@cm-test-app:/# cat /configfiles/logging.properties
handlers = 1catalina.org.apache.juli.AsyncFileHandler,
2localhost.org.apache.juli.AsyncFileHandler,
3manager.org.apache.juli.AsyncFileHandler,
4host-manager.org.apache.juli.AsyncFileHandler, java.util.logging.ConsoleHandler
......
```

如果在引用ConfigMap时不指定items，则使用volumeMount方式在容器内的目录下为每个item都生成一个文件名为key的文件。

Pod配置文件cm-test-app.yaml的内容如下：

```
apiVersion: v1
kind: Pod
metadata:
  name: cm-test-app
spec:
  containers:
  - name: cm-test-app
    image: kubeguide/tomcat-app:v1
    imagePullPolicy: Never
    ports:
    - containerPort: 8080
    volumeMounts:
    - name: serverxml              # 引用 Volume 的名称
      mountPath: /configfiles      # 挂载到容器内的目录
  volumes:
  - name: serverxml                # 定义 Volume 的名称
    configMap:
      name: cm-appconfigfiles      # 使用 ConfigMap "cm-appconfigfiles"
```

创建该 Pod：

```
# kubectl create -f cm-test-app.yaml
pod "cm-test-app" created
```

登录容器，查看到在/configfiles 目录下存在 key-loggingproperties 和 key-serverxml 文件，文件的名称来自在 ConfigMap cm-appconfigfiles 中定义的两个 key 的名称，文件的内容则为 value 的内容：

```
# ls /configfiles
key-loggingproperties  key-serverxml
```

3.5.4 使用 ConfigMap 的限制条件

使用 ConfigMap 的限制条件如下。

◎ ConfigMap 必须在 Pod 之前创建。
◎ ConfigMap 受 Namespace 限制，只有处于相同 Namespace 中的 Pod 才可以引用它。
◎ ConfigMap 中的配额管理还未能实现。
◎ kubelet 只支持可以被 API Server 管理的 Pod 使用 ConfigMap。kubelet 在本 Node 上通过 --manifest-url 或--config 自动创建的静态 Pod 将无法引用 ConfigMap。

◎ 在 Pod 对 ConfigMap 进行挂载（volumeMount）操作时，在容器内部只能挂载为"目录"，无法挂载为"文件"。在挂载到容器内部后，在目录下将包含 ConfigMap 定义的每个 item，如果在该目录下原来还有其他文件，则容器内的该目录将被挂载的 ConfigMap 覆盖。如果应用程序需要保留原来的其他文件，则需要进行额外的处理。可以将 ConfigMap 挂载到容器内部的临时目录，再通过启动脚本将配置文件复制或者链接到（cp 或 link 命令）应用所用的实际配置目录下。

3.6 在容器内获取 Pod 信息（Downward API）

我们知道，每个 Pod 在被成功创建出来之后，都会被系统分配唯一的名字、IP 地址，并且处于某个 Namespace 中，那么我们如何在 Pod 的容器内获取 Pod 的这些重要信息呢？答案就是使用 Downward API。

Downward API 可以通过以下两种方式将 Pod 信息注入容器内部。

（1）环境变量：用于单个变量，可以将 Pod 信息和 Container 信息注入容器内部。

（2）Volume 挂载：将数组类信息生成为文件并挂载到容器内部。

下面通过几个例子对 Downward API 的用法进行说明。

3.6.1 环境变量方式：将 Pod 信息注入为环境变量

下面的例子通过 Downward API 将 Pod 的 IP、名称和所在 Namespace 注入容器的环境变量中，容器应用使用 env 命令将全部环境变量打印到标准输出中：

```
dapi-test-pod.yaml
apiVersion: v1
kind: Pod
metadata:
  name: dapi-test-pod
spec:
  containers:
    - name: test-container
      image: busybox
      command: [ "/bin/sh", "-c", "env" ]
      env:
```

```
      - name: MY_POD_NAME
        valueFrom:
          fieldRef:
            fieldPath: metadata.name
      - name: MY_POD_NAMESPACE
        valueFrom:
          fieldRef:
            fieldPath: metadata.namespace
      - name: MY_POD_IP
        valueFrom:
          fieldRef:
            fieldPath: status.podIP
  restartPolicy: Never
```

注意到上面 valueFrom 这种特殊的语法是 Downward API 的写法。目前 Downward API 提供了以下变量。

- ◎ metadata.name：Pod 的名称，当 Pod 通过 RC 生成时，其名称是 RC 随机产生的唯一名称。
- ◎ status.podIP：Pod 的 IP 地址，之所以叫作 status.podIP 而非 metadata.IP，是因为 Pod 的 IP 属于状态数据，而非元数据。
- ◎ metadata.namespace：Pod 所在的 Namespace。

运行 kubectl create 命令创建 Pod：

```
# kubectl create -f dapi-test-pod.yaml
pod "dapi-test-pod" created
```

查看 dapi-test-pod 的日志：

```
# kubectl logs dapi-test-pod
......
MY_POD_NAMESPACE=default
MY_POD_IP=172.17.1.2
MY_POD_NAME=dapi-test-pod
......
```

从日志中我们可以看到 Pod 的 IP、Name 及 Namespace 等信息都被正确保存到了 Pod 的环境变量中。

3.6.2 环境变量方式：将容器资源信息注入为环境变量

下面的例子通过 Downward API 将 Container 的资源请求和限制信息注入容器的环境变量中，容器应用使用 printenv 命令将设置的资源请求和资源限制环境变量打印到标准输出中：

```yaml
dapi-test-pod-container-vars.yaml
apiVersion: v1
kind: Pod
metadata:
  name: dapi-test-pod-container-vars
spec:
  containers:
    - name: test-container
      image: busybox
      imagePullPolicy: Never
      command: [ "sh", "-c"]
      args:
      - while true; do
          echo -en '\n';
          printenv MY_CPU_REQUEST MY_CPU_LIMIT;
          printenv MY_MEM_REQUEST MY_MEM_LIMIT;
          sleep 3600;
        done;
      resources:
        requests:
          memory: "32Mi"
          cpu: "125m"
        limits:
          memory: "64Mi"
          cpu: "250m"
      env:
        - name: MY_CPU_REQUEST
          valueFrom:
            resourceFieldRef:
              containerName: test-container
              resource: requests.cpu
        - name: MY_CPU_LIMIT
          valueFrom:
            resourceFieldRef:
              containerName: test-container
```

```
        resource: limits.cpu
    - name: MY_MEM_REQUEST
      valueFrom:
        resourceFieldRef:
          containerName: test-container
          resource: requests.memory
    - name: MY_MEM_LIMIT
      valueFrom:
        resourceFieldRef:
          containerName: test-container
          resource: limits.memory
  restartPolicy: Never
```

注意 valueFrom 这种特殊的 Downward API 语法,目前 resourceFieldRef 可以将容器的资源请求和资源限制等配置设置为容器内部的环境变量。

- ◎ requests.cpu:容器的 CPU 请求值。
- ◎ limits.cpu:容器的 CPU 限制值。
- ◎ requests.memory:容器的内存请求值。
- ◎ limits.memory:容器的内存限制值。

运行 kubectl create 命令来创建 Pod:

```
# kubectl create -f dapi-test-pod-container-vars.yaml
pod "dapi-test-pod-container-vars" created

# kubectl get pods
NAME                          READY   STATUS    RESTARTS   AGE
dapi-test-pod-container-vars  1/1     Running   0          36s
```

查看 dapi-test-pod-container-vars 的日志:

```
# kubectl logs dapi-test-pod-container-vars
1
1
33554432
67108864
```

从日志中我们可以看到 Container 的 requests.cpu、limits.cpu、requests.memory、limits.memory 等信息都被正确保存到了 Pod 的环境变量中。

3.6.3 Volume 挂载方式

下面的例子通过 Downward API 将 Pod 的 Label、Annotation 列表通过 Volume 挂载为容器中的一个文件，容器应用使用 echo 命令将文件的内容打印到标准输出中：

```yaml
dapi-test-pod-volume.yaml
apiVersion: v1
kind: Pod
metadata:
  name: dapi-test-pod-volume
  labels:
    zone: us-est-coast
    cluster: test-cluster1
    rack: rack-22
  annotations:
    build: two
    builder: john-doe
spec:
  containers:
    - name: test-container
      image: busybox
      imagePullPolicy: Never
      command: ["sh", "-c"]
      args:
      - while true; do
          if [[ -e /etc/labels ]]; then
            echo -en '\n\n'; cat /etc/labels; fi;
          if [[ -e /etc/annotations ]]; then
            echo -en '\n\n'; cat /etc/annotations; fi;
          sleep 3600;
        done;
      volumeMounts:
        - name: podinfo
          mountPath: /etc
          readOnly: false
  volumes:
    - name: podinfo
      downwardAPI:
        items:
          - path: "labels"
            fieldRef:
```

```
            fieldPath: metadata.labels
        - path: "annotations"
          fieldRef:
            fieldPath: metadata.annotations
```

这里要注意 "volumes" 字段中 downwardAPI 的特殊语法，通过 items 的设置，系统会根据 path 的名称生成文件。根据上例的设置，系统将在容器内生成/etc/labels 和/etc/annotations 两个文件。在/etc/labels 文件中将包含 metadata.labels 的全部 Label 列表，在/etc/annotations 文件中将包含 metadata.annotations 的全部 Label 列表。

运行 kubectl create 命令创建 Pod：

```
# kubectl create -f dapi-test-pod-volume.yaml
pod "dapi-test-pod-volume" created

# kubectl get pods
NAME                    READY    STATUS     RESTARTS    AGE
dapi-test-pod-volume    1/1      Running    0           1m
```

查看 dapi-test-pod-volume 的日志：

```
# kubectl logs dapi-test-pod-volume
cluster="test-cluster1"
rack="rack-22"
zone="us-est-coast"

build="two"
builder="john-doe"
```

从日志中我们看到 Pod 的 Label 和 Annotation 信息都被保存到了容器内的/etc/labels 和/etc/annotations 文件中。

那么，Downward API 有什么价值呢？

在某些集群中，集群中的每个节点都需要将自身的标识（ID）及进程绑定的 IP 地址等信息事先写入配置文件中，进程在启动时会读取这些信息，然后将这些信息发布到某个类似服务注册中心的地方，以实现集群节点的自动发现功能。此时 Downward API 就可以派上用场了，具体做法是先编写一个预启动脚本或 Init Container，通过环境变量或文件方式获取 Pod 自身的名称、IP 地址等信息，然后将这些信息写入主程序的配置文件中，最后启动主程序。

3.7 Pod生命周期和重启策略

Pod在整个生命周期中被系统定义为各种状态,熟悉Pod的各种状态对于理解如何设置Pod的调度策略、重启策略是很有必要的。

Pod的状态如表3.2所示。

表3.2 Pod的状态

状态值	描述
Pending	API Server已经创建该Pod,但在Pod内还有一个或多个容器的镜像没有创建,包括正在下载镜像的过程
Running	Pod内所有容器均已创建,且至少有一个容器处于运行状态、正在启动状态或正在重启状态
Succeeded	Pod内所有容器均成功执行后退出,且不会再重启
Failed	Pod内所有容器均已退出,但至少有一个容器退出为失败状态
Unknown	由于某种原因无法获取该Pod的状态,可能由于网络通信不畅导致

Pod的重启策略(RestartPolicy)应用于Pod内的所有容器,并且仅在Pod所处的Node上由kubelet进行判断和重启操作。当某个容器异常退出或者健康检查(详见下节)失败时,kubelet将根据RestartPolicy的设置来进行相应的操作。

Pod的重启策略包括Always、OnFailure和Never,默认值为Always。

- ◎ Always:当容器失效时,由kubelet自动重启该容器。
- ◎ OnFailure:当容器终止运行且退出码不为0时,由kubelet自动重启该容器。
- ◎ Never:不论容器运行状态如何,kubelet都不会重启该容器。

kubelet重启失效容器的时间间隔以sync-frequency乘以$2n$来计算,例如1、2、4、8倍等,最长延时5min,并且在成功重启后的10min后重置该时间。

Pod的重启策略与控制方式息息相关,当前可用于管理Pod的控制器包括ReplicationController、Job、DaemonSet及直接通过kubelet管理(静态Pod)。每种控制器对Pod的重启策略要求如下。

- ◎ RC和DaemonSet:必须设置为Always,需要保证该容器持续运行。
- ◎ Job:OnFailure或Never,确保容器执行完成后不再重启。
- ◎ kubelet:在Pod失效时自动重启它,不论将RestartPolicy设置为什么值,也不会对 Pod 进行健康检查。

结合 Pod 的状态和重启策略，表 3.3 列出一些常见的状态转换场景。

表 3.3 一些常见的状态转换场景

Pod 包含的容器数	Pod 当前的状态	发生事件	Pod 的结果状态		
			RestartPolicy=Always	RestartPolicy=OnFailure	RestartPolicy=Never
包含 1 个容器	Running	容器成功退出	Running	Succeeded	Succeeded
包含 1 个容器	Running	容器失败退出	Running	Running	Failed
包含两个容器	Running	1 个容器失败退出	Running	Running	Running
包含两个容器	Running	容器被 OOM 杀掉	Running	Running	Failed

3.8 Pod 健康检查和服务可用性检查

Kubernetes 对 Pod 的健康状态可以通过两类探针来检查：LivenessProbe 和 ReadinessProbe，kubelet 定期执行这两类探针来诊断容器的健康状况。

（1）LivenessProbe 探针：用于判断容器是否存活（Running 状态），如果 LivenessProbe 探针探测到容器不健康，则 kubelet 将杀掉该容器，并根据容器的重启策略做相应的处理。如果一个容器不包含 LivenessProbe 探针，那么 kubelet 认为该容器的 LivenessProbe 探针返回的值永远是 Success。

（2）ReadinessProbe 探针：用于判断容器服务是否可用（Ready 状态），达到 Ready 状态的 Pod 才可以接收请求。对于被 Service 管理的 Pod，Service 与 Pod Endpoint 的关联关系也将基于 Pod 是否 Ready 进行设置。如果在运行过程中 Ready 状态变为 False，则系统自动将其从 Service 的后端 Endpoint 列表中隔离出去，后续再把恢复到 Ready 状态的 Pod 加回后端 Endpoint 列表。这样就能保证客户端在访问 Service 时不会被转发到服务不可用的 Pod 实例上。

LivenessProbe 和 ReadinessProbe 均可配置以下三种实现方式。

（1）ExecAction：在容器内部执行一个命令，如果该命令的返回码为 0，则表明容器健康。

在下面的例子中，通过执行"cat /tmp/health"命令来判断一个容器运行是否正常。在该 Pod 运行后，将在创建/tmp/health 文件 10s 后删除该文件，而 LivenessProbe 健康检查的初始探测时间（initialDelaySeconds）为 15s，探测结果是 Fail，将导致 kubelet 杀掉该容器并重启它：

```yaml
apiVersion: v1
kind: Pod
metadata:
  labels:
    test: liveness
  name: liveness-exec
spec:
  containers:
  - name: liveness
    image: gcr.io/google_containers/busybox
    args:
    - /bin/sh
    - -c
    - echo ok > /tmp/health; sleep 10; rm -rf /tmp/health; sleep 600
    livenessProbe:
      exec:
        command:
        - cat
        - /tmp/health
      initialDelaySeconds: 15
      timeoutSeconds: 1
```

（2）TCPSocketAction：通过容器的 IP 地址和端口号执行 TCP 检查，如果能够建立 TCP 连接，则表明容器健康。

在下面的例子中，通过与容器内的 localhost:80 建立 TCP 连接进行健康检查：

```yaml
apiVersion: v1
kind: Pod
metadata:
  name: pod-with-healthcheck
spec:
  containers:
  - name: nginx
    image: nginx
    ports:
```

```
      - containerPort: 80
    livenessProbe:
      tcpSocket:
        port: 80
      initialDelaySeconds: 30
      timeoutSeconds: 1
```

（3）HTTPGetAction：通过容器的 IP 地址、端口号及路径调用 HTTP Get 方法，如果响应的状态码大于等于 200 且小于 400，则认为容器健康。

在下面的例子中，kubelet 定时发送 HTTP 请求到 localhost:80/_status/healthz 来进行容器应用的健康检查：

```
apiVersion: v1
kind: Pod
metadata:
  name: pod-with-healthcheck
spec:
 containers:
 - name: nginx
   image: nginx
   ports:
   - containerPort: 80
   livenessProbe:
     httpGet:
       path: /_status/healthz
       port: 80
     initialDelaySeconds: 30
     timeoutSeconds: 1
```

对于每种探测方式，都需要设置 initialDelaySeconds 和 timeoutSeconds 两个参数，它们的含义分别如下。

- ◎ **initialDelaySeconds**：启动容器后进行首次健康检查的等待时间，单位为 s。
- ◎ **timeoutSeconds**：健康检查发送请求后等待响应的超时时间，单位为 s。当超时发生时，kubelet 会认为容器已经无法提供服务，将会重启该容器。

Kubernetes 的 ReadinessProbe 机制可能无法满足某些复杂应用对容器内服务可用状态的判断，所以 Kubernetes 从 1.11 版本开始，引入 Pod Ready++ 特性对 Readiness 探测机制进行扩展，在 1.14 版本时达到 GA 稳定版，称其为 Pod Readiness Gates。

通过 Pod Readiness Gates 机制，用户可以将自定义的 ReadinessProbe 探测方式设置在 Pod 上，辅助 Kubernetes 设置 Pod 何时达到服务可用状态（Ready）。为了使自定义的 ReadinessProbe 生效，用户需要提供一个外部的控制器（Controller）来设置相应的 Condition 状态。

Pod 的 Readiness Gates 在 Pod 定义中的 ReadinessGate 字段进行设置。下面的例子设置了一个类型为 www.example.com/feature-1 的新 Readiness Gate：

```
Kind: Pod
......
spec:
  readinessGates:
    - conditionType: "www.example.com/feature-1"
status:
  conditions:
    - type: Ready  # Kubernetes 系统内置的名为 Ready 的 Condition
      status: "True"
      lastProbeTime: null
      lastTransitionTime: 2018-01-01T00:00:00Z
    - type: "www.example.com/feature-1"   # 用户自定义 Condition
      status: "False"
      lastProbeTime: null
      lastTransitionTime: 2019-03-01T00:00:00Z
  containerStatuses:
    - containerID: docker://abcd...
      ready: true
......
```

新增的自定义 Condition 的状态（status）将由用户自定义的外部控制器设置，默认值为 False。Kubernetes 将在判断全部 readinessGates 条件都为 True 时，才设置 Pod 为服务可用状态（Ready 为 True）。

3.9　玩转 Pod 调度

在 Kubernetes 平台上，我们很少会直接创建一个 Pod，在大多数情况下会通过 RC、Deployment、DaemonSet、Job 等控制器完成对一组 Pod 副本的创建、调度及全生命周期的自动控制任务。

在最早的 Kubernetes 版本里是没有这么多 Pod 副本控制器的，只有一个 Pod 副本控制器 RC（Replication Controller），这个控制器是这样设计实现的：RC 独立于所控制的 Pod，并通过 Label 标签这个松耦合关联关系控制目标 Pod 实例的创建和销毁，随着 Kubernetes 的发展，RC 也出现了新的继任者——Deployment，用于更加自动地完成 Pod 副本的部署、版本更新、回滚等功能。

严谨地说，RC 的继任者其实并不是 Deployment，而是 ReplicaSet，因为 ReplicaSet 进一步增强了 RC 标签选择器的灵活性。之前 RC 的标签选择器只能选择一个标签，而 ReplicaSet 拥有集合式的标签选择器，可以选择多个 Pod 标签，如下所示：

```
selector:
  matchLabels:
    tier: frontend
  matchExpressions:
    - {key: tier, operator: In, values: [frontend]}
```

与 RC 不同，ReplicaSet 被设计成能控制多个不同标签的 Pod 副本。一种常见的应用场景是，应用 MyApp 目前发布了 v1 与 v2 两个版本，用户希望 MyApp 的 Pod 副本数保持为 3 个，可以同时包含 v1 和 v2 版本的 Pod，就可以用 ReplicaSet 来实现这种控制，写法如下：

```
selector:
  matchLabels:
    version: v2
  matchExpressions:
    - {key: version, operator: In, values: [v1,v2]}
```

其实，Kubernetes 的滚动升级就是巧妙运用 ReplicaSet 的这个特性来实现的，同时，Deployment 也是通过 ReplicaSet 来实现 Pod 副本自动控制功能的。我们不应该直接使用底层的 ReplicaSet 来控制 Pod 副本，而应该使用管理 ReplicaSet 的 Deployment 对象来控制副本，这是来自官方的建议。

在大多数情况下，我们希望 Deployment 创建的 Pod 副本被成功调度到集群中的任何一个可用节点，而不关心具体会调度到哪个节点。但是，在真实的生产环境中的确也存在一种需求：希望某种 Pod 的副本全部在指定的一个或者一些节点上运行，比如希望将 MySQL 数据库调度到一个具有 SSD 磁盘的目标节点上，此时 Pod 模板中的 NodeSelector 属性就开始发挥作用了，上述 MySQL 定向调度案例的实现方式可分为以下两步。

（1）把具有 SSD 磁盘的 Node 都打上自定义标签 "disk=ssd"。

（2）在Pod模板中设定NodeSelector的值为"disk: ssd"。

如此一来，Kubernetes在调度Pod副本的时候，就会先按照Node的标签过滤出合适的目标节点，然后选择一个最佳节点进行调度。

上述逻辑看起来既简单又完美，但在真实的生产环境中可能面临以下令人尴尬的问题。

（1）如果NodeSelector选择的Label不存在或者不符合条件，比如这些目标节点此时宕机或者资源不足，该怎么办？

（2）如果要选择多种合适的目标节点，比如SSD磁盘的节点或者超高速硬盘的节点，该怎么办？Kubernates引入了NodeAffinity（节点亲和性设置）来解决该需求。

在真实的生产环境中还存在如下所述的特殊需求。

（1）不同Pod之间的亲和性（Affinity）。比如MySQL数据库与Redis中间件不能被调度到同一个目标节点上，或者两种不同的Pod必须被调度到同一个Node上，以实现本地文件共享或本地网络通信等特殊需求，这就是PodAffinity要解决的问题。

（2）有状态集群的调度。对于ZooKeeper、Elasticsearch、MongoDB、Kafka等有状态集群，虽然集群中的每个Worker节点看起来都是相同的，但每个Worker节点都必须有明确的、不变的唯一ID（主机名或IP地址），这些节点的启动和停止次序通常有严格的顺序。此外，由于集群需要持久化保存状态数据，所以集群中的Worker节点对应的Pod不管在哪个Node上恢复，都需要挂载原来的Volume，因此这些Pod还需要捆绑具体的PV。针对这种复杂的需求，Kubernetes提供了StatefulSet这种特殊的副本控制器来解决问题，在Kubernetes 1.9版本发布后，StatefulSet才可用于正式生产环境中。

（3）在每个Node上调度并且仅仅创建一个Pod副本。这种调度通常用于系统监控相关的Pod，比如主机上的日志采集、主机性能采集等进程需要被部署到集群中的每个节点，并且只能部署一个副本，这就是DaemonSet这种特殊Pod副本控制器所解决的问题。

（4）对于批处理作业，需要创建多个Pod副本来协同工作，当这些Pod副本都完成自己的任务时，整个批处理作业就结束了。这种Pod运行且仅运行一次的特殊调度，用常规的RC或者Deployment都无法解决，所以Kubernates引入了新的Pod调度控制器Job来解决问题，并继续延伸了定时作业的调度控制器CronJob。

与单独的Pod实例不同，由RC、ReplicaSet、Deployment、DaemonSet等控制器创建

的 Pod 副本实例都是归属于这些控制器的，这就产生了一个问题：控制器被删除后，归属于控制器的 Pod 副本该何去何从？在 Kubernates 1.9 之前，在 RC 等对象被删除后，它们所创建的 Pod 副本都不会被删除；在 Kubernates 1.9 以后，这些 Pod 副本会被一并删除。如果不希望这样做，则可以通过 kubectl 命令的 --cascade=false 参数来取消这一默认特性：

```
kubectl delete replicaset my-repset --cascade=false
```

接下来深入理解和实践这些 Pod 调度控制器的各种功能和特性。

3.9.1 Deployment 或 RC：全自动调度

Deployment 或 RC 的主要功能之一就是自动部署一个容器应用的多份副本，以及持续监控副本的数量，在集群内始终维持用户指定的副本数量。

下面是一个 Deployment 配置的例子，使用这个配置文件可以创建一个 ReplicaSet，这个 ReplicaSet 会创建 3 个 Nginx 应用的 Pod：

```yaml
nginx-deployment.yaml
apiVersion: apps/v1
kind: Deployment
metadata:
  name: nginx-deployment
spec:
  selector:
    matchLabels:
      app: nginx
  replicas: 3
  template:
    metadata:
      labels:
        app: nginx
    spec:
      containers:
      - name: nginx
        image: nginx:1.7.9
        ports:
        - containerPort: 80
```

运行 kubectl create 命令创建这个 Deployment：

```
# kubectl create -f nginx-deployment.yaml
```

```
deployment "nginx-deployment" created
```

查看 Deployment 的状态：

```
# kubectl get deployments
NAME               DESIRED   CURRENT   UP-TO-DATE   AVAILABLE   AGE
nginx-deployment   3         3         3            3           18s
```

该状态说明 Deployment 已创建好所有 3 个副本，并且所有副本都是最新的可用的。

通过运行 kubectl get rs 和 kubectl get pods 可以查看已创建的 ReplicaSet（RS）和 Pod 的信息。

```
# kubectl get rs
NAME                          DESIRED   CURRENT   READY   AGE
nginx-deployment-4087004473   3         3         3       53s
# kubectl get pods
NAME                                READY   STATUS    RESTARTS   AGE
nginx-deployment-4087004473-9jqqs   1/1     Running   0          1m
nginx-deployment-4087004473-cq0cf   1/1     Running   0          1m
nginx-deployment-4087004473-vxn56   1/1     Running   0          1m
```

从调度策略上来说，这 3 个 Nginx Pod 由系统全自动完成调度。它们各自最终运行在哪个节点上，完全由 Master 的 Scheduler 经过一系列算法计算得出，用户无法干预调度过程和结果。

除了使用系统自动调度算法完成一组 Pod 的部署，Kubernetes 也提供了多种丰富的调度策略，用户只需在 Pod 的定义中使用 NodeSelector、NodeAffinity、PodAffinity、Pod 驱逐等更加细粒度的调度策略设置，就能完成对 Pod 的精准调度。下面对这些策略进行说明。

3.9.2 NodeSelector：定向调度

Kubernetes Master 上的 Scheduler 服务（kube-scheduler 进程）负责实现 Pod 的调度，整个调度过程通过执行一系列复杂的算法，最终为每个 Pod 都计算出一个最佳的目标节点，这一过程是自动完成的，通常我们无法知道 Pod 最终会被调度到哪个节点上。在实际情况下，也可能需要将 Pod 调度到指定的一些 Node 上，可以通过 Node 的标签（Label）和 Pod 的 nodeSelector 属性相匹配，来达到上述目的。

（1）首先通过 kubectl label 命令给目标 Node 打上一些标签：

```
kubectl label nodes <node-name> <label-key>=<label-value>
```

这里，我们为 k8s-node-1 节点打上一个 zone=north 标签，表明它是"北方"的一个节点：

```
$ kubectl label nodes k8s-node-1 zone=north
NAME            LABELS                                              STATUS
k8s-node-1      kubernetes.io/hostname=k8s-node-1,zone=north        Ready
```

上述命令行操作也可以通过修改资源定义文件的方式，并执行 kubectl replace -f xxx.yaml 命令来完成。

（2）然后，在 Pod 的定义中加上 nodeSelector 的设置，以 redis-master-controller.yaml 为例：

```
apiVersion: v1
kind: ReplicationController
metadata:
  name: redis-master
  labels:
    name: redis-master
spec:
  replicas: 1
  selector:
    name: redis-master
  template:
    metadata:
      labels:
        name: redis-master
    spec:
      containers:
      - name: master
        image: kubeguide/redis-master
        ports:
        - containerPort: 6379
      nodeSelector:
        zone: north
```

运行 kubectl create -f 命令创建 Pod，scheduler 就会将该 Pod 调度到拥有 zone=north 标签的 Node 上。

使用 kubectl get pods -o wide 命令可以验证 Pod 所在的 Node：

```
# kubectl get pods -o wide
```

```
NAME                READY   STATUS    RESTARTS   AGE   NODE
redis-master-f0rqj  1/1     Running   0          19s   k8s-node-1
```

如果我们给多个 Node 都定义了相同的标签（例如 zone=north），则 scheduler 会根据调度算法从这组 Node 中挑选一个可用的 Node 进行 Pod 调度。

通过基于 Node 标签的调度方式，我们可以把集群中具有不同特点的 Node 都贴上不同的标签，例如 "role=frontend" "role=backend" "role=database" 等标签，在部署应用时就可以根据应用的需求设置 NodeSelector 来进行指定 Node 范围的调度。

需要注意的是，如果我们指定了 Pod 的 nodeSelector 条件，且在集群中不存在包含相应标签的 Node，则即使在集群中还有其他可供使用的 Node，这个 Pod 也无法被成功调度。

除了用户可以自行给 Node 添加标签，Kubernetes 也会给 Node 预定义一些标签，包括：

◎ kubernetes.io/hostname
◎ beta.kubernetes.io/os（从 1.14 版本开始更新为稳定版，到 1.18 版本删除）
◎ beta.kubernetes.io/arch（从 1.14 版本开始更新为稳定版，到 1.18 版本删除）
◎ kubernetes.io/os（从 1.14 版本开始启用）
◎ kubernetes.io/arch（从 1.14 版本开始启用）

用户也可以使用这些系统标签进行 Pod 的定向调度。

NodeSelector 通过标签的方式，简单实现了限制 Pod 所在节点的方法。亲和性调度机制则极大扩展了 Pod 的调度能力，主要的增强功能如下。

◎ 更具表达力（不仅仅是"符合全部"的简单情况）。
◎ 可以使用软限制、优先采用等限制方式，代替之前的硬限制，这样调度器在无法满足优先需求的情况下，会退而求其次，继续运行该 Pod。
◎ 可以依据节点上正在运行的其他 Pod 的标签来进行限制，而非节点本身的标签。这样就可以定义一种规则来描述 Pod 之间的亲和或互斥关系。

亲和性调度功能包括节点亲和性（NodeAffinity）和 Pod 亲和性（PodAffinity）两个维度的设置。节点亲和性与 NodeSelector 类似，增强了上述前两点优势；Pod 的亲和与互斥限制则通过 Pod 标签而不是节点标签来实现，也就是上面第 4 点内容所陈述的方式，同时具有前两点提到的优点。

NodeSelector 将会继续使用，随着节点亲和性越来越能够表达 nodeSelector 的功能，最终 NodeSelector 会被废弃。

3.9.3　NodeAffinity：Node 亲和性调度

NodeAffinity 意为 Node 亲和性的调度策略，是用于替换 NodeSelector 的全新调度策略。目前有两种节点亲和性表达。

- RequiredDuringSchedulingIgnoredDuringExecution：必须满足指定的规则才可以调度 Pod 到 Node 上（功能与 nodeSelector 很像，但是使用的是不同的语法），相当于硬限制。
- PreferredDuringSchedulingIgnoredDuringExecution：强调优先满足指定规则，调度器会尝试调度 Pod 到 Node 上，但并不强求，相当于软限制。多个优先级规则还可以设置权重（weight）值，以定义执行的先后顺序。

IgnoredDuringExecution 的意思是：如果一个 Pod 所在的节点在 Pod 运行期间标签发生了变更，不再符合该 Pod 的节点亲和性需求，则系统将忽略 Node 上 Label 的变化，该 Pod 能继续在该节点运行。

下面的例子设置了 NodeAffinity 调度的如下规则。

- requiredDuringSchedulingIgnoredDuringExecution 要求只运行在 amd64 的节点上（beta.kubernetes.io/arch In amd64）。
- preferredDuringSchedulingIgnoredDuringExecution 的要求是尽量运行在磁盘类型为 ssd（disk-type In ssd）的节点上。

代码如下：

```
apiVersion: v1
kind: Pod
metadata:
  name: with-node-affinity
spec:
  affinity:
    nodeAffinity:
      requiredDuringSchedulingIgnoredDuringExecution:
        nodeSelectorTerms:
        - matchExpressions:
          - key: beta.kubernetes.io/arch
            operator: In
            values:
            - amd64
```

```
    preferredDuringSchedulingIgnoredDuringExecution:
    - weight: 1
      preference:
        matchExpressions:
        - key: disk-type
          operator: In
          values:
          - ssd
  containers:
  - name: with-node-affinity
    image: gcr.io/google_containers/pause:2.0
```

从上面的配置中可以看到 In 操作符，NodeAffinity 语法支持的操作符包括 In、NotIn、Exists、DoesNotExist、Gt、Lt。虽然没有节点排斥功能，但是用 NotIn 和 DoesNotExist 就可以实现排斥的功能了。

NodeAffinity 规则设置的注意事项如下。

◎ 如果同时定义了 nodeSelector 和 nodeAffinity，那么必须两个条件都得到满足，Pod 才能最终运行在指定的 Node 上。

◎ 如果 nodeAffinity 指定了多个 nodeSelectorTerms，那么其中一个能够匹配成功即可。

◎ 如果在 nodeSelectorTerms 中有多个 matchExpressions，则一个节点必须满足所有 matchExpressions 才能运行该 Pod。

3.9.4 PodAffinity：Pod 亲和与互斥调度策略

Pod 间的亲和与互斥从 Kubernetes 1.4 版本开始引入。这一功能让用户从另一个角度来限制 Pod 所能运行的节点：根据在节点上正在运行的 Pod 的标签而不是节点的标签进行判断和调度，要求对节点和 Pod 两个条件进行匹配。这种规则可以描述为：如果在具有标签 X 的 Node 上运行了一个或者多个符合条件 Y 的 Pod，那么 Pod 应该（如果是互斥的情况，那么就变成拒绝）运行在这个 Node 上。

这里 X 指的是一个集群中的节点、机架、区域等概念，通过 Kubernetes 内置节点标签中的 key 来进行声明。这个 key 的名字为 topologyKey，意为表达节点所属的 topology 范围。

◎ kubernetes.io/hostname

◎ failure-domain.beta.kubernetes.io/zone
◎ failure-domain.beta.kubernetes.io/region

与节点不同的是，Pod 是属于某个命名空间的，所以条件 Y 表达的是一个或者全部命名空间中的一个 Label Selector。

和节点亲和相同，Pod 亲和与互斥的条件设置也是 requiredDuringSchedulingIgnoredDuringExecution 和 preferredDuringSchedulingIgnoredDuringExecution。Pod 的亲和性被定义于 PodSpec 的 affinity 字段下的 podAffinity 子字段中。Pod 间的互斥性则被定义于同一层次的 podAntiAffinity 子字段中。

下面通过实例来说明 Pod 间的亲和性和互斥性策略设置。

1. 参照目标 Pod

首先，创建一个名为 pod-flag 的 Pod，带有标签 security=S1 和 app=nginx，后面的例子将使用 pod-flag 作为 Pod 亲和与互斥的目标 Pod：

```
apiVersion: v1
kind: Pod
metadata:
  name: pod-flag
  labels:
    security: "S1"
    app: "nginx"
spec:
  containers:
  - name: nginx
    image: nginx
```

2. Pod 的亲和性调度

下面创建第 2 个 Pod 来说明 Pod 的亲和性调度，这里定义的亲和标签是 security=S1，对应上面的 Pod "pod-flag"，topologyKey 的值被设置为 "kubernetes.io/hostname"：

```
apiVersion: v1
kind: Pod
metadata:
  name: pod-affinity
spec:
```

```
    affinity:
      podAffinity:
        requiredDuringSchedulingIgnoredDuringExecution:
        - labelSelector:
            matchExpressions:
            - key: security
              operator: In
              values:
              - S1
          topologyKey: kubernetes.io/hostname
    containers:
    - name: with-pod-affinity
      image: gcr.io/google_containers/pause:2.0
```

创建 Pod 之后，使用 kubectl get pods -o wide 命令可以看到，这两个 Pod 在同一个 Node 上运行。

有兴趣的读者还可以测试一下，在创建这个 Pod 之前，删掉这个节点的 kubernetes.io/hostname 标签，重复上面的创建步骤，将会发现 Pod 一直处于 Pending 状态，这是因为找不到满足条件的 Node 了。

3. Pod 的互斥性调度

创建第 3 个 Pod，我们希望它不与目标 Pod 运行在同一个 Node 上：

```
apiVersion: v1
kind: Pod
metadata:
  name: anti-affinity
spec:
  affinity:
    podAffinity:
      requiredDuringSchedulingIgnoredDuringExecution:
      - labelSelector:
          matchExpressions:
          - key: security
            operator: In
            values:
            - S1
        topologyKey: failure-domain.beta.kubernetes.io/zone
    podAntiAffinity:
```

```
        requiredDuringSchedulingIgnoredDuringExecution:
        - labelSelector:
            matchExpressions:
            - key: app
              operator: In
              values:
              - nginx
          topologyKey: kubernetes.io/hostname
  containers:
  - name: anti-affinity
    image: gcr.io/google_containers/pause:2.0
```

这里要求这个新 Pod 与 security=S1 的 Pod 为同一个 zone，但是不与 app=nginx 的 Pod 为同一个 Node。创建 Pod 之后，同样用 kubectl get pods -o wide 来查看，会看到新的 Pod 被调度到了同一 Zone 内的不同 Node 上。

与节点亲和性类似，Pod 亲和性的操作符也包括 In、NotIn、Exists、DoesNotExist、Gt、Lt。

原则上，topologyKey 可以使用任何合法的标签 Key 赋值，但是出于性能和安全方面的考虑，对 topologyKey 有如下限制。

- 在 Pod 亲和性和 RequiredDuringScheduling 的 Pod 互斥性的定义中，不允许使用空的 topologyKey。
- 如果 Admission controller 包含了 LimitPodHardAntiAffinityTopology，那么针对 Required DuringScheduling 的 Pod 互斥性定义就被限制为 kubernetes.io/hostname，要使用自定义的 topologyKey，就要改写或禁用该控制器。
- 在 PreferredDuringScheduling 类型的 Pod 互斥性定义中，空的 topologyKey 会被解释为 kubernetes.io/hostname、failure-domain.beta.kubernetes.io/zone 及 failure-domain.beta.kubernetes.io/region 的组合。
- 如果不是上述情况，就可以采用任意合法的 topologyKey 了。

PodAffinity 规则设置的注意事项如下。

- 除了设置 Label Selector 和 topologyKey，用户还可以指定 Namespace 列表来进行限制，同样，使用 Label Selector 对 Namespace 进行选择。Namespace 的定义和 Label Selector 及 topologyKey 同级。省略 Namespace 的设置，表示使用定义了 affinity/anti-affinity 的 Pod 所在的 Namespace。如果 Namespace 被设置为空值（""），

则表示所有 Namespace。
◎ 在所有关联 requiredDuringSchedulingIgnoredDuringExecution 的 matchExpressions 全都满足之后，系统才能将 Pod 调度到某个 Node 上。

关于 Pod 亲和性和互斥性调度的更多信息可以参考其设计文档，网址为 https://github.com/kubernetes/kubernetes/blob/master/docs/design/podaffinity.md。

3.9.5 Taints 和 Tolerations（污点和容忍）

前面介绍的 NodeAffinity 节点亲和性，是在 Pod 上定义的一种属性，使得 Pod 能够被调度到某些 Node 上运行（优先选择或强制要求）。Taint 则正好相反，它让 Node 拒绝 Pod 的运行。

Taint 需要和 Toleration 配合使用，让 Pod 避开那些不合适的 Node。在 Node 上设置一个或多个 Taint 之后，除非 Pod 明确声明能够容忍这些污点，否则无法在这些 Node 上运行。Toleration 是 Pod 的属性，让 Pod 能够（注意，只是能够，而非必须）运行在标注了 Taint 的 Node 上。

可以用 kubectl taint 命令为 Node 设置 Taint 信息：

```
$ kubectl taint nodes node1 key=value:NoSchedule
```

这个设置为 node1 加上了一个 Taint。该 Taint 的键为 key，值为 value，Taint 的效果是 NoSchedule。这意味着除非 Pod 明确声明可以容忍这个 Taint，否则就不会被调度到 node1 上。

然后，需要在 Pod 上声明 Toleration。下面的两个 Toleration 都被设置为可以容忍（Tolerate）具有该 Taint 的 Node，使得 Pod 能够被调度到 node1 上：

```
tolerations:
- key: "key"
  operator: "Equal"
  value: "value"
  effect: "NoSchedule"
```

或

```
tolerations:
- key: "key"
  operator: "Exists"
```

```
effect: "NoSchedule"
```

Pod 的 Toleration 声明中的 key 和 effect 需要与 Taint 的设置保持一致,并且满足以下条件之一。

◎ operator 的值是 Exists(无须指定 value)。
◎ operator 的值是 Equal 并且 value 相等。

如果不指定 operator,则默认值为 Equal。

另外,有如下两个特例。

◎ 空的 key 配合 Exists 操作符能够匹配所有的键和值。
◎ 空的 effect 匹配所有的 effect。

在上面的例子中,effect 的取值为 NoSchedule,还可以取值为 PreferNoSchedule,这个值的意思是优先,也可以算作 NoSchedule 的软限制版本——一个 Pod 如果没有声明容忍这个 Taint,则系统会尽量避免把这个 Pod 调度到这一节点上,但不是强制的。后面还会介绍另一个 effect "NoExecute"。

系统允许在同一个 Node 上设置多个 Taint,也可以在 Pod 上设置多个 Toleration。Kubernetes 调度器处理多个 Taint 和 Toleration 的逻辑顺序为:首先列出节点中所有的 Taint,然后忽略 Pod 的 Toleration 能够匹配的部分,剩下的没有忽略的 Taint 就是对 Pod 的效果了。下面是几种特殊情况。

◎ 如果在剩余的 Taint 中存在 effect=NoSchedule,则调度器不会把该 Pod 调度到这一节点上。
◎ 如果在剩余的 Taint 中没有 NoSchedule 效果,但是有 PreferNoSchedule 效果,则调度器会尝试不把这个 Pod 指派给这个节点。
◎ 如果在剩余的 Taint 中有 NoExecute 效果,并且这个 Pod 已经在该节点上运行,则会被驱逐;如果没有在该节点上运行,则也不会再被调度到该节点上。

例如,我们这样对一个节点进行 Taint 设置:

```
$ kubectl taint nodes node1 key1=value1:NoSchedule
$ kubectl taint nodes node1 key1=value1:NoExecute
$ kubectl taint nodes node1 key2=value2:NoSchedule
```

然后在 Pod 上设置两个 Toleration:

```
tolerations:
```

```
  - key: "key1"
    operator: "Equal"
    value: "value1"
    effect: "NoSchedule"
  - key: "key1"
    operator: "Equal"
    value: "value1"
    effect: "NoExecute"
```

这样的结果是该 Pod 无法被调度到 node1 上，这是因为第 3 个 Taint 没有匹配的 Toleration。但是如果该 Pod 已经在 node1 上运行了，那么在运行时设置第 3 个 Taint，它还能继续在 node1 上运行，这是因为 Pod 可以容忍前两个 Taint。

一般来说，如果给 Node 加上 effect=NoExecute 的 Taint，那么在该 Node 上正在运行的所有无对应 Toleration 的 Pod 都会被立刻驱逐，而具有相应 Toleration 的 Pod 永远不会被驱逐。不过，系统允许给具有 NoExecute 效果的 Toleration 加入一个可选的 tolerationSeconds 字段,这个设置表明 Pod 可以在 Taint 添加到 Node 之后还能在这个 Node 上运行多久（单位为 s）：

```
tolerations:
- key: "key1"
  operator: "Equal"
  value: "value1"
  effect: "NoExecute"
  tolerationSeconds: 3600
```

上述定义的意思是，如果 Pod 正在运行，所在节点都被加入一个匹配的 Taint，则这个 Pod 会持续在这个节点上存活 3600s 后被逐出。如果在这个宽限期内 Taint 被移除，则不会触发驱逐事件。

Taint 和 Toleration 是一种处理节点并且让 Pod 进行规避或者驱逐 Pod 的弹性处理方式，下面列举一些常见的用例。

1. 独占节点

如果想要拿出一部分节点专门给一些特定应用使用，则可以为节点添加这样的 Taint：

```
$ kubectl taint nodes nodename dedicated=groupName:NoSchedule
```

然后给这些应用的 Pod 加入对应的 Toleration。这样，带有合适 Toleration 的 Pod 就会被允许同使用其他节点一样使用有 Taint 的节点。

通过自定义 Admission Controller 也可以实现这一目标。如果希望让这些应用独占一批节点，并且确保它们只能使用这些节点，则还可以给这些 Taint 节点加入类似的标签 dedicated=groupName，然后 Admission Controller 需要加入节点亲和性设置，要求 Pod 只会被调度到具有这一标签的节点上。

2. 具有特殊硬件设备的节点

在集群里可能有一小部分节点安装了特殊的硬件设备（如 GPU 芯片），用户自然会希望把不需要占用这类硬件的 Pod 排除在外，以确保对这类硬件有需求的 Pod 能够被顺利调度到这些节点。

可以用下面的命令为节点设置 Taint：

```
$ kubectl taint nodes nodename special=true:NoSchedule
$ kubectl taint nodes nodename special=true:PreferNoSchedule
```

然后在 Pod 中利用对应的 Toleration 来保障特定的 Pod 能够使用特定的硬件。

和上面的独占节点的示例类似，使用 Admission Controller 来完成这一任务会更方便。例如，Admission Controller 使用 Pod 的一些特征来判断这些 Pod，如果可以使用这些硬件，就添加 Toleration 来完成这一工作。要保障需要使用特殊硬件的 Pod 只被调度到安装这些硬件的节点上，则还需要一些额外的工作，比如将这些特殊资源使用 opaque-int-resource 的方式对自定义资源进行量化，然后在 PodSpec 中进行请求；也可以使用标签的方式来标注这些安装有特别硬件的节点，然后在 Pod 中定义节点亲和性来实现这个目标。

3. 定义 Pod 驱逐行为，以应对节点故障（为 Alpha 版本的功能）

前面提到的 NoExecute 这个 Taint 效果对节点上正在运行的 Pod 有以下影响。

◎ 没有设置 Toleration 的 Pod 会被立刻驱逐。
◎ 配置了对应 Toleration 的 Pod，如果没有为 tolerationSeconds 赋值，则会一直留在这一节点中。
◎ 配置了对应 Toleration 的 Pod 且指定了 tolerationSeconds 值，则会在指定时间后驱逐。
◎ Kubernetes 从 1.6 版本开始引入一个 Alpha 版本的功能，即把节点故障标记为 Taint（目前只针对 node unreachable 及 node not ready，相应的 NodeCondition "Ready" 的值分别为 Unknown 和 False）。激活 TaintBasedEvictions 功能后（在 --feature-gates

参数中加入 TaintBasedEvictions=true)，NodeController 会自动为 Node 设置 Taint，而在状态为 Ready 的 Node 上，之前设置过的普通驱逐逻辑将会被禁用。注意，在节点故障的情况下，为了保持现存的 Pod 驱逐的限速（rate-limiting）设置，系统将会以限速的模式逐步给 Node 设置 Taint，这就能避免在一些特定情况下（比如 Master 暂时失联）大量的 Pod 被驱逐。这一功能兼容于 tolerationSeconds，允许 Pod 定义节点故障时持续多久才被逐出。

例如，一个包含很多本地状态的应用可能需要在网络发生故障时，还能持续在节点上运行，期望网络能够快速恢复，从而避免被从这个节点上驱逐。

Pod 的 Toleration 可以这样定义：

```
tolerations:
- key: "node.alpha.kubernetes.io/unreachable"
  operator: "Exists"
  effect: "NoExecute"
  tolerationSeconds: 6000
```

对于 Node 未就绪状态，可以把 Key 设置为 node.alpha.kubernetes.io/notReady。

如果没有为 Pod 指定 node.alpha.kubernetes.io/notReady 的 Toleration，那么 Kubernetes 会自动为 Pod 加入 tolerationSeconds=300 的 node.alpha.kubernetes.io/notReady 类型的 Toleration。

同样，如果 Pod 没有定义 node.alpha.kubernetes.io/unreachable 的 Toleration，那么系统会自动为其加入 tolerationSeconds=300 的 node.alpha.kubernetes.io/unreachable 类型的 Toleration。

这些系统自动设置的 toleration 在 Node 发现问题时，能够为 Pod 确保驱逐前再运行 5min。这两个默认的 Toleration 由 Admission Controller "DefaultTolerationSeconds" 自动加入。

3.9.6 Pod Priority Preemption：Pod 优先级调度

对于运行各种负载（如 Service、Job）的中等规模或者大规模的集群来说，出于各种原因，我们需要尽可能提高集群的资源利用率。而提高资源利用率的常规做法是采用优先级方案，即不同类型的负载对应不同的优先级，同时允许集群中的所有负载所需的资源总量超过集群可提供的资源，在这种情况下，当发生资源不足的情况时，系统可以选择释放一些不重要的负载（优先级最低的），保障最重要的负载能够获取足够的资源稳定运行。

在Kubernetes 1.8版本之前，当集群的可用资源不足时，在用户提交新的Pod创建请求后，该Pod会一直处于Pending状态，即使这个Pod是一个很重要（很有身份）的Pod，也只能被动等待其他Pod被删除并释放资源，才能有机会被调度成功。Kubernetes 1.8版本引入了基于Pod优先级抢占（Pod Priority Preemption）的调度策略，此时Kubernetes会尝试释放目标节点上低优先级的Pod，以腾出空间（资源）安置高优先级的Pod，这种调度方式被称为"抢占式调度"。在Kubernetes 1.11版本中，该特性升级为Beta版本，默认开启，在后继的Kubernetes 1.14版本中正式Release。如何声明一个负载相对其他负载"更重要"？我们可以通过以下几个维度来定义：

◎ Priority，优先级；
◎ QoS，服务质量等级；
◎ 系统定义的其他度量指标。

优先级抢占调度策略的核心行为分别是驱逐（Eviction）与抢占（Preemption），这两种行为的使用场景不同，效果相同。Eviction是kubelet进程的行为，即当一个Node发生资源不足（under resource pressure）的情况时，该节点上的kubelet进程会执行驱逐动作，此时Kubelet会综合考虑Pod的优先级、资源申请量与实际使用量等信息来计算哪些Pod需要被驱逐；当同样优先级的Pod需要被驱逐时，实际使用的资源量超过申请量最大倍数的高耗能Pod会被首先驱逐。对于QoS等级为"Best Effort"的Pod来说，由于没有定义资源申请（CPU/Memory Request），所以它们实际使用的资源可能非常大。Preemption则是Scheduler执行的行为，当一个新的Pod因为资源无法满足而不能被调度时，Scheduler可能（有权决定）选择驱逐部分低优先级的Pod实例来满足此Pod的调度目标，这就是Preemption机制。

需要注意的是，Scheduler可能会驱逐Node A上的一个Pod以满足Node B上的一个新Pod的调度任务。比如下面的这个例子：

一个低优先级的Pod A在Node A（属于机架R）上运行，此时有一个高优先级的Pod B等待调度，目标节点是同属机架R的Node B，他们中的一个或全部都定义了anti-affinity规则，不允许在同一个机架上运行，此时Scheduler只好"丢车保帅"，驱逐低优先级的Pod A以满足高优先级的Pod B的调度。

Pod优先级调度示例如下。

首先，由集群管理员创建 PriorityClasses，PriorityClass 不属于任何命名空间：

```
apiVersion: scheduling.k8s.io/v1beta1
kind: PriorityClass
metadata:
  name: high-priority
value: 1000000
globalDefault: false
description: "This priority class should be used for XYZ service pods only."
```

上述 YAML 文件定义了一个名为 high-priority 的优先级类别，优先级为 100000，数字越大，优先级越高，超过一亿的数字被系统保留，用于指派给系统组件。

我们可以在任意 Pod 中引用上述 Pod 优先级类别：

```
apiVersion: v1
kind: Pod
metadata:
  name: nginx
  labels:
    env: test
spec:
  containers:
  - name: nginx
    image: nginx
    imagePullPolicy: IfNotPresent
  priorityClassName: high-priority
```

如果发生了需要抢占的调度，高优先级 Pod 就可能抢占节点 N，并将其低优先级 Pod 驱逐出节点 N，高优先级 Pod 的 status 信息中的 nominatedNodeName 字段会记录目标节点 N 的名称。需要注意，高优先级 Pod 仍然无法保证最终被调度到节点 N 上，在节点 N 上低优先级 Pod 被驱逐的过程中，如果有新的节点满足高优先级 Pod 的需求，就会把它调度到新的 Node 上。而如果在等待低优先级的 Pod 退出的过程中，又出现了优先级更高的 Pod，调度器将会调度这个更高优先级的 Pod 到节点 N 上，并重新调度之前等待的高优先级 Pod。

优先级抢占的调度方式可能会导致调度陷入"死循环"状态。当 Kubernetes 集群配置了多个调度器（Scheduler）时，这一行为可能就会发生，比如下面这个例子：

Scheduler A 为了调度一个（批）Pod，特地驱逐了一些 Pod，因此在集群中有了空余的空间可以用来调度，此时 Scheduler B 恰好抢在 Scheduler A 之前调度了一个新的 Pod，消耗了相应的资源，因此，当 Scheduler A 清理完资源后正式发起 Pod 的调度时，却发现资源不足,被目标节点的 kubelet 进程拒绝了调度请求!这种情况的确无解,因此最好的做法是让多个 Scheduler 相互协作来共同实现一个目标。

最后要指出一点:使用优先级抢占的调度策略可能会导致某些 Pod 永远无法被成功调度。因此优先级调度不但增加了系统的复杂性,还可能带来额外不稳定的因素。因此,一旦发生资源紧张的局面,首先要考虑的是集群扩容,如果无法扩容,则再考虑有监管的优先级调度特性,比如结合基于 Namespace 的资源配额限制来约束任意优先级抢占行为。

3.9.7　DaemonSet:在每个 Node 上都调度一个 Pod

DaemonSet 是 Kubernetes 1.2 版本新增的一种资源对象,用于管理在集群中每个 Node 上仅运行一份 Pod 的副本实例,如图 3.3 所示。

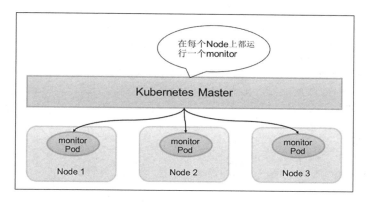

图 3.3　DaemonSet 示例

这种用法适合有这种需求的应用。

◎ 在每个 Node 上都运行一个 GlusterFS 存储或者 Ceph 存储的 Daemon 进程。
◎ 在每个 Node 上都运行一个日志采集程序,例如 Fluentd 或者 Logstach。
◎ 在每个 Node 上都运行一个性能监控程序,采集该 Node 的运行性能数据,例如 Prometheus Node Exporter、collectd、New Relic agent 或者 Ganglia gmond 等。

DaemonSet 的 Pod 调度策略与 RC 类似,除了使用系统内置的算法在每个 Node 上进行调度,也可以在 Pod 的定义中使用 NodeSelector 或 NodeAffinity 来指定满足条件的 Node 范围进行调度。

下面的例子定义为在每个 Node 上都启动一个 fluentd 容器,配置文件 fluentd-ds.yaml

的内容如下,其中挂载了物理机的两个目录"/var/log"和"/var/lib/docker/containers":

```yaml
apiVersion: apps/v1
kind: DaemonSet
metadata:
  name: fluentd-cloud-logging
  namespace: kube-system
  labels:
    k8s-app: fluentd-cloud-logging
spec:
  template:
    metadata:
      namespace: kube-system
      labels:
        k8s-app: fluentd-cloud-logging
    spec:
      containers:
      - name: fluentd-cloud-logging
        image: gcr.io/google_containers/fluentd-elasticsearch:1.17
        resources:
          limits:
            cpu: 100m
            memory: 200Mi
        env:
        - name: FLUENTD_ARGS
          value: -q
        volumeMounts:
        - name: varlog
          mountPath: /var/log
          readOnly: false
        - name: containers
          mountPath: /var/lib/docker/containers
          readOnly: false
      volumes:
      - name: containers
        hostPath:
          path: /var/lib/docker/containers
      - name: varlog
        hostPath:
          path: /var/log
```

使用 kubectl create 命令创建该 DaemonSet:

```
# kubectl create -f fluentd-ds.yaml
daemonset "fluentd-cloud-logging" created
```

查看创建好的 DaemonSet 和 Pod，可以看到在每个 Node 上都创建了一个 Pod：

```
# kubectl get daemonset --namespace=kube-system
NAME                    DESIRED   CURRENT   NODE-SELECTOR   AGE
fluentd-cloud-logging   2         2         <none>          3s

# kubectl get pods --namespace=kube-system
NAME                          READY   STATUS    RESTARTS   AGE
fluentd-cloud-logging-7tw9z   1/1     Running   0          1h
fluentd-cloud-logging-aqdn1   1/1     Running   0          1h
```

在 Kubernetes 1.6 以后的版本中，DaemonSet 也能执行滚动升级了，即在更新一个 DaemonSet 模板的时候，旧的 Pod 副本会被自动删除，同时新的 Pod 副本会被自动创建，此时 DaemonSet 的更新策略（updateStrategy）为 RollingUpdate，如下所示：

```
apiVersion: apps/v1
kind: DaemonSet
metadata:
  name: goldpinger
spec:
  updateStrategy:
    type: RollingUpdate
```

updateStrategy 的另外一个值是 OnDelete，即只有手工删除了 DaemonSet 创建的 Pod 副本，新的 Pod 副本才会被创建出来。如果不设置 updateStrategy 的值，则在 Kubernetes 1.6 之后的版本中会被默认设置为 RollingUpdate。

3.9.8 Job：批处理调度

Kubernetes 从 1.2 版本开始支持批处理类型的应用，我们可以通过 Kubernetes Job 资源对象来定义并启动一个批处理任务。批处理任务通常并行（或者串行）启动多个计算进程去处理一批工作项（work item），处理完成后，整个批处理任务结束。按照批处理任务实现方式的不同，批处理任务可以分为如图 3.4 所示的几种模式。

◎ Job Template Expansion 模式：一个 Job 对象对应一个待处理的 Work item，有几个 Work item 就产生几个独立的 Job，通常适合 Work item 数量少、每个 Work item 要处理的数据量比较大的场景，比如有一个 100GB 的文件作为一个 Work item，总共有 10 个文件需要处理。
◎ Queue with Pod Per Work Item 模式：采用一个任务队列存放 Work item，一个 Job 对象作为消费者去完成这些 Work item，在这种模式下，Job 会启动 N 个 Pod，每个 Pod 都对应一个 Work item。
◎ Queue with Variable Pod Count 模式：也是采用一个任务队列存放 Work item，一个 Job 对象作为消费者去完成这些 Work item，但与上面的模式不同，Job 启动的 Pod 数量是可变的。

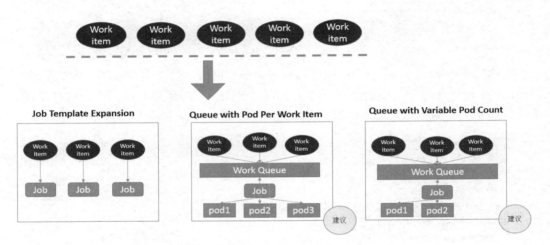

图 3.4　批处理任务的几种模式

还有一种被称为 Single Job with Static Work Assignment 的模式，也是一个 Job 产生多个 Pod，但它采用程序静态方式分配任务项，而不是采用队列模式进行动态分配。

如表 3.4 所示是这几种模式的一个对比。

表 3.4　批处理任务的模式对比

模 式 名 称	是否是一个 Job	Pod 的数量少于 Work item	用户程序是否要做相应的修改	Kubernetes 是否支持
Job Template Expansion	/	/	是	是
Queue with Pod Per Work Item	是	/	有时候需要	是
Queue with Variable Pod Count	是	/	/	是
Single Job with Static Work Assignment	是	/	是	/

考虑到批处理的并行问题，Kubernetes 将 Job 分以下三种类型。

1. Non-parallel Jobs

通常一个 Job 只启动一个 Pod，除非 Pod 异常，才会重启该 Pod，一旦此 Pod 正常结束，Job 将结束。

2. Parallel Jobs with a fixed completion count

并行 Job 会启动多个 Pod，此时需要设定 Job 的 .spec.completions 参数为一个正数，当正常结束的 Pod 数量达至此参数设定的值后，Job 结束。此外，Job 的 .spec.parallelism 参数用来控制并行度，即同时启动几个 Job 来处理 Work Item。

3. Parallel Jobs with a work queue

任务队列方式的并行 Job 需要一个独立的 Queue，Work item 都在一个 Queue 中存放，不能设置 Job 的 .spec.completions 参数，此时 Job 有以下特性。

- 每个 Pod 都能独立判断和决定是否还有任务项需要处理。
- 如果某个 Pod 正常结束，则 Job 不会再启动新的 Pod。
- 如果一个 Pod 成功结束，则此时应该不存在其他 Pod 还在工作的情况，它们应该都处于即将结束、退出的状态。
- 如果所有 Pod 都结束了，且至少有一个 Pod 成功结束，则整个 Job 成功结束。

下面分别讲解常见的三种批处理模型在 Kubernetes 中的应用例子。

首先是 Job Template Expansion 模式，由于在这种模式下每个 Work item 对应一个 Job 实例，所以这种模式首先定义一个 Job 模板，模板里的主要参数是 Work item 的标识，因为每个 Job 都处理不同的 Work item。如下所示的 Job 模板（文件名为 job.yaml.txt）中的 $ITEM 可以作为任务项的标识：

```
apiVersion: batch/v1
kind: Job
metadata:
  name: process-item-$ITEM
  labels:
    jobgroup: jobexample
spec:
  template:
```

```
      metadata:
        name: jobexample
        labels:
          jobgroup: jobexample
      spec:
        containers:
        - name: c
          image: busybox
          command: ["sh", "-c", "echo Processing item $ITEM && sleep 5"]
        restartPolicy: Never
```

通过下面的操作，生成了 3 个对应的 Job 定义文件并创建 Job：

```
# for i in apple banana cherry
> do
>   cat job.yaml.txt | sed "s/\$ITEM/$i/" > ./jobs/job-$i.yaml
> done
# ls jobs
job-apple.yaml  job-banana.yaml  job-cherry.yaml
# kubectl create -f jobs
job "process-item-apple" created
job "process-item-banana" created
job "process-item-cherry" created
```

首先，观察 Job 的运行情况：

```
# kubectl get jobs -l jobgroup=jobexample
NAME                  DESIRED   SUCCESSFUL   AGE
process-item-apple    1         1            4m
process-item-banana   1         1            4m
process-item-cherry   1         1            4m
```

其次，我们看看 Queue with Pod Per Work Item 模式，在这种模式下需要一个任务队列存放 Work item，比如 RabbitMQ，客户端程序先把要处理的任务变成 Work item 放入任务队列，然后编写 Worker 程序、打包镜像并定义成为 Job 中的 Work Pod。Worker 程序的实现逻辑是从任务队列中拉取一个 Work item 并处理，在处理完成后即结束进程。并行度为 2 的 Demo 示意图如图 3.5 所示。

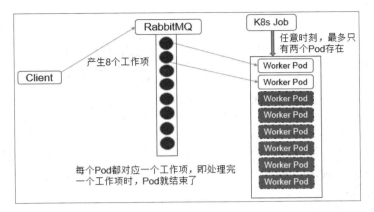

图 3.5　Queue with Pod Per Work Item 示例

最后，我们看看 Queue with Variable Pod Count 模式，如图 3.6 所示。由于这种模式下，Worker 程序需要知道队列中是否还有等待处理的 Work item，如果有就取出来处理，否则就认为所有工作完成并结束进程，所以任务队列通常要采用 Redis 或者数据库来实现。

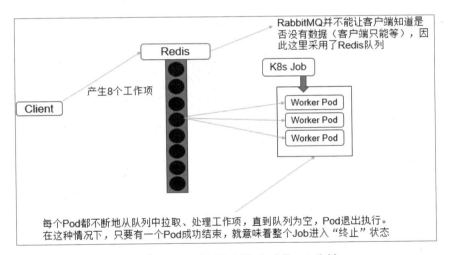

图 3.6　Queue with Variable Pod Count 示例

3.9.9　Cronjob：定时任务

Kubernetes 从 1.5 版本开始增加了一种新类型的 Job，即类似 Linux Cron 的定时任务 Cron Job，下面看看如何定义和使用这种类型的 Job。

首先，确保 Kubernetes 的版本为 1.8 及以上。

其次，需要掌握 Cron Job 的定时表达式，它基本上照搬了 Linux Cron 的表达式，格式如下：

```
Minutes Hours DayofMonth Month DayofWeek
```

其中每个域都可出现的字符如下。

- Minutes：可出现"," "-" "*" "/"这 4 个字符，有效范围为 0~59 的整数。
- Hours：可出现"," "-" "*" "/"这 4 个字符，有效范围为 0~23 的整数。
- DayofMonth：可出现"," "-" "*" "/" "?" "L" "W" "C"这 8 个字符，有效范围为 1~31 的整数。
- Month：可出现"," "-" "*" "/"这 4 个字符，有效范围为 1~12 的整数或 JAN~DEC。
- DayofWeek：可出现"," "-" "*" "/" "?" "L" "C" "#"这 8 个字符，有效范围为 1~7 的整数或 SUN~SAT。1 表示星期天，2 表示星期一，以此类推。

表达式中的特殊字符"*"与"/"的含义如下。

- *：表示匹配该域的任意值，假如在 Minutes 域使用"*"，则表示每分钟都会触发事件。
- /：表示从起始时间开始触发，然后每隔固定时间触发一次，例如在 Minutes 域设置为 5/20，则意味着第 1 次触发在第 5min 时，接下来每 20min 触发一次，将在第 25min、第 45min 等时刻分别触发。

比如，我们要每隔 1min 执行一次任务，则 Cron 表达式如下：

```
*/1 * * * *
```

掌握这些基本知识后，就可以编写一个 Cron Job 的配置文件了：

```
cron.yaml
apiVersion: batch/v1 beta
kind: CronJob
metadata:
  name: hello
spec:
  schedule: "*/1 * * * *"
  jobTemplate:
    spec:
```

```
      template:
        spec:
          containers:
          - name: hello
            image: busybox
            args:
            - /bin/sh
            - -c
            - date; echo Hello from the Kubernetes cluster
          restartPolicy: OnFailure
```

该例子定义了一个名为 hello 的 Cron Job，任务每隔 1min 执行一次，运行的镜像是 busybox，执行的命令是 Shell 脚本，脚本执行时会在控制台输出当前时间和字符串 "Hello from the Kubernetes cluster"。

接下来执行 kubectl create 命令完成创建：

```
# kubectl create -f cron.yaml
cronjob "hello" created
```

然后每隔 1min 执行 kubectl get cronjob hello 查看任务状态，发现的确每分钟调度了一次：

```
# kubectl get cronjob hello
NAME      SCHEDULE      SUSPEND   ACTIVE    LAST-SCHEDULE
hello     */1 * * * *   False     0         Thu, 29 Jun 2017 11:32:00 -0700
......
# kubectl get cronjob hello
NAME      SCHEDULE      SUSPEND   ACTIVE    LAST-SCHEDULE
hello     */1 * * * *   False     0         Thu, 29 Jun 2017 11:33:00 -0700
......
# kubectl get cronjob hello
NAME      SCHEDULE      SUSPEND   ACTIVE    LAST-SCHEDULE
hello     */1 * * * *   False     0         Thu, 29 Jun 2017 11:34:00 -0700
```

还可以通过查找 Cron Job 对应的容器，验证每隔 1min 产生一个容器的事实，如下所示：

```
# docker ps -a | grep busybox
83f7b86728ea   busybox@sha256:be3c11fdba7cfe299214e46edc642e09514dbb9bbefcd0d3836c05a1e0cd0642
"/bin/sh -c 'date; ec"   About a minute ago   Exited (0) About a minute ago
k8s_hello_hello-1498795860-qqwb4_default_207586cf-5d4a-11e7-86c1-000c2997487d_0
```

```
    36aa3b991980
busybox@sha256:be3c11fdba7cfe299214e46edc642e09514dbb9bbefcd0d3836c05a1e0cd0642
"/bin/sh -c 'date; ec"   2 minutes ago         Exited (0) 2 minutes ago
k8s_hello_hello-1498795800-g92vx_default_fca21ec0-5d49-11e7-86c1-000c2997487d_0
    3d762ae35172
busybox@sha256:be3c11fdba7cfe299214e46edc642e09514dbb9bbefcd0d3836c05a1e0cd0642
"/bin/sh -c 'date; ec"   3 minutes ago         Exited (0) 3 minutes ago
k8s_hello_hello-1498795740-3qxmd_default_d8c75d07-5d49-11e7-86c1-000c2997487d_0
    8ee5eefa8cd3
busybox@sha256:be3c11fdba7cfe299214e46edc642e09514dbb9bbefcd0d3836c05a1e0cd0642
"/bin/sh -c 'date; ec"   4 minutes ago         Exited (0) 4 minutes ago
k8s_hello_hello-1498795680-mgb7h_default_b4f7aec5-5d49-11e7-86c1-000c2997487d_0
```

查看任意一个容器的日志，结果如下：

```
# docker logs 83f7b86728ea
Thu Jun 29 18:33:07 UTC 2017
Hello from the Kubernetes cluster
```

运行下面的命令，可以更直观地了解 Cron Job 定期触发任务执行的历史和现状：

```
# kubectl get jobs --watch
NAME                 DESIRED    SUCCESSFUL    AGE
hello-1498761060     1          1             31m
hello-1498761120     1          1             30m
hello-1498761180     1          1             29m
hello-1498761240     1          1             28m
hello-1498761300     1          1             27m
hello-1498761360     1          1             26m
hello-1498761420     1          1             25m
```

其中 SUCCESSFUL 列为 1 的每一行都是一个调度成功的 Job，以第 1 行的 "hello-1498761060" 的 Job 为例，它对应的 Pod 可以通过下面的方式得到：

```
# kubectl get pods --show-all | grep hello-1498761060
hello-1498761060-shpwx    0/1    Completed    0    39m
```

查看该 Pod 的日志：

```
# kubectl logs hello-1498761060-shpwx
Thu Jun 29 18:31:07 UTC 2017
Hello from the Kubernetes cluster
```

最后，当不需要某个 Cron Job 时，可以通过下面的命令删除它：

```
# kubectl delete cronjob hello
cronjob "hello" deleted
```

在 Kubernetes 1.9 版本后,kubectl 命令增加了别名 cj 来表示 cronjob,同时 kubectl set image/env 命令也可以作用在 CronJob 对象上了。

3.9.10 自定义调度器

如果 Kubernetes 调度器的众多特性还无法满足我们的独特调度需求,则还可以用自己开发的调度器进行调度。从 1.6 版本开始,Kubernetes 的多调度器特性也进入了快速发展阶段。

一般情况下,每个新 Pod 都会由默认的调度器进行调度。但是如果在 Pod 中提供了自定义的调度器名称,那么默认的调度器会忽略该 Pod,转由指定的调度器完成 Pod 的调度。

在下面的例子中为 Pod 指定了一个名为 my-scheduler 的自定义调度器:

```
apiVersion: v1
kind: Pod
metadata:
  name: nginx
  labels:
    app: nginx
spec:
  schedulerName: my-scheduler
  containers:
  - name: nginx
    image: nginx
```

如果自定义的调度器还未在系统中部署,则默认的调度器会忽略这个 Pod,这个 Pod 将会永远处于 Pending 状态。

下面看看如何创建一个自定义的调度器。

可以用任何语言来实现简单或复杂的自定义调度器。下面的简单例子使用 Bash 脚本进行实现,调度策略为随机选择一个 Node(注意,这个调度器需要通过 kubectl proxy 来运行):

```
#!/bin/bash
SERVER='localhost:8001'
while true;
```

```
    do
        for PODNAME in $(kubectl --server $SERVER get pods -o json | jq '.items[] |
select(.spec.schedulerName == "my-scheduler") | select(.spec.nodeName == null)
| .metadata.name' | tr -d '"');
        do
            NODES=($(kubectl --server $SERVER get nodes -o json | jq
'.items[].metadata.name' | tr -d '"'))
            NUMNODES=${#NODES[@]}
            CHOSEN=${NODES[$[ $RANDOM % $NUMNODES ]]}
            curl --header "Content-Type:application/json" --request POST --data
'{"apiVersion":"v1", "kind": "Binding", "metadata": {"name": "'$PODNAME'"}, "target":
{"apiVersion": "v1", "kind": "Node", "name":"'$CHOSEN'"}}'
http://$SERVER/api/v1/namespaces/default/pods/$PODNAME/binding/
            echo "Assigned $PODNAME to $CHOSEN"
        done
        sleep 1
done
```

一旦这个自定义调度器成功启动，前面的 Pod 就会被正确调度到某个 Node 上。

3.10 Init Container（初始化容器）

在很多应用场景中，应用在启动之前都需要进行如下初始化操作。

◎ 等待其他关联组件正确运行（例如数据库或某个后台服务）。
◎ 基于环境变量或配置模板生成配置文件。
◎ 从远程数据库获取本地所需配置，或者将自身注册到某个中央数据库中。
◎ 下载相关依赖包，或者对系统进行一些预配置操作。

Kubernetes 1.3 引入了一个 Alpha 版本的新特性 init container（初始化容器，在 Kubernetes 1.5 时被更新为 Beta 版本），用于在启动应用容器（app container）之前启动一个或多个初始化容器，完成应用容器所需的预置条件，如图 3.7 所示。init container 与应用容器在本质上是一样的，但它们是仅运行一次就结束的任务，并且必须在成功执行完成后，系统才能继续执行下一个容器。根据 Pod 的重启策略（RestartPolicy），当 init container 执行失败，而且设置了 RestartPolicy=Never 时，Pod 将会启动失败；而设置 RestartPolicy=Always 时，Pod 将会被系统自动重启。

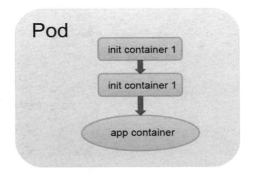

图 3.7　init container

下面以 Nginx 应用为例，在启动 Nginx 之前，通过初始化容器 busybox 为 Nginx 创建一个 index.html 主页文件。这里为 init container 和 Nginx 设置了一个共享的 Volume，以供 Nginx 访问 init container 设置的 index.html 文件：

```yaml
nginx-init-containers.yaml
apiVersion: v1
kind: Pod
metadata:
  name: nginx
  annotations:
spec:
  # These containers are run during pod initialization
  initContainers:
  - name: install
    image: busybox
    command:
    - wget
    - "-O"
    - "/work-dir/index.html"
    - http://kubernetes.io
    volumeMounts:
    - name: workdir
      mountPath: "/work-dir"
  containers:
  - name: nginx
    image: nginx
    ports:
    - containerPort: 80
    volumeMounts:
```

```yaml
    - name: workdir
      mountPath: /usr/share/nginx/html
  dnsPolicy: Default
  volumes:
  - name: workdir
    emptyDir: {}
```

创建这个 Pod：

```
# kubectl create -f nginx-init-containers.yaml
pod "nginx" created
```

在运行 init container 的过程中查看 Pod 的状态，可见 init 过程还未完成：

```
# kubectl get pods
NAME    READY    STATUS     RESTARTS   AGE
nginx   0/1      Init:0/1   0          1m
```

在 init container 成功执行完成后，系统继续启动 Nginx 容器，再次查看 Pod 的状态：

```
# kubectl get pods
NAME    READY    STATUS    RESTARTS   AGE
nginx   1/1      Running   0          7s
```

查看 Pod 的事件，可以看到系统首先创建并运行 init container 容器（名为 install），成功后继续创建和运行 Nginx 容器：

```
# kubectl describe pod nginx
Name:            nginx
Namespace:       default
......（略）
Events:
  FirstSeen     LastSeen        Count    From                              SubobjectPath
Type           Reason          Message
  ---------     --------        -----    ----                              --------------
--------       ------          -------
  3s            3s              1        default-scheduler
Normal         Scheduled       Successfully assigned init-demo to k8s-node-1
  3s            3s              1        kubelet, k8s-node-1
spec.initContainers{install}   Normal            Pulled           Container image
"busybox" already present on machine
  3s            3s              1        kubelet, k8s-node-1
spec.initContainers{install}   Normal            Created          Created container
with id 93d98cbc0251c60d43c2d8d0a6a9bb65f432344fe6f04561c4a940b79bcff74a
```

```
  3s          3s            1          kubelet, k8s-node-1
spec.initContainers{install}      Normal        Started       Started container
with id 93d98cbc0251c60d43c2d8d0a6a9bb65f432344fe6f04561c4a940b79bcff74a
  2s          2s            1          kubelet, k8s-node-1
spec.containers{nginx}            Normal        Pulled        Container image
"nginx" already present on machine
  2s          2s            1          kubelet, k8s-node-1
spec.containers{nginx}            Normal        Created       Created container with
id a388bbb9f1fe247cf42e61449328ab20f7c54a7c271590548d3d8610a28a6048
  1s          1s            1          kubelet, k8s-node-1
spec.containers{nginx}            Normal        Started       Started container with
id a388bbb9f1fe247cf42e61449328ab20f7c54a7c271590548d3d8610a28a6048
```

启动成功后，登录进 Nginx 容器，可以看到/usr/share/nginx/html 目录下的 index.html 文件为 init container 所生成，其内容为：

```
<html id="home" lang="en" class="">

<head>
......
<title>Kubernetes | Production-Grade Container Orchestration</title>
......
"url": "http://kubernetes.io/"}</script>
</head>

<body>
......
```

init container 与应用容器的区别如下。

（1）init container 的运行方式与应用容器不同，它们必须先于应用容器执行完成，当设置了多个 init container 时，将按顺序逐个运行，并且只有前一个 init container 运行成功后才能运行后一个 init container。当所有 init container 都成功运行后，Kubernetes 才会初始化 Pod 的各种信息，并开始创建和运行应用容器。

（2）在 init container 的定义中也可以设置资源限制、Volume 的使用和安全策略，等等。但资源限制的设置与应用容器略有不同。

◎ 如果多个 init container 都定义了资源请求/资源限制，则取最大的值作为所有 init container 的资源请求值/资源限制值。

- Pod 的有效（effective）资源请求值/资源限制值取以下二者中的较大值。
  a）所有应用容器的资源请求值/资源限制值之和。
  b）init container 的有效资源请求值/资源限制值。
- 调度算法将基于 Pod 的有效资源请求值/资源限制值进行计算，也就是说 init container 可以为初始化操作预留系统资源，即使后续应用容器无须使用这些资源。
- Pod 的有效 QoS 等级适用于 init container 和应用容器。
- 资源配额和限制将根据 Pod 的有效资源请求值/资源限制值计算生效。
- Pod 级别的 cgroup 将基于 Pod 的有效资源请求/限制，与调度机制一致。

（3）init container 不能设置 readinessProbe 探针，因为必须在它们成功运行后才能继续运行在 Pod 中定义的普通容器。

在 Pod 重新启动时，init container 将会重新运行，常见的 Pod 重启场景如下。

- init container 的镜像被更新时，init container 将会重新运行，导致 Pod 重启。仅更新应用容器的镜像只会使得应用容器被重启。
- Pod 的 infrastructure 容器更新时，Pod 将会重启。
- 若 Pod 中的所有应用容器都终止了，并且 RestartPolicy=Always，则 Pod 会重启。

3.11　Pod 的升级和回滚

下面说说 Pod 的升级和回滚问题。

当集群中的某个服务需要升级时，我们需要停止目前与该服务相关的所有 Pod，然后下载新版本镜像并创建新的 Pod。如果集群规模比较大，则这个工作变成了一个挑战，而且先全部停止然后逐步升级的方式会导致较长时间的服务不可用。Kubernetes 提供了滚动升级功能来解决上述问题。

如果 Pod 是通过 Deployment 创建的，则用户可以在运行时修改 Deployment 的 Pod 定义（spec.template）或镜像名称，并应用到 Deployment 对象上，系统即可完成 Deployment 的自动更新操作。如果在更新过程中发生了错误，则还可以通过回滚操作恢复 Pod 的版本。

3.11.1 Deployment 的升级

以 Deployment nginx 为例:

```yaml
nginx-deployment.yaml
apiVersion: apps/v1
kind: Deployment
metadata:
  name: nginx-deployment
spec:
  selector:
    matchLabels:
      app: nginx
  replicas: 3
  template:
    metadata:
      labels:
        app: nginx
    spec:
      containers:
      - name: nginx
        image: nginx:1.7.9
        ports:
        - containerPort: 80
```

已运行的 Pod 副本数量有 3 个:

```
# kubectl get pods
NAME                                READY   STATUS    RESTARTS   AGE
nginx-deployment-4087004473-9jqqs   1/1     Running   0          1m
nginx-deployment-4087004473-cq0cf   1/1     Running   0          1m
nginx-deployment-4087004473-vxn56   1/1     Running   0          1m
```

现在 Pod 镜像需要被更新为 Nginx:1.9.1,我们可以通过 kubectl set image 命令为 Deployment 设置新的镜像名称:

```
$ kubectl set image deployment/nginx-deployment nginx=nginx:1.9.1
deployment "nginx-deployment" image updated
```

另一种更新的方法是使用 kubectl edit 命令修改 Deployment 的配置,将 spec.template.spec.containers[0].image 从 Nginx:1.7.9 更改为 Nginx:1.9.1:

```
$ kubectl edit deployment/nginx-deployment
```

```
deployment "nginx-deployment" edited
```

一旦镜像名(或 Pod 定义)发生了修改,则将触发系统完成 Deployment 所有运行 Pod 的滚动升级操作。可以使用 kubectl rollout status 命令查看 Deployment 的更新过程:

```
$ kubectl rollout status deployment/nginx-deployment
Waiting for rollout to finish: 2 out of 3 new replicas have been updated...
Waiting for rollout to finish: 2 out of 3 new replicas have been updated...
Waiting for rollout to finish: 2 out of 3 new replicas have been updated...
Waiting for rollout to finish: 2 out of 3 new replicas have been updated...
Waiting for rollout to finish: 2 old replicas are pending termination...
Waiting for rollout to finish: 1 old replicas are pending termination...
Waiting for rollout to finish: 1 old replicas are pending termination...
Waiting for rollout to finish: 1 old replicas are pending termination...
Waiting for rollout to finish: 2 of 3 updated replicas are available...
deployment "nginx-deployment" successfully rolled out
```

查看当前运行的 Pod,名称已经更新了:

```
$ kubectl get pods
NAME                                READY   STATUS    RESTARTS   AGE
nginx-deployment-3599678771-01h26   1/1     Running   0          2m
nginx-deployment-3599678771-57thr   1/1     Running   0          2m
nginx-deployment-3599678771-s8p21   1/1     Running   0          2m
```

查看 Pod 使用的镜像,已经更新为 Nginx:1.9.1 了:

```
# kubectl describe pod/nginx-deployment-3599678771-s8p21
Name:           nginx-deployment-3599678771-s8p21
......
   Image:       nginx:1.9.1
......
```

那么,Deployment 是如何完成 Pod 更新的呢?

我们可以使用 kubectl describe deployments/nginx-deployment 命令仔细观察 Deployment 的更新过程。初始创建 Deployment 时,系统创建了一个 ReplicaSet(nginx-deployment-4087004473),并按用户的需求创建了 3 个 Pod 副本。当更新 Deployment 时,系统创建了一个新的 ReplicaSet(nginx-deployment-3599678771),并将其副本数量扩展到 1,然后将旧的 ReplicaSet 缩减为 2。之后,系统继续按照相同的更新策略对新旧两个 ReplicaSet 进行逐个调整。最后,新的 ReplicaSet 运行了 3 个新版本 Pod 副本,旧的 ReplicaSet 副本数量则缩减为 0。如图 3.8 所示。

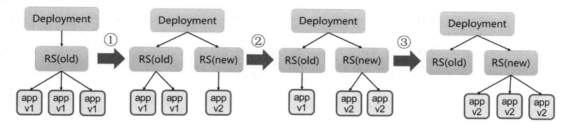

图 3.8　Pod 的滚动升级

下面列出 Deployment nginx-deployment 的详细事件信息：

```
$ kubectl describe deployments/nginx-deployment
Name:                   nginx-deployment
Namespace:              default
......
Replicas:               3 updated | 3 total | 3 available | 0 unavailable
StrategyType:           RollingUpdate
MinReadySeconds: 0
RollingUpdateStrategy:  1 max unavailable, 1 max surge
Conditions:
  Type          Status  Reason
  ----          ------  ------
  Available     TrueMinimumReplicasAvailable
OldReplicaSets: <none>
NewReplicaSet:  nginx-deployment-3599678771 (3/3 replicas created)
Events:
  FirstSeen  LastSeen Count   From                   SubObjectPath   Type        Reason
         Message
  ---------  -------------    ----                   --------------  ---------------
         -------
  55m        55m         1    {deployment-controller }               Normal
ScalingReplicaSet    Scaled up replica set nginx-deployment-4087004473 to 3
   4m         4m         1    {deployment-controller }               Normal
ScalingReplicaSet    Scaled up replica set nginx-deployment-3599678771 to 1
   4m         4m         1    {deployment-controller }               Normal
ScalingReplicaSet    Scaled down replica set nginx-deployment-4087004473 to 2
   4m         4m         1    {deployment-controller }               Normal
ScalingReplicaSet    Scaled up replica set nginx-deployment-3599678771 to 2
   4m         4m         1    {deployment-controller }               Normal
ScalingReplicaSet    Scaled down replica set nginx-deployment-4087004473 to 1
   4m         4m         1    {deployment-controller }               Normal
```

```
ScalingReplicaSet         Scaled up replica set nginx-deployment-3599678771 to 3
   4m      4m       1     {deployment-controller }             Normal
ScalingReplicaSet         Scaled down replica set nginx-deployment-4087004473 to 0
```

运行 kubectl get rs 命令，查看两个 ReplicaSet 的最终状态：

```
$ kubectl get rs
NAME                          DESIRED    CURRENT    READY    AGE
nginx-deployment-3599678771   3          3          3        1m
nginx-deployment-4087004473   0          0          0        52m
```

在整个升级的过程中，系统会保证至少有两个 Pod 可用，并且最多同时运行 4 个 Pod，这是 Deployment 通过复杂的算法完成的。Deployment 需要确保在整个更新过程中只有一定数量的 Pod 可能处于不可用状态。在默认情况下，Deployment 确保可用的 Pod 总数至少为所需的副本数量（DESIRED）减 1，也就是最多 1 个不可用（maxUnavailable=1）。Deployment 还需要确保在整个更新过程中 Pod 的总数量不会超过所需的副本数量太多。在默认情况下，Deployment 确保 Pod 的总数最多比所需的 Pod 数多 1 个，也就是最多 1 个浪涌值（maxSurge=1）。Kubernetes 从 1.6 版本开始，maxUnavailable 和 maxSurge 的默认值将从 1、1 更新为所需副本数量的 25%、25%。

这样，在升级过程中，Deployment 就能够保证服务不中断，并且副本数量始终维持为用户指定的数量（DESIRED）。

对更新策略的说明如下。

在 Deployment 的定义中，可以通过 spec.strategy 指定 Pod 更新的策略，目前支持两种策略：Recreate（重建）和 RollingUpdate（滚动更新），默认值为 RollingUpdate。在前面的例子中使用的就是 RollingUpdate 策略。

- Recreate：设置 spec.strategy.type=Recreate，表示 Deployment 在更新 Pod 时，会先杀掉所有正在运行的 Pod，然后创建新的 Pod。
- RollingUpdate：设置 spec.strategy.type=RollingUpdate，表示 Deployment 会以滚动更新的方式来逐个更新 Pod。同时，可以通过设置 spec.strategy.rollingUpdate 下的两个参数（maxUnavailable 和 maxSurge）来控制滚动更新的过程。

下面对滚动更新时两个主要参数的说明如下。

- spec.strategy.rollingUpdate.maxUnavailable：用于指定 Deployment 在更新过程中不可用状态的 Pod 数量的上限。该 maxUnavailable 的数值可以是绝对值（例如 5）

或 Pod 期望的副本数的百分比（例如 10%），如果被设置为百分比，那么系统会先以向下取整的方式计算出绝对值（整数）。而当另一个参数 maxSurge 被设置为 0 时，maxUnavailable 则必须被设置为绝对数值大于 0（从 Kubernetes 1.6 开始，maxUnavailable 的默认值从 1 改为 25%）。举例来说，当 maxUnavailable 被设置为 30% 时，旧的 ReplicaSet 可以在滚动更新开始时立即将副本数缩小到所需副本总数的 70%。一旦新的 Pod 创建并准备好，旧的 ReplicaSet 会进一步缩容，新的 ReplicaSet 又继续扩容，整个过程中系统在任意时刻都可以确保可用状态的 Pod 总数至少占 Pod 期望副本总数的 70%。

◎ spec.strategy.rollingUpdate.maxSurge：用于指定在 Deployment 更新 Pod 的过程中 Pod 总数超过 Pod 期望副本数部分的最大值。该 maxSurge 的数值可以是绝对值（例如 5）或 Pod 期望副本数的百分比（例如 10%）。如果设置为百分比，那么系统会先按照向上取整的方式计算出绝对数值（整数）。从 Kubernetes 1.6 开始，maxSurge 的默认值从 1 改为 25%。举例来说，当 maxSurge 的值被设置为 30% 时，新的 ReplicaSet 可以在滚动更新开始时立即进行副本数扩容，只需要保证新旧 ReplicaSet 的 Pod 副本数之和不超过期望副本数的 130% 即可。一旦旧的 Pod 被杀掉，新的 ReplicaSet 就会进一步扩容。在整个过程中系统在任意时刻都能确保新旧 ReplicaSet 的 Pod 副本总数之和不超过所需副本数的 130%。

这里需要注意多重更新（Rollover）的情况。如果 Deployment 的上一次更新正在进行，此时用户再次发起 Deployment 的更新操作，那么 Deployment 会为每一次更新都创建一个 ReplicaSet，而每次在新的 ReplicaSet 创建成功后，会逐个增加 Pod 副本数，同时将之前正在扩容的 ReplicaSet 停止扩容（更新），并将其加入旧版本 ReplicaSet 列表中，然后开始缩容至 0 的操作。

例如，假设我们创建一个 Deployment，这个 Deployment 开始创建 5 个 Nginx:1.7.9 的 Pod 副本，在这个创建 Pod 动作尚未完成时，我们又将 Deployment 进行更新，在副本数不变的情况下将 Pod 模板中的镜像修改为 Nginx:1.9.1，又假设此时 Deployment 已经创建了 3 个 Nginx:1.7.9 的 Pod 副本，则 Deployment 会立即杀掉已创建的 3 个 Nginx:1.7.9 Pod，并开始创建 Nginx:1.9.1 Pod。Deployment 不会在等待 Nginx:1.7.9 的 Pod 创建到 5 个之后再进行更新操作。

还需要注意更新 Deployment 的标签选择器（Label Selector）的情况。通常来说，不鼓励更新 Deployment 的标签选择器，因为这样会导致 Deployment 选择的 Pod 列表发生变化，也可能与其他控制器产生冲突。如果一定要更新标签选择器，那么请务必谨慎，确保不会出现其他问题。关于 Deployment 标签选择器的更新的注意事项如下。

（1）添加选择器标签时，必须同步修改 Deployment 配置的 Pod 的标签，为 Pod 添加新的标签，否则 Deployment 的更新会报验证错误而失败：

```
deployments "nginx-deployment" was not valid:
* spec.template.metadata.labels: Invalid value: {"app":"nginx"}: `selector` does not match template `labels`
```

添加标签选择器是无法向后兼容的，这意味着新的标签选择器不会匹配和使用旧选择器创建的 ReplicaSets 和 Pod，因此添加选择器将会导致所有旧版本的 ReplicaSets 和由旧 ReplicaSets 创建的 Pod 处于孤立状态（不会被系统自动删除，也不受新的 ReplicaSet 控制）。

为标签选择器和 Pod 模板添加新的标签（使用 kubectl edit deployment 命令）后，效果如下：

```
$ kubectl get rs
NAME                          DESIRED   CURRENT   READY   AGE
nginx-deployment-3661742516   3         3         3       2s
nginx-deployment-3599678771   3         3         3       1m
nginx-deployment-4087004473   0         0         0       52m
```

可以看到新 ReplicaSet（nginx-deployment-3661742516）创建的 3 个新 Pod：

```
$ kubectl get pods
NAME                                READY   STATUS    RESTARTS   AGE
nginx-deployment-3599678771-01h26   1/1     Running   0          2m
nginx-deployment-3599678771-57thr   1/1     Running   0          2m
nginx-deployment-3599678771-s8p21   1/1     Running   0          2m
nginx-deployment-3661742516-46djm   1/1     Running   0          52s
nginx-deployment-3661742516-kws84   1/1     Running   0          52s
nginx-deployment-3661742516-wq30s   1/1     Running   0          52s
```

（2）更新标签选择器，即更改选择器中标签的键或者值，也会产生与添加选择器标签类似的效果。

（3）删除标签选择器，即从 Deployment 的标签选择器中删除一个或者多个标签，该 Deployment 的 ReplicaSet 和 Pod 不会受到任何影响。但需要注意的是，被删除的标签仍会存在于现有的 Pod 和 ReplicaSets 上。

3.11.2 Deployment 的回滚

有时（例如新的 Deployment 不稳定时）我们可能需要将 Deployment 回滚到旧版本。在默认情况下，所有 Deployment 的发布历史记录都被保留在系统中，以便于我们随时进行回滚（可以配置历史记录数量）。

假设在更新 Deployment 镜像时，将容器镜像名误设置成 Nginx:1.91（一个不存在的镜像）：

```
$ kubectl set image deployment/nginx-deployment nginx=nginx:1.91
deployment "nginx-deployment" image updated
```

则这时 Deployment 的部署过程会卡住：

```
$ kubectl rollout status deployments nginx-deployment
Waiting for rollout to finish: 1 out of 3 new replicas have been updated...
```

由于执行过程卡住，所以需要执行 Ctrl-C 命令来终止这个查看命令。

查看 ReplicaSet，可以看到新建的 ReplicaSet（nginx-deployment-3660254150）：

```
$ kubectl get rs
NAME                          DESIRED   CURRENT   READY   AGE
nginx-deployment-3646295028   3         3         3       53s
nginx-deployment-3660254150   1         1         0       40s
nginx-deployment-4234284026   0         0         0       1m
```

再查看创建的 Pod，会发现新的 ReplicaSet 创建的 1 个 Pod 被卡在镜像拉取过程中。

```
$ kubectl get pods
NAME                                READY   STATUS            RESTARTS   AGE
nginx-deployment-3646295028-d5r6r   1/1     Running           0          1m
nginx-deployment-3646295028-jw22d   1/1     Running           0          59s
nginx-deployment-3646295028-tw6x7   1/1     Running           0          1m
nginx-deployment-3660254150-9kj51   0/1     ImagePullBackOff  0          49s
```

为了解决上面这个问题，我们需要回滚到之前稳定版本的 Deployment。

首先，用 kubectl rollout history 命令检查这个 Deployment 部署的历史记录：

```
$ kubectl rollout history deployment/nginx-deployment
deployments "nginx-deployment"
REVISION        CHANGE-CAUSE
1               kubectl create --filename=nginx-deployment.yaml --record=true
```

```
2           kubectl set image deployment/nginx-deployment nginx=nginx:1.9.1
3           kubectl set image deployment/nginx-deployment nginx=nginx:1.91
```

注意，在创建 Deployment 时使用--record 参数，就可以在 CHANGE-CAUSE 列看到每个版本使用的命令了。另外，Deployment 的更新操作是在 Deployment 进行部署（Rollout）时被触发的，这意味着当且仅当 Deployment 的 Pod 模板（即 spec.template）被更改时才会创建新的修订版本，例如更新模板标签或容器镜像。其他更新操作（如扩展副本数）将不会触发 Deployment 的更新操作，这也意味着我们将 Deployment 回滚到之前的版本时，只有 Deployment 的 Pod 模板部分会被修改。

如果需要查看特定版本的详细信息，则可以加上--revision=<N>参数：

```
$ kubectl rollout history deployment/nginx-deployment --revision=3
deployments "nginx-deployment" with revision #3
Pod Template:
  Labels:        app=nginx
        pod-template-hash=3660254150
  Annotations:   kubernetes.io/change-cause=kubectl set image deployment/nginx-deployment nginx=nginx:1.91
  Containers:
   nginx:
    Image:       nginx:1.91
    Port:        80/TCP
    Environment:        <none>
    Mounts:      <none>
  Volumes:       <none>
```

现在我们决定撤销本次发布并回滚到上一个部署版本：

```
$ kubectl rollout undo deployment/nginx-deployment
deployment "nginx-deployment" rolled back
```

当然，也可以使用--to-revision 参数指定回滚到的部署版本号：

```
$ kubectl rollout undo deployment/nginx-deployment --to-revision=2
deployment "nginx-deployment" rolled back
```

这样，该 Deployment 就回滚到之前的稳定版本了，可以从 Deployment 的事件信息中查看到回滚到版本 2 的操作过程：

```
$ kubectl describe deployment/nginx-deployment
Name:             nginx-deployment
......
```

```
  OldReplicaSets: <none>
  NewReplicaSet: nginx-deployment-3646295028 (3/3 replicas created)
  Events:
    FirstSeen      LastSeen        Count   From                    SubObjectPath
Type            Reason          Message
    ---------      --------        -----   ----                    -------------
--------        ------          -------
    4m             4m              1       deployment-controller
Normal          ScalingReplicaSet       Scaled up replica set nginx-deployment-4234284026
to 3
    4m             4m              1       deployment-controller
Normal          ScalingReplicaSet       Scaled up replica set nginx-deployment-3646295028
to 1
    4m             4m              1       deployment-controller
Normal          ScalingReplicaSet       Scaled down replica set
nginx-deployment-4234284026 to 2
    4m             4m              1       deployment-controller
Normal          ScalingReplicaSet       Scaled up replica set nginx-deployment-3646295028
to 2
    4m             4m              1       deployment-controller
Normal          ScalingReplicaSet       Scaled down replica set
nginx-deployment-4234284026 to 1
    4m             4m              1       deployment-controller
Normal          ScalingReplicaSet       Scaled up replica set nginx-deployment-3646295028
to 3
    4m             4m              1       deployment-controller
Normal          ScalingReplicaSet       Scaled down replica set
nginx-deployment-4234284026 to 0
    4m             4m              1       deployment-controller
Normal          ScalingReplicaSet       Scaled up replica set nginx-deployment-3660254150
to 1
    36s            36s             1       deployment-controller
Normal          DeploymentRollback      Rolled back deployment "nginx-deployment" to
revision 2
    36s            36s             1       deployment-controller
Normal          ScalingReplicaSet       Scaled down replica set
nginx-deployment-3660254150 to 0
```

3.11.3 暂停和恢复 Deployment 的部署操作，以完成复杂的修改

对于一次复杂的 Deployment 配置修改，为了避免频繁触发 Deployment 的更新操作，可以先暂停 Deployment 的更新操作，然后进行配置修改，再恢复 Deployment，一次性触发完整的更新操作，就可以避免不必要的 Deployment 更新操作了。

以之前创建的 Nginx 为例：

```
$ kubectl get deployments
NAME                 DESIRED    CURRENT    UP-TO-DATE    AVAILABLE    AGE
nginx-deployment     3          3          0             3            32s

$ kubectl get rs
NAME                            DESIRED    CURRENT    READY    AGE
nginx-deployment-4234284026     3          3          3        7s
```

通过 kubectl rollout pause 命令暂停 Deployment 的更新操作：

```
$ kubectl rollout pause deployment/nginx-deployment
deployment "nginx-deployment" paused
```

然后修改 Deployment 的镜像信息：

```
$ kubectl set image deploy/nginx-deployment nginx=nginx:1.9.1
deployment "nginx-deployment" image updated
```

查看 Deployment 的历史记录，发现并没有触发新的 Deployment 部署操作：

```
$ kubectl rollout history deploy/nginx-deployment
deployments "nginx-deployment"
REVISION        CHANGE-CAUSE
1               kubectl create --filename=nginx-deployment.yaml --record=true
```

在暂停 Deployment 部署之后，可以根据需要进行任意次数的配置更新。例如，再次更新容器的资源限制：

```
$ kubectl set resources deployment nginx-deployment -c=nginx --limits=cpu=200m,memory=512Mi
deployment "nginx-deployment" resource requirements updated
```

最后，恢复这个 Deployment 的部署操作：

```
$ kubectl rollout resume deploy nginx-deployment
deployment "nginx-deployment" resumed
```

可以看到一个新的 ReplicaSet 被创建出来了：

```
$ kubectl get rs
NAME                          DESIRED   CURRENT   READY   AGE
nginx-deployment-3133440882   3         3         3       6s
nginx-deployment-4234284026   0         0         0       49s
```

查看 Deployment 的事件信息，可以看到 Deployment 完成了更新：

```
# kubectl describe deployment/nginx-deployment
Name:           nginx-deployment
......
Events:
  FirstSeen   LastSeen    Count   From                    SubObjectPath   Type        Reason              Message
  ---------   --------    -----   ----                    -------------   --------    ------              -------
  1m          1m          1       deployment-controller                   Normal      ScalingReplicaSet   Scaled up replica set nginx-deployment-4234284026 to 3
  28s         28s         1       deployment-controller                   Normal      ScalingReplicaSet   Scaled up replica set nginx-deployment-3133440882 to 1
  27s         27s         1       deployment-controller                   Normal      ScalingReplicaSet   Scaled down replica set nginx-deployment-4234284026 to 2
  27s         27s         1       deployment-controller                   Normal      ScalingReplicaSet   Scaled up replica set nginx-deployment-3133440882 to 2
  26s         26s         1       deployment-controller                   Normal      ScalingReplicaSet   Scaled down replica set nginx-deployment-4234284026 to 1
  25s         25s         1       deployment-controller                   Normal      ScalingReplicaSet   Scaled up replica set nginx-deployment-3133440882 to 3
  23s         23s         1       deployment-controller                   Normal      ScalingReplicaSet   Scaled down replica set nginx-deployment-4234284026 to 0
```

注意，在恢复暂停的 Deployment 之前，无法回滚该 Deployment。

3.11.4 使用 kubectl rolling-update 命令完成 RC 的滚动升级

对于 RC 的滚动升级，Kubernetes 还提供了一个 kubectl rolling-update 命令进行实现。该命令创建了一个新的 RC，然后自动控制旧的 RC 中的 Pod 副本数量逐渐减少到 0，同时新的 RC 中的 Pod 副本数量从 0 逐步增加到目标值，来完成 Pod 的升级。需要注意的是，系统要求新的 RC 与旧的 RC 都在相同的命名空间内。

以 redis-master 为例，假设当前运行的 redis-master Pod 是 1.0 版本，现在需要升级到 2.0 版本。

创建 redis-master-controller-v2.yaml 的配置文件如下：

```yaml
apiVersion: v1
kind: ReplicationController
metadata:
  name: redis-master-v2
  labels:
    name: redis-master
    version: v2
spec:
  replicas: 1
  selector:
    name: redis-master
    version: v2
  template:
    metadata:
      labels:
        name: redis-master
        version: v2
    spec:
      containers:
      - name: master
        image: kubeguide/redis-master:2.0
        ports:
        - containerPort: 6379
```

在配置文件中需要注意以下两点。

◎ RC 的名字（name）不能与旧 RC 的名字相同。
◎ 在 selector 中应至少有一个 Label 与旧 RC 的 Label 不同，以标识其为新 RC。在本例中新增了一个名为 version 的 Label，以与旧 RC 进行区分。

运行 kubectl rolling-update 命令完成 Pod 的滚动升级：

```
kubectl rolling-update redis-master -f redis-master-controller-v2.yaml
```

kubectl 的执行过程如下：

```
Creating redis-master-v2
At beginning of loop: redis-master replicas: 2, redis-master-v2 replicas: 1
Updating redis-master replicas: 2, redis-master-v2 replicas: 1
At end of loop: redis-master replicas: 2, redis-master-v2 replicas: 1
At beginning of loop: redis-master replicas: 1, redis-master-v2 replicas: 2
Updating redis-master replicas: 1, redis-master-v2 replicas: 2
At end of loop: redis-master replicas: 1, redis-master-v2 replicas: 2
At beginning of loop: redis-master replicas: 0, redis-master-v2 replicas: 3
Updating redis-master replicas: 0, redis-master-v2 replicas: 3
At end of loop: redis-master replicas: 0, redis-master-v2 replicas: 3
Update succeeded. Deleting redis-master
redis-master-v2
```

等所有新的 Pod 都启动完成后，旧的 Pod 也被全部销毁，这样就完成了容器集群的更新工作。

另一种方法是不使用配置文件，直接用 kubectl rolling-update 命令，加上 --image 参数指定新版镜像名称来完成 Pod 的滚动升级：

```
kubectl rolling-update redis-master --image=redis-master:2.0
```

与使用配置文件的方式不同，执行的结果是旧 RC 被删除，新 RC 仍将使用旧 RC 的名称。

kubectl 的执行过程如下：

```
Creating redis-master-ea866a5d2c08588c3375b86fb253db75
  At beginning of loop: redis-master replicas: 2, redis-master-ea866a5d2c08588c
3375b86fb253db75 replicas: 1
  Updating redis-master replicas: 2, redis-master-ea866a5d2c08588c3375b86fb253db
75 replicas: 1
  At end of loop: redis-master replicas: 2, redis-master-ea866a5d2c08588c3375b86fb
253db75 replicas: 1
  At beginning of loop: redis-master replicas: 1, redis-master-ea866a5d2c08588c
3375b86fb253db75 replicas: 2
  Updating redis-master replicas: 1, redis-master-ea866a5d2c08588c3375b86fb
253db75 replicas: 2
```

```
    At end of loop: redis-master replicas: 1, redis-master-ea866a5d2c08588c3375b86fb
253db75 replicas: 2
    At beginning of loop: redis-master replicas: 0, redis-master-ea866a5d2c08588c
3375b86fb253db75 replicas: 3
    Updating redis-master replicas: 0, redis-master-ea866a5d2c08588c3375b86fb253db
75 replicas: 3
    At end of loop: redis-master replicas: 0, redis-master-ea866a5d2c08588c3375b86fb
253db75 replicas: 3
    Update succeeded. Deleting old controller: redis-master
    Renaming redis-master-ea866a5d2c08588c3375b86fb253db75 to redis-master
    redis-master
```

可以看到，kubectl 通过新建一个新版本 Pod，停掉一个旧版本 Pod，如此逐步迭代来完成整个 RC 的更新。

更新完成后，查看 RC：

```
    $ kubectl get rc
    CONTROLLER         CONTAINER(S)     IMAGE(S)              SELECTOR        REPLICAS
    redis-master       master           kubeguide/redis-master:2.0
deployment= ea866a5d2c08588c3375b86fb253db75,name=redis-master,version=v1    3
```

可以看到，kubectl 给 RC 增加了一个 key 为 "deployment" 的 Label（这个 key 的名字可通过 --deployment-label-key 参数进行修改），Label 的值是 RC 的内容进行 Hash 计算后的值，相当于签名，这样就能很方便地比较 RC 里的 Image 名字及其他信息是否发生了变化。

如果在更新过程中发现配置有误，则用户可以中断更新操作，并通过执行 kubectl rolling-update --rollback 完成 Pod 版本的回滚：

```
    $ kubectl rolling-update redis-master --image=kubeguide/redis-master:2.0
--rollback
    Found existing update in progress (redis-master-fefd9752aa5883ca4d53013a7b
583967), resuming.
    Found desired replicas.Continuing update with existing controller redis-master.
    At beginning of loop: redis-master-fefd9752aa5883ca4d53013a7b583967 replicas:
0, redis-master replicas: 3
    Updating redis-master-fefd9752aa5883ca4d53013a7b583967 replicas: 0, redis-master
replicas: 3
    At end of loop: redis-master-fefd9752aa5883ca4d53013a7b583967 replicas: 0,
redis-master replicas: 3
    Update succeeded. Deleting redis-master-fefd9752aa5883ca4d53013a7b583967
    redis-master
```

至此，可以看到 Pod 恢复到更新前的版本了。

可以看出，RC 的滚动升级不具有 Deployment 在应用版本升级过程中的历史记录、新旧版本数量的精细控制等功能，在 Kubernetes 的演进过程中，RC 将逐渐被 RS 和 Deployment 所取代，建议用户优先考虑使用 Deployment 完成 Pod 的部署和升级操作。

3.11.5 其他管理对象的更新策略

Kubernetes 从 1.6 版本开始，对 DaemonSet 和 StatefulSet 的更新策略也引入类似于 Deployment 的滚动升级，通过不同的策略自动完成应用的版本升级。

1. DaemonSet 的更新策略

目前 DaemonSet 的升级策略包括两种：OnDelete 和 RollingUpdate。

（1）OnDelete：DaemonSet 的默认升级策略，与 1.5 及以前版本的 Kubernetes 保持一致。当使用 OnDelete 作为升级策略时，在创建好新的 DaemonSet 配置之后，新的 Pod 并不会被自动创建，直到用户手动删除旧版本的 Pod，才触发新建操作。

（2）RollingUpdate：从 Kubernetes 1.6 版本开始引入。当使用 RollingUpdate 作为升级策略对 DaemonSet 进行更新时，旧版本的 Pod 将被自动杀掉，然后自动创建新版本的 DaemonSet Pod。整个过程与普通 Deployment 的滚动升级一样是可控的。不过有两点不同于普通 Pod 的滚动升级：一是目前 Kubernetes 还不支持查看和管理 DaemonSet 的更新历史记录；二是 DaemonSet 的回滚（Rollback）并不能如同 Deployment 一样直接通过 kubectl rollback 命令来实现，必须通过再次提交旧版本配置的方式实现。

2. StatefulSet 的更新策略

Kubernetes 从 1.6 版本开始，针对 StatefulSet 的更新策略逐渐向 Deployment 和 DaemonSet 的更新策略看齐，也将实现 RollingUpdate、Paritioned 和 OnDelete 这几种策略，以保证 StatefulSet 中各 Pod 有序地、逐个地更新，并且能够保留更新历史，也能回滚到某个历史版本。

3.12 Pod 的扩缩容

在实际生产系统中，我们经常会遇到某个服务需要扩容的场景，也可能会遇到由于资源紧张或者工作负载降低而需要减少服务实例数量的场景。此时可以利用 Deployment/RC 的 Scale 机制来完成这些工作。

Kubernetes 对 Pod 的扩缩容操作提供了手动和自动两种模式，手动模式通过执行 kubectl scale 命令或通过 RESTful API 对一个 Deployment/RC 进行 Pod 副本数量的设置，即可一键完成。自动模式则需要用户根据某个性能指标或者自定义业务指标，并指定 Pod 副本数量的范围，系统将自动在这个范围内根据性能指标的变化进行调整。

3.12.1 手动扩缩容机制

以 Deployment nginx 为例：

```yaml
nginx-deployment.yaml
apiVersion: apps/v1
kind: Deployment
metadata:
  name: nginx-deployment
spec:
  selector:
    matchLabels:
      app: nginx
  replicas: 3
  selector:
    matchLabels:
      app: nginx
  template:
    metadata:
      labels:
        app: nginx
    spec:
      containers:
      - name: nginx
        image: nginx:1.7.9
        ports:
        - containerPort: 80
```

已运行的 Pod 副本数量为 3 个：

```
$ kubectl get pods
NAME                                    READY   STATUS    RESTARTS   AGE
nginx-deployment-3973253433-scz37       1/1     Running   0          5s
nginx-deployment-3973253433-x8fsq       1/1     Running   0          5s
nginx-deployment-3973253433-x9z8z       1/1     Running   0          5s
```

通过 kubectl scale 命令可以将 Pod 副本数量从初始的 3 个更新为 5 个：

```
$ kubectl scale deployment nginx-deployment --replicas 5
deployment "nginx-deployment" scaled
$ kubectl get pods
NAME                                    READY   STATUS    RESTARTS   AGE
nginx-deployment-3973253433-3gt27       1/1     Running   0          4s
nginx-deployment-3973253433-7jls2       1/1     Running   0          4s
nginx-deployment-3973253433-scz37       1/1     Running   0          4m
nginx-deployment-3973253433-x8fsq       1/1     Running   0          4m
nginx-deployment-3973253433-x9z8z       1/1     Running   0          4m
```

将 --replicas 设置为比当前 Pod 副本数量更小的数字，系统将会"杀掉"一些运行中的 Pod，以实现应用集群缩容：

```
$ kubectl scale deployment nginx-deployment --replicas=1
deployment "nginx-deployment" scaled

$ kubectl get pods
NAME                                    READY   STATUS    RESTARTS   AGE
nginx-deployment-3973253433-x9z8z       1/1     Running   0          6m
```

3.12.2　自动扩缩容机制

Kubernetes 从 1.1 版本开始，新增了名为 Horizontal Pod Autoscaler（HPA）的控制器，用于实现基于 CPU 使用率进行自动 Pod 扩缩容的功能。HPA 控制器基于 Master 的 kube-controller-manager 服务启动参数 --horizontal-pod-autoscaler-sync-period 定义的探测周期（默认值为 15s），周期性地监测目标 Pod 的资源性能指标，并与 HPA 资源对象中的扩缩容条件进行对比，在满足条件时对 Pod 副本数量进行调整。

Kubernetes 在早期版本中，只能基于 Pod 的 CPU 使用率进行自动扩缩容操作，关于 CPU 使用率的数据来源于 Heapster 组件。Kubernetes 从 1.6 版本开始，引入了基于应用自定义性能指标的 HPA 机制，并在 1.9 版本之后逐步成熟。本节对 Kubernetes 的 HPA 的原理和实践进行详细说明。

1. HPA 的工作原理

Kubernetes 中的某个 Metrics Server（Heapster 或自定义 Metrics Server）持续采集所有 Pod 副本的指标数据。HPA 控制器通过 Metrics Server 的 API（Heapster 的 API 或聚合 API）获取这些数据，基于用户定义的扩缩容规则进行计算，得到目标 Pod 副本数量。当目标 Pod 副本数量与当前副本数量不同时，HPA 控制器就向 Pod 的副本控制器（Deployment、RC 或 ReplicaSet）发起 scale 操作，调整 Pod 的副本数量，完成扩缩容操作。图 3.9 描述了 HPA 体系中的关键组件和工作流程。

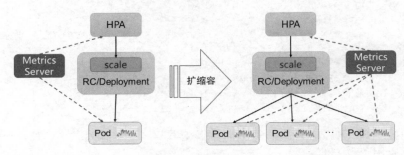

图 3.9　HPA 的工作原理

接下来首先对 HPA 能够管理的指标类型、扩缩容算法、HPA 对象的配置进行详细说明，然后通过一个完整的示例对如何搭建和使用基于自定义指标的 HPA 体系进行说明。

2. 指标的类型

Master 的 kube-controller-manager 服务持续监测目标 Pod 的某种性能指标，以计算是否需要调整副本数量。目前 Kubernetes 支持的指标类型如下。

◎ Pod 资源使用率：Pod 级别的性能指标，通常是一个比率值，例如 CPU 使用率。
◎ Pod 自定义指标：Pod 级别的性能指标，通常是一个数值，例如接收的请求数量。
◎ Object 自定义指标或外部自定义指标：通常是一个数值，需要容器应用以某种方式提供，例如通过 HTTP URL "/metrics" 提供，或者使用外部服务提供的指标采集 URL。

Kubernetes 从 1.11 版本开始，弃用基于 Heapster 组件完成 Pod 的 CPU 使用率采集的机制，全面转向基于 Metrics Server 完成数据采集。Metrics Server 将采集到的 Pod 性能指标数据通过聚合 API（Aggregated API）如 metrics.k8s.io、custom.metrics.k8s.io 和 external.metrics.k8s.io 提供给 HPA 控制器进行查询。关于聚合 API 和 API 聚合器（API

Aggregator）的概念详见 9.4 节的说明。

3. 扩缩容算法详解

Autoscaler 控制器从聚合 API 获取到 Pod 性能指标数据之后，基于下面的算法计算出目标 Pod 副本数量，与当前运行的 Pod 副本数量进行对比，决定是否需要进行扩缩容操作：

```
desiredReplicas = ceil[currentReplicas * ( currentMetricValue / desiredMetricValue )]
```

即当前副本数 ×（当前指标值/期望的指标值），将结果向上取整。

以 CPU 请求数量为例，如果用户设置的期望指标值为 100m，当前实际使用的指标值为 200m，则计算得到期望的 Pod 副本数量应为两个（200/100=2）。如果设置的期望指标值为 50m，计算结果为 0.5，则向上取整值为 1，得到目标 Pod 副本数量应为 1 个。

当计算结果与 1 非常接近时，可以设置一个容忍度让系统不做扩缩容操作。容忍度通过 kube-controller-manager 服务的启动参数 --horizontal-pod-autoscaler-tolerance 进行设置，默认值为 0.1（即 10%），表示基于上述算法得到的结果在[-10%-+10%]区间内，即[0.9-1.1]，控制器都不会进行扩缩容操作。

也可以将期望指标值（desiredMetricValue）设置为指标的平均值类型，例如 targetAverageValue 或 targetAverageUtilization，此时当前指标值（currentMetricValue）的算法为所有 Pod 副本当前指标值的总和除以 Pod 副本数量得到的平均值。

此外，存在几种 Pod 异常的情况，如下所述。

◎ Pod 正在被删除（设置了删除时间戳）：将不会计入目标 Pod 副本数量。
◎ Pod 的当前指标值无法获得：本次探测不会将这个 Pod 纳入目标 Pod 副本数量，后续的探测会被重新纳入计算范围。
◎ 如果指标类型是 CPU 使用率，则对于正在启动但是还未达到 Ready 状态的 Pod，也暂时不会纳入目标副本数量范围。可以通过 kube-controller-manager 服务的启动参数 --horizontal-pod-autoscaler-initial-readiness-delay 设置首次探测 Pod 是否 Ready 的延时时间，默认值为 30s。另一个启动参数 --horizontal-pod-autoscaler-cpu-initialization-period 设置首次采集 Pod 的 CPU 使用率的延时时间。

在计算"当前指标值/期望的指标值"（currentMetricValue / desiredMetricValue）时将不会包括上述这些异常 Pod。

当存在缺失指标的 Pod 时，系统将更保守地重新计算平均值。系统会假设这些 Pod 在需要缩容（Scale Down）时消耗了期望指标值的 100%，在需要扩容（Scale Up）时消耗了期望指标值的 0%，这样可以抑制潜在的扩缩容操作。

此外，如果存在未达到 Ready 状态的 Pod，并且系统原本会在不考虑缺失指标或 NotReady 的 Pod 情况下进行扩展，则系统仍然会保守地假设这些 Pod 消耗期望指标值的 0%，从而进一步抑制扩容操作。

如果在 HorizontalPodAutoscaler 中设置了多个指标，系统就会对每个指标都执行上面的算法，在全部结果中以期望副本数的最大值为最终结果。如果这些指标中的任意一个都无法转换为期望的副本数（例如无法获取指标的值），系统就会跳过扩缩容操作。

最后，在 HPA 控制器执行扩缩容操作之前，系统会记录扩缩容建议信息（Scale Recommendation）。控制器会在操作时间窗口（时间范围可以配置）中考虑所有的建议信息，并从中选择得分最高的建议。这个值可通过 kube-controller-manager 服务的启动参数 --horizontal-pod-autoscaler-downscale-stabilization-window 进行配置，默认值为 5min。这个配置可以让系统更为平滑地进行缩容操作，从而消除短时间内指标值快速波动产生的影响。

4. HorizontalPodAutoscaler 配置详解

Kubernetes 将 HorizontalPodAutoscaler 资源对象提供给用户来定义扩缩容的规则。

HorizontalPodAutoscaler 资源对象处于 Kubernetes 的 API 组 "autoscaling" 中，目前包括 v1 和 v2 两个版本。其中 autoscaling/v1 仅支持基于 CPU 使用率的自动扩缩容，autoscaling/v2 则用于支持基于任意指标的自动扩缩容配置，包括基于资源使用率、Pod 指标、其他指标等类型的指标数据，当前版本为 autoscaling/v2beta2。

下面对 HorizontalPodAutoscaler 的配置和用法进行说明。

（1）基于 autoscaling/v1 版本的 HorizontalPodAutoscaler 配置，仅可以设置 CPU 使用率：

```
apiVersion: autoscaling/v1
kind: HorizontalPodAutoscaler
metadata:
  name: php-apache
spec:
  scaleTargetRef:
```

```
    apiVersion: apps/v1
    kind: Deployment
    name: php-apache
minReplicas: 1
maxReplicas: 10
targetCPUUtilizationPercentage: 50
```

主要参数如下。

- scaleTargetRef：目标作用对象，可以是 Deployment、ReplicationController 或 ReplicaSet。
- targetCPUUtilizationPercentage：期望每个 Pod 的 CPU 使用率都为 50%，该使用率基于 Pod 设置的 CPU Request 值进行计算，例如该值为 200m，那么系统将维持 Pod 的实际 CPU 使用值为 100m。
- minReplicas 和 maxReplicas：Pod 副本数量的最小值和最大值，系统将在这个范围内进行自动扩缩容操作，并维持每个 Pod 的 CPU 使用率为 50%。

为了使用 autoscaling/v1 版本的 HorizontalPodAutoscaler，需要预先安装 Heapster 组件或 Metrics Server，用于采集 Pod 的 CPU 使用率。Heapster 从 Kubernetes 1.11 版本开始进入弃用阶段，本节不再对 Heapster 进行详细说明。关于 Metrics Server 的说明请参考 9.4 节的介绍，本节主要对基于自定义指标进行自动扩缩容的设置进行说明。

（2）基于 autoscaling/v2beta2 的 HorizontalPodAutoscaler 配置：

```
apiVersion: autoscaling/v2beta2
kind: HorizontalPodAutoscaler
metadata:
  name: php-apache
spec:
  scaleTargetRef:
    apiVersion: apps/v1
    kind: Deployment
    name: php-apache
  minReplicas: 1
  maxReplicas: 10
  metrics:
  - type: Resource
    resource:
      name: cpu
      target:
```

```
            type: Utilization
            averageUtilization: 50
```

主要参数如下。

- ◎ scaleTargetRef：目标作用对象，可以是 Deployment、ReplicationController 或 ReplicaSet。
- ◎ minReplicas 和 maxReplicas：Pod 副本数量的最小值和最大值，系统将在这个范围内进行自动扩缩容操作，并维持每个 Pod 的 CPU 使用率为 50%。
- ◎ metrics：目标指标值。在 metrics 中通过参数 type 定义指标的类型；通过参数 target 定义相应的指标目标值，系统将在指标数据达到目标值时（考虑容忍度的区间，见前面算法部分的说明）触发扩缩容操作。

可以将 metrics 中的 type（指标类型）设置为以下三种，可以设置一个或多个组合，如下所述。

（1）Resource：基于资源的指标值，可以设置的资源为 CPU 和内存。

（2）Pods：基于 Pod 的指标，系统将对全部 Pod 副本的指标值进行平均值计算。

（3）Object：基于某种资源对象（如 Ingress）的指标或应用系统的任意自定义指标。

Resource 类型的指标可以设置 CPU 和内存。对于 CPU 使用率，在 target 参数中设置 averageUtilization 定义目标平均 CPU 使用率。对于内存资源，在 target 参数中设置 AverageValue 定义目标平均内存使用值。指标数据可以通过 API "metrics.k8s.io" 进行查询，要求预先启动 Metrics Server 服务。

Pods 类型和 Object 类型都属于自定义指标类型，指标的数据通常需要搭建自定义 Metrics Server 和监控工具进行采集和处理。指标数据可以通过 API "custom.metrics.k8s.io" 进行查询，要求预先启动自定义 Metrics Server 服务。

类型为 Pods 的指标数据来源于 Pod 对象本身，其 target 指标类型只能使用 AverageValue，示例如下：

```
metrics:
- type: Pods
  pods:
    metric:
      name: packets-per-second
    target:
```

```
      type: AverageValue
      averageValue: 1k
```

其中,设置 Pod 的指标名为 packets-per-second,在目标指标平均值为 1000 时触发扩缩容操作。

类型为 Object 的指标数据来源于其他资源对象或任意自定义指标,其 target 指标类型可以使用 Value 或 AverageValue(根据 Pod 副本数计算平均值)进行设置。下面对几种常见的自定义指标给出示例和说明。

例 1,设置指标的名称为 requests-per-second,其值来源于 Ingress "main-route",将目标值(value)设置为 2000,即在 Ingress 的每秒请求数量达到 2000 个时触发扩缩容操作:

```
metrics:
- type: Object
  object:
    metric:
      name: requests-per-second
    describedObject:
      apiVersion: extensions/v1beta1
      kind: Ingress
      name: main-route
    target:
      type: Value
      value: 2k
```

例 2,设置指标的名称为 http_requests,并且该资源对象具有标签"verb=GET",在指标平均值达到 500 时触发扩缩容操作:

```
metrics:
- type: Object
  object:
    metric:
      name: 'http_requests'
      selector: 'verb=GET'
    target:
      type: AverageValue
      averageValue: 500
```

还可以在同一个 HorizontalPodAutoscaler 资源对象中定义多个类型的指标，系统将针对每种类型的指标都计算 Pod 副本的目标数量，以最大值为准进行扩缩容操作。例如：

```yaml
apiVersion: autoscaling/v2beta1
kind: HorizontalPodAutoscaler
metadata:
  name: php-apache
  namespace: default
spec:
  scaleTargetRef:
    apiVersion: apps/v1
    kind: Deployment
    name: php-apache
  minReplicas: 1
  maxReplicas: 10
  metrics:
  - type: Resource
    resource:
      name: cpu
      target:
        type: AverageUtilization
        averageUtilization: 50
  - type: Pods
    pods:
      metric:
        name: packets-per-second
      targetAverageValue: 1k
  - type: Object
    object:
      metric:
        name: requests-per-second
      describedObject:
        apiVersion: extensions/v1beta1
        kind: Ingress
        name: main-route
      target:
        kind: Value
        value: 10k
```

从 1.10 版本开始，Kubernetes 引入了对外部系统指标的支持。例如，用户使用了公有云服务商提供的消息服务或外部负载均衡器，希望基于这些外部服务的性能指标（如消息服务的队列长度、负载均衡器的 QPS）对自己部署在 Kubernetes 中的服务进行自动扩缩容操作。这时，就可以在 metrics 参数部分设置 type 为 External 来设置自定义指标，然后就可以通过 API "external.metrics.k8s.io" 查询指标数据了。当然，这同样要求自定义 Metrics Server 服务已正常工作。

例 3，设置指标的名称为 queue_messages_ready，具有 queue=worker_tasks 标签在目标指标平均值为 30 时触发自动扩缩容操作：

```
- type: External
  external:
    metric:
      name: queue_messages_ready
      selector: "queue=worker_tasks"
    target:
      type: AverageValue
      averageValue: 30
```

在使用外部服务的指标时，要安装、部署能够对接到 Kubernetes HPA 模型的监控系统，并且完全了解监控系统采集这些指标的机制，后续的自动扩缩容操作才能完成。

Kubernetes 推荐尽量使用 type 为 Object 的 HPA 配置方式，这可以通过使用 Operator 模式，将外部指标通过 CRD（自定义资源）定义为 API 资源对象来实现。

5. 基于自定义指标的 HPA 实践

下面通过一个完整的示例，对如何搭建和使用基于自定义指标的 HPA 体系进行说明。

基于自定义指标进行自动扩缩容时，需要预先部署自定义 Metrics Server，目前可以使用基于 Prometheus、Microsoft Azure、Datadog Cluster 等系统的 Adapter 实现自定义 Metrics Server，未来还将提供基于 Google Stackdriver 的实现自定义 Metrics Server。读者可以参考官网 https://github.com/kubernetes/metrics/blob/master/IMPLEMENTATIONS.md#custom-metrics-api 的说明。本节基于 Prometheus 监控系统对 HPA 的基础组件部署和 HPA 配置进行详细说明。

基于 Prometheus 的 HPA 架构如图 3.10 所示。

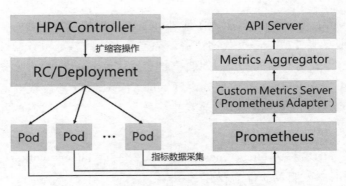

图3.10　基于Prometheus的HPA架构

关键组件包括如下。

◎ Prometheus：定期采集各Pod的性能指标数据。
◎ Custom Metrics Server：自定义Metrics Server，用Prometheus Adapter进行具体实现。它从Prometheus服务采集性能指标数据，通过Kubernetes的Metrics Aggregation层将自定义指标API注册到Master的API Server中，以/apis/custom.metrics.k8s.io路径提供指标数据。
◎ HPA Controller：Kubernetes的HPA控制器，基于用户定义的HorizontalPodAutoscaler进行自动扩缩容操作。

接下来对整个系统的部署过程进行说明。

（1）在Master的API Server启动Aggregation层，通过设置kube-apiserver服务的下列启动参数进行开启。

◎ --requestheader-client-ca-file=/etc/kubernetes/ssl_keys/ca.crt：客户端CA证书。
◎ --requestheader-allowed-names=：允许访问的客户端common names列表，通过header中由--requestheader-username-headers参数指定的字段获取。客户端common names的名称需要在client-ca-file中进行配置，将其设置为空值时，表示任意客户端都可以访问。
◎ --requestheader-extra-headers-prefix=X-Remote-Extra-：请求头中需要检查的前缀名。
◎ --requestheader-group-headers=X-Remote-Group：请求头中需要检查的组名。
◎ --requestheader-username-headers=X-Remote-User：请求头中需要检查的用户名。
◎ --proxy-client-cert-file=/etc/kubernetes/ssl_keys/kubelet_client.crt：在请求期间验证

Aggregator 的客户端 CA 证书。
◎ --proxy-client-key-file=/etc/kubernetes/ssl_keys/kubelet_client.key：在请求期间验证 Aggregator 的客户端私钥。

配置 kube-controller-manager 服务中 HPA 的相关启动参数（可选配置）如下。

◎ --horizontal-pod-autoscaler-sync-period=10s：HPA 控制器同步 Pod 副本数量的时间间隔，默认值为 15s。
◎ --horizontal-pod-autoscaler-downscale-stabilization=1m0s：执行缩容操作的等待时长，默认值为 5min。
◎ --horizontal-pod-autoscaler-initial-readiness-delay=30s：等待 Pod 达到 Ready 状态的时延，默认值为 30min。
◎ --horizontal-pod-autoscaler-tolerance=0.1：扩缩容计算结果的容忍度，默认值为 0.1，表示[-10%-+10%]。

（2）部署 Prometheus，这里使用 Operator 模式进行部署。

首先，使用下面的 YAML 配置文件部署 prometheus-operator：

```
apiVersion: apps/v1
kind: Deployment
metadata:
  labels:
    k8s-app: prometheus-operator
  name: prometheus-operator
spec:
  replicas: 1
  selector:
    matchLabels:
      k8s-app: prometheus-operator
  template:
    metadata:
      labels:
        k8s-app: prometheus-operator
    spec:
      containers:
      - image: quay.io/coreos/prometheus-operator:v0.17.0
        imagePullPolicy: IfNotPresent
        name: prometheus-operator
        ports:
```

```
        - containerPort: 8080
          name: http
        resources:
          limits:
            cpu: 200m
            memory: 100Mi
          requests:
            cpu: 100m
            memory: 50Mi
```

这个 prometheus-operator 会自动创建名为 monitoring.coreos.com 的 CRD 资源。

然后，通过 Operator 的配置部署 Prometheus 服务：

```
---
apiVersion: monitoring.coreos.com/v1
kind: Prometheus
metadata:
  name: prometheus
  labels:
    app: prometheus
    prometheus: prometheus
spec:
  replicas: 1
  baseImage: prom/prometheus
  version: v2.8.0
  serviceMonitorSelector:
    matchLabels:
      service-monitor: function
  resources:
    requests:
      memory: 300Mi

---
apiVersion: v1
kind: Service
metadata:
  name: prometheus
  labels:
    app: prometheus
    prometheus: prometheus
spec:
```

```
  selector:
    prometheus: prometheus
  ports:
  - name: http
    port: 9090
```

确认 Prometheus Operator 和 Prometheus 服务正常运行:

```
# kubectl get pods
NAME                                       READY   STATUS    RESTARTS   AGE
prometheus-operator-7c976597bc-xzdf5       1/1     Running   0          51m
prometheus-prometheus-0                    2/2     Running   0          42m
```

（3）部署自定义 Metrics Server，这里以 Prometheus Adapter 的实现进行部署。下面的 YAML 配置文件主要包含 Namespace、ConfigMap、Deployment、Service 和自定义 API 资源 custom.metrics.k8s.io/v1beta1，这里将它们部署在一个新的 Namespace "custom-metrics" 中。

```
---
kind: Namespace
apiVersion: v1
metadata:
  name: custom-metrics

---
apiVersion: v1
kind: ConfigMap
metadata:
  name: adapter-config
  namespace: custom-metrics
data:
  config.yaml: |
    rules:
    - seriesQuery:
'{__name__=~"^container_.*",container_name!="POD",namespace!="",pod_name!=""}'
      seriesFilters: []
      resources:
        overrides:
          namespace:
            resource: namespace
          pod_name:
            resource: pod
```

```yaml
        name:
          matches: ^container_(.*)_seconds_total$
          as: ""
        metricsQuery: sum(rate(<<.Series>>{<<.LabelMatchers>>,container_name!="POD"}[1m])) by (<<.GroupBy>>)

      - seriesQuery: '{__name__=~"^container_.*",container_name!="POD",namespace!="",pod_name!=""}'
        seriesFilters:
          - isNot: ^container_.*_seconds_total$
        resources:
          overrides:
            namespace:
              resource: namespace
            pod_name:
              resource: pod
        name:
          matches: ^container_(.*)_total$
          as: ""
        metricsQuery: sum(rate(<<.Series>>{<<.LabelMatchers>>,container_name!="POD"}[1m])) by (<<.GroupBy>>)

      - seriesQuery: '{__name__=~"^container_.*",container_name!="POD",namespace!="",pod_name!=""}'
        seriesFilters:
          - isNot: ^container_.*_total$
        resources:
          overrides:
            namespace:
              resource: namespace
            pod_name:
              resource: pod
        name:
          matches: ^container_(.*)$
          as: ""
        metricsQuery: sum(<<.Series>>{<<.LabelMatchers>>,container_name!="POD"}) by (<<.GroupBy>>)

      - seriesQuery: '{namespace!="",__name__!~"^container_.*"}'
        seriesFilters:
          - isNot: .*_total$
        resources:
```

```
        template: <<.Resource>>
      name:
        matches: ""
        as: ""
      metricsQuery: sum(<<.Series>>{<<.LabelMatchers>>}) by (<<.GroupBy>>)
    - seriesQuery: '{namespace!="",__name__!~"^container_.*"}'
      seriesFilters:
      - isNot: .*_seconds_total
      resources:
        template: <<.Resource>>
      name:
        matches: ^(.*)_total$
        as: ""
      metricsQuery: sum(rate(<<.Series>>{<<.LabelMatchers>>}[1m])) by (<<.GroupBy>>)
    - seriesQuery: '{namespace!="",__name__!~"^container_.*"}'
      seriesFilters: []
      resources:
        template: <<.Resource>>
      name:
        matches: ^(.*)_seconds_total$
        as: ""
      metricsQuery: sum(rate(<<.Series>>{<<.LabelMatchers>>}[1m])) by (<<.GroupBy>>)
  resourceRules:
    cpu:
      containerQuery: sum(rate(container_cpu_usage_seconds_total{<<.LabelMatchers>>}[1m])) by (<<.GroupBy>>)
      nodeQuery: sum(rate(container_cpu_usage_seconds_total{<<.LabelMatchers>>, id='/'}[1m])) by (<<.GroupBy>>)
      resources:
        overrides:
          instance:
            resource: node
          namespace:
            resource: namespace
          pod_name:
            resource: pod
      containerLabel: container_name
```

```yaml
        memory:
          containerQuery: sum(container_memory_working_set_bytes{<<.LabelMatchers>>}) by (<<.GroupBy>>)
          nodeQuery: sum(container_memory_working_set_bytes{<<.LabelMatchers>>,id='/'}) by (<<.GroupBy>>)
          resources:
            overrides:
              instance:
                resource: node
              namespace:
                resource: namespace
              pod_name:
                resource: pod
          containerLabel: container_name
        window: 1m
# 以上配置为针对应用自定义指标的计算逻辑

---
apiVersion: apps/v1
kind: Deployment
metadata:
  name: custom-metrics-server
  namespace: custom-metrics
  labels:
    app: custom-metrics-server
spec:
  replicas: 1
  selector:
    matchLabels:
      app: custom-metrics-server
  template:
    metadata:
      name: custom-metrics-server
      labels:
        app: custom-metrics-server
    spec:
      containers:
      - name: custom-metrics-server
        image: directxman12/k8s-prometheus-adapter-amd64
        imagePullPolicy: IfNotPresent
```

```
        args:
        - --prometheus-url=http://prometheus.default.svc:9090/
        - --metrics-relist-interval=30s
        - --v=10
        - --config=/etc/adapter/config.yaml
        - --logtostderr=true
        ports:
        - containerPort: 443
        securityContext:
          runAsUser: 0
        volumeMounts:
        - mountPath: /etc/adapter/
          name: config
          readOnly: true
      volumes:
      - name: config
        configMap:
          name: adapter-config
```

参数--prometheus-url 设置之前创建的 Prometheus 服务在 Kubernetes 中的 DNS 域名格式地址，例如 prometheus.default.svc
参数--metrics-relist-interval 设置从更新指标缓存的频率，应将其设置为大于或等于 Prometheus 的指标采集频率

```
---
apiVersion: v1
kind: Service
metadata:
  name: custom-metrics-server
  namespace: custom-metrics
spec:
  ports:
  - port: 443
    targetPort: 443
  selector:
    app: custom-metrics-server

---
apiVersion: apiregistration.k8s.io/v1beta1
kind: APIService
metadata:
```

```yaml
  name: v1beta1.custom.metrics.k8s.io
spec:
  service:
    name: custom-metrics-server
    namespace: custom-metrics
  group: custom.metrics.k8s.io
  version: v1beta1
  insecureSkipTLSVerify: true
  groupPriorityMinimum: 100
  versionPriority: 100
```

确认 custom-metrics-server 正常运行：

```
# kubectl -n custom-metrics get pods
NAME                                        READY   STATUS    RESTARTS   AGE
custom-metrics-server-594dd7c4db-z622f      1/1     Running   0          1m
```

（4）部署应用程序，它会在 HTTP URL "/metrics" 路径提供名为 http_requests_total 的指标值：

```yaml
---
apiVersion: apps/v1
kind: Deployment
metadata:
  name: sample-app
  labels:
    app: sample-app
spec:
  replicas: 1
  selector:
    matchLabels:
      app: sample-app
  template:
    metadata:
      labels:
        app: sample-app
    spec:
      containers:
      - image: luxas/autoscale-demo:v0.1.2
        imagePullPolicy: IfNotPresent
        name: metrics-provider
        ports:
```

```yaml
        - name: http
          containerPort: 8080

---
apiVersion: v1
kind: Service
metadata:
  name: sample-app
  labels:
    app: sample-app
spec:
  ports:
  - name: http
    port: 80
    targetPort: 8080
  selector:
    app: sample-app
```

部署成功之后,可以在应用的 URL "/metrics" 中查看指标 http_requests_total 的值:

```
# kubectl get service sample-app
NAME         TYPE        CLUSTER-IP        EXTERNAL-IP     PORT(S)     AGE
sample-app   ClusterIP   169.169.43.252    <none>          80/TCP      86m

# curl 169.169.43.252/metrics
# HELP http_requests_total The amount of requests served by the server in total
# TYPE http_requests_total counter
http_requests_total 1
```

(5) 创建一个 Prometheus 的 ServiceMonitor 对象,用于监控应用程序提供的指标:

```yaml
apiVersion: monitoring.coreos.com/v1
kind: ServiceMonitor
metadata:
  name: sample-app
  labels:
    service-monitor: function
spec:
  selector:
    matchLabels:
      app: sample-app
  endpoints:
  - port: http
```

关键配置参数如下。

- Selector：设置为 Pod 的 Label "app: sample-app"。
- Endpoints：设置为在 Service 中定义的端口名称 "http"。

（6）创建一个 HorizontalPodAutoscaler 对象，用于为 HPA 控制器提供用户期望的自动扩缩容配置。

```
apiVersion: autoscaling/v2beta2
kind: HorizontalPodAutoscaler
metadata:
  name: sample-app
spec:
  scaleTargetRef:
    apiVersion: apps/v1
    kind: Deployment
    name: sample-app
  minReplicas: 1
  maxReplicas: 10
  metrics:
  - type: Pods
    pods:
      metric:
        name: http_requests
      target:
        type: AverageValue
        averageValue: 500m
```

关键配置参数如下。

- scaleTargetRef：设置 HPA 的作用对象为之前部署的 Deployment "sample-app"。
- type=Pods：设置指标类型为 Pods，表示从 Pod 获取指标数据。
- metric.name=http_requests：将指标的名称设置为 "http_requests"，是自定义 Metrics Server 将应用程序提供的指标 "http_requests_total" 经过计算转换成的一个新比率值，即 sum(rate(http_requests_total{namespace="xx",pod="xx"}[1m])) by pod，指过去 1min 内全部 Pod 指标 http_requests_total 总和的每秒平均值。
- target：将指标 http_requests 的目标值设置为 500m，类型为 AverageValue，表示基于全部 Pod 副本数据计算平均值。目标 Pod 副本数量将使用公式 "http_requests 当前值/500m" 进行计算。

◎ minReplicas 和 maxReplicas：将扩缩容区间设置为 1～10（单位是 Pod 副本）。

此时可以通过查看自定义 Metrics Server 提供的 URL "custom.metrics.k8s.io/v1beta1" 查看 Pod 的指标是否已经被成功采集，并能够通过聚合 API 进行查询：

```
# kubectl get --raw "/apis/custom.metrics.k8s.io/v1beta1/namespaces/default/pods/*/http_requests?selector=app%3Dsample-app"
{"kind":"MetricValueList","apiVersion":"custom.metrics.k8s.io/v1beta1","metadata":{"selfLink":"/apis/custom.metrics.k8s.io/v1beta1/namespaces/default/pods/%2A/http_requests"},"items":[{"describedObject":{"kind":"Pod","namespace":"default","name":"sample-app-579f977995-jz98h","apiVersion":"/v1"},"metricName":"http_requests","timestamp":"2019-03-16T17:54:38Z","value":"33m"}]}
```

从结果中看到正确的 value 值，说明自定义 Metrics Server 工作正常。

查看 HorizontalPodAutoscaler 的详细信息，可以看到其成功从自定义 Metrics Server 处获取了应用的指标数据，可以进行扩缩容操作：

```
# kubectl describe hpa.v2beta2.autoscaling sample-app
Name:                                                   sample-app
Namespace:                                              default
Labels:                                                 <none>
Annotations:                                            <none>
CreationTimestamp:                                      Sun, 17 Mar 2019 01:05:33 +0800
Reference:                                              Deployment/sample-app
Metrics:                                                ( current / target )
  "http_requests" on pods:    33m / 500m
Min replicas:                 1
Max replicas:                 10
Deployment pods:              1 current / 1 desired
Conditions:
  Type            Status  Reason              Message
  ----            ------  ------              -------
  AbleToScale     True    ReadyForNewScale    recommended size matches current size
  ScalingActive   True    ValidMetricFound    the HPA was able to successfully calculate a replica count from pods metric http_requests
  ScalingLimited  False   DesiredWithinRange  the desired count is within the acceptable range
```

（7）对应用的服务地址发起 HTTP 访问请求，验证 HPA 自动扩容机制。例如，可以使用如下脚本对应用进行压力测试：

```
# for i in {1..100000}; do wget -q -O- 169.169.43.252 > /dev/null; done
```

一段时间之后，观察 HorizontalPodAutoscaler 和 Pod 数量的变化，可以看到自动扩容的过程：

```
# kubectl describe hpa.v2beta2.autoscaling sample-app
Name:                                        sample-app
Namespace:                                   default
Labels:                                      <none>
Annotations:                                 <none>
CreationTimestamp:                           Sun, 17 Mar 2019 02:01:30 +0800
Reference:                                   Deployment/sample-app
Metrics:                                     ( current / target )
  "http_requests" on pods:                   4296m / 500m
Min replicas:                                1
Max replicas:                                10
Deployment pods:                             10 current / 10 desired
Conditions:
  Type              Status  Reason
  ----              ------  ------
  AbleToScale       True    ScaleDownStabilized  recent recommendations were higher than current one, applying the highest recent recommendation
  ScalingActive     True    ValidMetricFound     the HPA was able to successfully calculate a replica count from pods metric http_requests
  ScalingLimited    True    TooManyReplicas      the desired replica count is more than the maximum replica count
Events:
  Type     Reason             Age   From                       Message
  ----     ------             ----  ----                       -------
  Normal   SuccessfulRescale  67s   horizontal-pod-autoscaler  New size: 4; reason: pods metric http_requests above target
  Normal   SuccessfulRescale  56s   horizontal-pod-autoscaler  New size: 8; reason: pods metric http_requests above target
  Normal   SuccessfulRescale  45s   horizontal-pod-autoscaler  New size: 10; reason: pods metric http_requests above target
```

发现 Pod 数量扩容到了 10 个（被 maxReplicas 参数限制的最大值）：

```
# kubectl get pods -l app=sample-app
NAME                            READY   STATUS    RESTARTS   AGE
sample-app-579f977995-dtgcw     1/1     Running   0          32s
```

```
sample-app-579f977995-hn5bd        1/1    Running    0    70s
sample-app-579f977995-jz98h        1/1    Running    0    75s
sample-app-579f977995-kllhq        1/1    Running    0    90s
sample-app-579f977995-p5d44        1/1    Running    0    85s
sample-app-579f977995-q6rxb        1/1    Running    0    70s
sample-app-579f977995-rhn5d        1/1    Running    0    70s
sample-app-579f977995-tjc8q        1/1    Running    0    86s
sample-app-579f977995-tzthf        1/1    Running    0    70s
sample-app-579f977995-wswcx        1/1    Running    0    32s
```

停止访问应用服务，等待一段时间后，观察 HorizontalPodAutoscaler 和 Pod 数量的变化，可以看到缩容操作：

```
# kubectl describe hpa.v2beta2.autoscaling sample-app
Name:                    sample-app
Namespace:               default
Labels:                  <none>
Annotations:             <none>
CreationTimestamp:       Sun, 17 Mar 2019 02:01:30 +0800
Reference:               Deployment/sample-app
Metrics:                 ( current / target )
  "http_requests" on pods:  33m / 500m
Min replicas:            1
Max replicas:            10
Deployment pods:         1 current / 1 desired
Conditions:
  Type            Status  Reason              Message
  ----            ------  ------              -------
  AbleToScale     True    ReadyForNewScale    recommended size matches current size
  ScalingActive   True    ValidMetricFound    the HPA was able to successfully calculate a replica count from pods metric http_requests
  ScalingLimited  False   DesiredWithinRange  the desired count is within the acceptable range
Events:
  Type    Reason             Age     From                       Message
  ----    ------             ----    ----                       -------
  Normal  SuccessfulRescale  6m48s   horizontal-pod-autoscaler  New size: 4; reason: pods metric http_requests above target
  Normal  SuccessfulRescale  6m37s   horizontal-pod-autoscaler  New size: 8; reason: pods metric http_requests above target
  Normal  SuccessfulRescale  6m26s   horizontal-pod-autoscaler  New size: 10;
```

```
reason: pods metric http_requests above target
         Normal   SuccessfulRescale   47s     horizontal-pod-autoscaler   New size: 1;
reason: All metrics below target
```

发现 Pod 的数量已经缩容到最小值 1 个：

```
# kubectl get pods -l app=sample-app
NAME                                READY   STATUS    RESTARTS   AGE
sample-app-579f977995-dtgcw         1/1     Running   0          10m
```

3.13 使用 StatefulSet 搭建 MongoDB 集群

本节以 MongoDB 为例，使用 StatefulSet 完成 MongoDB 集群的创建，为每个 MongoDB 实例在共享存储中（这里采用 GlusterFS）都申请一片存储空间，以实现一个无单点故障、高可用、可动态扩展的 MongoDB 集群。部署架构如图 3.11 所示。

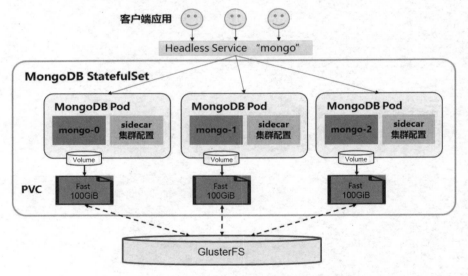

图 3.11 StatefulSet 部署 MongoDB 集群的架构

3.13.1 前提条件

在创建 StatefulSet 之前，需要确保在 Kubernetes 集群中管理员已经创建好共享存储，并能够与 StorageClass 对接，以实现动态存储供应的模式。本节的示例将使用 GlusterFS

作为共享存储（GlusterFS 的部署方法参见 8.6 节的说明）。

3.13.2 创建 StatefulSet

为了完成 MongoDB 集群的搭建，需要创建如下三个资源对象。

◎ 一个 StorageClass，用于 StatefulSet 自动为各个应用 Pod 申请 PVC。
◎ 一个 Headless Service，用于维护 MongoDB 集群的状态。
◎ 一个 StatefulSet。

首先，创建一个 StorageClass 对象。

storageclass-fast.yaml 文件的内容如下：

```
apiVersion: storage.k8s.io/v1
kind: StorageClass
metadata:
  name: fast
provisioner: kubernetes.io/glusterfs
parameters:
  resturl: "http://<heketi-rest-url>"
```

执行 kubectl create 命令创建该 StorageClass：

```
# kubectl create -f storageclass-fast.yaml
storageclass "fast" created
```

接下来，创建对应的 Headless Service。

mongo-sidecar 作为 MongoDB 集群的管理者，将使用此 Headless Service 来维护各个 MongoDB 实例之间的集群关系，以及集群规模变化时的自动更新。

mongo-headless-service.yaml 文件的内容如下：

```
apiVersion: v1
kind: Service
metadata:
  name: mongo
  labels:
    name: mongo
spec:
  ports:
```

```
    - port: 27017
      targetPort: 27017
  clusterIP: None
  selector:
    role: mongo
```

使用 kubectl create 命令创建该 Headless Service：

```
# kubectl create -f mongo-headless-service.yaml
service "mongo" created
```

最后，创建 MongoDB StatefulSet。

statefulset-mongo.yaml 文件的内容如下：

```
apiVersion: apps/v1
kind: StatefulSet
metadata:
  name: mongo
spec:
  serviceName: "mongo"
  replicas: 3
  template:
    metadata:
      labels:
        role: mongo
        environment: test
    spec:
      terminationGracePeriodSeconds: 10
      containers:
      - name: mongo
        image: mongo
        command:
        - mongod
        - "--replSet"
        - rs0
        - "--smallfiles"
        - "--noprealloc"
        ports:
        - containerPort: 27017
        volumeMounts:
        - name: mongo-persistent-storage
          mountPath: /data/db
```

```yaml
      - name: mongo-sidecar
        image: cvallance/mongo-k8s-sidecar
        env:
        - name: MONGO_SIDECAR_POD_LABELS
          value: "role=mongo,environment=test"
        - name: KUBERNETES_MONGO_SERVICE_NAME
          value: "mongo"
  volumeClaimTemplates:
  - metadata:
      name: mongo-persistent-storage
      annotations:
        volume.beta.kubernetes.io/storage-class: "fast"
    spec:
      accessModes: [ "ReadWriteOnce" ]
      resources:
        requests:
          storage: 100Gi
```

其中的主要配置说明如下。

（1）在该 StatefulSet 的定义中包括两个容器：mongo 和 mongo-sidecar。mongo 是主服务程序，mongo-sidecar 是将多个 mongo 实例进行集群设置的工具。mongo-sidecar 中的环境变量如下。

- MONGO_SIDECAR_POD_LABELS：设置为 mongo 容器的标签，用于 sidecar 查询它所要管理的 MongoDB 集群实例。
- KUBERNETES_MONGO_SERVICE_NAME：它的值为 mongo，表示 sidecar 将使用 mongo 这个服务名来完成 MongoDB 集群的设置。

（2）replicas=3 表示这个 MongoDB 集群由 3 个 mongo 实例组成。

（3）volumeClaimTemplates 是 StatefulSet 最重要的存储设置。在 annotations 段设置 volume.beta.kubernetes.io/storage-class="fast"表示使用名为 fast 的 StorageClass 自动为每个 mongo Pod 实例分配后端存储。resources.requests.storage=100Gi 表示为每个 mongo 实例都分配 100GiB 的磁盘空间。

使用 kubectl create 命令创建这个 StatefulSet：

```
# kubectl create -f statefulset-mongo.yaml
statefulset "mongo" created
```

最终可以看到 StatefulSet 依次创建并启动了 3 个 mongo Pod 实例，它们的名字依次为 mongo-0、mongo-1、mongo-2：

```
# kubectl get pods -l role=mongo
NAME       READY   STATUS    RESTARTS   AGE
mongo-0    2/2     Running   0          4m
mongo-1    2/2     Running   0          3m
mongo-2    2/2     Running   0          2m
```

StatefulSet 会用 volumeClaimTemplates 中的定义为每个 Pod 副本都创建一个 PVC 实例，每个 PVC 的名称由 StatefulSet 定义中 volumeClaimTemplates 的名称和 Pod 副本的名称组合而成，查看系统中的 PVC，可以验证这一点：

```
# kubectl get pvc
NAME                                                STATUS   VOLUME                                     CAPACITY   ACCESSMODES   STORAGECLASS   AGE
mongo-persistent-storage-mongo-0                    Bound    pvc-7d963fef-42b3-11e7-b4ca-000c291bc5fc   100Gi      RWO           fast           4m
mongo-persistent-storage-mongo-1                    Bound    pvc-8953f856-42b3-11e7-b4ca-000c291bc5fc   100Gi      RWO           fast           3m
mongo-persistent-storage-mongo-2                    Bound    pvc-a0fdc059-42b3-11e7-b4ca-000c291bc5fc   100Gi      RWO           fast           3m
```

下面是 mongo-0 这个 Pod 中的 Volume 设置，可以看到系统自动为其挂载了对应的 PVC：

```
# kubectl get pod mongo-0 -o yaml
apiVersion: v1
kind: Pod
metadata:
  name: mongo-0
......
  volumes:
  - name: mongo-persistent-storage
    persistentVolumeClaim:
      claimName: mongo-persistent-storage-mongo-0
......
```

至此，一个由 3 个实例组成的 MongoDB 集群就创建完成了，其中每个实例都拥有稳定的名称和独立的存储空间。

3.13.3 查看 MongoDB 集群的状态

登录任意一个 mongo Pod，在 mongo 命令行界面用 rs.status()命令查看 MongoDB 集群的状态，可以看到 mongo 集群已通过 sidecar 完成了创建。在集群中包含 3 个节点，每个节点的名称都是 StatefulSet 设置的 DNS 域名格式的网络标识名称：

◎ mongo-0.mongo.default.svc.cluster.local
◎ mongo-1.mongo.default.svc.cluster.local
◎ mongo-2.mongo.default.svc.cluster.local

同时，可以看到 3 个 mongo 实例各自的角色（PRIMARY 或 SECONDARY）也都进行了正确的设置：

```
# kubectl exec -ti mongo-0 -- mongo
MongoDB shell version v3.4.4
connecting to: mongodb://127.0.0.1:27017
MongoDB server version: 3.4.4
Welcome to the MongoDB shell.
......
rs0:PRIMARY>
rs0:PRIMARY> rs.status()
{
        "set" : "rs0",
        "date" : ISODate("2017-05-27T08:13:07.598Z"),
        "myState" : 2,
        "term" : NumberLong(1),
        "syncingTo" : "mongo-0.mongo.default.svc.cluster.local:27017",
        "heartbeatIntervalMillis" : NumberLong(2000),
        "optimes" : {
                "lastCommittedOpTime" : {
                        "ts" : Timestamp(1495872747, 1),
                        "t" : NumberLong(1)
                },
                "appliedOpTime" : {
                        "ts" : Timestamp(1495872747, 1),
                        "t" : NumberLong(1)
                },
                "durableOpTime" : {
                        "ts" : Timestamp(1495872747, 1),
                        "t" : NumberLong(1)
```

```
                    }
            },
            "members" : [
                    {
                            "_id" : 0,
                            "name" : "mongo-0.mongo.default.svc.cluster.local:27017",
                            "health" : 1,
                            "state" : 1,
                            "stateStr" : "PRIMARY",
                            "uptime" : 260,
                            "optime" : {
                                    "ts" : Timestamp(1495872747, 1),
                                    "t" : NumberLong(1)
                            },
                            "optimeDurable" : {
                                    "ts" : Timestamp(1495872747, 1),
                                    "t" : NumberLong(1)
                            },
                            "optimeDate" : ISODate("2017-05-27T08:12:27Z"),
                            "optimeDurableDate" : ISODate("2017-05-27T08:12:27Z"),
                            "lastHeartbeat" : ISODate("2017-05-27T08:13:05.777Z"),
                            "lastHeartbeatRecv" : ISODate("2017-05-27T08:13:05.776Z"),
                            "pingMs" : NumberLong(0),
                            "electionTime" : Timestamp(1495872445, 1),
                            "electionDate" : ISODate("2017-05-27T08:07:25Z"),
                            "configVersion" : 9
                    },
                    {
                            "_id" : 1,
                            "name" : "mongo-1.mongo.default.svc.cluster.local:27017",
                            "health" : 1,
                            "state" : 2,
                            "stateStr" : "SECONDARY",
                            "uptime" : 291,
                            "optime" : {
                                    "ts" : Timestamp(1495872747, 1),
                                    "t" : NumberLong(1)
                            },
                            "optimeDate" : ISODate("2017-05-27T08:12:27Z"),
                            "syncingTo" : "mongo-0.mongo.default.svc.cluster.local:
```

```
27017",
                    "configVersion" : 9,
                    "self" : true
            },
            {
                    "_id" : 2,
                    "name" : "mongo-2.mongo.default.svc.cluster.local:27017",
                    "health" : 1,
                    "state" : 2,
                    "stateStr" : "SECONDARY",
                    "uptime" : 164,
                    "optime" : {
                            "ts" : Timestamp(1495872747, 1),
                            "t" : NumberLong(1)
                    },
                    "optimeDurable" : {
                            "ts" : Timestamp(1495872747, 1),
                            "t" : NumberLong(1)
                    },
                    "optimeDate" : ISODate("2017-05-27T08:12:27Z"),
                    "optimeDurableDate" : ISODate("2017-05-27T08:12:27Z"),
                    "lastHeartbeat" : ISODate("2017-05-27T08:13:06.369Z"),
                    "lastHeartbeatRecv" : ISODate("2017-05-27T08:13:06.
635Z"),
                    "pingMs" : NumberLong(0),
                    "syncingTo" :
"mongo-0.mongo.default.svc.cluster.local:27017",
                    "configVersion" : 9
            }
    ],
    "ok" : 1
}
```

对于需要访问这个 mongo 集群的 Kubernetes 集群内部客户端来说，可以通过 Headless Service "mongo" 获取后端的所有 Endpoints 列表，并组合为数据库链接串，例如 "mongodb:// mongo-0.mongo, mongo-1.mongo, mongo-2.mongo:27017/dbname_?"。

3.13.4　StatefulSet 的常见应用场景

下面对 MongoDB 集群常见的两种场景进行操作，说明 StatefulSet 对有状态应用的自动化管理功能。

1. MongoDB 集群的扩容

假设在系统运行过程中，3 个 mongo 实例不足以满足业务的要求，这时就需要对 mongo 集群进行扩容。仅需要通过对 StatefulSet 进行 scale 操作，就能实现在 mongo 集群中自动添加新的 mongo 节点。

使用 kubectl scale 命令将 StatefulSet 设置为 4 个实例：

```
# kubectl scale --replicas=4 statefulset mongo
statefulset "mongo" scaled
```

等待一会儿，看到第 4 个实例 "mongo-3" 创建成功：

```
# kubectl get po -l role=mongo
NAME      READY  STATUS    RESTARTS  AGE
mongo-0   2/2    Running   0         1h
mongo-1   2/2    Running   0         2h
mongo-2   2/2    Running   0         2h
mongo-3   2/2    Running   0         1m
```

进入某个实例查看 mongo 集群的状态，可以看到第 4 个节点已经加入：

```
# kubectl exec -ti mongo-0 -- mongo
MongoDB shell version v3.4.4
connecting to: mongodb://127.0.0.1:27017
MongoDB server version: 3.4.4
Welcome to the MongoDB shell.
......
rs0:PRIMARY>
rs0:PRIMARY> rs.status()
{
......
        "members" : [
                {
                        "_id" : 0,
                        "name" : "mongo-0.mongo.default.svc.cluster.local:27017",
                        "health" : 1,
                        "state" : 1,
                        "stateStr" : "PRIMARY",
......
                {
```

```
                "_id" : 4,
                "name" : "mongo-3.mongo.default.svc.cluster.local:27017",
                "health" : 1,
                "state" : 2,
                "stateStr" : "SECONDARY",
                "uptime" : 102,
                "optime" : {
                        "ts" : Timestamp(1495880578, 1),
                        "t" : NumberLong(4)
                },
                "optimeDurable" : {
                        "ts" : Timestamp(1495880578, 1),
                        "t" : NumberLong(4)
                },
                "optimeDate" : ISODate("2017-05-27T10:22:58Z"),
                "optimeDurableDate" : ISODate("2017-05-27T10:22:58Z"),
                "lastHeartbeat" : ISODate("2017-05-27T10:23:00.049Z"),
                "lastHeartbeatRecv" :
ISODate("2017-05-27T10:23:00.049Z"),
                "pingMs" : NumberLong(0),
                "syncingTo" :
"mongo-1.mongo.default.svc.cluster.local:27017",
                "configVersion" : 100097
        }
    ],
    "ok" : 1
}
```

同时，系统也为 mongo-3 分配了一个新的 PVC 用于保存数据，此处不再赘述，有兴趣的读者可自行查看系统为 mongo-3 绑定的 Volume 设置和后端 GlusterFS 共享存储的资源分配情况。

2．自动故障恢复（MongoDB 集群的高可用）

假设在系统运行过程中，某个 mongo 实例或其所在主机发生故障，则 StatefulSet 将会自动重建该 mongo 实例，并保证其身份（ID）和使用的数据（PVC）不变。

以 mongo-0 实例发生故障为例，StatefulSet 将会自动重建 mongo-0 实例，并为其挂载之前分配的 PVC "mongo-persistent-storage-mongo-0"。服务 "mongo-0" 在重新启动后，原数据库中的数据不会丢失，可继续使用。

```
# kubectl get po -l role=mongo
NAME       READY   STATUS              RESTARTS   AGE
mongo-0    0/2     ContainerCreating   0          2h
mongo-1    2/2     Running             0          2h
mongo-2    2/2     Running             0          3s

# kubectl get pod mongo-0 -o yaml
apiVersion: v1
kind: Pod
metadata:
  name: mongo-0
......
  volumes:
  - name: mongo-persistent-storage
    persistentVolumeClaim:
      claimName: mongo-persistent-storage-mongo-0
......
```

进入某个实例查看 mongo 集群的状态，mongo-0 发生故障前在集群中的角色为 PRIMARY，在其脱离集群后，mongo 集群会自动选出一个 SECONDARY 节点提升为 PRIMARY 节点（本例中为 mongo-2）。重启后的 mongo-0 则会成为一个新的 SECONDARY 节点：

```
# kubectl exec -ti mongo-0 -- mongo
......
rs0:PRIMARY> rs.status()
{
......
        "members" : [
                {
                        "_id" : 1,
                        "name" : "mongo-1.mongo.default.svc.cluster.local:27017",
                        "health" : 1,
                        "state" : 2,
                        "stateStr" : "SECONDARY",
......
                {
                        "_id" : 2,
                        "name" : "mongo-2.mongo.default.svc.cluster.local:27017",
                        "health" : 1,
                        "state" : 1,
```

```
                "stateStr" : "PRIMARY",
                "uptime" : 6871,
......
        {
                "_id" : 3,
                "name" : "mongo-0.mongo.default.svc.cluster.local:27017",
                "health" : 1,
                "state" : 2,
                "stateStr" : "SECONDARY",
                "uptime" : 6806,
......
```

从上面的例子中可以看出，Kubernetes 使用 StatefulSet 来搭建有状态的应用集群（MongoDB、MySQL 等），同部署无状态的应用一样简便。Kubernetes 能够保证 StatefulSet 中各应用实例在创建和运行的过程中，都具有固定的身份标识和独立的后端存储；还支持在运行时对集群规模进行扩容、保障集群的高可用等非常重要的功能。

第 4 章
深入掌握 Service

第 4 章 深入掌握 Service

Service 是 Kubernetes 的核心概念，通过创建 Service，可以为一组具有相同功能的容器应用提供一个统一的入口地址，并且将请求负载分发到后端的各个容器应用上。本章对 Service 的使用进行详细说明，包括 Service 的负载均衡机制、如何访问 Service、Headless Service、DNS 服务的机制和实践、Ingress 7 层路由机制等。

4.1 Service 定义详解

YAML 格式的 Service 定义文件的完整内容如下：

```
apiVersion: v1              // Required
kind: Service               // Required
metadata:                   // Required
  name: string              // Required
  namespace: string         // Required
  labels:
    - name: string
  annotations:
    - name: string
spec:                       // Required
  selector: []              // Required
  type: string              // Required
  clusterIP: string
  sessionAffinity: string
  ports:
  - name: string
    protocol: string
    port: int
    targetPort: int
    nodePort: int
  status:
    loadBalancer:
      ingress:
        ip: string
        hostname: string
```

对各属性的说明如表 4.1 所示。

表 4.1　对 Service 的定义文件模板的各属性的说明

属性名称	取值类型	是否必选	取值说明
version	string	Required	v1
kind	string	Required	Service
metadata	object	Required	元数据
metadata.name	string	Required	Service 名称，需符合 RFC 1035 规范
metadata.namespace	string	Required	命名空间，不指定系统时将使用名为 default 的命名空间
metadata.labels[]	list		自定义标签属性列表
metadata.annotation[]	list		自定义注解属性列表
spec	object	Required	详细描述
spec.selector[]	list	Required	Label Selector 配置，将选择具有指定 Label 标签的 Pod 作为管理范围
spec.type	string	Required	Service 的类型，指定 Service 的访问方式，默认值为 ClusterIP。 （1）ClusterIP：虚拟的服务 IP 地址，该地址用于 Kubernetes 集群内部的 Pod 访问，在 Node 上 kube-proxy 通过设置的 iptables 规则进行转发。 （2）NodePort：使用宿主机的端口，使能够访问各 Node 的外部客户端通过 Node 的 IP 地址和端口号就能访问服务。 （3）LoadBalancer：使用外接负载均衡器完成到服务的负载分发，需要在 spec.status.loadBalancer 字段指定外部负载均衡器的 IP 地址，并同时定义 nodePort 和 clusterIP，用于公有云环境
spec.clusterIP	string		虚拟服务 IP 地址，当 type=ClusterIP 时，如果不指定，则系统进行自动分配，也可以手工指定；当 type=LoadBalancer 时，则需要指定
spec.sessionAffinity	string		是否支持 Session，可选值为 ClientIP，默认值为空。 ClientIP：表示将同一个客户端（根据客户端的 IP 地址决定）的访问请求都转发到同一个后端 Pod
spec.ports[]	list		Service 需要暴露的端口列表
spec.ports[].name	string		端口名称
spec.ports[].protocol	string		端口协议，支持 TCP 和 UDP，默认值为 TCP
spec.ports[].port	int		服务监听的端口号
spec.ports[].targetPort	int		需要转发到后端 Pod 的端口号
spec.ports[].nodePort	int		当 spec.type=NodePort 时，指定映射到物理机的端口号

续表

属性名称	取值类型	是否必选	取值说明
Status	object		当 spec.type=LoadBalancer 时，设置外部负载均衡器的地址，用于公有云环境
status.loadBalancer	object		外部负载均衡器
status.loadBalancer.ingress	object		外部负载均衡器
status.loadBalancer.ingress.ip	string		外部负载均衡器的 IP 地址
status.loadBalancer.ingress.hostname	string		外部负载均衡器的主机名

4.2　Service 的基本用法

一般来说，对外提供服务的应用程序需要通过某种机制来实现，对于容器应用最简便的方式就是通过 TCP/IP 机制及监听 IP 和端口号来实现。例如，定义一个提供 Web 服务的 RC，由两个 Tomcat 容器副本组成，每个容器都通过 containerPort 设置提供服务的端口号为 8080：

```yaml
webapp-rc.yaml
apiVersion: v1
kind: ReplicationController
metadata:
  name: webapp
spec:
  replicas: 2
  template:
    metadata:
      name: webapp
      labels:
        app: webapp
    spec:
      containers:
      - name: webapp
        image: tomcat
        ports:
        - containerPort: 8080
```

创建该 RC：

```
# kubectl create -f webapp-rc.yaml
replicationcontroller "webapp" created
```

获取 Pod 的 IP 地址：

```
# kubectl get pods -l app=webapp -o yaml | grep podIP
    podIP: 172.17.1.4
    podIP: 172.17.1.3
```

可以直接通过这两个 Pod 的 IP 地址和端口号访问 Tomcat 服务：

```
# curl 172.17.1.3:8080
<!DOCTYPE html>
<html lang="en">
    <head>
        <meta charset="UTF-8" />
        <title>Apache Tomcat/8.0.35</title>
......
# curl 172.17.1.4:8080
<!DOCTYPE html>
<html lang="en">
    <head>
        <meta charset="UTF-8" />
        <title>Apache Tomcat/8.0.35</title>
......
```

直接通过 Pod 的 IP 地址和端口号可以访问到容器应用内的服务，但是 Pod 的 IP 地址是不可靠的，例如当 Pod 所在的 Node 发生故障时，Pod 将被 Kubernetes 重新调度到另一个 Node，Pod 的 IP 地址将发生变化。更重要的是，如果容器应用本身是分布式的部署方式，通过多个实例共同提供服务，就需要在这些实例的前端设置一个负载均衡器来实现请求的分发。Kubernetes 中的 Service 就是用于解决这些问题的核心组件。

以前面创建的 webapp 应用为例，为了让客户端应用访问到两个 Tomcat Pod 实例，需要创建一个 Service 来提供服务。Kubernetes 提供了一种快速的方法，即通过 kubectl expose 命令来创建 Service：

```
# kubectl expose rc webapp
service "webapp" exposed
```

查看新创建的 Service，可以看到系统为它分配了一个虚拟的 IP 地址（ClusterIP），Service 所需的端口号则从 Pod 中的 containerPort 复制而来：

```
# kubectl get svc
```

```
NAME       CLUSTER-IP        EXTERNAL-IP     PORT(S)      AGE
webapp     169.169.235.79    <none>          8080/TCP     3s
```

接下来就可以通过 Service 的 IP 地址和 Service 的端口号访问该 Service 了：

```
# curl 169.169.235.79:8080
<!DOCTYPE html>
<html lang="en">
    <head>
        <meta charset="UTF-8" />
        <title>Apache Tomcat/8.0.35</title>
......
```

这里，对 Service 地址 169.169.235.79:8080 的访问被自动负载分发到了后端两个 Pod 之一：172.17.1.3:8080 或 172.17.1.4:8080。

除了使用 kubectl expose 命令创建 Service，我们也可以通过配置文件定义 Service，再通过 kubectl create 命令进行创建。例如对于前面的 webapp 应用，我们可以设置一个 Service，代码如下：

```
apiVersion: v1
kind: Service
metadata:
  name: webapp
spec:
  ports:
  - port: 8081
    targetPort: 8080
  selector:
    app: webapp
```

Service 定义中的关键字段是 ports 和 selector。本例中 ports 定义部分指定了 Service 所需的虚拟端口号为 8081，由于与 Pod 容器端口号 8080 不一样，所以需要再通过 targetPort 来指定后端 Pod 的端口号。selector 定义部分设置的是后端 Pod 所拥有的 label：app=webapp。

创建该 Service 并查看其 ClusterIP 地址：

```
# kubectl create -f webapp-svc.yaml
service "webapp" created

# kubectl get svc
NAME       CLUSTER-IP        EXTERNAL-IP     PORT(S)      AGE
webapp     169.169.28.190    <none>          8081/TCP     3s
```

通过 Service 的 IP 地址和 Service 的端口号进行访问：

```
# curl 169.169.28.190:8081
<!DOCTYPE html>
<html lang="en">
    <head>
        <meta charset="UTF-8" />
        <title>Apache Tomcat/8.0.35</title>
......
```

同样，对 Service 地址 169.169.28.190:8081 的访问被自动负载分发到了后端两个 Pod 之一：172.17.1.3:8080 或 172.17.1.4:8080。目前 Kubernetes 提供了两种负载分发策略：RoundRobin 和 SessionAffinity，具体说明如下。

◎ RoundRobin：轮询模式，即轮询将请求转发到后端的各个 Pod 上。
◎ SessionAffinity：基于客户端 IP 地址进行会话保持的模式，即第 1 次将某个客户端发起的请求转发到后端的某个 Pod 上，之后从相同的客户端发起的请求都将被转发到后端相同的 Pod 上。

在默认情况下，Kubernetes 采用 RoundRobin 模式对客户端请求进行负载分发，但我们也可以通过设置 service.spec.sessionAffinity=ClientIP 来启用 SessionAffinity 策略。这样，同一个客户端 IP 发来的请求就会被转发到后端固定的某个 Pod 上了。

通过 Service 的定义，Kubernetes 实现了一种分布式应用统一入口的定义和负载均衡机制。Service 还可以进行其他类型的设置，例如设置多个端口号、直接设置为集群外部服务，或实现为 Headless Service（无头服务）模式（将在 4.3 节介绍）。

4.2.1 多端口 Service

有时一个容器应用也可能提供多个端口的服务，那么在 Service 的定义中也可以相应地设置为将多个端口对应到多个应用服务。在下面的例子中，Service 设置了两个端口号，并且为每个端口号都进行了命名：

```
apiVersion: v1
kind: Service
metadata:
  name: webapp
spec:
  ports:
```

```
    - port: 8080
      targetPort: 8080
      name: web
    - port: 8005
      targetPort: 8005
      name: management
  selector:
    app: webapp
```

另一个例子是两个端口号使用了不同的 4 层协议——TCP 和 UDP：

```
apiVersion: v1
kind: Service
metadata:
  name: kube-dns
  namespace: kube-system
  labels:
    k8s-app: kube-dns
    kubernetes.io/cluster-service: "true"
    kubernetes.io/name: "KubeDNS"
spec:
  selector:
    k8s-app: kube-dns
  clusterIP: 169.169.0.100
  ports:
  - name: dns
    port: 53
    protocol: UDP
  - name: dns-tcp
    port: 53
    protocol: TCP
```

4.2.2 外部服务 Service

在某些环境中，应用系统需要将一个外部数据库作为后端服务进行连接，或将另一个集群或 Namespace 中的服务作为服务的后端，这时可以通过创建一个无 Label Selector 的 Service 来实现：

```
kind: Service
apiVersion: v1
metadata:
```

```
  name: my-service
spec:
  ports:
  - protocol: TCP
    port: 80
    targetPort: 80
```

通过该定义创建的是一个不带标签选择器的 Service，即无法选择后端的 Pod，系统不会自动创建 Endpoint，因此需要手动创建一个和该 Service 同名的 Endpoint，用于指向实际的后端访问地址。创建 Endpoint 的配置文件内容如下：

```
kind: Endpoints
apiVersion: v1
metadata:
  name: my-service
subsets:
- addresses:
  - ip: 1.2.3.4
  ports:
  - port: 80
```

如图 4.1 所示，访问没有标签选择器的 Service 和带有标签选择器的 Service 一样，请求将会被路由到由用户手动定义的后端 Endpoint 上。

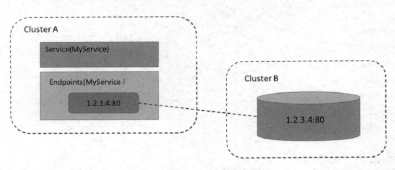

图 4.1　Service 指向外部服务

4.3　Headless Service

在某些应用场景中，开发人员希望自己控制负载均衡的策略，不使用 Service 提供的默认负载均衡的功能，或者应用程序希望知道属于同组服务的其他实例。Kubernetes 提供了 Headless Service 来实现这种功能，即不为 Service 设置 ClusterIP（入口 IP 地址），仅通过 Label Selector 将后端的 Pod 列表返回给调用的客户端。例如：

```yaml
apiVersion: v1
kind: Service
metadata:
  name: nginx
  labels:
    app: nginx
spec:
  ports:
  - port: 80
  clusterIP: None
  selector:
    app: nginx
```

这样，Service 就不再具有一个特定的 ClusterIP 地址，对其进行访问将获得包含 Label "app=nginx" 的全部 Pod 列表，然后客户端程序自行决定如何处理这个 Pod 列表。例如，StatefulSet 就是使用 Headless Service 为客户端返回多个服务地址的。

对于"去中心化"类的应用集群，Headless Service 将非常有用。下面以搭建 Cassandra 集群为例，看看如何通过对 Headless Service 的巧妙使用，自动实现应用集群的创建。

Apache Cassandra 是一套开源分布式 NoSQL 数据库系统，主要特点为它不是单个数据库，而是由一组数据库节点共同构成的一个分布式的集群数据库。由于 Cassandra 使用的是"去中心化"模式，所以在集群里的一个节点启动之后，需要一个途径获知集群中新节点的加入。Cassandra 使用了 Seed（种子）来完成集群中节点之间的相互查找和通信。

通过对 Headless Service 的使用，实现了 Cassandra 各节点之间的相互查找和集群的自动搭建。主要步骤包括：自定义 SeedProvider；通过 Headless Service 自动查找后端 Pod；自动添加新 Cassandra 节点。

4.3.1 自定义 SeedProvider

在本例中使用了一个自定义的 SeedProvider 类来完成新节点的查询和添加，类名为 io.k8s.cassandra.KubernetesSeedProvider。

KubernetesSeedProvider.java 类的源代码节选如下：

```java
......
    public List<InetAddress> getSeeds() {
        List<InetAddress> list = new ArrayList<InetAddress>();
        String host = "https://kubernetes.default.cluster.local";
        String serviceName = getEnvOrDefault("CASSANDRA_SERVICE", "cassandra");
        String podNamespace = getEnvOrDefault("POD_NAMESPACE", "default");
        String path = String.format("/api/v1/namespaces/%s/endpoints/", podNamespace);
......
    public static void main(String[] args) {
        SeedProvider provider = new KubernetesSeedProvider(new HashMap<String, String>());
        System.out.println(provider.getSeeds());
    }
}
```

完整的源代码可以从 https://github.com/kubernetes/kubernetes/blob/master/examples/storage/cassandra/java/src/main/java/io/k8s/cassandra/KubernetesSeedProvider.java 获取。

定制的 KubernetesSeedProvider 类将使用 REST API 来访问 Kubernetes Master，然后通过查询 name=cassandra 的服务（Headless Service 将返回 Pod 列表）完成对其他"节点"的查找。

4.3.2 通过 Service 动态查找 Pod

在 KubernetesSeedProvider 类中，通过查询环境变量 CASSANDRA_SERVICE 的值来获得服务的名称。这样就要求 Service 在 Pod 之前创建。如果我们已经创建好 DNS 服务（详见 4.5 节的说明），那么也可以直接使用服务的名称而无须使用环境变量。

回顾一下 Service 的概念。Service 通常用作一个负载均衡器，供 Kubernetes 集群中其他应用（Pod）对属于该 Service 的一组 Pod 进行访问。由于 Pod 的创建和销毁都会实时更新 Service 的 Endpoint 数据，所以可以动态地对 Service 的后端 Pod 进行查询。Cassandra 的"去中心化"设计使得 Cassandra 集群中的一个 Cassandra 实例（节点）只需要查询到其他节点，即可自动组成一个集群，正好可以使用 Service 的这个特性查询到新增的节点。图 4.2 描述了 Cassandra 新节点加入集群的过程。

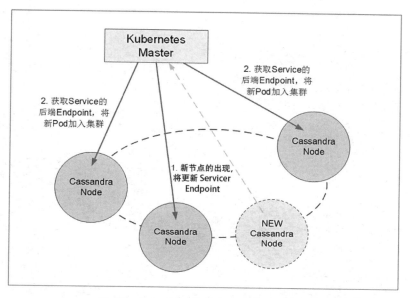

图4.2　Cassandra 新节点加入集群的过程

在 Kubernetes 系统中，首先需要为 Cassandra 集群定义一个 Service：

cassandra-service.yaml
```
apiVersion: v1
kind: Service
metadata:
  labels:
    name: cassandra
  name: cassandra
spec:
  ports:
    - port: 9042
  selector:
    name: cassandra
```

在 Service 的定义中指定 Label Selector 为 name=cassandra。

使用 kubectl create 命令创建这个 Service：

```
$ kubectl create -f cassandra-service.yaml
service "cassandra" created
```

然后，创建一个 Cassandra Pod：

cassandra-rc.yaml
```yaml
apiVersion: v1
kind: ReplicationController
metadata:
  labels:
    name: cassandra
  name: cassandra
spec:
  replicas: 1
  selector:
    name: cassandra
  template:
    metadata:
      labels:
        name: cassandra
    spec:
      containers:
        - command:
            - /run.sh
          resources:
            limits:
              cpu: 0.5
          env:
            - name: MAX_HEAP_SIZE
              value: 512M
            - name: HEAP_NEWSIZE
              value: 100M
            - name: POD_NAMESPACE
              valueFrom:
                fieldRef:
                  fieldPath: metadata.namespace
          image: gcr.io/google_containers/cassandra:v5
          name: cassandra
          ports:
            - containerPort: 9042
              name: cql
            - containerPort: 9160
              name: thrift
          volumeMounts:
```

```
            - mountPath: /cassandra_data
              name: data
      volumes:
        - name: data
          emptyDir: {}

$ kubectl create -f cassandra-rc.yaml
replicationcontroller "cassandra" created
```

现在，一个 Cassandra Pod 运行起来了，但还没有组成 Cassandra 集群。

4.3.3　Cassandra 集群中新节点的自动添加

现在，我们使用 Kubernetes 提供的 Scale（扩容和缩容）机制对 Cassandra 集群进行扩容：

```
$ kubectl scale rc cassandra --replicas=2
replicationcontroller "cassandra" scaled
```

查看 Pod，可以看到 RC 创建并启动了一个新的 Pod：

```
$ kubectl get pods -l="name=cassandra"
NAME              READY   STATUS    RESTARTS   AGE
cassandra         1/1     Running   0          5m
cassandra-g52t3   1/1     Running   0          50s
```

使用 Cassandra 提供的 nodetool 工具对任一 cassandra 实例（Pod）进行访问来验证 Cassandra 集群的状态。下面的命令将访问名为 cassandra 的 Pod（访问 cassandra-g52t3 也能获得相同的结果）：

```
$ kubectl exec -ti cassandra -- nodetool status
Datacenter: datacenter1
=======================
Status=Up/Down
|/ State=Normal/Leaving/Joining/Moving
--  Address       Load       Tokens  Owns (effective)  Host ID                               Rack
    UN  10.1.20.16   51.58 KB   256     100.0%            1625c65d-b5b6-40f4-a794-
6f5a12322d86  rack1
    UN  10.1.10.11   51.51 KB   256     100.0%            cdfcbf1a-795c-4412-9d3f-
e8fe50bb8deb  rack1
```

可以看到在 Cassandra 集群中有两个节点处于正常运行状态（Up and Normal，UN）。

该结果中的两个 IP 地址为两个 Cassandra Pod 的 IP 地址。

内部的过程为：每个 Cassandra 节点（Pod）都通过 API 访问 Kubernetes Master，查询名为 cassandra 的 Service 的 Endpoint（即 Cassandra 节点），若发现有新节点加入，就进行添加操作，最后成功组成一个 Cassandra 集群。

我们再增加两个 Cassandra 实例：

```
$ kubectl scale rc cassandra --replicas=4
```

用 nodetool 工具查看 Cassandra 集群的状态：

```
$ kubectl exec -ti cassandra -- nodetool status
Datacenter: datacenter1
=======================
Status=Up/Down
|/ State=Normal/Leaving/Joining/Moving
--  Address      Load       Tokens  Owns (effective)  Host ID                               Rack
UN  10.1.20.16   51.58 KB   256     50.5%             1625c65d-b5b6-40f4-a794-6f5a12322d86  rack1
UN  10.1.10.12   52.03 KB   256     47.0%             8bcc1c3e-44ec-46a7-b981-4090b206f14e  rack1
UN  10.1.20.17   68.05 KB   256     50.6%             579b6493-e92a-47f5-91f2-9313198a24c9  rack1
UN  10.1.10.11   51.51 KB   256     51.9%             cdfcbf1a-795c-4412-9d3f-e8fe50bb8deb  rack1
```

可以看到 4 个 Cassandra 节点都加入 Cassandra 集群中了。

另外，可以通过查看 Cassandra Pod 的日志来看到新节点加入集群的记录：

```
$ kubectl logs cassandra-g52t3
......
INFO 18:05:36 Handshaking version with /10.1.20.17
INFO 18:05:36 Node /10.1.20.17 is now part of the cluster
INFO 18:05:36 InetAddress /10.1.20.17 is now UP
INFO 18:05:38 Handshaking version with /10.1.10.12
INFO 18:05:39 Node /10.1.10.12 is now part of the cluster
INFO 18:05:39 InetAddress /10.1.10.12 is now UP
```

本例描述了一种通过 API 查询 Service 来完成动态 Pod 发现的应用场景。对于类似于 Cassandra 的去中心化集群类应用，都可以使用 Headless Service 查询后端 Endpoint 这种巧妙的方法来实现对应用集群（属于同一 Service）中新加入节点的查找。

4.4 从集群外部访问 Pod 或 Service

由于 Pod 和 Service 都是 Kubernetes 集群范围内的虚拟概念，所以集群外的客户端系统无法通过 Pod 的 IP 地址或者 Service 的虚拟 IP 地址和虚拟端口号访问它们。为了让外部客户端可以访问这些服务，可以将 Pod 或 Service 的端口号映射到宿主机，以使客户端应用能够通过物理机访问容器应用。

4.4.1 将容器应用的端口号映射到物理机

（1）通过设置容器级别的 hostPort，将容器应用的端口号映射到物理机上：

```
pod-hostport.yaml
apiVersion: v1
kind: Pod
metadata:
  name: webapp
  labels:
    app: webapp
spec:
  containers:
  - name: webapp
    image: tomcat
    ports:
    - containerPort: 8080
      hostPort: 8081
```

通过 kubectl create 命令创建这个 Pod：

```
# kubectl create -f pod-hostport.yaml
pod "webapp" created
```

通过物理机的 IP 地址和 8081 端口号访问 Pod 内的容器服务：

```
# curl 192.168.18.3:8081
<!DOCTYPE html>
<html lang="en">
    <head>
        <meta charset="UTF-8" />
        <title>Apache Tomcat/8.0.35</title>
```

......

（2）通过设置 Pod 级别的 hostNetwork=true，该 Pod 中所有容器的端口号都将被直接映射到物理机上。在设置 hostNetwork=true 时需要注意，在容器的 ports 定义部分如果不指定 hostPort，则默认 hostPort 等于 containerPort，如果指定了 hostPort，则 hostPort 必须等于 containerPort 的值：

```yaml
pod-hostnetwork.yaml
apiVersion: v1
kind: Pod
metadata:
  name: webapp
  labels:
    app: webapp
spec:
  hostNetwork: true
  containers:
  - name: webapp
    image: tomcat
    imagePullPolicy: Never
    ports:
    - containerPort: 8080
```

创建这个 Pod：

```
# kubectl create -f pod-hostnetwork.yaml
pod "webapp" created
```

通过物理机的 IP 地址和 8080 端口号访问 Pod 内的容器服务：

```
# curl 192.168.18.4:8080
<!DOCTYPE html>
<html lang="en">
    <head>
        <meta charset="UTF-8" />
        <title>Apache Tomcat/8.0.35</title>
......
```

4.4.2　将 Service 的端口号映射到物理机

（1）通过设置 nodePort 映射到物理机，同时设置 Service 的类型为 NodePort：

```
apiVersion: v1
kind: Service
metadata:
  name: webapp
spec:
  type: NodePort
  ports:
  - port: 8080
    targetPort: 8080
    nodePort: 8081
  selector:
    app: webapp
```

创建这个 Service：

```
# kubectl create -f webapp-svc-nodeport.yaml
You have exposed your service on an external port on all nodes in your
cluster.  If you want to expose this service to the external internet, you may
need to set up firewall rules for the service port(s) (tcp:8081) to serve traffic.

See http://releases.k8s.io/release-1.3/docs/user-guide/services-firewalls.md for more details.
service "webapp" created
```

系统提示信息说明：由于要使用物理机的端口号，所以需要在防火墙上做好相应的配置，以使外部客户端能够访问到该端口。

通过物理机的 IP 地址和 nodePort 8081 端口号访问服务：

```
# curl 192.168.18.3:8081
<!DOCTYPE html>
<html lang="en">
    <head>
        <meta charset="UTF-8" />
        <title>Apache Tomcat/8.0.35</title>
......
```

同样，对该 Service 的访问也将被负载分发到后端的多个 Pod 上。

（2）通过设置 LoadBalancer 映射到云服务商提供的 LoadBalancer 地址。这种用法仅用于在公有云服务提供商的云平台上设置 Service 的场景。在下面的例子中，status.loadBalancer.ingress.ip 设置的 146.148.47.155 为云服务商提供的负载均衡器的 IP 地址。对该 Service 的访问请求将会通过 LoadBalancer 转发到后端 Pod 上，负载分发的实现方式则依赖于云服务商提供的 LoadBalancer 的实现机制：

```
kind: Service
apiVersion: v1
metadata:
  name: my-service
spec:
  selector:
    app: MyApp
  ports:
  - protocol: TCP
    port: 80
    targetPort: 9376
    nodePort: 30061
  clusterIP: 10.0.171.239
  loadBalancerIP: 78.11.24.19
  type: LoadBalancer
status:
  loadBalancer:
    ingress:
    - ip: 146.148.47.155
```

4.5 DNS 服务搭建和配置指南

作为服务发现机制的基本功能，在集群内需要能够通过服务名对服务进行访问，这就需要一个集群范围内的 DNS 服务来完成从服务名到 ClusterIP 的解析。

DNS 服务在 Kubernetes 的发展过程中经历了 3 个阶段，接下来会进行讲解。

在 Kubernetes 1.2 版本时，DNS 服务是由 SkyDNS 提供的，它由 4 个容器组成：kube2sky、skydns、etcd 和 healthz。kube2sky 容器监控 Kubernetes 中 Service 资源的变化，根据 Service 的名称和 IP 地址信息生成 DNS 记录，并将其保存到 etcd 中；skydns 容器从 etcd 中读取 DNS 记录，并为客户端容器应用提供 DNS 查询服务；healthz 容器提供对 skydns 服务的健康检查功能。

图 4.3 展现了 SkyDNS 的总体架构。

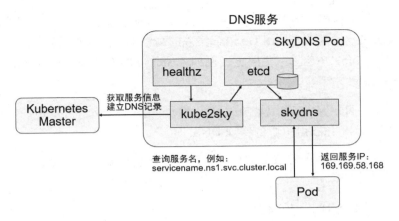

图 4.3 SkyDNS 的总体架构

从 Kubernetes 1.4 版本开始，SkyDNS 组件便被 KubeDNS 替换，主要考虑是 SkyDNS 组件之间通信较多，整体性能不高。KubeDNS 由 3 个容器组成：kubedns、dnsmasq 和 sidecar，去掉了 SkyDNS 中的 etcd 存储，将 DNS 记录直接保存在内存中，以提高查询性能。kubedns 容器监控 Kubernetes 中 Service 资源的变化，根据 Service 的名称和 IP 地址生成 DNS 记录，并将 DNS 记录保存在内存中；dnsmasq 容器从 kubedns 中获取 DNS 记录，提供 DNS 缓存，为客户端容器应用提供 DNS 查询服务；sidecar 提供对 kubedns 和 dnsmasq 服务的健康检查功能。

图 4.4 展现了 KubeDNS 的总体架构。

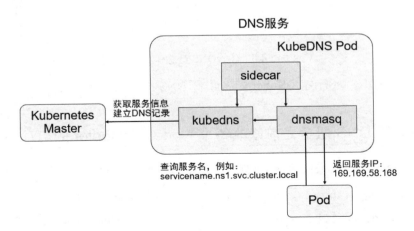

图 4.4 KubeDNS 的总体架构

从 Kubernetes 1.11 版本开始，Kubernetes 集群的 DNS 服务由 CoreDNS 提供。CoreDNS 是 CNCF 基金会的一个项目，是用 Go 语言实现的高性能、插件式、易扩展的 DNS 服务端。CoreDNS 解决了 KubeDNS 的一些问题，例如 dnsmasq 的安全漏洞、externalName 不能使用 stubDomains 设置，等等。

CoreDNS 支持自定义 DNS 记录及配置 upstream DNS Server，可以统一管理 Kubernetes 基于服务的内部 DNS 和数据中心的物理 DNS。

CoreDNS 没有使用多个容器的架构，只用一个容器便实现了 KubeDNS 内 3 个容器的全部功能。

图 4.5 展现了 CoreDNS 的总体架构

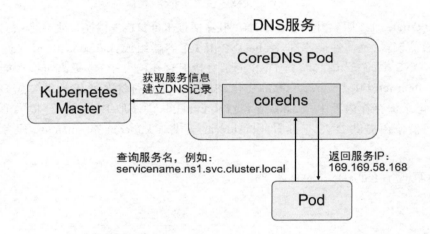

图 4.5　CoreDNS 的总体架构

下面以 CoreDNS 为例说明 Kubernetes 集群 DNS 服务的搭建过程。

4.5.1　在创建 DNS 服务之前修改每个 Node 上 kubelet 的启动参数

修改每个 Node 上 kubelet 的启动参数，加上以下两个参数。

◎ --cluster-dns=169.169.0.100：为 DNS 服务的 ClusterIP 地址。
◎ --cluster-domain=cluster.local：为在 DNS 服务中设置的域名。

然后重启 kubelet 服务。

4.5.2 创建 CoreDNS 应用

在部署 CoreDNS 应用前，至少需要创建一个 ConfigMap、一个 Deployment 和一个 Service 共 3 个资源对象。在启用了 RBAC 的集群中，还可以设置 ServiceAccount、ClusterRole、ClusterRoleBinding 对 CoreDNS 容器进行权限设置。关于 RBAC 的内容详见 6.2.3 节的说明。

ConfigMap "coredns" 主要设置 CoreDNS 的主配置文件 Corefile 的内容，其中可以定义各种域名的解析方式和使用的插件，示例如下（Corefile 的详细配置说明参见 4.5.4 节）：

```
apiVersion: v1
kind: ConfigMap
metadata:
  name: coredns
  namespace: kube-system
  labels:
      addonmanager.kubernetes.io/mode: EnsureExists
data:
  Corefile: |
    cluster.local {
        errors
        health
        kubernetes cluster.local in-addr.arpa ip6.arpa {
            pods insecure
            upstream
            fallthrough in-addr.arpa ip6.arpa
        }
        prometheus :9153
        forward . /etc/resolv.conf
        cache 30
        loop
        reload
        loadbalance
    }
    . {
        cache 30
        loadbalance
        forward . /etc/resolv.conf
    }
```

Deployment "coredns" 主要设置 CoreDNS 容器应用的内容，示例如下。

其中，replicas 副本的数量通常应该根据集群的规模和服务数量确定，如果单个 CoreDNS 进程不足以支撑整个集群的 DNS 查询，则可以通过水平扩展提高查询能力。由于 DNS 服务是 Kubernetes 集群的关键核心服务，所以建议为其 Deployment 设置自动扩缩容控制器，自动管理其副本数量。

另外，对资源限制部分（CPU 限制和内存限制）的设置也应根据实际环境进行调整：

```yaml
apiVersion: apps/v1
kind: Deployment
metadata:
  name: coredns
  namespace: kube-system
  labels:
    k8s-app: kube-dns
    kubernetes.io/cluster-service: "true"
    addonmanager.kubernetes.io/mode: Reconcile
    kubernetes.io/name: "CoreDNS"
spec:
  replicas: 1
  strategy:
    type: RollingUpdate
    rollingUpdate:
      maxUnavailable: 1
  selector:
    matchLabels:
      k8s-app: kube-dns
  template:
    metadata:
      labels:
        k8s-app: kube-dns
      annotations:
        seccomp.security.alpha.kubernetes.io/pod: 'docker/default'
    spec:
      priorityClassName: system-cluster-critical
      tolerations:
        - key: "CriticalAddonsOnly"
          operator: "Exists"
      nodeSelector:
        beta.kubernetes.io/os: linux
      containers:
      - name: coredns
```

```yaml
        image: coredns/coredns:1.3.1
        imagePullPolicy: IfNotPresent
        resources:
          limits:
            memory: 170Mi
          requests:
            cpu: 100m
            memory: 70Mi
        args: [ "-conf", "/etc/coredns/Corefile" ]
        volumeMounts:
        - name: config-volume
          mountPath: /etc/coredns
          readOnly: true
        ports:
        - containerPort: 53
          name: dns
          protocol: UDP
        - containerPort: 53
          name: dns-tcp
          protocol: TCP
        - containerPort: 9153
          name: metrics
          protocol: TCP
        livenessProbe:
          httpGet:
            path: /health
            port: 8080
            scheme: HTTP
          initialDelaySeconds: 60
          timeoutSeconds: 5
          successThreshold: 1
          failureThreshold: 5
        securityContext:
          allowPrivilegeEscalation: false
          capabilities:
            add:
            - NET_BIND_SERVICE
            drop:
            - all
          readOnlyRootFilesystem: true
      dnsPolicy: Default
```

```
        volumes:
          - name: config-volume
            configMap:
              name: coredns
              items:
              - key: Corefile
                path: Corefile
```

Service "kube-dns" 是 DNS 服务的配置,示例如下。

这个服务需要设置固定的 ClusterIP,也需要将所有 Node 上的 kubelet 启动参数 --cluster-dns 设置为这个 ClusterIP:

```
apiVersion: v1
kind: Service
metadata:
  name: kube-dns
  namespace: kube-system
  annotations:
    prometheus.io/port: "9153"
    prometheus.io/scrape: "true"
  labels:
    k8s-app: kube-dns
    kubernetes.io/cluster-service: "true"
    addonmanager.kubernetes.io/mode: Reconcile
    kubernetes.io/name: "CoreDNS"
spec:
  selector:
    k8s-app: kube-dns
  clusterIP: 169.169.0.100
  ports:
  - name: dns
    port: 53
    protocol: UDP
  - name: dns-tcp
    port: 53
    protocol: TCP
  - name: metrics
    port: 9153
    protocol: TCP
```

通过 kubectl create 完成 CoreDNS 服务的创建:

```
# kubectl create -f coredns.yaml
```

查看 RC、Pod 和 Service,确保容器成功启动:

```
# kubectl get deployment --namespace=kube-system
NAME                          DESIRED         CURRENT         AGE
kube-dns-v11                  1               1               1d

# kubectl get pods --namespace=kube-system
NAME                          READY           STATUS          RESTARTS        AGE
kube-dns-v11-6d1wu            4/4             Running         0               1d

# kubectl get services --namespace=kube-system
NAME              CLUSTER-IP          EXTERNAL-IP     PORT(S)             AGE
kube-dns          169.169.0.100       <none>          53/UDP,53/TCP       1d
```

4.5.3 服务名的 DNS 解析

接下来使用一个带有 nslookup 工具的 Pod 来验证 DNS 服务能否正常工作:

```
busybox.yaml
apiVersion: v1
kind: Pod
metadata:
  name: busybox
  namespace: default
spec:
  containers:
  - name: busybox
    image: gcr.io/google_containers/busybox
    command:
      - sleep
      - "3600"
```

运行 kubectl create -f busybox.yaml 即可完成创建。

在该容器成功启动后,通过 kubectl exec <container_id> nslookup 进行测试:

```
# kubectl exec busybox -- nslookup redis-master
Server:      169.169.0.100
Address 1:   169.169.0.100
```

```
Name:         redis-master
Address 1: 169.169.8.10
```

可以看到,通过 DNS 服务器 169.169.0.100 成功找到了 redis-master 服务的 IP 地址:169.169.8.10。

如果某个 Service 属于不同的命名空间,那么在进行 Service 查找时,需要补充 Namespace 的名称,组合成完整的域名。下面以查找 kube-dns 服务为例,将其所在的 Namespace "kube-system" 补充在服务名之后,用 "." 连接为 "kube-dns.kube-system",即可查询成功:

```
# kubectl exec busybox -- nslookup kube-dns.kube-system
Server:    169.169.0.100
Address 1: 169.169.0.100

Name:      kube-dns.kube-system
Address 1: 169.169.0.100
```

如果仅使用 "kube-dns" 进行查找,则会失败:

```
nslookup: can't resolve 'kube-dns'
```

4.5.4 CoreDNS 的配置说明

CoreDNS 的主要功能是通过插件系统实现的。CoreDNS 实现了一种链式插件结构,将 DNS 的逻辑抽象成了一个个插件,能够灵活组合使用。

常用的插件如下。

- loadbalance:提供基于 DNS 的负载均衡功能。
- loop:检测在 DNS 解析过程中出现的简单循环问题。
- cache:提供前端缓存功能。
- health:对 Endpoint 进行健康检查。
- kubernetes:从 Kubernetes 中读取 zone 数据。
- etcd:从 etcd 读取 zone 数据,可以用于自定义域名记录。
- file:从 RFC1035 格式文件中读取 zone 数据。
- hosts:使用/etc/hosts 文件或者其他文件读取 zone 数据,可以用于自定义域名记录。
- auto:从磁盘中自动加载区域文件。

- reload：定时自动重新加载 Corefile 配置文件的内容。
- forward：转发域名查询到上游 DNS 服务器。
- proxy：转发特定的域名查询到多个其他 DNS 服务器，同时提供到多个 DNS 服务器的负载均衡功能。
- prometheus：为 Prometheus 系统提供采集性能指标数据的 URL。
- pprof：在 URL 路径/debug/pprof 下提供运行时的性能数据。
- log：对 DNS 查询进行日志记录。
- errors：对错误信息进行日志记录。

在下面的示例中为域名"cluster.local"设置了一系列插件，包括 errors、health、kubernetes、prometheus、forward、cache、loop、reload 和 loadbalance，在进行域名解析时，这些插件将以从上到下的顺序依次执行：

```
cluster.local {
    errors
    health
    kubernetes cluster.local in-addr.arpa ip6.arpa {
        pods insecure
        upstream
        fallthrough in-addr.arpa ip6.arpa
    }
    prometheus :9153
    forward . /etc/resolv.conf
    cache 30
    loop
    reload
    loadbalance
}
```

另外，etcd 和 hosts 插件都可以用于用户自定义域名记录。

下面是使用 etcd 插件的配置示例，将以".com"结尾的域名记录配置为从 etcd 中获取，并将域名记录保存在/skydns 路径下：

```
{
    etcd com {
        path /skydns
        endpoint http://192.168.18.3:2379
        upstream /etc/resolv.conf
    }
```

```
    cache 160 com
    loadbalance
    proxy . /etc/resolv.conf
}
```

如果用户在 etcd 中插入一条 "10.1.1.1 mycompany.com" DNS 记录：

```
# etcdctl put /skydns/com/mycompany '{"host":"10.1.1.1", "ttl":60}'
```

客户端应用就能访问域名 "mycompany.com" 了：

```
# nslookup mycompany.com
nslookup mycompany.com
Server:         169.169.0.100
Address:        169.169.0.100#53

Name:   mycompany.com
Address: 10.1.1.1
```

forward 和 proxy 插件都可以用于配置上游 DNS 服务器或其他 DNS 服务器，当在 CoreDNS 中查询不到域名时，会到其他 DNS 服务器上进行查询。在实际环境中，可以将 Kubernetes 集群外部的 DNS 纳入 CoreDNS，进行统一的 DNS 管理。

4.5.5 Pod 级别的 DNS 配置说明

除了使用集群范围的 DNS 服务（如 CoreDNS），在 Pod 级别也能设置 DNS 的相关策略和配置。

在 Pod 的 YAML 配置文件中通过 spec.dnsPolicy 字段设置 DNS 策略，例如：

```
apiVersion: v1
kind: Pod
metadata:
  name: nginx
spec:
  containers:
  - name: nginx
    image: nginx
  dnsPolicy: Default
```

目前可以设置的 DNS 策略如下。

- Default:继承Pod所在宿主机的DNS设置。
- ClusterFirst:优先使用Kubernetes环境的DNS服务(如CoreDNS提供的域名解析服务),将无法解析的域名转发到从宿主机继承的DNS服务器。
- ClusterFirstWithHostNet:与ClusterFirst相同,对于以hostNetwork模式运行的Pod,应明确指定使用该策略。
- None:忽略Kubernetes环境的DNS配置,通过spec.dnsConfig自定义DNS配置。这个选项从Kubernetes 1.9版本开始引入,到Kubernetes 1.10版本升级为Beta版,到Kubernetes 1.14版本升级为稳定版。

自定义DNS配置可以通过spec.dnsConfig字段进行设置,可以设置下列信息。

- nameservers:一组DNS服务器的列表,最多可以设置3个。
- searches:一组用于域名搜索的DNS域名后缀,最多可以设置6个。
- options:配置其他可选DNS参数,例如ndots、timeout等,以name或name/value对的形式表示。

以下面的dnsConfig为例:

```
apiVersion: v1
kind: Pod
metadata:
  namespace: default
  name: dns-example
spec:
  containers:
  - name: test
    image: nginx
  dnsPolicy: "None"
  dnsConfig:
    nameservers:
      - 1.2.3.4
    searches:
      - ns1.svc.cluster.local
      - my.dns.search.suffix
    options:
      - name: ndots
        value: "2"
      - name: edns0
```

该Pod被成功创建之后,容器内的DNS配置文件/etc/resolv.conf的内容将被系统设置为：

```
nameserver 1.2.3.4
search ns1.svc.cluster.local my.dns.search.suffix
options ndots:2 edns0
```

表示该Pod完全使用自定义的DNS配置，不再使用Kubernetes环境的DNS服务。

4.6 Ingress：HTTP 7层路由机制

根据前面对Service的使用说明，我们知道Service的表现形式为IP:Port，即工作在TCP/IP层。而对于基于HTTP的服务来说，不同的URL地址经常对应到不同的后端服务或者虚拟服务器（Virtual Host），这些应用层的转发机制仅通过Kubernetes的Service机制是无法实现的。从Kubernetes 1.1版本开始新增Ingress资源对象，用于将不同URL的访问请求转发到后端不同的Service，以实现HTTP层的业务路由机制。Kubernetes使用了一个Ingress策略定义和一个具体的Ingress Controller，两者结合并实现了一个完整的Ingress负载均衡器。

使用Ingress进行负载分发时，Ingress Controller基于Ingress规则将客户端请求直接转发到Service对应的后端Endpoint（Pod）上，这样会跳过kube-proxy的转发功能，kube-proxy不再起作用。如果Ingress Controller提供的是对外服务，则实际上实现的是边缘路由器的功能。

图4.6显示了一个典型的HTTP层路由的例子。

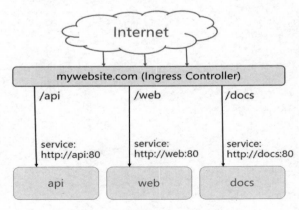

图4.6 一个典型的HTTP层路由的例子

其中：

- 对 http://mywebsite.com/api 的访问将被路由到后端名为 api 的 Service；
- 对 http://mywebsite.com/web 的访问将被路由到后端名为 web 的 Service；
- 对 http://mywebsite.com/docs 的访问将被路由到后端名为 docs 的 Service。

为使用 Ingress，需要创建 Ingress Controller（带一个默认 backend 服务）和 Ingress 策略设置来共同完成。下面通过一个例子分三步说明 Ingress Controller 和 Ingress 策略的配置方法，以及客户端如何访问 Ingress 提供的服务。

4.6.1 创建 Ingress Controller 和默认的 backend 服务

在定义 Ingress 策略之前，需要先部署 Ingress Controller，以实现为所有后端 Service 都提供一个统一的入口。Ingress Controller 需要实现基于不同 HTTP URL 向后转发的负载分发规则，并可以灵活设置 7 层负载分发策略。如果公有云服务商能够提供该类型的 HTTP 路由 LoadBalancer，则也可设置其为 Ingress Controller。

在 Kubernetes 中，Ingress Controller 将以 Pod 的形式运行，监控 API Server 的 /ingress 接口后端的 backend services，如果 Service 发生变化，则 Ingress Controller 应自动更新其转发规则。

下面的例子使用 Nginx 来实现一个 Ingress Controller，需要实现的基本逻辑如下。

（1）监听 API Server，获取全部 Ingress 的定义。

（2）基于 Ingress 的定义，生成 Nginx 所需的配置文件 /etc/nginx/nginx.conf。

（3）执行 nginx -s reload 命令，重新加载 nginx.conf 配置文件的内容。

基于 Go 语言的核心代码实现如下：

```go
for {
    rateLimiter.Accept()
    ingresses, err := ingClient.List(labels.Everything(), fields.Everything())
    if err != nil || reflect.DeepEqual(ingresses.Items, known.Items) {
        continue
    }
    if w, err := os.Create("/etc/nginx/nginx.conf"); err != nil {
        log.Fatalf("Failed to open %v: %v", nginxConf, err)
    } else if err := tmpl.Execute(w, ingresses); err != nil {
```

```
        log.Fatalf("Failed to write template %v", err)
    }
    shellOut("nginx -s reload")
}
```

本例使用谷歌提供的 nginx-ingress-controller 镜像来创建 Ingress Controller。该 Ingress Controller 以 daemonset 的形式进行创建，在每个 Node 上都将启动一个 Nginx 服务。

这里为 Nginx 容器设置了 hostPort，将容器应用监听的 80 和 443 端口号映射到物理机上，使得客户端应用可以通过 URL 地址"http://物理机 IP:80"或"https://物理机 IP:443"来访问该 Ingress Controller。这使得 Nginx 类似于通过 NodePort 映射到物理机的 Service，成为代替 kube-proxy 的 HTTP 层的 Load Balancer：

```
nginx-ingress-daemonset.yaml
apiVersion: extensions/v1beta1
kind: DaemonSet
metadata:
  name: nginx-ingress-lb
  labels:
    name: nginx-ingress-lb
  namespace: kube-system
spec:
  template:
    metadata:
      labels:
        name: nginx-ingress-lb
    spec:
      terminationGracePeriodSeconds: 60
      containers:
      - image: gcr.io/google_containers/nginx-ingress-controller:0.9.0-beta.2
        name: nginx-ingress-lb
        readinessProbe:
          httpGet:
            path: /healthz
            port: 10254
            scheme: HTTP
        livenessProbe:
          httpGet:
            path: /healthz
            port: 10254
            scheme: HTTP
```

```
        initialDelaySeconds: 10
        timeoutSeconds: 1
      ports:
      - containerPort: 80
        hostPort: 80
      - containerPort: 443
        hostPort: 443
      env:
        - name: POD_NAME
          valueFrom:
            fieldRef:
              fieldPath: metadata.name
        - name: POD_NAMESPACE
          valueFrom:
            fieldRef:
              fieldPath: metadata.namespace
      args:
      - /nginx-ingress-controller
      - --default-backend-service=$(POD_NAMESPACE)/default-http-backend
```

为了让 Ingress Controller 正常启动，还需要为它配置一个默认的 backend，用于在客户端访问的 URL 地址不存在时，返回一个正确的 404 应答。这个 backend 服务用任何应用实现都可以，只要满足对根路径"/"的访问返回 404 应答，并且提供/healthz 路径以使 kubelet 完成对它的健康检查。另外，由于 Nginx 通过 default-backend-service 的服务名称（Service Name）去访问它，所以需要 DNS 服务正确运行：

```
default-backend.yaml
apiVersion: extensions/v1beta1
kind: Deployment
metadata:
  name: default-http-backend
  labels:
    k8s-app: default-http-backend
  namespace: kube-system
spec:
  replicas: 1
  template:
    metadata:
      labels:
        k8s-app: default-http-backend
    spec:
```

```yaml
      terminationGracePeriodSeconds: 60
      containers:
      - name: default-http-backend
        image: gcr.io/google_containers/defaultbackend:1.0
        livenessProbe:
          httpGet:
            path: /healthz
            port: 8080
            scheme: HTTP
          initialDelaySeconds: 30
          timeoutSeconds: 5
        ports:
        - containerPort: 8080
        resources:
          limits:
            cpu: 10m
            memory: 20Mi
          requests:
            cpu: 10m
            memory: 20Mi
---
apiVersion: v1
kind: Service
metadata:
  name: default-http-backend
  namespace: kube-system
  labels:
    k8s-app: default-http-backend
spec:
  ports:
  - port: 80
    targetPort: 8080
  selector:
    k8s-app: default-http-backend
```

通过 kubectl create 命令创建 backend 服务：

```
# kubectl create -f default-backend.yaml
deployment "default-http-backend" created
service "default-http-backend" created
```

创建 nginx-ingress-controller：

```
# kubectl create -f nginx-ingress-daemonset.yaml
daemonset "nginx-ingress-lb" created
```

查看 default-http-backend 和 nginx-ingress-controller 容器是否正确运行：

```
# kubectl get po --namespace=kube-system
NAME                                       READY   STATUS    RESTARTS   AGE
default-http-backend-1132503640-84lnv      1/1     Running   0          3m
kube-dns-v11-z3cb0                         4/4     Running   0          10m
nginx-ingress-lb-5jbwv                     1/1     Running   0          3m
nginx-ingress-lb-60j7h                     1/1     Running   0          3m
nginx-ingress-lb-dttr9                     1/1     Running   0          3m
```

用 curl 访问任意 Node 的 80 端口号，验证 nginx-ingress-controller 和 default-http-backend 服务正常工作：

```
# curl k8s-node-2
default backend - 404
```

4.6.2 定义 Ingress 策略

本例对 mywebsite.com 网站的访问设置 Ingress 策略，定义对其/demo 路径的访问转发到后端 webapp Service 的规则：

```
apiVersion: extensions/v1beta1
kind: Ingress
metadata:
  name: mywebsite-ingress
spec:
  rules:
  - host: mywebsite.com
    http:
      paths:
      - path: /demo
        backend:
          serviceName: webapp
          servicePort: 8080
```

这个 Ingress 的定义，说明对目标地址 http://mywebsite.com/demo 的访问将被转发到集群中的 Service webapp 即 webapp:8080/demo 上。

在 Ingress 生效之前，需要先将 webapp 服务部署完成。同时需要注意 Ingress 中 path 的定义，需要与后端真实 Service 提供的 path 一致，否则将被转发到一个不存在的 path 上，引发错误。这里以第 1 章的例子为例，假设 myweb 服务已经部署完毕并正常运行，myweb 提供的 Web 服务的路径也为/demo。

创建该 Ingress：

```
# kubectl create -f ingress.yaml
ingress "mywebsite-ingress" created

# kubectl get ingress -o wide
NAME                HOSTS           ADDRESS                                    PORTS   AGE
mywebsite-ingress   mywebsite.com   192.168.18.3,192.168.18.4,192.168.18.5     80      59s
```

在成功创建该 Ingress 后，查看其 ADDRESS 列，如果显示了所有 nginx-ingress-controller Pod 的 IP 地址，则表示 Nginx 已经设置好后端 Service 的 Endpoint，该 Ingress 可以正常工作了。如果 ADDRESS 列为空，则通常说明 Nginx 未能正确连接到后端 Service，需要排错。

登录任一 nginx-ingress-controller Pod，查看其自动生成的 nginx.conf 配置文件的内容，可以看到对 mywebsite.com/demo 的转发规则的正确配置：

```
daemon off;
worker_processes 1;
......
http {
    real_ip_header      X-Forwarded-For;
    set_real_ip_from    0.0.0.0/0;
    real_ip_recursive   on;
......
    upstream default-myweb-8080 {
        least_conn;
        server 172.17.1.5:8080 max_fails=0 fail_timeout=0;
    }
......
    server {
        server_name mywebsite.com;
        listen [::]:80;
```

```
        location /demo {
            set $proxy_upstream_name "default-myweb-8080";

            port_in_redirect off;
            client_max_body_size                    "1m";

            proxy_set_header Host                   $host;

            # Pass Real IP
            proxy_set_header X-Real-IP              $remote_addr;

            # Allow websocket connections
            proxy_set_header                        Upgrade           $http_upgrade;
            proxy_set_header                        Connection        $connection_upgrade;

            proxy_set_header X-Forwarded-For        $proxy_add_x_forwarded_for;
            proxy_set_header X-Forwarded-Host       $host;
            proxy_set_header X-Forwarded-Port       $pass_port;
            proxy_set_header X-Forwarded-Proto      $pass_access_scheme;
......
```

4.6.3　客户端访问 http://mywebsite.com/demo

由于 Ingress Controller 容器通过 hostPort 将服务端口号 80 映射到了所有 Node 上，所以客户端可以通过任意 Node 访问 mywebsite.com 提供的服务。

需要说明的是，客户端只能通过域名 mywebsite.com 访问服务，这时要求客户端或者 DNS 将 mywebsite.com 域名解析到后端多个 Node 的真实 IP 地址上。

通过 curl 访问 mywebsite.com 提供的服务（可以用 --resolve 参数模拟 DNS 解析，目标地址为域名；也可以用 -H 'Host:mywebsite.com' 参数设置在 HTTP 头中要访问的域名，目标地址为 IP 地址），可以得到 myweb 服务返回的网页内容。

```
# curl --resolve mywebsite.com:80:192.168.18.3 http://mywebsite.com/demo/
```

或

```
# curl -H 'Host:mywebsite.com' http://192.168.18.3/demo/
<!DOCTYPE html PUBLIC "-//W3C//DTD HTML 4.01 Transitional//EN"
"http://www.w3.org/TR/html4/loose.dtd">
```

```html
<html>
<head>
<meta http-equiv="Content-Type" content="text/html; charset=utf-8">
<title>HPE University Docker&Kubernetes Learning</title>
</head>
<body align="center">

    <h2>Congratulations!!</h2>
    <br></br>
       <input type="button" value="Add..." onclick="location.href='input.html'" >
           <br></br>
     <TABLE align="center"  border="1" width="600px">
   <TR>
     <TD>Name</TD>
     <TD>Level(Score)</TD>
   </TR>

   <TR>
      <TD>google</TD>
      <TD>100</TD>
   </TR>

   <TR>
      <TD>docker</TD>
      <TD>100</TD>
   </TR>

   <TR>
      <TD>teacher</TD>
      <TD>100</TD>
   </TR>

   <TR>
      <TD>HPE</TD>
      <TD>100</TD>
   </TR>

   <TR>
      <TD>our team</TD>
      <TD>100</TD>
```

```
        </TR>

    <TR>
        <TD>me</TD>
        <TD>100</TD>
    </TR>

    </TABLE>

</body>
</html>
```

如果通过浏览器访问,那么需要先在本机上设置域名 mywebsite.com 对应的 IP 地址,再到浏览器上进行访问。以 Windows 为例,修改 C:\Windows\System32\drivers\etc\hosts 文件,加入一行记录:

```
192.168.18.3 mywebsite.com
```

然后在浏览器的地址栏输入 http://mywebsite.com/demo/,就能够访问 Ingress 提供的服务了,如图 4.7 所示。

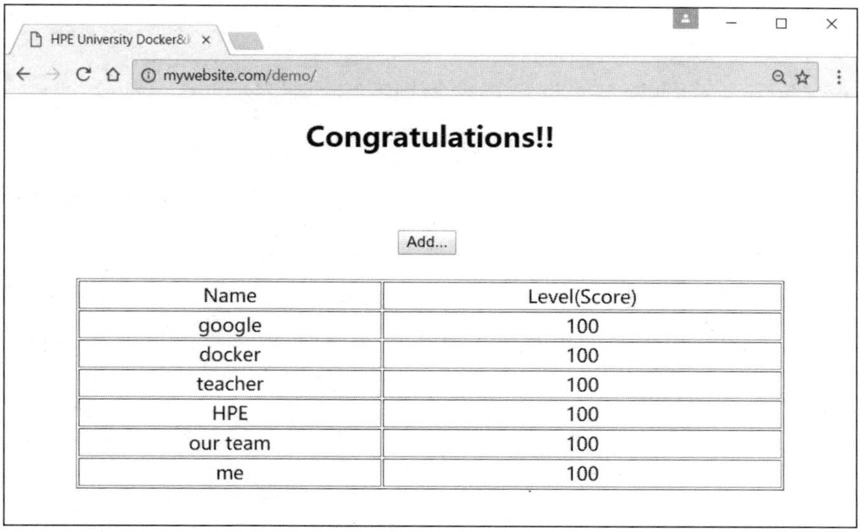

图 4.7 通过浏览器访问 Ingress 服务

4.6.4 Ingress 的策略配置技巧

为了实现灵活的负载分发策略，Ingress 策略可以按多种方式进行配置，下面对几种常见的 Ingress 转发策略进行说明。

1. 转发到单个后端服务上

基于这种设置，客户端到 Ingress Controller 的访问请求都将被转发到后端的唯一 Service 上，在这种情况下 Ingress 无须定义任何 rule。

通过如下所示的设置，对 Ingress Controller 的访问请求都将被转发到"myweb:8080"这个服务上。

```
apiVersion: extensions/v1beta1
kind: Ingress
metadata:
  name: test-ingress
spec:
  backend:
    serviceName: myweb
    servicePort: 8080
```

2. 同一域名下，不同的 URL 路径被转发到不同的服务上

这种配置常用于一个网站通过不同的路径提供不同的服务的场景，例如 /web 表示访问 Web 页面，/api 表示访问 API 接口，对应到后端的两个服务，通过 Ingress 的设置很容易就能将基于 URL 路径的转发规则定义出来。

通过如下所示的设置，对 "mywebsite.com/web" 的访问请求将被转发到 "web-service:80" 服务上；对 "mywebsite.com/api" 的访问请求将被转发到 "api-service:80" 服务上：

```
apiVersion: extensions/v1beta1
kind: Ingress
metadata:
  name: test-ingress
spec:
  rules:
  - host: mywebsite.com
    http:
```

```
    paths:
    - path: /web
      backend:
        serviceName: web-service
        servicePort: 80
    - path: /api
      backend:
        serviceName: api-service
        servicePort: 8081
```

3. 不同的域名（虚拟主机名）被转发到不同的服务上

这种配置常用于一个网站通过不同的域名或虚拟主机名提供不同服务的场景，例如 foo.bar.com 域名由 service1 提供服务，bar.foo.com 域名由 service2 提供服务。

通过如下所示的设置，对 "foo.bar.com" 的访问请求将被转发到 "service1:80" 服务上，对 "bar.foo.com" 的访问请求将被转发到 "service2:80" 服务上：

```
apiVersion: extensions/v1beta1
kind: Ingress
metadata:
  name: test
spec:
  rules:
  - host: foo.bar.com
    http:
      paths:
      - backend:
          serviceName: service1
          servicePort: 80
  - host: bar.foo.com
    http:
      paths:
      - backend:
          serviceName: service2
          servicePort: 80
```

4. 不使用域名的转发规则

这种配置用于一个网站不使用域名直接提供服务的场景，此时通过任意一台运行 ingress-controller 的 Node 都能访问到后端的服务。

以上节的后端服务 webapp 为例，下面的配置为将 "<ingress-controller-ip>/demo" 的访问请求转发到 "webapp:8080/demo" 服务上：

```
apiVersion: extensions/v1beta1
kind: Ingress
metadata:
  name: test-ingress
spec:
  rules:
  - http:
      paths:
      - path: /demo
        backend:
          serviceName: webapp
          servicePort: 8080
```

注意，使用无域名的 Ingress 转发规则时，将默认禁用非安全 HTTP，强制启用 HTTPS。例如，当使用 Nginx 作为 Ingress Controller 时，在其配置文件/etc/nginx/nginx.conf 中将会自动设置下面的规则，将全部 HTTP 的访问请求直接返回 301 错误：

```
......
if ($pass_access_scheme = http) {
    return 301 https://$best_http_host$request_uri;
}
......
```

客户端使用 HTTP 访问将得到 301 的错误应答：

```
# curl http://192.168.18.3/demo/
<html>
<head><title>301 Moved Permanently</title></head>
<body bgcolor="white">
<center><h1>301 Moved Permanently</h1></center>
<hr><center>nginx/1.13.0</center>
</body>
</html>
```

使用 HTTPS 能够访问成功：

```
# curl -k https://192.168.18.3/demo/
......
    <h2>Congratulations!!</h2>
    <br></br>
```

```
            <input type="button" value="Add..."
onclick="location.href='input.html'" >
                <br></br>
        <TABLE align="center" border="1" width="600px">
      <TR>
        <TD>Name</TD>
        <TD>Level(Score)</TD>
      </TR>
......
```

可以在 Ingress 的定义中设置一个 annotation "ingress.kubernetes.io/ssl-redirect=false" 来关闭强制启用 HTTPS 的设置：

```
apiVersion: extensions/v1beta1
kind: Ingress
metadata:
  name: test-ingress
  annotations:
    ingress.kubernetes.io/ssl-redirect: "false"
spec:
  rules:
  - http:
      paths:
      - path: /
        backend:
          serviceName: web-service
          servicePort: 8080
```

这样，到 Ingress Controller 的访问就可以使用 HTTP 了：

```
# curl http://192.168.18.3/demo/
......
    <h2>Congratulations!!</h2>
    <br></br>
        <input type="button" value="Add..."
onclick="location.href='input.html'" >
                <br></br>
        <TABLE align="center" border="1" width="600px">
      <TR>
        <TD>Name</TD>
        <TD>Level(Score)</TD>
      </TR>
```

......

4.6.5 Ingress 的 TLS 安全设置

为了 Ingress 提供 HTTPS 的安全访问，可以为 Ingress 中的域名进行 TLS 安全证书的设置。设置的步骤如下。

（1）创建自签名的密钥和 SSL 证书文件。

（2）将证书保存到 Kubernetes 中的一个 Secret 资源对象上。

（3）将该 Secret 对象设置到 Ingress 中。

根据提供服务的网站域名是一个还是多个，可以使用不同的操作完成前两步 SSL 证书和 Secret 对象的创建，在只有一个域名的情况下设置相对简单。第 3 步对于这两种场景来说是相同的。

对于只有一个域名的场景来说，可以通过 OpenSSL 工具直接生成密钥和证书文件，将命令行参数-subj 中的/CN 设置为网站域名：

```
# openssl req -x509 -nodes -days 5000 -newkey rsa:2048 -keyout tls.key -out tls.crt -subj "/CN=mywebsite.com"
Generating a 2048 bit RSA private key
..................+++
.................................................+++
writing new private key to 'tls.key'
-----
```

上述命令将生成 tls.key 和 tls.crt 两个文件。

然后根据 tls.key 和 tls.crt 文件创建 secret 资源对象，有以下两种方法。

方法一：使用 kubectl create secret tls 命令直接通过 tls.key 和 tls.crt 文件创建 secret 对象。

```
# kubectl create secret tls mywebsite-ingress-secret --key tls.key --cert tls.crt
secret "mywebsite-ingress-secret" created
```

方法二：编辑 mywebsite-ingress-secret.yaml 文件，将 tls.key 和 tls.crt 文件的内容经过 BASE64 编码的结果复制进去，使用 kubectl create 命令进行创建。

mywebsite-ingress-secret.yaml

```yaml
apiVersion: v1
kind: Secret
metadata:
  name: mywebsite-ingress-secret
type: kubernetes.io/tls
data:
  tls.crt:
MIIDAzCCAeugAwIBAgIJALrTg9VLmFgdMA0GCSqGSIb3DQEBCwUAMBgxFjAUBgNVBAMMDW15d2Vic2l0
ZS5jb20wHhcNMTcwNDIzMTMwMjA1WhcNMzAxMjMxMTMwMjA1WjAYMRYwFAYDVQQDDA1teXdlYnNpdGUu
Y29tMIIBIjANBgkqhkiG9w0BAQEFAAOCAQ8AMIIBCgKCAQEApLly1rq1I3EQ5E0PjzW8Lc3heW4WYTyk
POisDT9Zgyc+TLPGj/YF4QnAuoIUAUNtXPlmINKuD9Fxzmh6q0oSBVb42BU0RzOTtvaCVOU+uoJ9MgJp
d7Bao5higTZMyvj5a1M9iwb7k4xRAsuGCh/jDO8fj6tgJW4WfzawO5w1pDd2fFDxYn34Ma1pg0xFebVa
iqBu9FL0JbiEimsV9y7V+g6jjfGffu2xl06X3svqAdfGhvS+uCTArAXiZgS279se1Xp834CG0MJeP7ta
mD44IfA2wkkmD+uCVjSEcNFsveY5cJevjf0PSE9g5wohSXphd1sIGyjEy2APeIJBP8bQ+wIDAQABo1Aw
TjAdBgNVHQ4EFgQUjmpxpmdFPKWkr+A2XLF7oqro2GkwHwYDVR0jBBgwFoAUjmpxpmdFPKWkr+A2XLF7
oqro2GkwDAYDVR0TBAUwAwEB/zANBgkqhkiG9w0BAQsFAAOCAQEAAVXPyfagP1AIov3kXRhI3WfyCOIN
/sgNSqKM3FuykboSBN6c1w4UhrpF71Hd4nt0myeyX/o69o2Oc9a9dIS2FEGKvfxZQ4sa99iI3qjoMAuu
f/Q9fDYIZ+k0YvY4pbcCqqOyICFBCMLlAct/aB0K1GBvC5k06vD4Rn2fOdVMkloW+Zf41cxVIRZe/tQG
nZoEhtM6FQADrv1+jM5gjIKRX3s2/Jcxy5g2XLPqtSpzYA0F7FJyuFJXEG+P9X466xPi9ia1Uri66vkb
UVT6uLXGhhunsu6bZ/qwsm2HzdPo4WRQ3z2VhgFzHEzHVVX+CEyZ8fJGoSi7njapHb08lRiztQ==
  tls.key:
MIIEvQIBADANBgkqhkiG9w0BAQEFAASCBKcwggSjAgEAAoIBAQCkvXLWurUjcRDkTQ+PNbwtzeF5bhZh
PKQ86KwNP1mDJz5Ms8aP9gXhCcC6ghQBQ21c+WYg0q4P0XHOaHqrShIFVvjYFTRHM5O29oJU5T66gn0y
Aml3sFqjmGKBNkzK+PlrUz2LBvuTjFECy4YKH+MM7x+Pq2AlbhZ/NrA7nDWkN3Z8UPFiffgxrWmDTEV5
tVqKoG70UvQluISKaxX3LtX6DqON8Z9+7bGXTpfey+oB18aG9L64JMCsBeJmBLbv2x7VenzfgIbQwl4/
u1qYPjgh8DbCSSYP64JWNIRw0Wy95jlwl6+N/Q9IT2DnCiFJemF3WwgbKMTLYA94gkE/xtD7AgMBAAEC
ggEAUftNePq1RgvwYgzPX29YVFsOiAV28bDh8sW/SWBrRU90O2uDtwSx7EmUNbyiA/bwJ8KdRlxR7uFG
B3gLA876pNmhQLdcqspKClUmiuUCkIJ7lzWIEt4aXStqae8BzEiWpwhnqhYxgD3l2sQ50jQII9mkFTUt
xbLBU1F95kxYjX2XmFTrrvwroDLZEHCPcbY9hNUFhZaCdBBYKADmWo9eV/xZJ97ZAFpbpWyONrFjNwMj
jqCmxMx3HwOI/tLbhpvob6RT1UG1QUPlbB8aXR1FeSgt0NYhYwWKF7JSXcYBiyQubtd3T6RBtNjFk4b/
zuEUhdFN1lKJLcsVDVQZgMsO4QKBgQDajXAq4hMKPH3CKdieAialj4rVAPyrAFYDMokW+7buZZAgZO1a
rRtqFWLTtp6hwHqwTySHFyiRsK2Ikfct1H16hRn6FXbiPrFDP8gpYveu31Cd1qqYUYI7xaodWUiLldrt
eun9sLr3YYR7kaXYRenWZFjZbbUkq3KJfoh+uArPwwKBgQDA95Y4xhcL0F5pE/TLEdj33WjRXMkXMCHX
Gl3fTnBImoRf7jF9e5fRK/v4YIHaMCOn+6drwwMv9KHFL0nvxPbgbECW1F2OfzmNgm6l7jkpcsCQOVtuu
1+4gK+B2geQYRA2LhBk+9MtGQFmwSPgwSg+VHUrm28qhzUmTCN1etdpeaQKBgAFqHSO44Kp1S8Lp6q0
kzpGeN7hEiIngaLh/y1j5pmTceFptocSa2sOfl86azPyF3WDMC9SU3a/Q18vkoRGSeMcu68O4y7AEK3V
RiI4402nvAm9GTLXDPsp+3XtllwNuSSBznCxx1ONOuH3uf/tp7GUYR0WgHHeCfKy71GNluJ1AoGAKhHQ
XnBRdfHno2EGbX9mniNXRs3DyZpkx1CpRpYDRNDrKz7y6ziW0LOWK4BezwLPwz/KMGPIFV1L2gv5mY6r
JLtQfTqsLZsBb36AZL+Q1sRQGBA3tNa+w6TNOwj2gZPUoCYcmu0jpB1DcHt4II8E9q18NviUJNJsx/GW
0Z80DIECgYEAxzQBh/ckRvRaprN0v8w9GRq3wTYYD9y15U+3ecEIZrr1g9bLOi/rktXy3vqL6kj6CFlp
wwRVLj8R3u1QPy3MpJNXYR1Bua+/FVn2xKwyYDuXaqs0vW3xLONVO7z44gAKmEQyDq2sir+vpayuY4ps
```

```
fXXK06uifz6ELfVyY6XZvRA=
```

```
# kubectl create -f mywebsite-ingress-secret.yaml
secret "mywebsite-ingress-secret" created
```

如果提供服务的网站不止一个域名，例如前面第3种Ingress策略配置方式，则SSL证书需要使用额外的一个x509 v3配置文件辅助完成，在[alt_names]段中完成多个DNS域名的设置。

首先编写openssl.cnf文件，内容为：

```
[req]
req_extensions = v3_req
distinguished_name = req_distinguished_name
[req_distinguished_name]
[ v3_req ]
basicConstraints = CA:FALSE
keyUsage = nonRepudiation, digitalSignature, keyEncipherment
subjectAltName = @alt_names
[alt_names]
DNS.1 = mywebsite.com
DNS.2 = mywebsite2.com
```

接着使用OpenSSL工具完成密钥和证书的创建。

生成自签名CA证书：

```
# openssl genrsa -out ca.key 2048
Generating RSA private key, 2048 bit long modulus
..+++
.........................+++
e is 65537 (0x10001)
# openssl req -x509 -new -nodes -key ca.key -days 5000 -out ca.crt -subj "/CN=mywebsite.com"
```

基于openssl.cnf和CA证书生成ingress SSL证书：

```
# openssl genrsa -out ingress.key 2048
Generating RSA private key, 2048 bit long modulus
..........................+++
......................................................+++
e is 65537 (0x10001)
# openssl req -new -key ingress.key -out ingress.csr -subj "/CN=mywebsite.com"
```

```
-config openssl.cnf
   # openssl x509 -req -in ingress.csr -CA ca.crt -CAkey ca.key -CAcreateserial -out
ingress.crt -days 5000 -extensions v3_req -extfile openssl.cnf
   Signature ok
   subject=/CN=mywebsite.com
   Getting CA Private Key
```

然后根据 ingress.key 和 ingress.crt 文件创建 secret 资源对象，同样可以通过 kubectl create secret tls 命令或 YAML 配置文件生成。这里通过命令行直接生成：

```
   # kubectl create secret tls mywebsite-ingress-secret --key ingress.key --cert
ingress.crt
   secret "mywebsite-ingress-secret" created
```

至此，Ingress 的 TLS 证书就被成功创建到 Secret 对象中了。

下面创建 Ingress 对象，在 tls 段引用刚刚创建好的 Secret 对象：

```
mywebsite-ingress-tls.yaml
apiVersion: extensions/v1beta1
kind: Ingress
metadata:
  name: mywebsite-ingress-tls
spec:
  tls:
  - hosts:
    - mywebsite.com
    secretName: mywebsite-ingress-secret
  rules:
  - host: mywebsite.com
    http:
      paths:
      - path: /demo
        backend:
          serviceName: myweb
          servicePort: 8080
```

之后，就可以通过 HTTPS 来访问 mywebsite.com 了。

以 curl 为例，访问 https://192.168.18.3/demo/：

```
   # curl -H 'Host:mywebsite.com' -k https://192.168.18.3/demo/
   <!DOCTYPE html PUBLIC "-//W3C//DTD HTML 4.01 Transitional//EN"
"http://www.w3.org/TR/html4/loose.dtd">
```

```html
<html>
......
    <h2>Congratulations!!</h2>
    <br></br>
       <input type="button" value="Add..." onclick="location.href='input.html'" >
         <br></br>
    <TABLE align="center"  border="1" width="600px">
  <TR>
    <TD>Name</TD>
    <TD>Level(Score)</TD>
  </TR>

  <TR>
    <TD>google</TD>
    <TD>100</TD>
  </TR>
......
</html>
```

如果是通过浏览器访问的，则在浏览器的地址栏输入 https://mywebsite.com/demo/ 来访问 Ingress 提供的服务，浏览器会提示不安全，如图 4.8 所示。

图 4.8　通过浏览器访问 Ingress HTTPS 服务的警告信息

单击"继续前往 mywebsite.com（不安全）"标签，访问后可看到 Ingress 后端服务提供的页面，如图 4.9 所示。

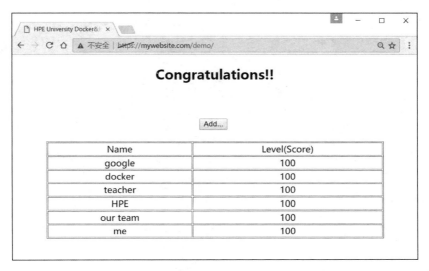

图 4.9　使用浏览器访问 Ingress HTTPS 服务

第 5 章
核心组件运行机制

5.1 Kubernetes API Server 原理解析

总体来看，Kubernetes API Server 的核心功能是提供 Kubernetes 各类资源对象（如 Pod、RC、Service 等）的增、删、改、查及 Watch 等 HTTP Rest 接口，成为集群内各个功能模块之间数据交互和通信的中心枢纽，是整个系统的数据总线和数据中心。除此之外，它还有以下一些功能特性。

（1）是集群管理的 API 入口。

（2）是资源配额控制的入口。

（3）提供了完备的集群安全机制。

5.1.1 Kubernetes API Server 概述

Kubernetes API Server 通过一个名为 kube-apiserver 的进程提供服务，该进程运行在 Master 上。在默认情况下，kube-apiserver 进程在本机的 8080 端口（对应参数--insecure-port）提供 REST 服务。我们可以同时启动 HTTPS 安全端口（--secure-port=6443）来启动安全机制，加强 REST API 访问的安全性。

我们通常可以通过命令行工具 kubectl 来与 Kubernetes API Server 交互，它们之间的接口是 RESTful API。为了测试和学习 Kubernetes API Server 所提供的接口，我们也可以使用 curl 命令行工具进行快速验证。

比如，登录 Master 并运行下面的 curl 命令，得到以 JSON 方式返回的 Kubernetes API 的版本信息：

```
# curl localhost:8080/api
{
  "kind": "APIVersions",
  "versions": [
    "v1"
  ],
  "serverAddressByClientCIDRs": [
    {
      "clientCIDR": "0.0.0.0/0",
      "serverAddress": "192.168.18.131:6443"
```

```
    }
  ]
}
```

可以运行下面的命令查看 Kubernetes API Server 目前支持的资源对象的种类：

```
# curl localhost:8080/api/v1
```

根据以上命令的输出，我们可以运行下面的 curl 命令，分别返回集群中的 Pod 列表、Service 列表、RC 列表等：

```
# curl localhost:8080/api/v1/pods
# curl localhost:8080/api/v1/services
# curl localhost:8080/api/v1/replicationcontrollers
```

如果只想对外暴露部分 REST 服务，则可以在 Master 或其他节点上运行 kubectl proxy 进程启动一个内部代理来实现。

运行下面的命令，在 8001 端口启动代理，并且拒绝客户端访问 RC 的 API：

```
# kubectl proxy --reject-paths="^/api/v1/replicationcontrollers" --port=8001 --v=2
Starting to serve on 127.0.0.1:8001
```

运行下面的命令进行验证：

```
# curl localhost:8001/api/v1/replicationcontrollers
<h3>Unauthorized</h3>
```

kubectl proxy 具有很多特性，最实用的一个特性是提供简单有效的安全机制，比如在采用白名单限制非法客户端访问时，只需增加下面这个参数即可：

```
--accept-hosts="^localhost$,^127\\.0\\.0\\.1$,^\\[::1\\]$"
```

最后一种方式是通过编程方式调用 Kubernetes API Server。具体使用场景又细分为以下两种。

第 1 种使用场景：运行在 Pod 里的用户进程调用 Kubernetes API，通常用来实现分布式集群搭建的目标。比如下面这段来自谷歌官方的 Elasticsearch 集群例子中的代码，Pod 在启动的过程中通过访问 Endpoints 的 API，找到属于 elasticsearch-logging 这个 Service 的所有 Pod 副本的 IP 地址，用来构建集群，如图 5.1 所示。

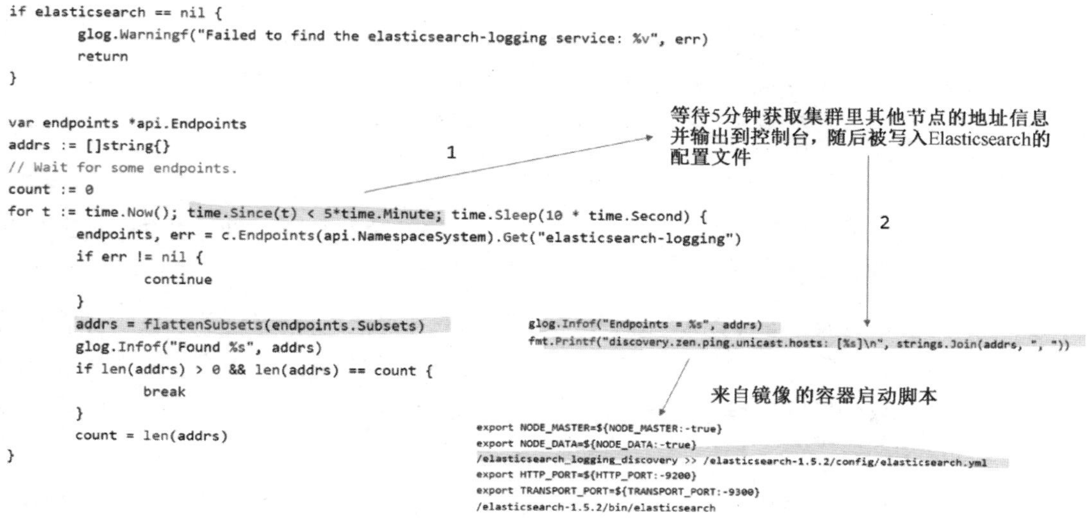

```
if elasticsearch == nil {
        glog.Warningf("Failed to find the elasticsearch-logging service: %v", err)
        return
}

var endpoints *api.Endpoints
addrs := []string{}
// Wait for some endpoints.
count := 0
for t := time.Now(); time.Since(t) < 5*time.Minute; time.Sleep(10 * time.Second) {
    endpoints, err = c.Endpoints(api.NamespaceSystem).Get("elasticsearch-logging")
    if err != nil {
        continue
    }
    addrs = flattenSubsets(endpoints.Subsets)
    glog.Infof("Found %s", addrs)
    if len(addrs) > 0 && len(addrs) == count {
        break
    }
    count = len(addrs)
}
```

1 → 等待5分钟获取集群里其他节点的地址信息并输出到控制台，随后被写入Elasticsearch的配置文件

2 →
```
glog.Infof("Endpoints = %s", addrs)
fmt.Printf("discovery.zen.ping.unicast.hosts: [%s]\n", strings.Join(addrs, ", "))
```
来自镜像的容器启动脚本
```
export NODE_MASTER=${NODE_MASTER:-true}
export NODE_DATA=${NODE_DATA:-true}
/elasticsearch_logging_discovery >> /elasticsearch-1.5.2/config/elasticsearch.yml
export HTTP_PORT=${HTTP_PORT:-9200}
export TRANSPORT_PORT=${TRANSPORT_PORT:-9300}
/elasticsearch-1.5.2/bin/elasticsearch
```

图 5.1　应用程序编程访问 API Server

在上述使用场景中，Pod 中的进程如何知道 API Server 的访问地址呢？答案很简单：Kubernetes API Server 本身也是一个 Service，它的名称就是 kubernetes，并且它的 Cluster IP 地址是 Cluster IP 地址池里的第 1 个地址！另外，它所服务的端口是 HTTPS 端口 443，通过 kubectl get service 命令可以确认这一点：

```
# kubectl get service
NAME              CLUSTER-IP        EXTERNAL-IP       PORT(S)           AGE
kubernetes        169.169.0.1       <none>            443/TCP           30d
```

第 2 种使用场景：开发基于 Kubernetes 的管理平台。比如调用 Kubernetes API 来完成 Pod、Service、RC 等资源对象的图形化创建和管理界面，此时可以使用 Kubernetes 及各开源社区为开发人员提供的各种语言版本的 Client Library。后面会介绍通过编程方式访问 API Server 的一些细节技术。

由于 API Server 是 Kubernetes 集群数据的唯一访问入口，因此安全性与高性能就成为 API Server 设计和实现的两大核心目标。通过采用 HTTPS 安全传输通道与 CA 签名数字证书强制双向认证的方式，API Server 的安全性得以保障。此外，为了更细粒度地控制用户或应用对 Kubernetes 资源对象的访问权限，Kubernetes 启用了 RBAC 访问控制策略，之后会深入讲解这一安全策略。

API Server 的性能是决定 Kubernetes 集群整体性能的关键因素，因此 Kubernetes 的设

计者综合运用以下方式来最大程度地保证 API Server 的性能。

（1）API Server 拥有大量高性能的底层代码。在 API Server 源码中使用协程（Coroutine）+队列（Queue）这种轻量级的高性能并发代码，使得单进程的 API Server 具备了超强的多核处理能力，从而以很快的速度并发处理大量的请求。

（2）普通 List 接口结合异步 Watch 接口，不但完美解决了 Kubernetes 中各种资源对象的高性能同步问题，也极大提升了 Kubernetes 集群实时响应各种事件的灵敏度。

（3）采用了高性能的 etcd 数据库而非传统的关系数据库，不仅解决了数据的可靠性问题，也极大提升了 API Server 数据访问层的性能。在常见的公有云环境中，一个 3 节点的 etcd 集群在轻负载环境中处理一个请求的时间可以低于 1ms，在重负载环境中可以每秒处理超过 30000 个请求。

正是由于采用了上述提升性能的方法，API Server 可以支撑很大规模的 Kubernetes 集群。截至 1.13 版本时，Kubernetes 已经可以支持最多 5000 节点规模的集群，同时 Kubernetes API Server 也淘汰了 etcd 2.0，只支持 etcd 3.0 以上版本。

5.1.2 API Server 架构解析

API Server 的架构从上到下可以分为以下几层，如图 5.2 所示。

（1）API 层：主要以 REST 方式提供各种 API 接口，除了有 Kubernetes 资源对象的 CRUD 和 Watch 等主要 API，还有健康检查、UI、日志、性能指标等运维监控相关的 API。Kubernetes 从 1.11 版本开始废弃 Heapster 监控组件，转而使用 Metrics Server 提供 Metrics API 接口，进一步完善了自身的监控能力。

（2）访问控制层：当客户端访问 API 接口时，访问控制层负责对用户身份鉴权，验明用户身份，核准用户对 Kubernetes 资源对象的访问权限，然后根据配置的各种资源访问许可逻辑（Admission Control），判断是否允许访问。

（3）注册表层：Kubernetes 把所有资源对象都保存在注册表（Registry）中，针对注册表中的各种资源对象都定义了：资源对象的类型、如何创建资源对象、如何转换资源的不同版本，以及如何将资源编码和解码为 JSON 或 ProtoBuf 格式进行存储。

（4）etcd 数据库：用于持久化存储 Kubernetes 资源对象的 KV 数据库。etcd 的 watch API 接口对于 API Server 来说至关重要，因为通过这个接口，API Server 创新性地设计了

List-Watch 这种高性能的资源对象实时同步机制,使 Kubernetes 可以管理超大规模的集群,及时响应和快速处理集群中的各种事件。

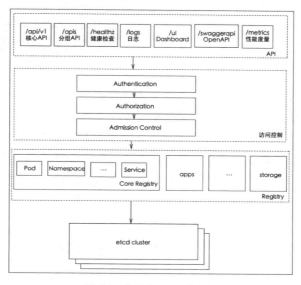

图 5.2 API Server 的架构

从本质上看,API Server 与常见的 MIS 或 ERP 系统中的 DAO 模块类似,可以将主要处理逻辑视作对数据库表的 CRUD 操作。这里解读 API Server 中资源对象的 List-Watch 机制。图 5.3 以一个完整的 Pod 调度过程为例,对 API Server 的 List-Watch 机制进行说明。

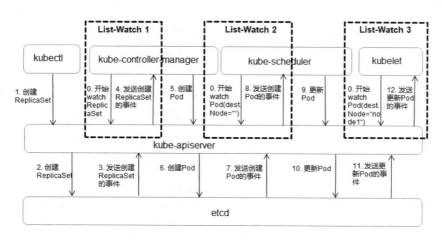

图 5.3 Pod 调度过程中的 List-Watch 机制

首先，借助 etcd 提供的 Watch API 接口，API Server 可以监听（Watch）在 etcd 上发生的数据操作事件，比如 Pod 创建事件、更新事件、删除事件等，在这些事件发生后，etcd 会及时通知 API Server。图 5.3 中 API Server 与 etcd 之间的交互箭头表明了这个过程：当一个 ReplicaSet 对象被创建并被保存到 etcd 中后（图中的 2. Create RepliatSet 箭头），etcd 会立即发送一个对应的 Create 事件给 API Server（图中的 3. Send RepliatSet Create Event 箭头），与其类似的 6、7、10、11 箭头都是针对 Pod 的创建、更新事件的。

然后，为了让 Kubernetes 中的其他组件在不访问底层 etcd 数据库的情况下，也能及时获取资源对象的变化事件，API Server 模仿 etcd 的 Watch API 接口提供了自己的 Watch 接口，这样一来，这些组件就能近乎实时地获取它们感兴趣的任意资源对象的相关事件通知了。图 5.3 中 controller-manager、scheduler、kublet 等组件与 API Server 之间的 3 个标记有 List-Watch 的虚框表明了这个过程。同时，在监听自己感兴趣的资源的时候，客户端可以增加过滤条件，以 List-Watch 3 为例，node1 节点上的 kubelet 进程只对自己节点上的 Pod 事件感兴趣。

最后，Kubernetes List-Watch 用于实现数据同步的代码逻辑。客户端首先调用 API Server 的 List 接口获取相关资源对象的全量数据并将其缓存到内存中，然后启动对应资源对象的 Watch 协程，在接收到 Watch 事件后，再根据事件的类型（比如新增、修改或删除）对内存中的全量资源对象列表做出相应的同步修改，从实现上来看，这是一种全量结合增量的、高性能的、近乎实时的数据同步方式。

接下来说说 API Server 中的另一处精彩设计。我们知道，对于不断迭代更新的系统，对象的属性一定是在不断变化的，API 接口的版本也在不断升级，此时就会面临版本问题，即同一个对象不同版本之间的数据转换问题及 API 接口版本的兼容问题。后面这个问题解决起来比较容易，即定义不同的 API 版本号（比如 v1alpha1、v1beta1）来加以区分，但前面的问题就有点麻烦了，比如数据对象经历 v1alpha1、v1beta1、v1beta1、v1beta2 等变化后最终变成 v1 版本，此时该数据对象就存在 5 个版本，如果这 5 个版本之间的数据两两直接转换，就存在很多种逻辑组合，变成一种典型的网状网络，如图 5.4 所示，为此我们不得不增加很多重复的转换代码。

上述直接转换的设计模式还存在另一个不可控的变数，即每增加一个新的对象版本，之前每个版本的对象就都需要增加一个到新版本对象的转换逻辑。如此一来，对直接转换的实现就更难了。于是，API Server 针对每种资源对象都引入了一个相对不变的 internal 版本，每个版本只要支持转换为 internal 版本，就能够与其他版本进行间接转换。于是对象版本转换的拓扑图就简化成了如图 5.5 所示的星状图。

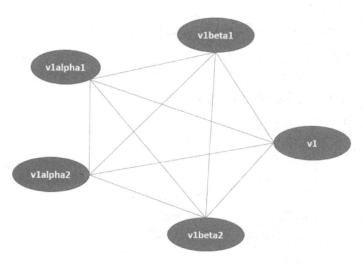

图 5.4 对象版本转换的拓扑图

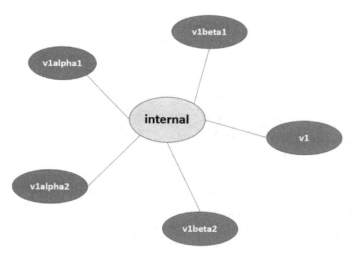

图 5.5 星状图

本节最后简单说说 Kubernetes 中的 CRD 在 API Server 中的设计和实现机制。根据 Kubernetes 的设计，每种官方内建的资源对象如 Node、Pod、Service 等的实现都包含以下主要功能。

（1）资源对象的元数据（Schema）的定义：可以将其理解为数据库 Table 的定义，定义了对应资源对象的数据结构，官方内建资源对象的元数据定义是固化在源码中的。

（2）资源对象的校验逻辑：确保用户提交的资源对象的属性的合法性。

（3）资源对象的 CRUD 操作代码：可以将其理解为数据库表的 CRUD 代码，但比后者更难，因为 API Server 对资源对象的 CRUD 操作都会保存到 etcd 数据库中，对处理性能的要求也更高，还要考虑版本兼容性和版本转换等复杂问题。

（4）资源对象相关的"自动控制器"（如 RC、Deployment 等资源对象背后的控制器）：这是很重要的一个功能。因为 Kubernetes 是一个以自动化为核心目标的平台，用户给出期望的资源对象声明，运行过程中则由资源背后的"自动控制器"负责，确保对应资源对象的数量、状态、行为都始终符合用户的预期。

类似地，每个自定义 CRD 的开发人员都需要实现上面这些功能。为了减小编程的难度与工作量，API Server 的设计者们做出了大量的努力，使得上面前 3 个功能无须编程实现，直接编写 YAML 定义文件即可实现。对于唯一需要编程的第 4 个功能来说，由于 API Server 提供了大量的基础 API 库，特别是易用的 List-Watch 的编程框架，也使得 CRD 自动控制器的编程难度大大减小。

5.1.3 独特的 Kubernetes Proxy API 接口

前面讲到，Kubernetes API Server 最主要的 REST 接口是资源对象的增、删、改、查接口，除此之外，它还提供了一类很特殊的 REST 接口——Kubernetes Proxy API 接口，这类接口的作用是代理 REST 请求，即 Kubernetes API Server 把收到的 REST 请求转发到某个 Node 上的 kubelet 守护进程的 REST 端口，由该 kubelet 进程负责响应。

首先来说说 Kubernetes Proxy API 里关于 Node 的相关接口。该接口的 REST 路径为 /api/v1/nodes/{name}/proxy，其中 {name} 为节点的名称或 IP 地址，包括以下几个具体接口：

```
/api/v1/nodes/{name}/proxy/pods      # 列出指定节点内所有 Pod 的信息
/api/v1/nodes/{name}/proxy/stats     # 列出指定节点内物理资源的统计信息
/api/v1/nodes/{name}/proxy/spec      # 列出指定节点的概要信息
```

例如，当前 Node 的名称为 k8s-node-1，用下面的命令即可获取该节点上所有运行中的 Pod：

```
# curl localhost:8080/api/v1/nodes/k8s-node-1/proxy/pods
```

需要说明的是：这里获取的 Pod 的信息数据来自 Node 而非 etcd 数据库，所以两者可能在某些时间点有所偏差。此外，如果 kubelet 进程在启动时包含 --enable-debugging-handlers =true 参数，那么 Kubernetes Proxy API 还会增加下面的接口：

```
/api/v1/nodes/{name}/proxy/run           # 在节点上运行某个容器
/api/v1/nodes/{name}/proxy/exec          # 在节点上的某个容器中运行某条命令
/api/v1/nodes/{name}/proxy/attach        # 在节点上 attach 某个容器
/api/v1/nodes/{name}/proxy/portForward   # 实现节点上的 Pod 端口转发
/api/v1/nodes/{name}/proxy/logs          # 列出节点的各类日志信息，例如 tallylog、
                                         # lastlog、wtmp、ppp/、rhsm/、audit/、
                                         # tuned/和 anaconda/等
/api/v1/nodes/{name}/proxy/metrics       # 列出和该节点相关的 Metrics 信息
/api/v1/nodes/{name}/proxy/runningpods   # 列出节点内运行中的 Pod 信息
/api/v1/nodes/{name}/proxy/debug/pprof   # 列出节点内当前 Web 服务的状态
                                         # 包括 CPU 占用情况和内存使用情况等
```

接下来说说 Kubernetes Proxy API 里关于 Pod 的相关接口，通过这些接口，我们可以访问 Pod 里某个容器提供的服务（如 Tomcat 在 8080 端口的服务）：

```
/api/v1/namespaces/{namespace}/pods/{name}/proxy              # 访问 Pod
/api/v1/namespaces/{namespace}/pods/{name}/proxy/{path:*}     # 访问 Pod 服务的 URL 路径
```

下面用第 1 章 Java Web 例子中的 Tomcat Pod 来说明上述 Proxy 接口的用法。

首先，得到 Pod 的名称：

```
# kubectl get pods
NAME           READY   STATUS    RESTARTS   AGE
mysql-c95jc    1/1     Running   0          8d
myweb-g9pmm    1/1     Running   0          8d
```

然后，运行下面的命令，会输出 Tomcat 的首页，即相当于访问 http://localhost:8080/：

```
# curl http://localhost:8080/api/v1/namespaces/default/pods/myweb-g9pmm/proxy
```

我们也可以在浏览器中访问上面的地址，比如 Master 的 IP 地址是 192.168.18.131，我们在浏览器中输入 http://192.168.18.131:8080/api/v1/namespaces/default/pods/myweb-g9pmm/proxy，就能够访问 Tomcat 首页了；而如果输入 /api/v1/namespaces/default/pods/myweb-g9pmm/proxy/demo，就能访问 Tomcat 中 Demo 应用的页面了。

看到这里，你可能明白 Pod 的 Proxy 接口的作用和意义了：在 Kubernetes 集群之外访问某个 Pod 容器的服务（HTTP 服务）时，可以用 Proxy API 实现，这种场景多用于管理目的，比如逐一排查 Service 的 Pod 副本，检查哪些 Pod 的服务存在异常。

最后说说 Service。Kubernetes Proxy API 也有 Service 的 Proxy 接口，其接口定义与 Pod

的接口定义基本一样：/api/v1/namespaces/{namespace}/services/{name}/proxy。比如，若我们想访问 myweb 服务的/demo 页面，则可以在浏览器里输入 http://192.168.18.131:8080/api/v1/namespaces/default/services/myweb/proxy/demo/。

5.1.4 集群功能模块之间的通信

从图 5.6 中可以看出，Kubernetes API Server 作为集群的核心，负责集群各功能模块之间的通信。集群内的各个功能模块通过 API Server 将信息存入 etcd，当需要获取和操作这些数据时，则通过 API Server 提供的 REST 接口（用 GET、LIST 或 WATCH 方法）来实现，从而实现各模块之间的信息交互。

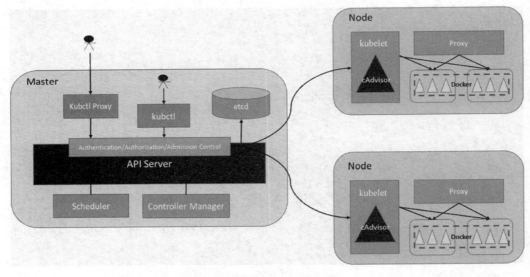

图 5.6　Kubernetes 结构图

常见的一个交互场景是 kubelet 进程与 API Server 的交互。每个 Node 上的 kubelet 每隔一个时间周期，就会调用一次 API Server 的 REST 接口报告自身状态，API Server 在接收到这些信息后，会将节点状态信息更新到 etcd 中。此外，kubelet 也通过 API Server 的 Watch 接口监听 Pod 信息，如果监听到新的 Pod 副本被调度绑定到本节点，则执行 Pod 对应的容器创建和启动逻辑；如果监听到 Pod 对象被删除，则删除本节点上相应的 Pod 容器；如果监听到修改 Pod 的信息，kubelet 就会相应地修改本节点的 Pod 容器。

另一个交互场景是 kube-controller-manager 进程与 API Server 的交互。

kube-controller-manager 中的 Node Controller 模块通过 API Server 提供的 Watch 接口实时监控 Node 的信息，并做相应处理。

还有一个比较重要的交互场景是 kube-scheduler 与 API Server 的交互。Scheduler 通过 API Server 的 Watch 接口监听到新建 Pod 副本的信息后，会检索所有符合该 Pod 要求的 Node 列表，开始执行 Pod 调度逻辑，在调度成功后将 Pod 绑定到目标节点上。

为了缓解集群各模块对 API Server 的访问压力，各功能模块都采用缓存机制来缓存数据。各功能模块定时从 API Server 获取指定的资源对象信息（通过 List-Watch 方法），然后将这些信息保存到本地缓存中，功能模块在某些情况下不直接访问 API Server，而是通过访问缓存数据来间接访问 API Server。

5.2 Controller Manager 原理解析

一般来说，智能系统和自动系统通常会通过一个"操作系统"来不断修正系统的工作状态。在 Kubernetes 集群中，每个 Controller 都是这样的一个"操作系统"，它们通过 API Server 提供的（List-Watch）接口实时监控集群中特定资源的状态变化，当发生各种故障导致某资源对象的状态发生变化时，Controller 会尝试将其状态调整为期望的状态。比如当某个 Node 意外宕机时，Node Controller 会及时发现此故障并执行自动化修复流程，确保集群始终处于预期的工作状态。Controller Manager 是 Kubernetes 中各种操作系统的管理者，是集群内部的管理控制中心，也是 Kubernetes 自动化功能的核心。

如图 5.7 所示，Controller Manager 内部包含 Replication Controller、Node Controller、ResourceQuota Controller、Namespace Controller、ServiceAccount Controller、Token Controller、Service Controller 及 Endpoint Controller 这 8 种 Controller，每种 Controller 都负责一种特定资源的控制流程，而 Controller Manager 正是这些 Controller 的核心管理者。

由于 ServiceAccount Controller 与 Token Controller 是与安全相关的两个控制器，并且与 Service Account、Token 密切相关，所以我们将对它们的分析放到后面讲解。

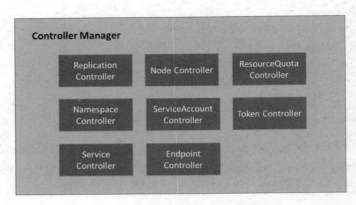

图 5.7　Controller Manager 结构图

在 Kubernetes 集群中与 Controller Manager 并重的另一个组件是 Kubernetes Scheduler，它的作用是将待调度的 Pod（包括通过 API Server 新创建的 Pod 及 RC 为补足副本而创建的 Pod 等）通过一些复杂的调度流程计算出最佳目标节点，然后绑定到该节点上。本章最后会介绍 Kubernetes Scheduler 调度器的基本原理。

5.2.1　Replication Controller

为了区分 Controller Manager 中的 Replication Controller（副本控制器）和资源对象 Replication Controller，我们将资源对象 Replication Controller 简写为 RC，而本节中的 Replication Controller 是指"副本控制器"，以便于后续分析。

Replication Controller 的核心作用是确保在任何时候集群中某个 RC 关联的 Pod 副本数量都保持预设值。如果发现 Pod 的副本数量超过预期值，则 Replication Controller 会销毁一些 Pod 副本；反之，Replication Controller 会自动创建新的 Pod 副本，直到符合条件的 Pod 副本数量达到预设值。需要注意：只有当 Pod 的重启策略是 Always 时（RestartPolicy=Always），Replication Controller 才会管理该 Pod 的操作（例如创建、销毁、重启等）。在通常情况下，Pod 对象被成功创建后不会消失，唯一的例外是当 Pod 处于 succeeded 或 failed 状态的时间过长（超时参数由系统设定）时，该 Pod 会被系统自动回收，管理该 Pod 的副本控制器将在其他工作节点上重新创建、运行该 Pod 副本。

RC 中的 Pod 模板就像一个模具，模具制作出来的东西一旦离开模具，它们之间就再也没关系了。同样，一旦 Pod 被创建完毕，无论模板如何变化，甚至换成一个新的模板，也不会影响到已经创建的 Pod 了。此外，Pod 可以通过修改它的标签来脱离 RC 的管控。

该方法可以用于将 Pod 从集群中迁移、数据修复等调试。对于被迁移的 Pod 副本，RC 会自动创建一个新的副本替换被迁移的副本。需要注意的是，删除一个 RC 不会影响它所创建的 Pod。如果想删除一个被 RC 所控制的 Pod，则需要将该 RC 的副本数（Replicas）属性设置为 0，这样所有的 Pod 副本就都会被自动删除。

最好不要越过 RC 直接创建 Pod，因为 Replication Controller 会通过 RC 管理 Pod 副本，实现自动创建、补足、替换、删除 Pod 副本，这样能提高系统的容灾能力，减少由于节点崩溃等意外状况造成的损失。即使你的应用程序只用到一个 Pod 副本，我们也强烈建议你使用 RC 来定义 Pod。

总结一下 Replication Controller 的职责，如下所述。

（1）确保在当前集群中有且仅有 N 个 Pod 实例，N 是在 RC 中定义的 Pod 副本数量。

（2）通过调整 RC 的 spec.replicas 属性值来实现系统扩容或者缩容。

（3）通过改变 RC 中的 Pod 模板（主要是镜像版本）来实现系统的滚动升级。

最后总结一下 Replication Controller 的典型使用场景，如下所述。

（1）重新调度（Rescheduling）。如前面所述，不管想运行 1 个副本还是 1000 个副本，副本控制器都能确保指定数量的副本存在于集群中，即使发生节点故障或 Pod 副本被终止运行等意外状况。

（2）弹性伸缩（Scaling）。手动或者通过自动扩容代理修改副本控制器的 spec.replicas 属性值，非常容易实现增加或减少副本的数量。

（3）滚动更新（Rolling Updates）。副本控制器被设计成通过逐个替换 Pod 的方式来辅助服务的滚动更新。推荐的方式是创建一个只有一个副本的新 RC，若新 RC 副本数量加 1，则旧 RC 的副本数量减 1，直到这个旧 RC 的副本数量为 0，然后删除该旧 RC。通过上述模式，即使在滚动更新的过程中发生了不可预料的错误，Pod 集合的更新也都在可控范围内。在理想情况下，滚动更新控制器需要将准备就绪的应用考虑在内，并保证在集群中任何时刻都有足够数量的可用 Pod。

5.2.2　Node Controller

kubelet 进程在启动时通过 API Server 注册自身的节点信息，并定时向 API Server 汇报状态信息，API Server 在接收到这些信息后，会将这些信息更新到 etcd 中。在 etcd 中存储的节点信息包括节点健康状况、节点资源、节点名称、节点地址信息、操作系统版本、Docker版本、kubelet版本等。节点健康状况包含"就绪"(True)"未就绪"(False)和"未知"(Unknown)三种。

Node Controller 通过 API Server 实时获取 Node 的相关信息，实现管理和监控集群中的各个 Node 的相关控制功能，Node Controller 的核心工作流程如图 5.8 所示。

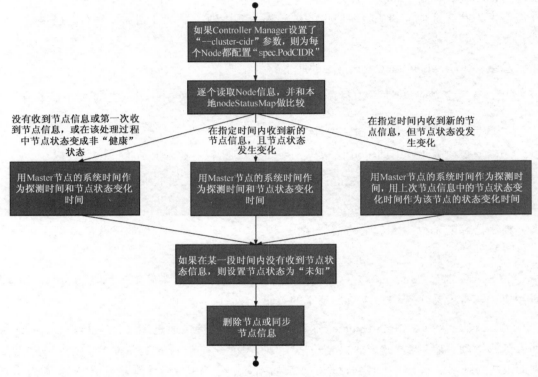

图 5.8　Node Controller 的核心工作流程

对流程中关键点的解释如下。

（1）Controller Manager 在启动时如果设置了 --cluster-cidr 参数，那么为每个没有设置 Spec.PodCIDR 的 Node 都生成一个 CIDR 地址，并用该 CIDR 地址设置节点的 Spec.PodCIDR 属性，这样做的目的是防止不同节点的 CIDR 地址发生冲突。

（2）逐个读取 Node 信息，多次尝试修改 nodeStatusMap 中的节点状态信息，将该节点信息和 Node Controller 的 nodeStatusMap 中保存的节点信息做比较。如果判断出没有收到

kubelet 发送的节点信息、第 1 次收到节点 kubelet 发送的节点信息，或在该处理过程中节点状态变成非"健康"状态，则在 nodeStatusMap 中保存该节点的状态信息，并用 Node Controller 所在节点的系统时间作为探测时间和节点状态变化时间。如果判断出在指定时间内收到新的节点信息，且节点状态发生变化，则在 nodeStatusMap 中保存该节点的状态信息，并用 Node Controller 所在节点的系统时间作为探测时间和节点状态变化时间。如果判断出在指定时间内收到新的节点信息，但节点状态没发生变化，则在 nodeStatusMap 中保存该节点的状态信息，并用 Node Controller 所在节点的系统时间作为探测时间，将上次节点信息中的节点状态变化时间作为该节点的状态变化时间。如果判断出在某段时间（gracePeriod）内没有收到节点状态信息，则设置节点状态为"未知"，并且通过 API Server 保存节点状态。

（3）逐个读取节点信息，如果节点状态变为非"就绪"状态，则将节点加入待删除队列，否则将节点从该队列中删除。如果节点状态为非"就绪"状态，且系统指定了 Cloud Provider，则 Node Controller 调用 Cloud Provider 查看节点，若发现节点故障，则删除 etcd 中的节点信息，并删除和该节点相关的 Pod 等资源的信息。

5.2.3　ResourceQuota Controller

作为完备的企业级的容器集群管理平台，Kubernetes 也提供了 ResourceQuota Controller（资源配额管理）这一高级功能，资源配额管理确保了指定的资源对象在任何时候都不会超量占用系统物理资源，避免了由于某些业务进程的设计或实现的缺陷导致整个系统运行紊乱甚至意外宕机，对整个集群的平稳运行和稳定性有非常重要的作用。

目前 Kubernetes 支持如下三个层次的资源配额管理。

（1）容器级别，可以对 CPU 和 Memory 进行限制。

（2）Pod 级别，可以对一个 Pod 内所有容器的可用资源进行限制。

（3）Namespace 级别，为 Namespace（多租户）级别的资源限制，包括：

◎　Pod 数量；
◎　Replication Controller 数量；
◎　Service 数量；
◎　ResourceQuota 数量；
◎　Secret 数量；

◎ 可持有的 PV 数量。

Kubernetes 的配额管理是通过 Admission Control（准入控制）来控制的，Admission Control 当前提供了两种方式的配额约束，分别是 LimitRanger 与 ResourceQuota。其中 LimitRanger 作用于 Pod 和 Container，ResourceQuota 则作用于 Namespace，限定一个 Namespace 里的各类资源的使用总额。

如图 5.9 所示，如果在 Pod 定义中同时声明了 LimitRanger，则用户通过 API Server 请求创建或修改资源时，Admission Control 会计算当前配额的使用情况，如果不符合配额约束，则创建对象失败。对于定义了 ResourceQuota 的 Namespace，ResourceQuota Controller 组件则负责定期统计和生成该 Namespace 下的各类对象的资源使用总量，统计结果包括 Pod、Service、RC、Secret 和 Persistent Volume 等对象实例个数，以及该 Namespace 下所有 Container 实例所使用的资源量（目前包括 CPU 和内存），然后将这些统计结果写入 etcd 的 resourceQuotaStatusStorage 目录（resourceQuotas/status）下。写入 resourceQuotaStatusStorage 的内容包含 Resource 名称、配额值（ResourceQuota 对象中 spec.hard 域下包含的资源的值）、当前使用值（ResourceQuota Controller 统计出来的值）。随后这些统计信息被 Admission Control 使用，以确保相关 Namespace 下的资源配额总量不会超过 ResourceQuota 中的限定值。

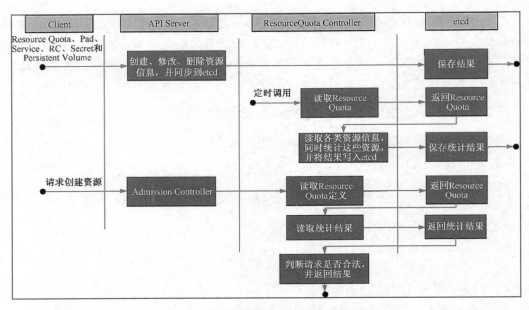

图 5.9　ResourceQuota Controller 流程图

5.2.4　Namespace Controller

用户通过 API Server 可以创建新的 Namespace 并将其保存在 etcd 中，Namespace Controller 定时通过 API Server 读取这些 Namespace 的信息。如果 Namespace 被 API 标识为优雅删除（通过设置删除期限实现，即设置 DeletionTimestamp 属性），则将该 NameSpace 的状态设置成 Terminating 并保存到 etcd 中。同时 Namespace Controller 删除该 Namespace 下的 ServiceAccount、RC、Pod、Secret、PersistentVolume、ListRange、ResourceQuota 和 Event 等资源对象。

在 Namespace 的状态被设置成 Terminating 后，由 Admission Controller 的 NamespaceLifecycle 插件来阻止为该 Namespace 创建新的资源。同时，在 Namespace Controller 删除该 Namespace 中的所有资源对象后，Namespace Controller 对该 Namespace 执行 finalize 操作，删除 Namespace 的 spec.finalizers 域中的信息。

如果 Namespace Controller 观察到 Namespace 设置了删除期限，同时 Namespace 的 spec.finalizers 域值是空的，那么 Namespace Controller 将通过 API Server 删除该 Namespace 资源。

5.2.5　Service Controller 与 Endpoints Controller

本节讲解 Endpoints Controller，在这之前，让我们先看看 Service、Endpoints 与 Pod 的关系。如图 5.10 所示，Endpoints 表示一个 Service 对应的所有 Pod 副本的访问地址，Endpoints Controller 就是负责生成和维护所有 Endpoints 对象的控制器。

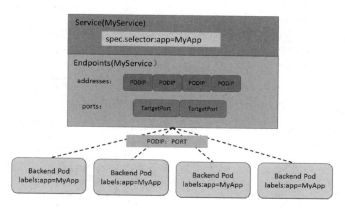

图 5.10　Service、Endpoints 与 Pod 的关系

它负责监听 Service 和对应的 Pod 副本的变化，如果监测到 Service 被删除，则删除和该 Service 同名的 Endpoints 对象。如果监测到新的 Service 被创建或者修改，则根据该 Service 信息获得相关的 Pod 列表，然后创建或者更新 Service 对应的 Endpoints 对象。如果监测到 Pod 的事件，则更新它所对应的 Service 的 Endpoints 对象（增加、删除或者修改对应的 Endpoint 条目）。

那么，Endpoints 对象是在哪里被使用的呢？答案是每个 Node 上的 kube-proxy 进程，kube-proxy 进程获取每个 Service 的 Endpoints，实现了 Service 的负载均衡功能。在后面的章节中会深入讲解这部分内容。

接下来说说 Service Controller 的作用，它其实是属于 Kubernetes 集群与外部的云平台之间的一个接口控制器。Service Controller 监听 Service 的变化，如果该 Service 是一个 LoadBalancer 类型的 Service（externalLoadBalancers=true），则 Service Controller 确保在外部的云平台上该 Service 对应的 LoadBalancer 实例被相应地创建、删除及更新路由转发表（根据 Endpoints 的条目）。

5.3 Scheduler 原理解析

前面深入分析了 Controller Manager 及它所包含的各个组件的运行机制，本节将继续对 Kubernetes 中负责 Pod 调度的重要功能模块——Kubernetes Scheduler 的工作原理和运行机制做深入分析。

Kubernetes Scheduler 在整个系统中承担了"承上启下"的重要功能，"承上"是指它负责接收 Controller Manager 创建的新 Pod，为其安排一个落脚的"家"——目标 Node；"启下"是指安置工作完成后，目标 Node 上的 kubelet 服务进程接管后继工作，负责 Pod 生命周期中的"下半生"。

具体来说，Kubernetes Scheduler 的作用是将待调度的 Pod（API 新创建的 Pod、Controller Manager 为补足副本而创建的 Pod 等）按照特定的调度算法和调度策略绑定（Binding）到集群中某个合适的 Node 上，并将绑定信息写入 etcd 中。在整个调度过程中涉及三个对象，分别是待调度 Pod 列表、可用 Node 列表，以及调度算法和策略。简单地说，就是通过调度算法调度为待调度 Pod 列表中的每个 Pod 从 Node 列表中选择一个最适合的 Node。

随后，目标节点上的 kubelet 通过 API Server 监听到 Kubernetes Scheduler 产生的 Pod 绑定事件，然后获取对应的 Pod 清单，下载 Image 镜像并启动容器。完整的流程如图 5.11

所示。

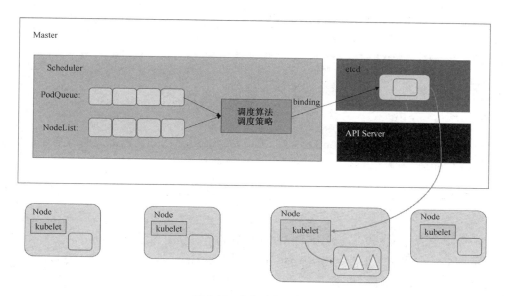

图 5.11 Scheduler 流程

Kubernetes Scheduler 当前提供的默认调度流程分为以下两步。

（1）预选调度过程，即遍历所有目标 Node，筛选出符合要求的候选节点。为此，Kubernetes 内置了多种预选策略（xxx Predicates）供用户选择。

（2）确定最优节点，在第 1 步的基础上，采用优选策略（xxx Priority）计算出每个候选节点的积分，积分最高者胜出。

Kubernetes Scheduler 的调度流程是通过插件方式加载的"调度算法提供者"（AlgorithmProvider）具体实现的。一个 AlgorithmProvider 其实就是包括了一组预选策略与一组优先选择策略的结构体，注册 AlgorithmProvider 的函数如下：

```
func RegisterAlgorithmProvider(name string, predicateKeys, priorityKeys util.StringSet)
```

它包含三个参数：name string 为算法名；predicateKeys 为算法用到的预选策略集合；priorityKeys 为算法用到的优选策略集合。

Scheduler 中可用的预选策略包含：NoDiskConflict、PodFitsResources、PodSelectorMatches、PodFitsHost、CheckNodeLabelPresence、CheckServiceAffinity 和

PodFitsPorts策略等。其默认的AlgorithmProvider加载的预选策略Predicates包括：PodFitsPorts（PodFitsPorts）、PodFitsResources（PodFitsResources）、NoDiskConflict（NoDiskConflict）、MatchNodeSelector（PodSelectorMatches）和HostName（PodFitsHost），即每个节点只有通过前面提及的5个默认预选策略后，才能初步被选中，进入下一个流程。

下面列出的是对所有预选策略的详细说明。

1）NoDiskConflict

判断备选Pod的gcePersistentDisk或AWSElasticBlockStore和备选的节点中已存在的Pod是否存在冲突。检测过程如下。

（1）首先，读取备选Pod的所有Volume的信息（即pod.Spec.Volumes），对每个Volume执行以下步骤进行冲突检测。

（2）如果该Volume是gcePersistentDisk，则将Volume和备选节点上的所有Pod的每个Volume都进行比较，如果发现相同的gcePersistentDisk，则返回false，表明存在磁盘冲突，检查结束，反馈给调度器该备选节点不适合作为备选Pod；如果该Volume是AWSElasticBlockStore，则将Volume和备选节点上的所有Pod的每个Volume都进行比较，如果发现相同的AWSElasticBlockStore，则返回false，表明存在磁盘冲突，检查结束，反馈给调度器该备选节点不适合备选Pod。

（3）如果检查完备选Pod的所有Volume均未发现冲突，则返回true，表明不存在磁盘冲突，反馈给调度器该备选节点适合备选Pod。

2）PodFitsResources

判断备选节点的资源是否满足备选Pod的需求，检测过程如下。

（1）计算备选Pod和节点中已存在Pod的所有容器的需求资源（内存和CPU）的总和。

（2）获得备选节点的状态信息，其中包含节点的资源信息。

（3）如果在备选Pod和节点中已存在Pod的所有容器的需求资源（内存和CPU）的总和，超出了备选节点拥有的资源，则返回false，表明备选节点不适合备选Pod，否则返回true，表明备选节点适合备选Pod。

3）PodSelectorMatches

判断备选节点是否包含备选Pod的标签选择器指定的标签。

（1）如果 Pod 没有指定 spec.nodeSelector 标签选择器，则返回 true。

（2）否则，获得备选节点的标签信息，判断节点是否包含备选 Pod 的标签选择器（spec.nodeSelector）所指定的标签，如果包含，则返回 true，否则返回 false。

4）PodFitsHost

判断备选 Pod 的 spec.nodeName 域所指定的节点名称和备选节点的名称是否一致，如果一致，则返回 true，否则返回 false。

5）CheckNodeLabelPresence

如果用户在配置文件中指定了该策略，则 Scheduler 会通过 RegisterCustomFitPredicate 方法注册该策略。该策略用于判断策略列出的标签在备选节点中存在时，是否选择该备选节点。

（1）读取备选节点的标签列表信息。

（2）如果策略配置的标签列表存在于备选节点的标签列表中，且策略配置的 presence 值为 false，则返回 false，否则返回 true；如果策略配置的标签列表不存在于备选节点的标签列表中，且策略配置的 presence 值为 true，则返回 false，否则返回 true。

6）CheckServiceAffinity

如果用户在配置文件中指定了该策略，则 Scheduler 会通过 RegisterCustomFitPredicate 方法注册该策略。该策略用于判断备选节点是否包含策略指定的标签，或包含和备选 Pod 在相同 Service 和 Namespace 下的 Pod 所在节点的标签列表。如果存在，则返回 true，否则返回 false。

7）PodFitsPorts

判断备选 Pod 所用的端口列表中的端口是否在备选节点中已被占用，如果被占用，则返回 false，否则返回 true。

Scheduler 中的优选策略包含：LeastRequestedPriority、CalculateNodeLabelPriority 和 BalancedResourceAllocation 等。每个节点通过优先选择策略时都会算出一个得分，计算各项得分，最终选出得分值最大的节点作为优选的结果（也是调度算法的结果）。

下面是对所有优选策略的详细说明。

1) LeastRequestedPriority

该优选策略用于从备选节点列表中选出资源消耗最小的节点。

（1）计算出在所有备选节点上运行的 Pod 和备选 Pod 的 CPU 占用量 totalMilliCPU。

（2）计算出在所有备选节点上运行的 Pod 和备选 Pod 的内存占用量 totalMemory。

（3）计算每个节点的得分，计算规则大致如下，其中，NodeCpuCapacity 为节点 CPU 计算能力，NodeMemoryCapacity 为节点内存大小：

```
score=int(((nodeCpuCapacity-totalMilliCPU)*10)/ nodeCpuCapacity+((nodeMemory
Capacity-totalMemory)*10)/ nodeCpuMemory)/2)
```

2) CalculateNodeLabelPriority

如果用户在配置文件中指定了该策略，则 scheduler 会通过 RegisterCustomPriorityFunction 方法注册该策略。该策略用于判断策略列出的标签在备选节点中存在时，是否选择该备选节点。如果备选节点的标签在优选策略的标签列表中且优选策略的 presence 值为 true，或者备选节点的标签不在优选策略的标签列表中且优选策略的 presence 值为 false，则备选节点 score=10，否则备选节点 score=0。

3) BalancedResourceAllocation

该优选策略用于从备选节点列表中选出各项资源使用率最均衡的节点。

（1）计算出在所有备选节点上运行的 Pod 和备选 Pod 的 CPU 占用量 totalMilliCPU。

（2）计算出在所有备选节点上运行的 Pod 和备选 Pod 的内存占用量 totalMemory。

（3）计算每个节点的得分，计算规则大致如下，其中，NodeCpuCapacity 为节点的 CPU 计算能力，NodeMemoryCapacity 为节点的内存大小：

```
score= int(10-math.Abs(totalMilliCPU/nodeCpuCapacity-totalMemory/
nodeMemoryCapacity)*10)
```

5.4 kubelet 运行机制解析

在 Kubernetes 集群中，在每个 Node（又称 Minion）上都会启动一个 kubelet 服务进程。该进程用于处理 Master 下发到本节点的任务，管理 Pod 及 Pod 中的容器。每个 kubelet 进程都会在 API Server 上注册节点自身的信息，定期向 Master 汇报节点资源的使用情况，并通过 cAdvisor 监控容器和节点资源。

5.4.1 节点管理

节点通过设置 kubelet 的启动参数"--register-node",来决定是否向 API Server 注册自己。如果该参数的值为 true,那么 kubelet 将试着通过 API Server 注册自己。在自注册时,kubelet 启动时还包含下列参数。

- --api-servers:API Server 的位置。
- --kubeconfig:kubeconfig 文件,用于访问 API Server 的安全配置文件。
- --cloud-provider:云服务商(IaaS)地址,仅用于公有云环境。

当前每个 kubelet 都被授予创建和修改任何节点的权限。但是在实践中,它仅仅创建和修改自己。将来,我们计划限制 kubelet 的权限,仅允许它修改和创建所在节点的权限。如果在集群运行过程中遇到集群资源不足的情况,用户就很容易通过添加机器及运用 kubelet 的自注册模式来实现扩容。

在某些情况下,Kubernetes 集群中的某些 kubelet 没有选择自注册模式,用户需要自己去配置 Node 的资源信息,同时告知 Node 上 Kubelet API Server 的位置。集群管理者能够创建和修改节点信息。如果管理者希望手动创建节点信息,则通过设置 kubelet 的启动参数"--register- node=false"即可完成。

kubelet 在启动时通过 API Server 注册节点信息,并定时向 API Server 发送节点的新消息,API Server 在接收到这些信息后,将这些信息写入 etcd。通过 kubelet 的启动参数"--node-status- update-frequency"设置 kubelet 每隔多长时间向 API Server 报告节点状态,默认为 10s。

5.4.2 Pod 管理

kubelet 通过以下几种方式获取自身 Node 上要运行的 Pod 清单。

(1)文件:kubelet 启动参数"--config"指定的配置文件目录下的文件(默认目录为"/etc/ kubernetes/manifests/")。通过--file-check-frequency 设置检查该文件目录的时间间隔,默认为 20s。

（2）HTTP端点（URL）：通过"--manifest-url"参数设置。通过--http-check-frequency设置检查该HTTP端点数据的时间间隔，默认为20s。

（3）API Server：kubelet通过API Server监听etcd目录，同步Pod列表。

所有以非API Server方式创建的Pod都叫作Static Pod。kubelet将Static Pod的状态汇报给API Server，API Server为该Static Pod创建一个Mirror Pod和其相匹配。Mirror Pod的状态将真实反映Static Pod的状态。当Static Pod被删除时，与之相对应的Mirror Pod也会被删除。在本章中只讨论通过API Server获得Pod清单的方式。kubelet通过API Server Client使用Watch加List的方式监听"/registry/nodes/$"当前节点的名称和"/registry/pods"目录，将获取的信息同步到本地缓存中。

kubelet监听etcd，所有针对Pod的操作都会被kubelet监听。如果发现有新的绑定到本节点的Pod，则按照Pod清单的要求创建该Pod。

如果发现本地的Pod被修改，则kubelet会做出相应的修改，比如在删除Pod中的某个容器时，会通过Docker Client删除该容器。

如果发现删除本节点的Pod，则删除相应的Pod，并通过Docker Client删除Pod中的容器。

kubelet读取监听到的信息，如果是创建和修改Pod任务，则做如下处理。

（1）为该Pod创建一个数据目录。

（2）从API Server读取该Pod清单。

（3）为该Pod挂载外部卷（External Volume）。

（4）下载Pod用到的Secret。

（5）检查已经运行在节点上的Pod，如果该Pod没有容器或Pause容器（"kubernetes/pause"镜像创建的容器）没有启动，则先停止Pod里所有容器的进程。如果在Pod中有需要删除的容器，则删除这些容器。

（6）用"kubernetes/pause"镜像为每个Pod都创建一个容器。该Pause容器用于接管Pod中所有其他容器的网络。每创建一个新的Pod，kubelet都会先创建一个Pause容器，然后创建其他容器。"kubernetes/pause"镜像大概有200KB，是个非常小的容器镜像。

（7）为 Pod 中的每个容器做如下处理。

◎ 为容器计算一个 Hash 值，然后用容器的名称去查询对应 Docker 容器的 Hash 值。若查找到容器，且二者的 Hash 值不同，则停止 Docker 中容器的进程，并停止与之关联的 Pause 容器的进程；若二者相同，则不做任何处理。
◎ 如果容器被终止了，且容器没有指定的 restartPolicy（重启策略），则不做任何处理。
◎ 调用 Docker Client 下载容器镜像，调用 Docker Client 运行容器。

5.4.3 容器健康检查

Pod 通过两类探针来检查容器的健康状态。一类是 LivenessProbe 探针，用于判断容器是否健康并反馈给 kubelet。如果 LivenessProbe 探针探测到容器不健康，则 kubelet 将删除该容器，并根据容器的重启策略做相应的处理。如果一个容器不包含 LivenessProbe 探针，那么 kubelet 认为该容器的 LivenessProbe 探针返回的值永远是 Success；另一类是 ReadinessProbe 探针，用于判断容器是否启动完成，且准备接收请求。如果 ReadinessProbe 探针检测到容器启动失败，则 Pod 的状态将被修改，Endpoint Controller 将从 Service 的 Endpoint 中删除包含该容器所在 Pod 的 IP 地址的 Endpoint 条目。

kubelet 定期调用容器中的 LivenessProbe 探针来诊断容器的健康状况。LivenessProbe 包含以下 3 种实现方式。

（1）ExecAction：在容器内部执行一个命令，如果该命令的退出状态码为 0，则表明容器健康。

（2）TCPSocketAction：通过容器的 IP 地址和端口号执行 TCP 检查，如果端口能被访问，则表明容器健康。

（3）HTTPGetAction：通过容器的 IP 地址和端口号及路径调用 HTTP Get 方法，如果响应的状态码大于等于 200 且小于等于 400，则认为容器状态健康。

LivenessProbe 探针被包含在 Pod 定义的 spec.containers.{某个容器}中。下面的例子展示了两种 Pod 中容器健康检查的方式：HTTP 检查和容器命令执行检查。下面所列的内容实现了通过容器命令执行检查：

```
livenessProbe:
  exec:
```

```
        command:
        - cat
        - /tmp/health
      initialDelaySeconds: 15
      timeoutSeconds: 1
```

kubelet 在容器中执行 "cat /tmp/health" 命令，如果该命令返回的值为 0，则表明容器处于健康状态，否则表明容器处于不健康状态。

下面所列的内容实现了容器的 HTTP 检查：

```
    livenessProbe:
      httpGet:
        path: /healthz
        port: 8080
      initialDelaySeconds: 15
      timeoutSeconds: 1
```

kubelet 发送一个 HTTP 请求到本地主机、端口及指定的路径，来检查容器的健康状况。

5.4.4 cAdvisor 资源监控

在 Kubernetes 集群中，应用程序的执行情况可以在不同的级别上监测到，这些级别包括：容器、Pod、Service 和整个集群。作为 Kubernetes 集群的一部分，Kubernetes 希望提供给用户详细的各个级别的资源使用信息，这将使用户深入地了解应用的执行情况，并找到应用中可能的瓶颈。

cAdvisor 是一个开源的分析容器资源使用率和性能特性的代理工具，它是因为容器而产生的，因此自然支持 Docker 容器，在 Kubernetes 项目中，cAdvisor 被集成到 Kubernetes 代码中，kubelet 则通过 cAdvisor 获取其所在节点及容器的数据。cAdvisor 自动查找所有在其所在 Node 上的容器，自动采集 CPU、内存、文件系统和网络使用的统计信息。在大部分 Kubernetes 集群中，cAdvisor 通过它所在 Node 的 4194 端口暴露一个简单的 UI。

如图 5.12 所示是 cAdvisor 的一个截图。

kubelet 作为连接 Kubernetes Master 和各 Node 之间的桥梁，管理运行在 Node 上的 Pod 和容器。kubelet 将每个 Pod 都转换成它的成员容器，同时从 cAdvisor 获取单独的容器使用统计信息，然后通过该 REST API 暴露这些聚合后的 Pod 资源使用的统计信息。

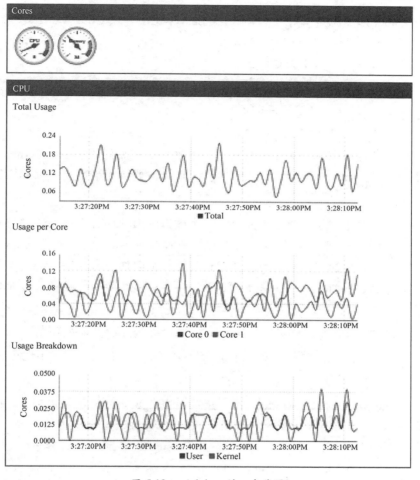

图 5.12 cAdvisor 的一个截图

cAdvisor 只能提供 2～3min 的监控数据，对性能数据也没有持久化，因此在 Kubernetes 早期版本中需要依靠 Heapster 来实现集群范围内全部容器性能指标的采集和查询功能。从 Kubernetes 1.8 版本开始，性能指标数据的查询接口升级为标准的 Metrics API，后端服务则升级为全新的 Metrics Server。因此，cAdvisor 在 4194 端口提供的 UI 和 API 服务从 Kubernetes 1.10 版本开始进入弃用流程，并于 1.12 版本完全关闭。如果还希望使用 cAdvisor 的这个特性，则从 1.13 版本开始可以通过部署一个 DaemonSet 在每个 Node 上启动一个 cAdvisor 来提供 UI 和 API，请参考 cAdvisor 在 GitHub 上的说明

（https://github.com/google/cadvisor）。

在新的Kubernetes监控体系中，Metrics Server用于提供Core Metrics（核心指标），包括Node和Pod的CPU和内存使用数据。其他Custom Metrics（自定义指标）则由第三方组件（如Prometheus）采集和存储。

5.5 kube-proxy 运行机制解析

我们在前面已经了解到，为了支持集群的水平扩展、高可用性，Kubernetes抽象出了Service的概念。Service是对一组Pod的抽象，它会根据访问策略（如负载均衡策略）来访问这组Pod。

Kubernetes在创建服务时会为服务分配一个虚拟的IP地址，客户端通过访问这个虚拟的IP地址来访问服务，服务则负责将请求转发到后端的Pod上。这不就是一个反向代理吗？没错，这就是一个反向代理。但是，它和普通的反向代理有一些不同：首先，它的IP地址是虚拟的，想从外面访问还需要一些技巧；其次，它的部署和启停是由Kubernetes统一自动管理的。

在很多情况下，Service 只是一个概念，而真正将 Service 的作用落实的是它背后的kube-proxy 服务进程。只有理解了kube-proxy 的原理和机制，我们才能真正理解Service背后的实现逻辑。

在Kubernetes集群的每个Node上都会运行一个kube-proxy服务进程，我们可以把这个进程看作Service的透明代理兼负载均衡器，其核心功能是将到某个Service的访问请求转发到后端的多个Pod实例上。此外，Service的Cluster IP与NodePort等概念是kube-proxy服务通过iptables的NAT转换实现的，kube-proxy在运行过程中动态创建与Service相关的iptables规则，这些规则实现了将访问服务（Cluster IP或NodePort）的请求负载分发到后端Pod的功能。由于iptables机制针对的是本地的kube-proxy端口，所以在每个Node上都要运行kube-proxy组件，这样一来，在Kubernetes集群内部，我们可以在任意Node上发起对Service的访问请求。综上所述，由于kube-proxy的作用，在Service的调用过程中客户端无须关心后端有几个Pod，中间过程的通信、负载均衡及故障恢复都是透明的。

起初，kube-proxy进程是一个真实的TCP/UDP代理，类似HA Proxy，负责从Service到Pod 的访问流量的转发，这种模式被称为 userspace（用户空间代理）模式。如图 5.13所示，当某个Pod以Cluster IP方式访问某个Service的时候，这个流量会被Pod所在本机的iptables转发到本机的kube-proxy进程,然后由kube-proxy建立起到后端Pod的TCP/UDP连接,随后将请求转发到某个后端Pod上,并在这个过程中实现负载均衡功能。

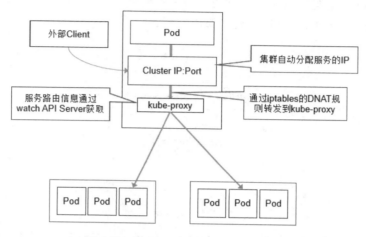

图5.13　Service的负载均衡转发规则

关于Cluster IP与Node Port的实现原理,以及kube-proxy与API Server的交互过程,图5.14给出了较为详细的说明,由于这是最古老的kube-proxy的实现方式,所以不再赘述。

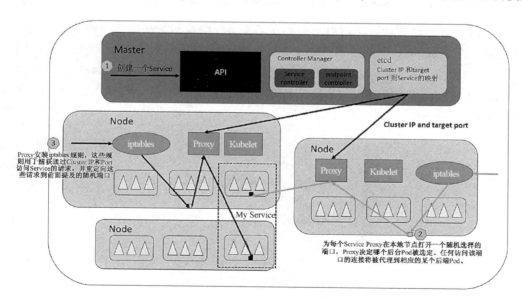

图5.14　kube-proxy工作原理示意

如图 5.15 所示，Kubernetes 从 1.2 版本开始，将 iptables 作为 kube-proxy 的默认模式。iptables 模式下的 kube-proxy 不再起到 Proxy 的作用，其核心功能：通过 API Server 的 Watch 接口实时跟踪 Service 与 Endpoint 的变更信息，并更新对应的 iptables 规则，Client 的请求流量则通过 iptables 的 NAT 机制"直接路由"到目标 Pod。

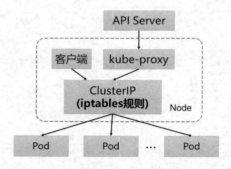

图 5.15　应用程序编程访问 API Server

根据 Kubernetes 的网络模型，一个 Node 上的 Pod 与其他 Node 上的 Pod 应该能够直接建立双向的 TCP/IP 通信通道，所以如果直接修改 iptables 规则，则也可以实现 kube-proxy 的功能，只不过后者更加高端，因为是全自动模式的。与第 1 代的 userspace 模式相比，iptables 模式完全工作在内核态，不用再经过用户态的 kube-proxy 中转，因而性能更强。

iptables 模式虽然实现起来简单，但存在无法避免的缺陷：在集群中的 Service 和 Pod 大量增加以后，iptables 中的规则会急速膨胀，导致性能显著下降，在某些极端情况下甚至会出现规则丢失的情况，并且这种故障难以重现与排查，于是 Kubernetes 从 1.8 版本开始引入第 3 代的 IPVS（IP Virtual Server）模式，如图 5.16 所示。IPVS 在 Kubernetes 1.11 中升级为 GA 稳定版。

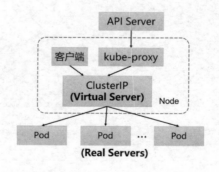

图 5.16　应用程序编程访问 API Server

iptables 与 IPVS 虽然都是基于 Netfilter 实现的，但因为定位不同，二者有着本质的差别：iptables 是为防火墙而设计的；IPVS 则专门用于高性能负载均衡，并使用更高效的数据结构（Hash 表），允许几乎无限的规模扩张，因此被 kube-proxy 采纳为第三代模式。

与 iptables 相比，IPVS 拥有以下明显优势：

◎ 为大型集群提供了更好的可扩展性和性能；
◎ 支持比 iptables 更复杂的复制均衡算法（最小负载、最少连接、加权等）；
◎ 支持服务器健康检查和连接重试等功能；
◎ 可以动态修改 ipset 的集合，即使 iptables 的规则正在使用这个集合。

由于 IPVS 无法提供包过滤、airpin-masquerade tricks（地址伪装）、SNAT 等功能，因此在某些场景（如 NodePort 的实现）下还要与 iptables 搭配使用。在 IPVS 模式下，kube-proxy 又做了重要的升级，即使用 iptables 的扩展 ipset，而不是直接调用 iptables 来生成规则链。

iptables 规则链是一个线性的数据结构，ipset 则引入了带索引的数据结构，因此当规则很多时，也可以很高效地查找和匹配。我们可以将 ipset 简单理解为一个 IP（段）的集合，这个集合的内容可以是 IP 地址、IP 网段、端口等，iptables 可以直接添加规则对这个"可变的集合"进行操作，这样做的好处在于可以大大减少 iptables 规则的数量，从而减少性能损耗。

假设要禁止上万个 IP 访问我们的服务器，则用 iptables 的话，就需要一条一条地添加规则，会在 iptables 中生成大量的规则；但是用 ipset 的话，只需将相关的 IP 地址（网段）加入 ipset 集合中即可，这样只需设置少量的 iptables 规则即可实现目标。

kube-proxy 针对 Service 和 Pod 创建的一些主要的 iptables 规则如下。

◎ KUBE-CLUSTER-IP：在 masquerade-all=true 或 clusterCIDR 指定的情况下对 Service Cluster IP 地址进行伪装，以解决数据包欺骗问题。
◎ KUBE-EXTERNAL-IP：将数据包伪装成 Service 的外部 IP 地址。
◎ KUBE-LOAD-BALANCER、KUBE-LOAD-BALANCER-LOCAL：伪装 Load Balancer 类型的 Service 流量。
◎ KUBE-NODE-PORT-TCP、KUBE-NODE-PORT-LOCAL-TCP、KUBE-NODE-PORT-UDP、KUBE-NODE-PORT-LOCAL-UDP：伪装 NodePort 类型的 Service 流量。

第 6 章

深入分析集群安全机制

Kubernetes 通过一系列机制来实现集群的安全控制,其中包括 API Server 的认证授权、准入控制机制及保护敏感信息的 Secret 机制等。集群的安全性必须考虑如下几个目标。

(1)保证容器与其所在宿主机的隔离。

(2)限制容器给基础设施或其他容器带来的干扰。

(3)最小权限原则——合理限制所有组件的权限,确保组件只执行它被授权的行为,通过限制单个组件的能力来限制它的权限范围。

(4)明确组件间边界的划分。

(5)划分普通用户和管理员的角色。

(6)在必要时允许将管理员权限赋给普通用户。

(7)允许拥有 Secret 数据(Keys、Certs、Passwords)的应用在集群中运行。

下面分别从 Authentication、Authorization、Admission Control、Secret 和 Service Account 等方面来说明集群的安全机制。

6.1 API Server 认证管理

我们知道,Kubernetes 集群中所有资源的访问和变更都是通过 Kubernetes API Server 的 REST API 来实现的,所以集群安全的关键点就在于如何识别并认证客户端身份(Authentication),以及随后访问权限的授权(Authorization)这两个关键问题,本节对认证管理进行说明。

我们知道,Kubernetes 集群提供了 3 种级别的客户端身份认证方式。

◎ 最严格的 HTTPS 证书认证:基于 CA 根证书签名的双向数字证书认证方式。
◎ HTTP Token 认证:通过一个 Token 来识别合法用户。
◎ HTTP Base 认证:通过用户名+密码的方式认证。

首先说说 HTTPS 证书认证的原理。

这里需要有一个 CA 证书,我们知道 CA 是 PKI 系统中通信双方都信任的实体,被称为可信第三方(Trusted Third Party,TTP)。CA 作为可信第三方的重要条件之一就是 CA 的行为具有非否认性。作为第三方而不是简单的上级,就必须能让信任者有追究自己责任的能力。CA 通过证书证实他人的公钥信息,证书上有 CA 的签名。用户如果因为信任证书而有了损失，则证书可以作为有效的证据用于追究 CA 的法律责任。正是因为 CA 承担责任的承诺，所以 CA 也被称为可信第三方。在很多情况下，CA 与用户是相互独立的实体，CA 作为服务提供方，有可能因为服务质量问题（例如，发布的公钥数据有错误）而给用户带来损失。在证书中绑定了公钥数据和相应私钥拥有者的身份信息，并带有 CA 的数字签名；在证书中也包含了 CA 的名称，以便于依赖方找到 CA 的公钥，验证证书上的数字签名。

CA 认证涉及诸多概念，比如根证书、自签名证书、密钥、私钥、加密算法及 HTTPS 等，本书大致讲述 SSL 协议的流程，有助于理解 CA 认证和 Kubernetes CA 认证的配置过程。

如图 6.1 所示，CA 认证大概包含下面几个步骤。

（1）HTTPS 通信双方的服务器端向 CA 机构申请证书，CA 机构是可信的第三方机构，它可以是一个公认的权威企业，也可以是企业自身。企业内部系统一般都用企业自身的认证系统。CA 机构下发根证书、服务端证书及私钥给申请者。

（2）HTTPS 通信双方的客户端向 CA 机构申请证书，CA 机构下发根证书、客户端证书及私钥给申请者。

（3）客户端向服务器端发起请求，服务端下发服务端证书给客户端。客户端接收到证书后，通过私钥解密证书，并利用服务器端证书中的公钥认证证书信息比较证书里的消息，例如，比较域名和公钥与服务器刚刚发送的相关消息是否一致，如果一致，则客户端认可这个服务器的合法身份。

（4）客户端发送客户端证书给服务器端，服务端在接收到证书后，通过私钥解密证书，获得客户端证书公钥，并用该公钥认证证书信息，确认客户端是否合法。

（5）客户端通过随机密钥加密信息，并发送加密后的信息给服务端。在服务器端和客户端协商好加密方案后，客户端会产生一个随机的密钥，客户端通过协商好的加密方案加密该随机密钥，并发送该随机密钥到服务器端。服务器端接收这个密钥后，双方通信的所有内容都通过该随机密钥加密。

上述是双向认证 SSL 协议的具体通信过程，这种情况要求服务器和用户双方都有证书。单向认证 SSL 协议则不需要客户端拥有 CA 证书，对于上面的步骤，只需将服务器端验证客户证书的过程去掉，之后协商对称密码方案和对称通话密钥时，服务器发送给客户的密码没被加密即可。

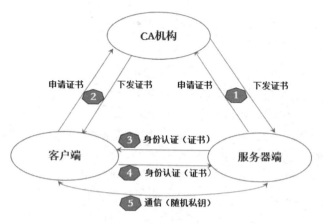

图 6.1　CA 认证流程

然后说说 HTTP Token 的认证原理。

HTTP Token 的认证是用一个很长的特殊编码方式的并且难以被模仿的字符串——Token 来表明客户身份的一种方式。在通常情况下，Token 是一个很复杂的字符串，比如我们用私钥签名一个字符串后的数据就可以被当作一个 Token。此外，每个 Token 对应一个用户名，存储在 API Server 能访问的一个文件中。当客户端发起 API 调用请求时，需要在 HTTP Header 里放入 Token，这样一来，API Server 就能识别合法用户和非法用户了。

最后说说 HTTP Base 认证。

我们知道，HTTP 是无状态的，浏览器和 Web 服务器之间可以通过 Cookie 来进行身份识别。桌面应用程序（比如新浪桌面客户端、SkyDrive 客户端、命令行程序）一般不会使用 Cookie，那么它们与 Web 服务器之间是如何进行身份识别的呢？这就用到了 HTTP Base 认证，这种认证方式是把"用户名+冒号+密码"用 BASE64 算法进行编码后的字符串放在 HTTP Request 中的 Header Authorization 域里发送给服务端，服务端在收到后进行解码，获取用户名及密码，然后进行用户身份鉴权。

6.2　API Server 授权管理

当客户端发起 API Server 调用时，API Server 内部要先进行用户认证，然后执行用户授权流程，即通过授权策略来决定一个 API 调用是否合法。对合法用户进行授权并且随后在用户访问时进行鉴权，是权限与安全系统的重要一环。简单地说，授权就是授予不同的用户不同的访问权限。API Server 目前支持以下几种授权策略（通过 API Server 的启动参数 "--authorization-mode" 设置）。

- AlwaysDeny：表示拒绝所有请求，一般用于测试。
- AlwaysAllow：允许接收所有请求，如果集群不需要授权流程，则可以采用该策略，这也是 Kubernetes 的默认配置。
- ABAC（Attribute-Based Access Control）：基于属性的访问控制，表示使用用户配置的授权规则对用户请求进行匹配和控制。
- Webhook：通过调用外部 REST 服务对用户进行授权。
- RBAC：Role-Based Access Control，基于角色的访问控制。
- Node：是一种专用模式，用于对 kubelet 发出的请求进行访问控制。

API Server 在接收到请求后，会读取该请求中的数据，生成一个访问策略对象，如果在该请求中不带某些属性（如 Namespace），则这些属性的值将根据属性类型的不同，设置不同的默认值（例如，为字符串类型的属性设置一个空字符串；为布尔类型的属性设置 false；为数值类型的属性设置 0）。然后将这个访问策略对象和授权策略文件中的所有访问策略对象逐条匹配，如果至少有一个策略对象被匹配，则该请求被鉴权通过，否则终止 API 调用流程，并返回客户端的错误调用码。

6.2.1　ABAC 授权模式详解

在 API Server 启用 ABAC 模式时，需要指定授权策略文件的路径和名称（--authorization-policy-file=SOME_FILENAME），授权策略文件里的每一行都以一个 Map 类型的 JSON 对象进行设置，这被称为"访问策略对象"。通过设置访问策略对象中的 apiVersion、kind、spec 属性来确定具体的授权策略，其中，apiVersion 当前版本为 abac.authorization.kubernetes.io/v1beta1；kind 被设置为 Policy；spec 指详细的策略设置，包括主题属性、资源属性、非资源属性这三个字段，如下所述。

（1）主体属性

- user（用户名）：字符串类型，该字符串类型的用户名来源于 Token 文件（--token-auth-file 参数设置的文件）或基本认证文件中用户名称段的值。
- group（用户组）：在被设置为 "system:authenticated" 时表示匹配所有已认证的请求，在被设置为 "system:unauthenticated" 时表示匹配所有未认证的请求。

（2）资源属性

- apiGroup（API 组）：字符串类型，表明匹配哪些 API Group，例如 extensions 或*（表示匹配所有 API Group）。
- namespace（命名空间）：字符串类型，表明该策略允许访问某个 Namespace 的资源，例如 kube-system 或*（表示匹配所有 Namespace）。
- resource（资源）：字符串类型，API 资源对象，例如 pods 或*（表示匹配所有资源对象）。

（3）非资源属性

- nonResourcePath（非资源对象类路径）：非资源对象类的 URL 路径，例如/version 或/apis，*表示匹配所有非资源对象类的请求路径，也可以设置为子路径，/foo/*表示匹配所有/foo 路径下的所有子路径。
- readonly（只读标识）：布尔类型，当它的值为 true 时，表明仅允许 GET 请求通过。

下面对 ABAC 授权算法、使用 kubectl 时的授权机制、常见 ABAC 授权示例、以及如何对 Service Account 进行授权进行说明。

1. ABAC 授权算法

API Server 进行 ABAC 授权的算法为：在 API Server 收到请求之后，首先识别出请求携带的策略对象的属性，然后根据在策略文件中定义的策略对这些属性进行逐条匹配，以判定是否允许授权。如果有至少一条匹配成功，那么这个请求就通过了授权（不过还是可能在后续其他授权校验中失败）。常见的策略配置如下。

- 要允许所有认证用户做某件事，可以写一个策略，将 group 属性设置为 system:authenticated。
- 要允许所有未认证用户做某件事，可以把策略的 group 属性设置为 system:unauthenticated。
- 要允许一个用户做任何事，将策略的 apiGroup、namespace、resource 和 nonResourcePath 属性设置为 "*" 即可。

2. 使用 kubectl 时的授权机制

kubectl 使用 API Server 的/api 和/apis 端点来获取版本信息。要验证 kubectl create/update 命令发送给服务器的对象，kubectl 需要向 OpenAPI 进行查询，对应的 URL

路径为/openapi/v2。

当使用 ABAC 授权模式时，下列特殊资源必须显式地通过 nonResourcePath 属性进行设置。

- ◎ API 版本协商过程中的/api、/api/*、/apis、和/apis/*。
- ◎ 使用 kubectl version 命令从服务器获取版本时的/version。
- ◎ create/update 操作过程中的/swaggerapi/*。

在使用 kubectl 操作时，如果需要查看发送到 API Server 的 HTTP 请求，则可以将日志级别设置为 8，例如：

```
# kubectl --v=8 version
```

3. 常见的 ABAC 授权示例

下面通过几个授权策略文件（JSON 格式）示例说明 ABAC 的访问控制用法。

（1）允许用户 alice 对所有资源做任何操作：

```
{"apiVersion": "abac.authorization.kubernetes.io/v1beta1", "kind": "Policy", "spec": {"user": "alice", "namespace": "*", "resource": "*", "apiGroup": "*"}}
```

（2）kubelet 可以读取任意 Pod：

```
{"apiVersion": "abac.authorization.kubernetes.io/v1beta1", "kind": "Policy", "spec": {"user": "`", "namespace": "*", "resource": "pods", "readonly": true}}
```

（3）kubelet 可以读写 Event 对象：

```
{"apiVersion": "abac.authorization.kubernetes.io/v1beta1", "kind": "Policy", "spec": {"user": "kubelet", "namespace": "*", "resource": "events"}}
```

（4）用户 bob 只能读取 projectCaribou 中的 Pod：

```
{"apiVersion": "abac.authorization.kubernetes.io/v1beta1", "kind": "Policy", "spec": {"user": "bob", "namespace": "projectCaribou", "resource": "pods", "readonly": true}}
```

（5）任何用户都可以对非资源类路径进行只读请求：

```
{"apiVersion": "abac.authorization.kubernetes.io/v1beta1", "kind": "Policy", "spec": {"group": "system:authenticated", "readonly": true, "nonResourcePath": "*"}}
{"apiVersion": "abac.authorization.kubernetes.io/v1beta1", "kind": "Policy", "spec": {"group": "system:unauthenticated", "readonly": true, "nonResourcePath":
```

```
"*"}}
```

如果添加了新的 ABAC 策略,则需要重启 API Server 以使其生效。

4. 对 Service Account 进行授权

Service Account 会自动生成一个 ABAC 用户名(username),用户名按照以下命名规则生成:

```
system:serviceaccount:<namespace>:<serviceaccountname>
```

创建新的命名空间时,会产生一个如下名称的 Service Account:

```
system:serviceaccount:<namespace>:default
```

如果希望 kube-system 命名空间中的 Service Account "default" 具有全部权限,就要在策略文件中加入如下内容:

```
{"apiVersion":"abac.authorization.kubernetes.io/v1beta1","kind":"Policy","spec":{"user":"system:serviceaccount:kube-system:default","namespace":"*","resource":"*","apiGroup":"*"}}
```

6.2.2 Webhook 授权模式详解

Webhook 定义了一个 HTTP 回调接口,实现 Webhook 的应用会在指定事件发生时,向一个 URL 地址发送(POST)通知信息。启用 Webhook 授权模式后,Kubernetes 会调用外部 REST 服务对用户进行授权。

Webhook 模式用参数 --authorization-webhook-config-file=SOME_FILENAME 来设置远端授权服务的信息。

配置文件使用的是 kubeconfig 文件的格式。文件里 user 一节的内容指的是 API Server。相对于远程授权服务来说,API Server 是客户端,也就是用户;cluster 一节的内容指的是远程授权服务器的配置。下面的例子为设置一个使用 HTTPS 客户端认证的配置:

```
clusters:                 # 指向远端服务
  - name: name-of-remote-authz-service
    cluster:
      certificate-authority: /path/to/ca.pem      # 验证远端服务的 CA
      server: https://authz.example.com/authorize # 远端服务的 URL,必须使用 HTTPS
```

```
users:                   # API Server 的 Webhook 配置
  - name: name-of-api-server
    user:
      client-certificate: /path/to/cert.pem # Webhook 插件使用的证书
      client-key: /path/to/key.pem          # 证书的 key
current-context: webhook     # kubeconfig 文件需要设置 context
contexts:
- context:
    cluster: name-of-remote-authz-service
    user: name-of-api-server
  name: webhook
```

在授权开始时，API Server 会生成一个 api.authorization.v1beta1.SubjectAccessReview 对象，用于描述操作信息，在进行 JSON 序列化之后 POST 出来。在这个对象中包含用户尝试访问资源的请求动作的描述，以及被访问资源的属性。

Webhook API 对象和其他 API 对象一样，遵循同样的版本兼容性规则，在实现时要注意 apiVersion 字段的版本，以实现正确的反序列化操作。另外，API Server 必须启用 authorization.k8s.io/v1beta1 API 扩展（--runtime-config=authorization.k8s.io/v1beta1=true）。

下面是一个希望获取 Pod 列表的请求报文示例：

```
{
  "apiVersion": "authorization.k8s.io/v1beta1",
  "kind": "SubjectAccessReview",
  "spec": {
    "resourceAttributes": {
      "namespace": "kittensandponies",
      "verb": "get",
      "group": "unicorn.example.org",
      "resource": "pods"
    },
    "user": "jane",
    "group": [
      "group1",
      "group2"
    ]
  }
}
```

远端服务需要填充请求中的 SubjectAccessReviewStatus 字段，并返回允许或不允许访问的结果。应答报文中的 spec 字段是无效的,也可以省略。

一个返回"运行访问"的应答报文示例如下:

```
{
  "apiVersion": "authorization.k8s.io/v1beta1",
  "kind": "SubjectAccessReview",
  "status": {
    "allowed": true
  }
}
```

一个返回"不允许访问"的应答报文示例如下:

```
{
  "apiVersion": "authorization.k8s.io/v1beta1",
  "kind": "SubjectAccessReview",
  "status": {
    "allowed": false,
    "reason": "user does not have read access to the namespace"
  }
}
```

非资源的访问请求路径包括/api、/apis、/metrics、/resetMetrics、/logs、/debug、/healthz、/swagger-ui/、/swaggerapi/、/ui 和/version。通常可以对/api、/api/*、/apis、/apis/*和/version 对于客户端发现服务器提供的资源和版本信息给予"允许"授权,对于其他非资源的访问一般可以禁止,以限制客户端对 API Server 进行没有必要的查询。

查询/debug 的请求报文示例如下:

```
{
  "apiVersion": "authorization.k8s.io/v1beta1",
  "kind": "SubjectAccessReview",
  "spec": {
    "nonResourceAttributes": {
      "path": "/debug",
      "verb": "get"
    },
    "user": "jane",
    "group": [
      "group1",
      "group2"
    ]
```

```
    }
}
```

6.2.3 RBAC 授权模式详解

RBAC（Role-Based Access Control，基于角色的访问控制）在 Kubernetes 的 1.5 版本中引入，在 1.6 版本时升级为 Beta 版本，在 1.8 版本时升级为 GA。作为 kubeadm 安装方式的默认选项，足见其重要程度。相对于其他访问控制方式，新的 RBAC 具有如下优势。

◎ 对集群中的资源和非资源权限均有完整的覆盖。
◎ 整个 RBAC 完全由几个 API 对象完成，同其他 API 对象一样，可以用 kubectl 或 API 进行操作。
◎ 可以在运行时进行调整，无须重新启动 API Server。

要使用 RBAC 授权模式，需要在 API Server 的启动参数中加上--authorization-mode=RBAC。

下面对 RBAC 的原理和用法进行说明。

1. RBAC 的 API 资源对象说明

RBAC 引入了 4 个新的顶级资源对象：Role、ClusterRole、RoleBinding 和 ClusterRoleBinding。同其他 API 资源对象一样，用户可以使用 kubectl 或者 API 调用等方式操作这些资源对象。

1）角色（Role）

一个角色就是一组权限的集合，这里的权限都是许可形式的，不存在拒绝的规则。在一个命名空间中，可以用角色来定义一个角色，如果是集群级别的，就需要使用 ClusterRole 了。

角色只能对命名空间内的资源进行授权，在下面例子中定义的角色具备读取 Pod 的权限：

```
kind: Role
apiVersion: rbac.authorization.k8s.io/v1
metadata:
  namespace: default
  name: pod-reader
```

```
rules:
- apiGroups: [""]      # "" 空字符串,表示核心 API 群
  resources: ["pods"]
  verbs: ["get", "watch", "list"]
```

rules 中的参数说明如下。

- apiGroups:支持的 API 组列表,例如 "apiVersion: batch/v1" "apiVersion: extensions: v1beta1" "apiVersion: apps/v1beta1" 等,详细的 API 组说明参见第 9 章的说明。
- resources:支持的资源对象列表,例如 pods、deployments、jobs 等。
- verbs:对资源对象的操作方法列表,例如 get、watch、list、delete、replace、patch 等,详细的操作方法说明参见第 9 章的说明。

2)集群角色(ClusterRole)

集群角色除了具有和角色一致的命名空间内资源的管理能力,因其集群级别的范围,还可以用于以下特殊元素的授权。

- 集群范围的资源,例如 Node。
- 非资源型的路径,例如 "/healthz"。
- 包含全部命名空间的资源,例如 pods(用于 kubectl get pods --all-namespaces 这样的操作授权)。

下面的集群角色可以让用户有权访问任意一个或所有命名空间的 secrets(视其绑定方式而定):

```
kind: ClusterRole
apiVersion: rbac.authorization.k8s.io/v1
metadata:
  # ClusterRole 不受限于命名空间,所以无须设置 Namespace 的名称
rules:
- apiGroups: [""]
  resources: ["secrets"]
  verbs: ["get", "watch", "list"]
```

3)角色绑定(RoleBinding)和集群角色绑定(ClusterRoleBinding)

角色绑定或集群角色绑定用来把一个角色绑定到一个目标上,绑定目标可以是 User(用户)、Group(组)或者 Service Account。使用 RoleBinding 为某个命名空间授权,使用 ClusterRoleBinding 为集群范围内授权。

RoleBinding 可以引用 Role 进行授权。下面的例子中的 RoleBinding 将在 default 命名空间中把 pod-reader 角色授予用户 jane，这一操作可以让 jane 读取 default 命名空间中的 Pod：

```
kind: RoleBinding
apiVersion: rbac.authorization.k8s.io/v1
metadata:
  name: read-pods
  namespace: default
subjects:
- kind: User
  name: jane
  apiGroup: rbac.authorization.k8s.io
roleRef:
  kind: Role
  name: pod-reader
  apiGroup: rbac.authorization.k8s.io
```

RoleBinding 也可以引用 ClusterRole，对属于同一命名空间内 ClusterRole 定义的资源主体进行授权。一种常见的做法是集群管理员为集群范围预先定义好一组 ClusterRole，然后在多个命名空间中重复使用这些 ClusterRole。

例如，在下面的例子中，虽然 secret-reader 是一个集群角色，但是因为使用了 RoleBinding，所以 dave 只能读取 development 命名空间中的 secret：

```
kind: RoleBinding
apiVersion: rbac.authorization.k8s.io/v1
metadata:
  name: read-secrets
  namespace: development # 集群角色中，只有development命名空间中的权限才能赋予dave
subjects:
- kind: User
  name: dave
  apiGroup: rbac.authorization.k8s.io
roleRef:
  kind: ClusterRole
  name: secret-reader
  apiGroup: rbac.authorization.k8s.io
```

集群角色绑定中的角色只能是集群角色，用于进行集群级别或者对所有命名空间都生效的授权。下面的例子允许 manager 组的用户读取任意 Namespace 中的 secret：

```
kind: ClusterRoleBinding
apiVersion: rbac.authorization.k8s.io/v1
metadata:
  name: read-secrets-global
subjects:
- kind: Group
  name: manager
  apiGroup: rbac.authorization.k8s.io
roleRef:
  kind: ClusterRole
  name: secret-reader
  apiGroup: rbac.authorization.k8s.io
```

图 6.2 展示了上述对 Pod 的 get/watch/list 操作进行授权的 Role 和 RoleBinding 的逻辑关系。

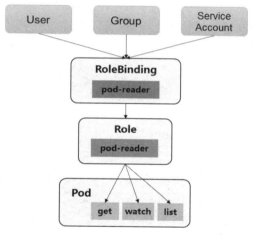

图 6.2　RoleBinding、Role 与对 Pod 的操作授权

2. 对资源的引用方式

多数资源可以用其名称的字符串来表达，也就是 Endpoint 中的 URL 相对路径，例如 pods。然而，某些 Kubernetes API 包含下级资源，例如 Pod 的日志（logs）。Pod 日志的 Endpoint 是 GET/ api/v1/namespaces/{namespace}/pods/{name}/log。

在这个例子中，Pod 是一个命名空间内的资源，log 就是一个下级资源。要在一个 RBAC 角色中体现，就需要用斜线"/"来分隔资源和下级资源。若想授权让某个主体同时能够读取 Pod 和 Pod log，则可以配置 resources 为一个数组：

```
kind: Role
apiVersion: rbac.authorization.k8s.io/v1
metadata:
  namespace: default
  name: pod-and-pod-logs-reader
rules:
- apiGroups: [""]
  resources: ["pods", "pods/log"]
  verbs: ["get", "list"]
```

资源还可以通过名称（ResourceName）进行引用。在指定 ResourceName 后，使用 get、delete、update 和 patch 动词的请求，就会被限制在这个资源实例范围内。例如，下面的声明让一个主体只能对一个 ConFigmap 进行 get 和 update 操作：

```
kind: Role
apiVersion: rbac.authorization.k8s.io/v1
metadata:
  namespace: default
  name: configmap-updater
rules:
- apiGroups: [""]
  resources: ["configmap"]
  resourceNames: ["my-configmap"]
  verbs: ["update", "get"]
```

可想而知，resourceName 这种用法对 list、watch、create 或 deletecollection 操作是无效的，这是因为必须要通过 URL 进行鉴权，而资源名称在 list、watch、create 或 deletecollection 请求中只是请求 Body 数据的一部分。

下面对常见的角色和角色绑定给出示例，提供参考用法。

3. 常用的角色示例

注意，下面的例子只展示了 rules 部分的内容。

（1）允许读取核心 API 组中的 Pod 资源：

```
rules:
- apiGroups: [""]
  resources: ["pods"]
```

```
verbs: ["get", "list", "watch"]
```

（2）允许读写 extensions 和 apps 两个 API 组中的 deployment 资源：

```
rules:
- apiGroups: ["extensions", "apps"]
  resources: ["deployments"]
  verbs: ["get", "list", "watch", "create", "update", "patch", "delete"]
```

（3）允许读取 pods 及读写 jobs：

```
rules:
- apiGroups: [""]
  resources: ["pods"]
  verbs: ["get", "list", "watch"]
- apiGroups: ["batch", "extensions"]
  resources: ["jobs"]
  verbs: ["get", "list", "watch", "create", "update", "patch", "delete"]
```

（4）允许读取一个名为 my-config 的 ConfigMap（必须绑定到一个 RoleBinding 来限制到一个 Namespace 下的 ConfigMap）：

```
rules:
- apiGroups: [""]
  resources: ["configmaps"]
  resourceNames: ["my-config"]
  verbs: ["get"]
```

（5）读取核心组的 Node 资源（Node 属于集群级的资源，所以必须存在于 ClusterRole 中，并使用 ClusterRoleBinding 进行绑定）：

```
rules:
- apiGroups: [""]
  resources: ["nodes"]
  verbs: ["get", "list", "watch"]
```

（6）允许对非资源端点"/healthz"及其所有子路径进行 GET 和 POST 操作（必须使用 ClusterRole 和 ClusterRoleBinding）：

```
rules:
- nonResourceURLs: ["/healthz", "/healthz/*"]
  verbs: ["get", "post"]
```

4. 常见的角色绑定示例

注意，在下面的例子中只包含 subjects 部分的内容。

（1）用户名 alice@example.com：

```
subjects:
- kind: User
  name: "alice@example.com"
  apiGroup: rbac.authorization.k8s.io
```

（2）组名 frontend-admins：

```
subjects:
- kind: Group
  name: "frontend-admins"
  apiGroup: rbac.authorization.k8s.io
```

（3）kube-system 命名空间中的默认 Service Account：

```
subjects:
- kind: ServiceAccount
  name: default
  namespace: kube-system
```

（4）qa 命名空间中的所有 Service Account：

```
subjects:
- kind: Group
  name: system:serviceaccounts:qa
  apiGroup: rbac.authorization.k8s.io
```

（5）所有 Service Account：

```
subjects:
- kind: Group
  name: system:serviceaccounts
  apiGroup: rbac.authorization.k8s.io
```

（6）所有认证用户（Kubernetes 1.5 以上版本）：

```
subjects:
- kind: Group
  name: system:authenticated
  apiGroup: rbac.authorization.k8s.io
```

（7）所有未认证用户（Kubernetes 1.5 以上版本）：

```
subjects:
- kind: Group
  name: system:unauthenticated
  apiGroup: rbac.authorization.k8s.io
```

（8）全部用户（Kubernetes 1.5 以上版本）：

```
subjects:
- kind: Group
  name: system:authenticated
  apiGroup: rbac.authorization.k8s.io
- kind: Group
  name: system:unauthenticated
  apiGroup: rbac.authorization.k8s.io
```

5. 默认的角色和角色绑定

API Server 会创建一套默认的 ClusterRole 和 ClusterRoleBinding 对象，其中很多是以"system:"为前缀的，以表明这些资源属于基础架构，对这些对象的改动可能造成集群故障。举例来说，system:node 这个 ClusterRole 为 kubelet 定义了权限，如果这个集群角色被改动了，kubelet 就会停止工作。

所有默认的 ClusterRole 和 RoleBinding 都会用标签 kubernetes.io/bootstrapping=rbac-defaults 进行标记。

下面对一些常见的默认 ClusterRole 和 ClusterRoleBinding 对象进行说明。

对系统角色的说明如表 6.1 所示。

表 6.1 系统角色

默认的 ClusterRole	默认的 ClusterRoleBinding	描述
system:basic-user	system:authenticated	让用户能够读取自身的信息（在 Kubernetes 1.14 版本之前，还绑定了 system:unauthenticated 组）
system:discovery	system:authenticated	对 API 发现 Endpoint 的只读访问，用于 API 级别的发现和协商（在 Kubernetes 1.14 版本之前还绑定了 system:unauthenticated 组）
system:public-info-viewer	system:authenticated 和 system:unauthenticated 组	允许读取集群的非敏感信息（从 Kubernetes 1.14 开始引入）

对用户角色的说明如表 6.2 所示。有些默认角色不是以 "system:" 为前缀的，这部分角色是针对用户的。其中包含超级用户角色（cluster-admin），有的用于集群一级的角色（cluster-status），还有针对 Namespace 的角色（admin、edit、view）。

表 6.2 用户角色

默认的 ClusterRole	默认的 ClusterRoleBinding	描述
cluster-admin	system:masters 组	让超级用户可以对任何资源执行任何操作。如果在 ClusterRoleBinding 中使用，则影响的是整个集群所有 Namespace 中的任何资源；如果使用的是 RoleBinding，则能控制这一绑定的 Namespace 中的资源，还包括 Namespace 本身
admin	None	允许 admin 访问，可以限制在一个 Namespace 中使用 RoleBinding。如果在 RoleBinding 中使用，则允许对 Namespace 中的大多数资源进行读写访问，其中包含创建角色和角色绑定的能力。这一角色不允许操作 Namespace 本身，也不能写入资源限额
edit	None	允许对命名空间中的大多数资源进行读写操作，不允许查看或修改角色，以及角色绑定
view	None	允许对多数对象进行只读操作，但是对角色、角色绑定及 secret 是不可访问的

对核心 Master 组件角色的说明如表 6.3 所示。

表 6.3 核心组件角色

默认的 ClusterRole	默认的 ClusterRoleBinding	描述
system:kube-scheduler	system:kube-scheduler 用户	能够访问 kube-scheduler 组件所需的资源
system:volume-scheduler	system:kube-scheduler 用户	允许 kube-scheduler 组件访问 Volume 资源
system:kube-controller-manager	system:kube-controller-manager 用户	能够访问 kube-controller-manager 组件所需的资源。不同的控制所需的权限参见表 6.4
system:node	None	允许访问 kubelet 组件所需的资源，包括对所有 secret 的读取，以及对所有 Pod 状态的写访问。从 Kubernetes 1.8 版本开始，就没有自动绑定过程了（在 Kubernetes 1.7 及以前的版本中绑定了 system:nodes 组）
system:node-proxier	system:kube-proxy 用户	允许访问 kube-proxy 所需的资源

对其他组件角色的说明如表 6.4 所示。

表 6.4 其他组件角色

默认的 ClusterRole	默认的 ClusterRoleBinding	描述
system:auth-delegator	None	允许对授权和认证进行托管,通常用于附加的 API 服务器
system:heapster	None	Heapster 组件的角色
system:kube-aggregator	None	kube-aggregator 的角色
system:kube-dns	在 kube-system namespace 中 kube-dns 的 Service Account	kube-dns 的角色
system:kubelet-api-admin	None	允许对 kubelet API 的完全访问
system:node-bootstrapper	None	允许访问 kubelet TLS 启动所需的资源
system:node-problem-detector	None	允许访问 node-problem-detector 组件所需的资源
system:persistent-volume-provisioner	None	允许访问多数动态卷供给所需的资源

Controller 角色如表 6.5 所示。

表 6.5 Controller 角色

需要赋予的角色
system:controller:attachdetach-controller
system:controller:certificate-controller
system:controller:clusterrole-aggregation-controller
system:controller:cronjob-controller
system:controller:daemon-set-controller
system:controller:deployment-controller
system:controller:disruption-controller
system:controller:endpoint-controller
system:controller:expand-controller
system:controller:generic-garbage-collector
system:controller:horizontal-pod-autoscaler
system:controller:job-controller
system:controller:namespace-controller
system:controller:node-controller
system:controller:persistent-volume-binder

续表

需要赋予的角色
system:controller:pod-garbage-collector
system:controller:pv-protection-controller
system:controller:pvc-protection-controller
system:controller:replicaset-controller
system:controller:replication-controller
system:controller:resourcequota-controller
system:controller:root-ca-cert-publisher
system:controller:route-controller
system:controller:service-account-controller
system:controller:service-controller
system:controller:statefulset-controller
system:controller:ttl-controller

Kubernetes Controller Manager 负责的是核心控制流。如果用 --use-service-account-credentials 调用，则每个控制过程都会使用不同的 Service Account 启动，因此就有了对应各个控制过程的角色，前缀是 system:controller。如果 Controller Manager 没有用 --use-service-account-credentials 启动参数，则将使用自己的凭据运行各个控制流程，这就需要为该凭据授予所有相关角色。

6. 授权注意事项：预防提升权限和授权初始化

RBAC API 拒绝用户通过编辑角色或者角色绑定进行提升权限。这一限制是在 API 层面做出的，因此即使 RBAC 没有启用也仍然有效。

用户要对角色进行创建或更新操作，需要满足下列至少一个条件：

（1）拥有一个角色的所有权限，且与该角色的生效范围一致（如果是集群角色，则是集群范围；如果是普通角色，则可能是同一个命名空间或者整个集群）；

（2）为用户显式授予针对该角色或集群角色的提权（escalate）操作的权限（要求 Kubernetes 1.12 及以上版本）。

例如，用户 user-1 没有列出集群中所有 secret 的权限，就不能创建具有这一权限的集群角色。要让一个用户能够创建或更新角色，需要：

（1）为其授予一个允许创建或更新 Role 或 ClusterRole 资源对象的角色；

（2）为其授予绑定某一角色的权限，有隐式或显式两种方法。

- ◎ 隐式：为用户授予权限，要覆盖该用户所能控制的所有权限范围。用户如果尝试创建超出其自身权限的角色或者集群角色，则该 API 调用会被禁止。
- ◎ 显式：为用户显式授予针对该 Role 或 ClusterRole 的提权（Escalate）操作的权限（要求 Kubernetes 1.12 及以上版本）。

如果一个用户的权限包含一个角色的所有权限，就可以为其创建和更新角色绑定（要求同样的作用范围）；或者如果被授予了针对某个角色的绑定授权，则也有权完成此操作。例如，如果用户 user-1 没有列出集群内所有 secret 的权限，就无法为一个具有这样权限的角色创建集群角色绑定。要使用户能够创建、更新这一角色绑定，则需要这样做，如下所述。

（1）为其授予一个允许创建和更新 RoleBinding 或 ClusterRoleBinding 的角色。

（2）为其授予绑定某一角色的权限，有隐式或显式两种方法。

- ◎ 隐式：让其具有该角色的所有权限。
- ◎ 显式：为用户授予针对该角色（或集群角色）的绑定操作的权限。

例如，下面的集群角色和角色绑定能让 user-1 为其他用户在 user-1-namespace 命名空间中授予 admin、edit 及 view 角色：

```
apiVersion: rbac.authorization.k8s.io/v1
kind: ClusterRole
metadata:
  name: role-grantor
rules:
- apiGroups: ["rbac.authorization.k8s.io"]
  resources: ["rolebindings"]
  verbs: ["create"]
- apiGroups: ["rbac.authorization.k8s.io"]
  resources: ["clusterroles"]
  verbs: ["bind"]
  resourceNames: ["admin","edit","view"]
---
apiVersion: rbac.authorization.k8s.io/v1
```

```
kind: RoleBinding
metadata:
  name: role-grantor-binding
  namespace: user-1-namespace
roleRef:
  apiGroup: rbac.authorization.k8s.io
  kind: ClusterRole
  name: role-grantor
subjects:
- apiGroup: rbac.authorization.k8s.io
  kind: User
  name: user-1
```

在进行第 1 个角色和角色绑定时,必须让初始用户具备其尚未被授予的权限。要进行初始的角色和角色绑定设置,有以下两种办法。

(1)使用属于 system:masters 组的身份,这一群组默认具有 cluster-admin 这一超级角色的绑定。

(2)如果 API Server 以 --insecure-port 参数运行,则客户端通过这个非安全端口进行接口调用,这一端口没有认证鉴权的限制。

7. 对 Service Account 的授权管理

默认的 RBAC 策略为控制平台组件、节点和控制器授予有限范围的权限,但是除 kube-system 外的 Service Account 是没有任何权限的(除了所有认证用户都具有的 discovery 权限)。

这就要求用户为 Service Account 赋予所需的权限。细粒度的角色分配能够提高安全性,但也会提高管理成本。粗放的授权方式可能会给 Service Account 多余的权限,但更易于管理。

下面的实践以安全性递减的方式排序。

(1)为一个应用专属的 Service Account 赋权(最佳实践)。

这个应用需要在 Pod 的 Spec 中指定一个 serviceAccountName,用 API、Application Manifest、kubectl create serviceaccount 命令等创建 Service Account,例如为 my-namespace 中的 "my-sa" Service Account 授予只读权限:

```
$ kubectl create rolebinding my-sa-view \
```

```
  --clusterrole=view \
  --serviceaccount=my-namespace:my-sa \
  --namespace=my-namespace
```

（2）为一个命名空间中名为 default 的 Service Account 授权。

如果一个应用没有指定 serviceAccountName，则会使用名为 default 的 Service Account。注意，赋给 Service Account "default" 的权限会让所有没有指定 serviceAccountName 的 Pod 都具有这些权限。

例如，在 my-namespace 命名空间中为 Service Account "default" 授予只读权限：

```
$ kubectl create rolebinding default-view \
  --clusterrole=view \
  --serviceaccount=my-namespace:default \
  --namespace=my-namespace
```

另外，许多系统级 Add-Ons 都需要在 kube-system 命名空间中运行。要让这些 Add-Ons 能够使用超级用户权限，则可以把 cluster-admin 权限赋予 kube-system 命名空间中名为 default 的 Service Account。注意，这一操作意味着 kube-system 命名空间包含了通向 API 超级用户的捷径：

```
$ kubectl create clusterrolebinding add-on-cluster-admin \
  --clusterrole=cluster-admin \
  --serviceaccount=kube-system:default
```

（3）为命名空间中的所有 Service Account 都授予一个角色。

如果希望在一个命名空间中，任何 Service Account 的应用都具有一个角色，则可以为这一命名空间的 Service Account 群组进行授权。

例如，为 my-namespace 命名空间中的所有 Service Account 赋予只读权限：

```
$ kubectl create rolebinding serviceaccounts-view \
  --clusterrole=view \
  --group=system:serviceaccounts:my-namespace \
  --namespace=my-namespace
```

（4）为集群范围内的所有 Service Account 都授予一个低权限角色（不推荐）。

如果不想为每个命名空间管理授权，则可以把一个集群级别的角色赋给所有 Service Account。

例如，为所有命名空间中的所有 Service Account 授予只读权限：

```
$ kubectl create clusterrolebinding serviceaccounts-view \
  --clusterrole=view \
  --group=system:serviceaccounts
```

（5）为所有 Service Account 授予超级用户权限（强烈不建议这样设置）。

如果完全不在意权限，则可以把超级用户权限分配给每个 Service Account。

注意，这让所有具有读取 Secret 权限的用户都可以创建 Pod 来访问超级用户的专属权限：

```
$ kubectl create clusterrolebinding serviceaccounts-cluster-admin \
  --clusterrole=cluster-admin \
  --group=system:serviceaccounts
```

8. 使用 kubectl 命令行工具创建资源对象

除了使用 YAML 配置文件来创建这些资源对象，也可以直接使用 kubectl 命令行工具对它们进行创建。下面通过几个例子进行说明。

（1）在命名空间 acme 中为用户 bob 授权 admin ClusterRole：

```
kubectl create rolebinding bob-admin-binding --clusterrole=admin --user=bob --namespace=acme
```

（2）在命名空间 acme 中为名为 myapp 的 Service Account 授予 view ClusterRole：

```
kubectl create rolebinding myapp-view-binding --clusterrole=view --serviceaccount=acme:myapp --namespace=acme
```

（3）在全集群范围内为用户 root 授予 cluster-admin ClusterRole：

```
kubectl create clusterrolebinding root-cluster-admin-binding --clusterrole=cluster-admin --user=root
```

（4）在全集群范围内为用户 kubelet 授予 system:node ClusterRole：

```
kubectl create clusterrolebinding kubelet-node-binding --clusterrole=system:node --user=kubelet
```

（5）在全集群范围内为名为 myapp 的 Service Account 授予 view ClusterRole：

```
kubectl create clusterrolebinding myapp-view-binding --clusterrole=view --serviceaccount=acme:myapp
```

可以在 kubectl --help 的帮助中查看更详细的说明。

9. RBAC 的 Auto-reconciliation（自动恢复）功能

自动恢复从 Kubernetes 1.6 版本开始引入。每次启动时，API Server 都会更新默认的集群角色的缺失权限，也会刷新在默认的角色绑定中缺失的主体，这样就防止了一些破坏性的修改，也保证了在集群升级的情况下相关内容能够及时更新。

如果不希望使用这一功能，则可以将一个默认的集群角色或者角色绑定的 Annotation 注解 "rbac.authorization.kubernetes.io/autoupdate" 值设置为 false。

10. 从旧版本的授权策略升级到 RBAC

在 Kubernetes 1.6 之前，很多 Deployment 都试用了比较宽松的 ABAC 策略，包含为所有 Service Account 都开放完全 API 访问权限。

默认的 RBAC 策略为控制台组件、节点和控制器授予了范围受限的权限，但是不会为 kube-system 以外的 Service Account 授予任何权限。

这样一来，可能会对现有的一些工作负载造成影响，这时有两种办法来解决这一问题。

（1）并行认证。RBAC 和 ABAC 同时运行，并包含传统的 ABAC 策略：

```
--authorization-mode=RBAC,ABAC --authorization-policy-file=mypolicy.jsonl
```

首先会由 RBAC 尝试对请求进行鉴权，如果得到的结果是拒绝，就轮到 ABAC 生效。这样，所有应用只要满足 RBAC 或 ABAC 之一即可工作。

使用 5 或者更详细的日志级别（--v=5 或--vmodule=rbac*=5），则可以在 API Server 日志中看到 RBAC 的拒绝行为（前缀：RBAC DENY）。可以利用这一信息来确定需要授予何种权限给用户、组或 Service Account。等到集群管理员按照 RBAC 的方式对相关组件进行了授权，并且在日志中不再出现 RBAC 的拒绝信息，就可以移除 ABAC 认证方式了。

（2）粗放管理。可以使用 RBAC 的角色绑定，复制一个粗放的策略。

警告：下面的策略让所有 Service Account 都具备了集群管理员权限，所有容器运行的应用都会自动接收 Service Account 的认证，能够对任何 API 做任何事情，包括查看 secret 和修改授权。注意，这不是一个值得推荐的策略。

```
$ kubectl create clusterrolebinding permissive-binding \
  --clusterrole=cluster-admin \
  --user=admin \
  --user=kubelet \
  --group=system:serviceaccounts
```

6.3 Admission Control

突破了之前所说的认证和鉴权两道关卡之后,客户端的调用请求就能够得到 API Server 的真正响应了吗?答案是:不能!这个请求还需要通过 Admission Control(准入控制)所控制的一个准入控制链的层层考验,才能获得成功的响应。Kubernetes 官方标准的"关卡"有 30 多个,还允许用户自定义扩展。

Admission Control 配备了一个准入控制器的插件列表,发送给 API Server 的任何请求都需要通过列表中每个准入控制器的检查,检查不通过,则 API Server 拒绝此调用请求。此外,准入控制器插件能够修改请求参数以完成一些自动化任务,比如 ServiceAccount 这个控制器插件。当前可配置的准入控制器插件如下。

◎ AlwaysAdmit:已弃用,允许所有请求。
◎ AlwaysPullImages:在启动容器之前总是尝试重新下载镜像。这对于多租户共享一个集群的场景非常有用,系统在启动容器之前可以保证总是使用租户的密钥去下载镜像。如果不设置这个控制器,则在 Node 上下载的镜像的安全性将被削弱,只要知道该镜像的名称,任何人便都可以使用它们了。
◎ AlwaysDeny:已弃用,禁止所有请求,用于测试。
◎ DefaultStorageClass:会关注 PersistentVolumeClaim 资源对象的创建,如果其中没有包含任何针对特定 Storage class 的请求,则为其指派指定的 Storage class。在这种情况下,用户无须在 PVC 中设置任何特定的 Storage class 就能完成 PVC 的创建了。如果没有设置默认的 Storage class,该控制器就不会进行任何操作;如果设置了超过一个的默认 Storage class,该控制器就会拒绝所有 PVC 对象的创建申请,并返回错误信息。管理员必须检查 StorageClass 对象的配置,确保只有一个默认值。该控制器仅关注 PVC 的创建过程,对更新过程无效。
◎ DefaultTolerationSeconds:针对没有设置容忍 node.kubernetes.io/not-ready:NoExecute 或者 node.alpha.kubernetes.io/unreachable:NoExecute 的 Pod,设置 5min 的默认容忍时间。

- DenyExecOnPrivileged：已弃用，拦截所有想在 Privileged Container 上执行命令的请求。如果你的集群支持 Privileged Container，又希望限制用户在这些 Privileged Container 上执行命令，那么强烈推荐使用它。其功能已被合并到 DenyEscalatingExec 中。
- DenyEscalatingExec：拦截所有 exec 和 attach 到具有特权的 Pod 上的请求。如果你的集群支持运行有 escalated privilege 权限的容器，又希望限制用户在这些容器内执行命令，那么强烈推荐使用它。
- EventReateLimit：Alpha 版本，用于应对事件密集情况下对 API Server 造成的洪水攻击。
- ExtendedResourceToleration：如果运维人员要创建带有特定资源（例如 GPU、FPGA 等）的独立节点，则可能会对节点进行 Taint 处理来进行特别配置。该控制器能够自动为申请这些特别资源的 Pod 加入 Toleration 定义，无须人工干预。
- ImagePolicyWebhook：这个插件将允许后端的一个 Webhook 程序来完成 admission controller 的功能。ImagePolicyWebhook 需要使用一个配置文件（通过 kube-apiserver 的启动参数 --admission-control-config-file 设置）定义后端 Webhook 的参数。目前是 Alpha 版本的功能。
- Initializers：Alpha。用于为动态准入控制提供支持，通过修改待创建资源的元数据来完成对该资源的修改。
- LimitPodHardAntiAffinityTopology：该插件启用了 Pod 的反亲和性调度策略设置，在设置亲和性策略参数 requiredDuringSchedulingRequiredDuringExecution 时要求将 topologyKey 的值设置为 "kubernetes.io/hostname"，否则 Pod 会被拒绝创建。
- LimitRanger：这个插件会监控进入的请求，确保请求的内容符合在 Namespace 中定义的 LimitRange 对象里的资源限制。如果要在 Kubernetes 集群中使用 LimitRange 对象，则必须启用该插件才能实施这一限制。LimitRanger 还能用于为没有设置资源请求的 Pod 自动设置默认的资源请求，该插件会为 default 命名空间中的所有 Pod 设置 0.1CPU 的资源请求。请参考第 10 章对 LimitRange 原理和用法的详细说明。
- MutatingAdmissionWebhook：Beta。这一插件会变更符合要求的请求的内容，Webhook 以串行的方式顺序执行。
- NamespaceAutoProvision：这一插件会检测所有进入的具备命名空间的资源请求，如果其中引用的命名空间不存在，就会自动创建命名空间。
- NamespaceExists：这一插件会检测所有进入的具备命名空间的资源请求，如果其中引用的命名空间不存在，就会拒绝这一创建过程。
- NamespaceLifecycle：如果尝试在一个不存在的 Namespace 中创建资源对象，则该创建请求将被拒绝。当删除一个 Namespace 时，系统将会删除该 Namespace 中的所有对象，包括 Pod、Service 等，并阻止删除 default、kube-system 和 kube-public 这三个命名空间。
- NodeRestriction：该插件会限制 kubelet 对 Node 和 Pod 的修改行为。为了实现这一限制，kubelet 必须使用 system:nodes 组中用户名为 system:node:<nodeName> 的 Token 来运行。符合条件的 kubelet 只能修改自己的 Node 对象，也只能修改分配到各自 Node 上的 Pod 对象。在 Kubernetes 1.11 以后的版本中，kubelet 无法修改或者更新自身 Node 的 taint 属性。在 Kubernetes 1.13 以后，这一插件还会阻止 kubelet 删除自己的 Node 资源，并限制对有 kubernetes.io/ 或 k8s.io/ 前缀的标签的修改。
- OwnerReferencesPermissionEnforcement：在该插件启用后，一个用户要想修改对象的 metadata.ownerReferences，就必须具备 delete 权限。该插件还会保护对象的 metadata.ownerReferences[x].blockOwnerDeletion 字段，用户只有在对 finalizers 子资源拥有 update 权限的时候才能进行修改。
- PersistentVolumeLabel：弃用。这一插件自动根据云供应商（例如 GCE 或 AWS）的定义，为 PersistentVolume 对象加入 region 或 zone 标签，以此来保障 PersistentVolume 和 Pod 同处一区。如果插件不为 PV 自动设置标签，则需要用户手动保证 Pod 和其加载卷的相对位置。该插件正在被 Cloud controller manager 替换，从 Kubernetes 1.11 版本开始默认被禁止。
- PodNodeSelector：该插件会读取命名空间的 annotation 字段及全局配置，来对一个命名空间中对象的节点选择器设置默认值或限制其取值。
- PersistentVolumeClaimResize：该插件实现了对 PersistentVolumeClaim 发起的 resize 请求的额外校验。
- PodPreset：该插件会使用 PodSelector 选择 Pod，为符合条件的 Pod 进行注入。
- PodSecurityPolicy：在创建或修改 Pod 时决定是否根据 Pod 的 security context 和可用的 PodSecurityPolicy 对 Pod 的安全策略进行控制。
- PodTolerationRestriction：该插件首先会在 Pod 和其命名空间的 Toleration 中进行冲突检测，如果其中存在冲突，则拒绝该 Pod 的创建。它会把命名空间和 Pod 的 Toleration 进行合并，然后将合并的结果与命名空间中的白名单进行比较，如果合并的结果不在白名单内，则拒绝创建。如果不存在命名空间级的默认 Toleration 和白名单，则会采用集群级别的默认 Toleration 和白名单。

- Priority：这一插件使用 priorityClassName 字段来确定优先级，如果没有找到对应的 Priority Class，该 Pod 就会被拒绝。
- ResourceQuota：用于资源配额管理目的，作用于 Namespace。该插件拦截所有请求，以确保在 Namespace 上的资源配额使用不会超标。推荐在 Admission Control 参数列表中将这个插件排最后一个，以免可能被其他插件拒绝的 Pod 被过早分配资源。在 10.4 节将详细介绍 ResourceQuota 的原理和用法。
- SecurityContextDeny：这个插件将在 Pod 中定义的 SecurityContext 选项全部失效。SecurityContext 在 Container 中定义了操作系统级别的安全设定（uid、gid、capabilities、SELinux 等）。在未设置 PodSecurityPolicy 的集群中建议启用该插件，以禁用容器设置的非安全访问权限。
- ServiceAccount：这个插件将 ServiceAccount 实现了自动化，如果想使用 ServiceAccount 对象，那么强烈推荐使用它，在后面讲述 ServiceAccount 的章节会详细说明其作用。
- StorageObjectInUseProtection：这一插件会在新创建的 PVC 或 PV 中加入 kubernetes.io/pvc-protection 或 kubernetes.io/pv-protection 的 finalizer。如果想要删除 PVC 或者 PV，则直到所有 finalizer 的工作都完成，删除动作才会执行。
- ValidatingAdmissionWebhook：在 Kubernetes 1.8 中为 Alpha 版本，在 Kubernetes 1.9 中为 Beta 版本。该插件会针对符合其选择要求的请求调用校验 Webhook。目标 Webhook 会以并行方式运行；如果其中任何一个 Webhook 拒绝了该请求，该请求就会失败。

在 API Server 上设置参数即可定制我们需要的准入控制链，如果启用多种准入控制选项，则建议设置：在 Kubernetes 1.9 及之前的版本中使用的参数是--admission-control，并且其中的内容是顺序相关的；在 Kubernetes 1.10 及之后的版本中，该参数为--enable-admission-plugins，并且与顺序无关。

对 Kubernetes 1.10 及以上版本设置如下：

```
--enable-admission-plugins=NamespaceLifecycle,LimitRanger,ServiceAccount,DefaultStorageClass,DefaultTolerationSeconds,MutatingAdmissionWebhook,ValidatingAdmissionWebhook,ResourceQuota
```

对 Kubernetes 1.9 及以下版本设置如下：

```
--admission-control=NamespaceLifecycle,LimitRanger,ServiceAccount,DefaultStorageClass,DefaultTolerationSeconds,MutatingAdmissionWebhook,ValidatingAdmissionWebhook,ResourceQuota
```

6.4 Service Account

Service Account 也是一种账号，但它并不是给 Kubernetes 集群的用户（系统管理员、运维人员、租户用户等）用的，而是给运行在 Pod 里的进程用的，它为 Pod 里的进程提供了必要的身份证明。

在继续学习之前，请回忆 6.1 节 API Server 认证的内容。

在正常情况下，为了确保 Kubernetes 集群的安全，API Server 都会对客户端进行身份认证，认证失败的客户端无法进行 API 调用。此外，在 Pod 中访问 Kubernetes API Server 服务时，是以 Service 方式访问名为 Kubernetes 的这个服务的，而 Kubernetes 服务又只在 HTTPS 安全端口 443 上提供，那么如何进行身份认证呢？这的确是个谜，因为 Kubernetes 的官方文档并没有清楚说明这个问题。

通过查看官方源码，我们发现这是在用一种类似 HTTP Token 的新认证方式——Service Account Auth，Pod 中的客户端调用 Kubernetes API 时，在 HTTP Header 中传递了一个 Token 字符串，这类似于之前提到的 HTTP Token 认证方式，但有以下几个不同之处。

◎ 这个 Token 的内容来自 Pod 里指定路径下的一个文件（/run/secrets/kubernetes.io/serviceaccount/token），这种 Token 是动态生成的，确切地说，是由 Kubernetes Controller 进程用 API Server 的私钥（--service-account-private-key-file 指定的私钥）签名生成的一个 JWT Secret。

◎ 在官方提供的客户端 REST 框架代码里，通过 HTTPS 方式与 API Server 建立连接后，会用 Pod 里指定路径下的一个 CA 证书（/run/secrets/kubernetes.io/serviceaccount/ca.crt）验证 API Server 发来的证书，验证是否为 CA 证书签名的合法证书。

◎ API Server 在收到这个 Token 以后，采用自己的私钥（实际上是使用 service-account-key-file 参数指定的私钥，如果没有设置此参数，则默认采用 tls-private-key-file 指定的参数，即自己的私钥）对 Token 进行合法性验证。

明白了认证原理，我们接下来继续分析在上面的认证过程中所涉及的 Pod 中的以下三个文件。

◎ /run/secrets/kubernetes.io/serviceaccount/token。

- /run/secrets/kubernetes.io/serviceaccount/ca.crt。
- /run/secrets/kubernetes.io/serviceaccount/namespace（客户端采用这里指定的 namespace 作为参数调用 Kubernetes API）。

这三个文件由于参与到 Pod 进程与 API Server 认证的过程中，起到了类似 secret（私密凭据）的作用，所以它们被称为 Kubernetes Secret 对象。Secret 从属于 Service Account 资源对象，属于 Service Account 的一部分，在一个 Service Account 对象里面可以包括多个不同的 Secret 对象，分别用于不同目的的认证活动。

下面通过运行一些命令来加深我们对 Service Account 与 Secret 的直观认识。

首先，查看系统中的 Service Account 对象，看到有一个名为 default 的 Service Account 对象，包含一个名为 default-token-77oyg 的 Secret，这个 Secret 同时是 Mountable secrets，表明它是需要被挂载到 Pod 上的：

```
# kubectl describe serviceaccounts
Name:                default
Namespace:           default
Labels:              <none>
Image pull secrets:  <none>
Mountable secrets:   default-token-77oyg
Tokens:              default-token-77oyg
```

接下来看看 default-token-77oyg 都有什么内容：

```
# kubectl describe secrets default-token-77oyg
Name:         default-token-77oyg
Namespace:    default
Labels:       <none>
Annotations:  kubernetes.io/service-account.name=default
              kubernetes.io/service-account.uid=3e5b99c0-432c-11e6-b45c-000c29dc2102

Type:   kubernetes.io/service-account-token

Data
====
token:
eyJhbGciOiJSUzI1NiIsInR5cCI6IkpXVCJ9.eyJpc3MiOiJrdWJlcm5ldGVzL3NlcnZpY2VhY2N
vdW50Iiwia3ViZXJuZXRlcy5pby9zZXJ2aWNlYWNjb3VudC9uYW1lc3BhY2UiOiJkZWZhdWx0Iiwia3
iZXJuZXRlcy5pby9zZXJ2aWNlYWNjb3VudC9zZWNyZXQubmFtZSI6ImRlZmF1bHQtdG9rZW4tNzdveWc
```

```
iLCJrdWJlcm5ldGVzLmlvL3NlcnZpY2VhY2NvdW50L3NlcnZpY2UtYWNjb3VudC5uYW1lIjoiZGVmYXV
sdCIsImt1YmVybmV0ZXMuaW8vc2VydmljZWFjY291bnQvc2VydmljZS1hY2NvdW50LnVpZCI6IjNlNWI
5OWMwLTQzMmMtMTFlNi1iNDVjLTAwMGMyOWRjMjEwMiIsInN1YiI6InN5c3RlbTpzZXJ2aWNlYWNjb3V
udDpkZWZhdWx0OmRlZmF1bHQifQ.MFsBrYmTLMB55X3UGfO_pADP6FSsQgHb0SxGJtTsJnY-ze2vFc8Q
dO7bVdmQfFbnkHgLWht1KIpR_EyvJTRP538uovgcA_QGN9yIMEdqIfQC2wfnLFuk10a8OdSH4uzayBb5
0yI7gJWXWbXn6u0wAGMneiTKtCvzGfR4q-p19Jjh5qNPiUdJ0NhjsJJSAc1hdNK40XtOgMHdNNyPEmPg
k6Ow2cM7DRb6ifiSOs05cTeLYv1TpIBMvcQy4sYedCEL2cJ20BwcSo4-1Dev9rdxr5OdtgCvo6OxbPF7
RcWwjjgUMLYO3YCi07WmQNdmxWHJkwvBtkWZhzdvuFCpHeWANA
    ca.crt:      1115 bytes
    namespace:   7 bytes
```

从上面的输出信息中可以看到，default-token-77oyg包含三个数据项，分别是token、ca.crt、namespace。联想到Mountable secrets的标记，以及之前看到的Pod中的三个文件的文件名，我们恍然大悟：在每个Namespace下都有一个名为default的默认Service Account对象，在这个Service Account里面有一个名为Tokens的可以当作Volume被挂载到Pod里的Secret，当Pod启动时，这个Secret会自动被挂载到Pod的指定目录下，用来协助完成Pod中的进程访问API Server时的身份鉴权。

如图6.3所示，一个Service Account可以包括多个Secret对象。

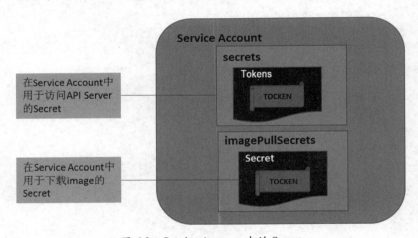

图6.3 Service Account中的Secret

其中：

（1）名为Tokens的Secret用于访问API Server的Secret，也被称为Service Account Secret。

（2）名为 imagePullSecrets 的 Secret 用于下载容器镜像时的认证过程，通常镜像库运行在 Insecure 模式下，所以这个 Secret 为空。

（3）用户自定义的其他 Secret，用于用户的进程。

如果一个 Pod 在定义时没有指定 spec.serviceAccountName 属性，则系统会自动为其赋值为 default，即大家都使用同一个 Namespace 下的默认 Service Account。如果某个 Pod 需要使用非 default 的 Service Account，则需要在定义时指定：

```
apiVersion: v1
kind: Pod
metadata:
  name: mypod
spec:
  containers:
    - name: mycontainter
      image: nginx:v1
  serviceAccountName: myserviceaccount
```

Kubernetes 之所以要创建两套独立的账号系统，原因如下。

- User 账号是给人用的，Service Account 是给 Pod 里的进程使用的，面向的对象不同。
- User 账号是全局性的，Service Account 则属于某个具体的 Namespace。
- 通常来说，User 账号是与后端的用户数据库同步的，创建一个新用户通常要走一套复杂的业务流程才能实现，Service Account 的创建则需要极轻量级的实现方式，集群管理员可以很容易地为某些特定任务创建一个 Service Account。
- 对于这两种不同的账号，其审计要求通常不同。
- 对于一个复杂的系统来说，多个组件通常拥有各种账号的配置信息，Service Account 是 Namespace 隔离的，可以针对组件进行一对一的定义，同时具备很好的"便携性"。

接下来深入分析 Service Account 与 Secret 相关的一些运行机制。

根据 5.2 节的内容，我们知道 Controller manager 创建了 ServiceAccount Controller 与 Token Controller 这两个安全相关的控制器。其中 ServiceAccount Controller 一直监听 Service Account 和 Namespace 的事件，如果在一个 Namespace 中没有 default Service Account，那么 ServiceAccount Controller 会为该 Namespace 创建一个默认（default）的 Service Account，

这就是我们之前看到在每个Namespace下都有一个名为default的Service Account的原因。

如果Controller manager进程在启动时指定了API Server私钥（service-account-private-key-file参数），那么Controller manager会创建Token Controller。Token Controller也监听Service Account的事件，如果发现在新创建的Service Account里没有对应的Service Account Secret，则会用API Server私钥创建一个Token（JWT Token），并用该Token、CA证书及Namespace名称等三个信息产生一个新的Secret对象，然后放入刚才的Service Account中；如果监听到的事件是删除Service Account事件，则自动删除与该Service Account相关的所有Secret。此外，Token Controller对象同时监听Secret的创建、修改和删除事件，并根据事件做不同的处理。

当我们在API Server的鉴权过程中启用了Service Account类型的准入控制器，即在kube-apiserver启动参数中包括下面的内容时：

```
--admission_control=ServiceAccount
```

则针对Pod新增或修改的请求，Service Account准入控制器会验证Pod里的Service Account是否合法。

（1）如果spec.serviceAccount域没有被设置，则Kubernetes默认为其指定名称为default的Service accout。

（2）如果Pod的spec.serviceAccount域指定了default以外的Service Account，而该Service Account没有被事先创建，则该Pod操作失败。

（3）如果在Pod中没有指定ImagePullSecrets，那么这个spec.serviceAccount域指定的Service Account的ImagePullSecrets会被加入该Pod中。

（4）给Pod添加一个特殊的Volume，在该Volume中包含Service Account Secret中的Token，并将Volume挂载到Pod中所有容器的指定目录下（/var/run/secrets/kubernetes.io/serviceaccount）。

综上所述，Service Account的正常工作离不开以下几个控制器。

（1）Admission Controller。

（2）Token Controller。

（3）ServiceAccount Controller。

6.5 Secret 私密凭据

上一节提到 Secret 对象，Secret 的主要作用是保管私密数据，比如密码、OAuth Tokens、SSH Keys 等信息。将这些私密信息放在 Secret 对象中比直接放在 Pod 或 Docker Image 中更安全，也更便于使用和分发。

下面的例子用于创建一个 Secret：

```
secrets.yaml:
apiVersion: v1
kind: Secret
metadata:
  name: mysecret
type: Opaque
data:
  password: dmFsdWUtMg0K
  username: dmFsdWUtMQ0K

# kubectl create -f secrets.yaml
```

在上面的例子中，data 域的各子域的值必须为 BASE64 编码值，其中 password 域和 username 域 BASE64 编码前的值分别为 value-1 和 value-2。

一旦 Secret 被创建，就可以通过下面三种方式使用它。

（1）在创建 Pod 时，通过为 Pod 指定 Service Account 来自动使用该 Secret。

（2）通过挂载该 Secret 到 Pod 来使用它。

（3）在 Docker 镜像下载时使用，通过指定 Pod 的 spc.ImagePullSecrets 来引用它。

第 1 种使用方式主要用在 API Server 鉴权方面，之前提到过。下面的例子展示了第 2 种使用方式：将一个 Secret 通过挂载的方式添加到 Pod 的 Volume 中：

```
apiVersion: v1
kind: Pod
metadata:
  name: mypod
  namespace: myns
spec:
  containers:
```

```
    - name: mycontainer
      image: redis
      volumeMounts:
      - name: foo
        mountPath: "/etc/foo"
        readOnly: true
    volumes:
    - name: foo
      secret:
        secretName: mysecret
```

其结果如图 6.4 所示。

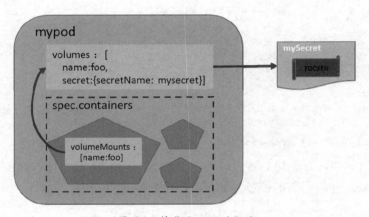

图 6.4　挂载 Secret 到 Pod

第 3 种使用方式的使用流程如下。

（1）执行 login 命令，登录私有 Registry：

```
# docker login localhost:5000
```

输入用户名和密码，如果是第 1 次登录系统，则会创建新用户，相关信息被会写入 ~/.dockercfg 文件中。

（2）用 BASE64 编码 dockercfg 的内容：

```
# cat ~/.dockercfg | base64
```

（3）将上一步命令的输出结果作为 Secret 的 data.dockercfg 域的内容，由此来创建一个 Secret：

```
image-pull-secret.yaml:
apiVersion: v1
kind: Secret
metadata:
  name: myregistrykey
data:
  .dockercfg: eyAiaHR0cHM6Ly9pbmRleC5kb2NrZXIuaW8vdjEvIjogeyAiYXV0aCI6ICJab
UZyWlhCaGMzTjNiM0prTVRJSyIsICJlbWFpbCI6ICJqZG9lQGV4YW1wbGUuY29tIiB9IH0K
type: kubernetes.io/dockercfg

# kubectl create -f image-pull-secret.yaml
```

（4）在创建 Pod 时引用该 Secret：

```
pods.yaml:
apiVersion: v1
kind: Pod
metadata:
  name: mypod2
spec:
  containers:
    - name: foo
      image: janedoe/awesomeapp:v1
  imagePullSecrets:
    - name: myregistrykey

$ kubectl create -f pods.yaml
```

其结果如图 6.5 所示。

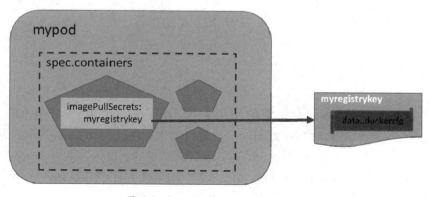

图 6.5　imagePullSecret 引用 Secret

每个单独的 Secret 大小不能超过 1MB，Kubernetes 不鼓励创建大的 Secret，因为如果使用大的 Secret，则将大量占用 API Server 和 kubelet 的内存。当然，创建许多小的 Secret 也能耗尽 API Server 和 kubelet 的内存。

在使用 Mount 方式挂载 Secret 时，Container 中 Secret 的 data 域的各个域的 Key 值作为目录中的文件，Value 值被 BASE64 编码后存储在相应的文件中。在前面的例子中创建的 Secret，被挂载到一个叫作 mycontainer 的 Container 中，在该 Container 中可通过相应的查询命令查看所生成的文件和文件中的内容，如下所示：

```
$ ls /etc/foo/
username
password
$ cat /etc/foo/username
value-1
$ cat /etc/foo/password
value-2
```

通过上面的例子可以得出如下结论：我们可以通过 Secret 保管其他系统的敏感信息（比如数据库的用户名和密码），并以 Mount 的方式将 Secret 挂载到 Container 中，然后通过访问目录中文件的方式获取该敏感信息。当 Pod 被 API Server 创建时，API Server 不会校验该 Pod 引用的 Secret 是否存在。一旦这个 Pod 被调度，则 kubelet 将试着获取 Secret 的值。如果 Secret 不存在或暂时无法连接到 API Server，则 kubelet 按一定的时间间隔定期重试获取该 Secret，并发送一个 Event 来解释 Pod 没有启动的原因。一旦 Secret 被 Pod 获取，则 kubelet 将创建并挂载包含 Secret 的 Volume。只有所有 Volume 都挂载成功，Pod 中的 Container 才会被启动。在 kubelet 启动 Pod 中的 Container 后，Container 中和 Secret 相关的 Volume 将不会被改变，即使 Secret 本身被修改。为了使用更新后的 Secret，必须删除旧 Pod，并重新创建一个新 Pod。

6.6 Pod 的安全策略配置

为了更精细地控制 Pod 对资源的使用方式，Kubernetes 从 1.4 版本开始引入了 PodSecurityPolicy 资源对象对 Pod 的安全策略进行管理，并在 1.10 版本中升级为 Beta 版，到 1.14 版本时趋于成熟。目前 PodSecurityPolicy 资源对象的 API 版本为 extensions/v1beta1，从 1.10 版本开始更新为 policy/v1beta1，并计划于 1.16 版本时弃用 extensions/v1beta1。

6.6.1 PodSecurityPolicy 的工作机制

若想启用 PodSecurityPolicy 机制，则需要在 kube-apiserver 服务的启动参数 --enable-admission-plugins 中进行设置：

```
--enable-admission-plugins=PodSecurityPolicy
```

在开启 PodSecurityPolicy 准入控制器后，Kubernetes 默认不允许创建任何 Pod，需要创建 PodSecurityPolicy 策略和相应的 RBAC 授权策略（Authorizing Policies），Pod 才能创建成功。

例如，尝试创建如下 Pod：

```
apiVersion: v1
kind: Pod
metadata:
  name: nginx
spec:
  containers:
  - name: nginx
    image: nginx
```

使用 kubectl 命令创建时，系统将提示"禁止创建"的报错信息：

```
# kubectl create -f pod.yaml
Error from server (Forbidden): error when creating "pod.yaml": pods "nginx" is forbidden: no providers available to validate pod request
```

接下来创建一个 PodSecurityPolicy，配置文件 psp-non-privileged.yaml 的内容如下：

```
apiVersion: policy/v1beta1
kind: PodSecurityPolicy
metadata:
  name: psp-non-privileged
spec:
  privileged: false    # 不允许特权模式的 Pod
  seLinux:
    rule: RunAsAny
  supplementalGroups:
    rule: RunAsAny
  runAsUser:
    rule: RunAsAny
  fsGroup:
```

```
    rule: RunAsAny
  volumes:
  - '*'
```

使用 kubectl create 命令创建该 PodSecurityPolicy：

```
# kubectl create -f psp-non-privileged.yaml
podsecuritypolicy.policy/psp-non-privileged created
```

查看 PodSecurityPolicy：

```
# kubectl get psp psp-non-privileged
NAME                  PRIV    CAPS    SELINUX    RUNASUSER    FSGROUP    SUPGROUP    READONLYROOTFS    VOLUMES
psp-non-privileged    false           RunAsAny   RunAsAny     RunAsAny   RunAsAny    false             *
```

再次创建 Pod 就能成功：

```
# kubectl create -f pod.yaml
pod/nginx created
```

上面的 PodSecurityPolicy "psp-non-privileged" 设置了 privileged: false，表示不允许创建特权模式的 Pod。在下面的 YAML 配置文件 pod-privileged.yaml 中为 Pod 设置了特权模式：

```
apiVersion: v1
kind: Pod
metadata:
  name: nginx
spec:
  containers:
  - name: nginx
    image: nginx
    securityContext:
      privileged: true
```

创建 Pod 时，系统将提示"禁止创建特权模式的 Pod"的报错信息：

```
# kubectl create -f pod-privileged.yaml
Error from server (Forbidden): error when creating "pod-pri.yaml": pods "nginx" is forbidden: unable to validate against any pod security policy: [spec.containers[0].securityContext.privileged: Invalid value: true: Privileged containers are not allowed]
```

6.6.2 PodSecurityPolicy 配置详解

在 PodSecurityPolicy 对象中可以设置下列字段来控制 Pod 运行时的各种安全策略。

1. 特权模式相关配置

privileged：是否允许 Pod 以特权模式运行。

2. 宿主机资源相关配置

（1）hostPID：是否允许 Pod 共享宿主机的进程空间。

（2）hostIPC：是否允许 Pod 共享宿主机的 IPC 命名空间。

（3）hostNetwork：是否允许 Pod 使用宿主机网络的命名空间。

（4）hostPorts：是否允许 Pod 使用宿主机的端口号，可以通过 hostPortRange 字段设置允许使用的端口号范围，以[min, max]设置最小端口号和最大端口号。

（5）Volumes：允许 Pod 使用的存储卷 Volume 类型，设置为"*"表示允许使用任意 Volume 类型，建议至少允许 Pod 使用下列 Volume 类型。

- ◎ configMap
- ◎ downwardAPI
- ◎ emptyDir
- ◎ persistentVolumeClaim
- ◎ secret
- ◎ projected

（6）AllowedHostPaths：允许 Pod 使用宿主机的 hostPath 路径名称，可以通过 pathPrefix 字段设置路径的前缀，并可以设置是否为只读属性，例子如下。

```
apiVersion: policy/v1beta1
kind: PodSecurityPolicy
metadata:
  name: allow-hostpath-volumes
spec:
  volumes:
    - hostPath
  allowedHostPaths:
```

```
      - pathPrefix: "/foo"
        readOnly: true   # 只读属性
```

结果为允许 Pod 访问宿主机上以 "/foo" 为前缀的路径，包括 "/foo" "/foo/" "/foo/bar" 等，但不能访问 "/fool" "/etc/foo" 等路径，也不允许通过 "/foo/../" 表达式访问 /foo 的上层目录。

（7）FSGroup：设置允许访问某些 Volume 的 Group ID 范围，可以将规则（rule 字段）设置为 MustRunAs、MayRunAs 或 RunAsAny。

- MustRunAs：需要设置 Group ID 的范围，例如 1～65535，要求 Pod 的 securityContext.fsGroup 设置的值必须属于该 Group ID 的范围。
- MayRunAs：需要设置 Group ID 的范围，例如 1～65535，不强制要求 Pod 设置 securityContext.fsGroup。
- RunAsAny：不限制 Group ID 的范围，任何 Group 都可以访问 Volume。

（8）ReadOnlyRootFilesystem：要求容器运行的根文件系统（root filesystem）必须是只读的。

（9）allowedFlexVolumes：对于类型为 flexVolume 的存储卷，设置允许使用的驱动类型，例子如下。

```
apiVersion: policy/v1beta1
kind: PodSecurityPolicy
metadata:
  name: allow-flex-volumes
spec:
  volumes:
    - flexVolume
  allowedFlexVolumes:
    - driver: example/lvm
    - driver: example/cifs
```

3. 用户和组相关配置

（1）RunAsUser：设置运行容器的用户 ID（User ID）范围，规则字段（rule）的值可以被设置为 MustRunAs、MustRunAsNonRoot 或 RunAsAny。

- MustRunAs：需要设置 User ID 的范围，要求 Pod 的 securityContext.runAsUser 设置的值必须属于该 User ID 的范围。

- MustRunAsNonRoot：必须以非 root 用户运行容器，要求 Pod 的 securityContext.runAsUser 设置一个非 0 的用户 ID，或者镜像中在 USER 字段设置了用户 ID，建议同时设置 allowPrivilegeEscalation=false 以避免不必要的提升权限操作。
- RunAsAny：不限制 User ID 的范围，任何 User 都可以运行。

（2）RunAsGroup：设置运行容器的 Group ID 范围，规则字段的值可以被设置为 MustRunAs、MustRunAsNonRoot 或 RunAsAny。

- MustRunAs：需要设置 Group ID 的范围，要求 Pod 的 securityContext.runAsGroup 设置的值必须属于该 Group ID 的范围。
- MustRunAsNonRoot：必须以非 root 组运行容器，要求 Pod 的 securityContext.runAsUser 设置一个非 0 的用户 ID，或者镜像中在 USER 字段设置了用户 ID，建议同时设置 allowPrivilegeEscalation=false 以避免不必要的提升权限操作。
- RunAsAny：不限制 Group ID 的范围，任何 Group 的用户都可以运行。

（3）SupplementalGroups：设置容器可以额外添加的 Group ID 范围，可以将规则（rule 字段）设置为 MustRunAs、MayRunAs 或 RunAsAny。

- MustRunAs：需要设置 Group ID 的范围，要求 Pod 的 securityContext.supplementalGroups 设置的值必须属于该 Group ID 范围。
- MayRunAs：需要设置 Group ID 的范围，不强制要求 Pod 设置 securityContext.supplementalGroups。
- RunAsAny：不限制 Group ID 的范围，任何 supplementalGroups 的用户都可以运行。

4. 提升权限相关配置

（1）AllowPrivilegeEscalation：设置容器内的子进程是否可以提升权限，通常在设置非 root 用户（MustRunAsNonRoot）时进行设置。

（2）DefaultAllowPrivilegeEscalation：设置 AllowPrivilegeEscalation 的默认值，设置为 disallow 时，管理员还可以显式设置 AllowPrivilegeEscalation 来指定是否允许提升权限。

5. Linux 能力相关配置

（1）AllowedCapabilities：设置容器可以使用的 Linux 能力列表，设置为 "*" 表示允许使用 Linux 的所有能力（如 NET_ADMIN、SYS_TIME 等）。

（2）RequiredDropCapabilities：设置不允许容器使用的 Linux 能力列表。

（3）DefaultAddCapabilities：设置默认为容器添加的 Linux 能力列表，例如 SYS_TIME 等，Docker 建议默认设置的 Linux 能力请查看 https://docs.docker.com/engine/reference/run/#runtime-privilege-and-linux-capabilities。

6. SELinux 相关配置

seLinux：设置 SELinux 参数，可以将规则字段（rule）的值设置为 MustRunAs 或 RunAsAny。

- ◎ MustRunAs：要求设置 seLinuxOptions，系统将对 Pod 的 securityContext.seLinuxOptions 设置的值进行校验。
- ◎ RunAsAny：不限制 seLinuxOptions 的设置。

7. 其他 Linux 相关配置

（1）AllowedProcMountTypes：设置允许的 ProcMountTypes 类型列表，可以设置 allowedProcMountTypes 或 DefaultProcMount。

（2）AppArmor：设置对容器可执行程序的访问控制权限，详情请参考 https://kubernetes.io/docs/tutorials/clusters/apparmor/#podsecuritypolicy-annotations。

（3）Seccomp：设置允许容器使用的系统调用（System Calls）的 profile。

（4）Sysctl：设置允许调整的内核参数，详情请参考 https://kubernetes.io/docs/concepts/cluster-administration/sysctl-cluster/#podsecuritypolicy。

下面列举两种常用的 PodSecurityPolicy 安全策略配置。

例 1：基本没有限制的安全策略，允许创建任意安全设置的 Pod。

```
apiVersion: policy/v1beta1
kind: PodSecurityPolicy
metadata:
  name: privileged
  annotations:
    seccomp.security.alpha.kubernetes.io/allowedProfileNames: '*'
spec:
```

```yaml
    privileged: true
    allowPrivilegeEscalation: true
    allowedCapabilities:
    - '*'
    volumes:
    - '*'
    hostNetwork: true
    hostPorts:
    - min: 0
      max: 65535
    hostIPC: true
    hostPID: true
    runAsUser:
      rule: 'RunAsAny'
    seLinux:
      rule: 'RunAsAny'
    supplementalGroups:
      rule: 'RunAsAny'
    fsGroup:
      rule: 'RunAsAny'
```

例2：要求 Pod 运行用户为非特权用户；禁止提升权限；不允许使用宿主机网络、端口号、IPC 等资源；限制可以使用的 Volume 类型，等等。

```yaml
apiVersion: policy/v1beta1
kind: PodSecurityPolicy
metadata:
  name: restricted
  annotations:
    seccomp.security.alpha.kubernetes.io/allowedProfileNames: 'docker/default'
    apparmor.security.beta.kubernetes.io/allowedProfileNames: 'runtime/default'
    seccomp.security.alpha.kubernetes.io/defaultProfileName: 'docker/default'
    apparmor.security.beta.kubernetes.io/defaultProfileName: 'runtime/default'
spec:
  privileged: false
  allowPrivilegeEscalation: false
  requiredDropCapabilities:
    - ALL
```

```
  volumes:
    - 'configMap'
    - 'emptyDir'
    - 'projected'
    - 'secret'
    - 'downwardAPI'
    - 'persistentVolumeClaim'
  hostNetwork: false
  hostIPC: false
  hostPID: false
  runAsUser:
    rule: 'MustRunAsNonRoot'
  seLinux:
     rule: 'RunAsAny'
  supplementalGroups:
    rule: 'MustRunAs'
    ranges:
      - min: 1
        max: 65535
  fsGroup:
    rule: 'MustRunAs'
    ranges:
      - min: 1
        max: 65535
  readOnlyRootFilesystem: false
```

此外,Kubernetes 建议使用 RBAC 授权机制来设置针对 Pod 安全策略的授权,通常应该对 Pod 的 ServiceAccount 进行授权。

例如,可以创建如下 ClusterRole(也可以创建 Role)并将其设置为允许使用 PodSecurityPolicy:

```
kind: ClusterRole
apiVersion: rbac.authorization.k8s.io/v1
metadata:
  name: <role name>
rules:
- apiGroups: ['policy']
  resources: ['podsecuritypolicies']
  verbs:     ['use']
  resourceNames:
  - <list of policies to authorize>   #允许使用的 PodSecurityPolicy 列表
```

然后创建一个 ClusterRoleBinding 与用户和 ServiceAccount 进行绑定：

```
kind: ClusterRoleBinding
apiVersion: rbac.authorization.k8s.io/v1
metadata:
  name: <binding name>
roleRef:
  kind: ClusterRole
  name: <role name>   # 之前创建的 ClusterRole 名称
  apiGroup: rbac.authorization.k8s.io
subjects:
# 对特定 Namespace 中的 ServiceAccount 进行授权
- kind: ServiceAccount
  name: <authorized service account name>   # ServiceAccount 的名称
  namespace: <authorized pod namespace>   # Namespace 的名称
# 对特定用户进行授权（不推荐）
- kind: User
  apiGroup: rbac.authorization.k8s.io
  name: <authorized user name>   # 用户名
```

也可以创建 RoleBinding 对与该 RoleBinding 相同的 Namespace 中的 Pod 进行授权，通常可以与某个系统级别的 Group 关联配置，例如：

```
kind: RoleBinding
apiVersion: rbac.authorization.k8s.io/v1
metadata:
  name: <binding name>
  namespace: <binding namespace>   # 该 RoleBinding 所属的 Namespace
roleRef:
  kind: Role
  name: <role name>
  apiGroup: rbac.authorization.k8s.io
subjects:
# 授权该 Namespace 中的全部 ServiceAccount
- kind: Group
  apiGroup: rbac.authorization.k8s.io
  name: system:serviceaccounts
# 授权该 Namespace 中的全部用户
- kind: Group
  apiGroup: rbac.authorization.k8s.io
  name: system:authenticated
```

6.6.3 Pod 的安全设置详解

在系统管理员对 Kubernetes 集群中设置了 PodSecurityPolicy 策略之后，系统将对 Pod 和 Container 级别的安全设置进行校验，对于不满足 PodSecurityPolicy 安全策略的 Pod，系统将拒绝创建。

Pod 和容器的安全策略可以在 Pod 或 Container 的 securityContext 字段中进行设置，如果在 Pod 和 Container 级别都设置了相同的安全类型字段，容器将使用 Container 级别的设置。

在 Pod 级别可以设置的安全策略类型如下。

◎ runAsUser：容器内运行程序的用户 ID。
◎ runAsGroup：容器内运行程序的用户组 ID。
◎ runAsNonRoot：是否必须以非 root 用户运行程序。
◎ fsGroup：SELinux 相关设置。
◎ seLinuxOptions：SELinux 相关设置。
◎ supplementalGroups：允许容器使用的其他用户组 ID。
◎ sysctls：设置允许调整的内核参数。

在 Container 级别可以设置的安全策略类型如下。

◎ runAsUser：容器内运行程序的用户 ID。
◎ runAsGroup：容器内运行程序的用户组 ID。
◎ runAsNonRoot：是否必须以非 root 用户运行程序。
◎ privileged：是否以特权模式运行。
◎ allowPrivilegeEscalation：是否允许提升权限。
◎ readOnlyRootFilesystem：根文件系统是否为只读属性。
◎ capabilities：Linux 能力列表。
◎ seLinuxOptions：SELinux 相关设置。

下面通过几个例子对 Pod 的安全设置进行说明。

例 1：Pod 级别的安全设置，作用于该 Pod 内的全部容器。

```
apiVersion: v1
kind: Pod
metadata:
```

```yaml
  name: security-context-demo
spec:
  securityContext:
    runAsUser: 1000
    runAsGroup: 3000
    fsGroup: 2000
  volumes:
  - name: sec-ctx-vol
    emptyDir: {}
  containers:
  - name: sec-ctx-demo
    image: gcr.io/google-samples/node-hello:1.0
    volumeMounts:
    - name: sec-ctx-vol
      mountPath: /data/demo
    securityContext:
      allowPrivilegeEscalation: false
```

在 spec.securityContext 中设置了如下参数。

- runAsUser=1000：所有容器都将以 User ID 1000 运行程序，所有新生成文件的 User ID 也被设置为 1000。
- runAsGroup=3000：所有容器都将以 Group ID 3000 运行程序，所有新生成文件的 Group ID 也被设置为 3000。
- fsGroup=2000：挂载的卷 "/data/demo" 及其中创建的文件都将属于 Group ID 2000。

创建该 Pod 之后进入容器环境，查看到运行进程的用户 ID 为 1000：

```
# ps aux
USER    PID %CPU %MEM    VSZ   RSS TTY      STAT START   TIME COMMAND
1000      1  0.0  0.0   4336   724 ?        Ss   18:16   0:00 /bin/sh -c node server.js
1000      5  0.2  0.6 772124 22768 ?        Sl   18:16   0:00 node server.js
......
```

查看从 Volume 挂载到容器的 /data/demo 目录，其 Group ID 为 2000：

```
# ls -l /data
drwxrwsrwx 2 root 2000 4096 Jun  6 20:08 demo
```

在该目录下创建一个新文件，可见其用户 ID 为 1000，组 ID 为 2000：

```
# cd demo
# touch hello
```

```
# ls -l
-rw-r--r-- 1 1000 2000 6 Mar  6 10:10 hello
```

例 2：Container 级别的安全设置，作用于特定的容器。

```
apiVersion: v1
kind: Pod
metadata:
  name: security-context-demo-2
spec:
  securityContext:
    runAsUser: 1000
  containers:
  - name: sec-ctx-demo-2
    image: gcr.io/google-samples/node-hello:1.0
    securityContext:
      runAsUser: 2000
      allowPrivilegeEscalation: false
```

创建该 Pod 之后进入容器环境，查看到运行进程的用户 ID 为 2000：

```
# ps aux
USER       PID %CPU %MEM    VSZ   RSS TTY      STAT START   TIME COMMAND
2000         1  0.0  0.0   4336   764 ?        Ss   20:36   0:00 /bin/sh -c node server.js
2000         8  0.1  0.5 772124 22604 ?        Sl   20:36   0:00 node server.js
......
```

例 3：为 Container 设置可用的 Linux 能力，为容器设置允许使用的 Linux 能力包括 NET_ADMIN 和 SYS_TIME。

```
apiVersion: v1
kind: Pod
metadata:
  name: security-context-demo-4
spec:
  containers:
  - name: sec-ctx-4
    image: gcr.io/google-samples/node-hello:1.0
    securityContext:
      capabilities:
        add: ["NET_ADMIN", "SYS_TIME"]
```

创建该 Pod 之后进入容器环境，查看 1 号进程的 Linux 能力设置：

```
# cd /proc/1
# cat status
...
CapPrm: 00000000aa0435fb
CapEff: 00000000aa0435fb
```

未设置这两个能力的配置为：

```
CapPrm: 00000000a80425fb
CapEff: 00000000a80425fb
```

可以看到系统在第 12 位和第 25 位添加了 CAP_NET_ADMIN 和 CAP_SYS_TIME 这两个 Linux 能力。

第 7 章

网络原理

关于Kubernetes网络，我们通常有如下问题需要回答。

◎ Kubernetes的网络模型是什么？
◎ Docker背后的网络基础是什么？
◎ Docker自身的网络模型和局限是什么？
◎ Kubernetes的网络组件之间是怎么通信的？
◎ 外部如何访问Kubernetes集群？
◎ 有哪些开源组件支持Kubernetes的网络模型？

本章分别回答这些问题，然后通过一个具体的实验将这些相关的知识点串联成一个整体。

7.1 Kubernetes网络模型

Kubernetes网络模型设计的一个基础原则是：每个Pod都拥有一个独立的IP地址，并假定所有Pod都在一个可以直接连通的、扁平的网络空间中。所以不管它们是否运行在同一个Node（宿主机）中，都要求它们可以直接通过对方的IP进行访问。设计这个原则的原因是，用户不需要额外考虑如何建立Pod之间的连接，也不需要考虑如何将容器端口映射到主机端口等问题。

实际上，在Kubernetes的世界里，IP是以Pod为单位进行分配的。一个Pod内部的所有容器共享一个网络堆栈（相当于一个网络命名空间，它们的IP地址、网络设备、配置等都是共享的）。按照这个网络原则抽象出来的为每个Pod都设置一个IP地址的模型也被称作IP-per-Pod模型。

由于Kubernetes的网络模型假设Pod之间访问时使用的是对方Pod的实际地址，所以一个Pod内部的应用程序看到的自己的IP地址和端口与集群内其他Pod看到的一样。它们都是Pod实际分配的IP地址。将IP地址和端口在Pod内部和外部都保持一致，也就不需要使用NAT来进行地址转换了。Kubernetes的网络之所以这么设计，主要原因就是可以兼容过去的应用。当然，我们使用Linux命令"ip addr show"也能看到这些地址，和程序看到的没有什么区别。所以这种IP-per-Pod的方案很好地利用了现有的各种域名解析和发现机制。

为每个Pod都设置一个IP地址的模型还有另外一层含义，那就是同一个Pod内的不同容器会共享同一个网络命名空间，也就是同一个Linux网络协议栈。这就意味着同一个Pod

内的容器可以通过 localhost 来连接对方的端口。这种关系和同一个 VM 内的进程之间的关系是一样的，看起来 Pod 内容器之间的隔离性减小了，而且 Pod 内不同容器之间的端口是共享的，就没有所谓的私有端口的概念了。如果你的应用必须要使用一些特定的端口范围，那么你也可以为这些应用单独创建一些 Pod。反之，对那些没有特殊需要的应用，由于 Pod 内的容器是共享部分资源的，所以可以通过共享资源互相通信，这显然更加容易和高效。针对这些应用，虽然损失了可接受范围内的部分隔离性，却也是值得的。

IP-per-Pod 模式和 Docker 原生的通过动态端口映射方式实现的多节点访问模式有什么区别呢？主要区别是后者的动态端口映射会引入端口管理的复杂性，而且访问者看到的 IP 地址和端口与服务提供者实际绑定的不同（因为 NAT 的缘故，它们都被映射成新的地址或端口了），这也会引起应用配置的复杂化。同时，标准的 DNS 等名字解析服务也不适用了，甚至服务注册和发现机制都将迎来挑战，因为在端口映射情况下，服务自身很难知道自己对外暴露的真实的服务 IP 和端口，外部应用也无法通过服务所在容器的私有 IP 地址和端口来访问服务。

总的来说，IP-per-Pod 模型是一个简单的兼容性较好的模型。从该模型的网络的端口分配、域名解析、服务发现、负载均衡、应用配置和迁移等角度来看，Pod 都能够被看作一台独立的虚拟机或物理机。

按照这个网络抽象原则，Kubernetes 对网络有什么前提和要求呢？

Kubernetes 对集群网络有如下要求。

（1）所有容器都可以在不用 NAT 的方式下同别的容器通信。

（2）所有节点都可以在不用 NAT 的方式下同所有容器通信，反之亦然。

（3）容器的地址和别人看到的地址是同一个地址。

这些基本要求意味着并不是只要两台机器都运行 Docker，Kubernetes 就可以工作了。具体的集群网络实现必须满足上述基本要求，原生的 Docker 网络目前还不能很好地支持这些要求。

实际上，这些对网络模型的要求并没有降低整个网络系统的复杂度。如果你的程序原来在 VM 上运行，而那些 VM 拥有独立 IP，并且它们之间可以直接透明地通信，那么 Kubernetes 的网络模型就和 VM 使用的网络模型一样。所以使用这种模型可以很容易地将已有的应用程序从 VM 或者物理机迁移到容器上。

当然，谷歌设计 Kubernetes 的一个主要运行基础就是其公有云 GCE，GCE 默认支持这些网络要求。另外，常见的其他公有云服务商如亚马逊等，其公有云环境也支持这些网络要求。

由于部署私有云的场景也非常普遍，所以在私有云中运行 Kubernetes+Docker 集群之前，需要自己搭建出符合 Kubernetes 要求的网络环境。有很多开源组件可以帮助我们打通 Docker 容器和容器之间的网络，实现满足 Kubernetes 要求的网络模型。当然，每种方案都有适合的场景，我们要根据自己的实际需要进行选择。在后面的章节中会对常见的开源方案进行介绍。

Kubernetes 的网络依赖于 Docker，Docker 的网络又离不开 Linux 操作系统内核特性的支持，所以我们有必要先深入了解 Docker 背后的网络原理和基础知识。接下来一起深入学习必要的 Linux 网络知识。

7.2 Docker 网络基础

Docker 本身的技术依赖于近年来 Linux 内核虚拟化技术的发展，所以 Docker 对 Linux 内核的特性有很强的依赖。这里将 Docker 使用到的与 Linux 网络有关的主要技术进行简要介绍，这些技术有：网络命名空间（Network Namespace）、Veth 设备对、网桥、ipatables 和路由。

7.2.1 网络命名空间

为了支持网络协议栈的多个实例，Linux 在网络栈中引入了网络命名空间，这些独立的协议栈被隔离到不同的命名空间中。处于不同命名空间中的网络栈是完全隔离的，彼此之间无法通信，就好像两个"平行宇宙"。通过对网络资源的隔离，就能在一个宿主机上虚拟多个不同的网络环境。Docker 正是利用了网络的命名空间特性，实现了不同容器之间的网络隔离。

在 Linux 的网络命名空间中可以有自己独立的路由表及独立的 iptables 设置来提供包转发、NAT 及 IP 包过滤等功能。

为了隔离出独立的协议栈，需要纳入命名空间的元素有进程、套接字、网络设备等。进程创建的套接字必须属于某个命名空间，套接字的操作也必须在命名空间中进行。同样，网络设备也必须属于某个命名空间。因为网络设备属于公共资源，所以可以通过修改属性实现在命名空间之间移动。当然，是否允许移动与设备的特征有关。

让我们稍微深入 Linux 操作系统内部，看它是如何实现网络命名空间的，这也会对理解后面的概念有帮助。

1. 网络命名空间的实现

Linux 的网络协议栈是十分复杂的，为了支持独立的协议栈，相关的这些全局变量都必须被修改为协议栈私有。最好的办法就是让这些全局变量成为一个 Net Namespace 变量的成员，然后为协议栈的函数调用加入一个 Namespace 参数。这就是 Linux 实现网络命名空间的核心。

同时，为了保证对已经开发的应用程序及内核代码的兼容性，内核代码隐式地使用了命名空间中的变量。程序如果没有对命名空间有特殊需求，就不需要编写额外的代码，网络命名空间对应用程序而言是透明的。

在建立了新的网络命名空间，并将某个进程关联到这个网络命名空间后，就出现了类似于如图 7.1 所示的内核数据结构，所有网站栈变量都被放入了网络命名空间的数据结构中。这个网络命名空间是其进程组私有的，和其他进程组不冲突。

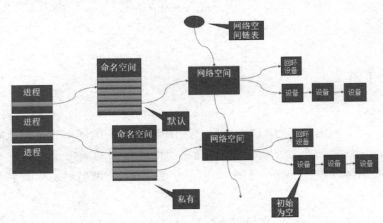

图 7.1　命名空间的内核数据结构

在新生成的私有命名空间中只有回环设备（名为"lo"且是停止状态），其他设备默认都不存在，如果我们需要，则要一一手工建立。Docker 容器中的各类网络栈设备都是 Docker Daemon 在启动时自动创建和配置的。

所有的网络设备（物理的或虚拟接口、桥等在内核里都叫作 Net Device）都只能属于一个命名空间。当然，物理设备（连接实际硬件的设备）通常只能关联到 root 这个命名空间中。虚拟的网络设备（虚拟的以太网接口或者虚拟网口对）则可以被创建并关联到一个给定的命名空间中，而且可以在这些命名空间之间移动。

前面提到，由于网络命名空间代表的是一个独立的协议栈，所以它们之间是相互隔离的，彼此无法通信，在协议栈内部都看不到对方。那么有没有办法打破这种限制，让处于不同命名空间的网络相互通信，甚至和外部的网络进行通信呢？答案就是"有，应用 Veth 设备对即可"。Veth 设备对的一个重要作用就是打通互相看不到的协议栈之间的壁垒，它就像一条管子，一端连着这个网络命名空间的协议栈，一端连着另一个网络命名空间的协议栈。所以如果想在两个命名空间之间通信，就必须有一个 Veth 设备对。后面会介绍如何操作 Veth 设备对来打通不同命名空间之间的网络。

2. 网络命名空间的操作

下面列举网络命名空间的一些操作。

我们可以使用 Linux iproute2 系列配置工具中的 IP 命令来操作网络命名空间。注意，这个命令需要由 root 用户运行。

创建一个命名空间：

```
ip netns add <name>
```

在命名空间中执行命令：

```
ip netns exec <name> <command>
```

也可以先通过 bash 命令进入内部的 shell 界面，然后执行各种命令：

```
ip netns exec <name> bash
```

退出到外面的命名空间时，请输入"exit"。

3. 网络命名空间的实用技巧

操作网络命名空间时的一些实用技巧如下。

我们可以在不同的网络命名空间之间转移设备，例如下面会提到的 Veth 设备对的转移。因为一个设备只能属于一个命名空间，所以转移后在这个命名空间中就看不到这个设备了。具体哪些设备能被转移到不同的命名空间呢？在设备里面有一个重要的属性：

NETIF_F_NETNS_LOCAL，如果这个属性为 on，就不能被转移到其他命名空间中。Veth 设备属于可以转移的设备，而很多其他设备如 lo 设备、vxlan 设备、ppp 设备、bridge 设备等都是不可以转移的。将无法转移的设备移动到别的命名空间时，会得到无效参数的错误提示。

```
# ip link set br0 netns ns1
RTNETLINK answers: Invalid argument
```

如何知道这些设备是否可以转移呢？可以使用 ethtool 工具查看：

```
# ethtool -k br0
netns-local: on [fixed]
```

netns-local 的值是 on，说明不可以转移，否则可以转移。

7.2.2　Veth 设备对

引入 Veth 设备对是为了在不同的网络命名空间之间通信，利用它可以直接将两个网络命名空间连接起来。由于要连接两个网络命名空间，所以 Veth 设备都是成对出现的，很像一对以太网卡，并且中间有一根直连的网线。既然是一对网卡，那么我们将其中一端称为另一端的 peer。在 Veth 设备的一端发送数据时，它会将数据直接发送到另一端，并触发另一端的接收操作。

整个 Veth 的实现非常简单，有兴趣的读者可以参考源代码 "drivers/net/veth.c" 的实现。如图 7.2 所示是 Veth 设备对的示意图。

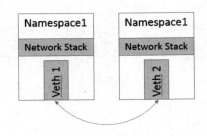

图 7.2　Veth 设备对的示意图

1. Veth 设备对的操作命令

接下来看看如何创建 Veth 设备对，如何连接到不同的命名空间，并设置它们的地址，让它们通信。

创建 Veth 设备对：

```
ip link add veth0 type veth peer name veth1
```

创建后，可以查看 Veth 设备对的信息。使用 ip link show 命令查看所有网络接口：

```
# ip link show
1: lo: <LOOPBACK,UP,LOWER_UP> mtu 65536 qdisc noqueue state UNKNOWN mode DEFAULT
    Link/loopback: 00:00:00:00:00:00 brd 00:00:00:00:00:00
2: eno16777736: <BROADCAST,MULTICAST,UP,LOWER_UP> mtu 1500 qdisc pfifo_fast state UP mode DEFAULT qlen 1000
    link/ether 00:0c:29:cf:1a:2e brd ff:ff:ff:ff:ff:ff
3: docker0: <NO-CARRIER,BROADCAST,MULTICAST,UP> mtu 1500 qdisc noqueue state UP mode DEFAULT
    link/ether 56:84:7a:fe:97:99 brd ff:ff:ff:ff:ff:ff
19: veth1: <BROADCAST,MULTICAST> mtu 1500 qdisc noop state DOWN mode DEFAULT qlen 1000
    link/ether 7e:4a:ae:41:a3:65 brd ff:ff:ff:ff:ff:ff
20: veth0: <BROADCAST,MULTICAST> mtu 1500 qdisc noop state DOWN mode DEFAULT qlen 1000
    link/ether ea:da:85:a3:75:8a brd ff:ff:ff:ff:ff:ff
```

看到了吧，有两个设备生成了，一个是 veth0，它的 peer 是 veth1。

现在这两个设备都在自己的命名空间中，那怎么能行呢？好了，如果将 Veth 看作有两个头的网线，那么我们将另一个头甩给另一个命名空间：

```
ip link set veth1 netns netns1
```

这时可在外面这个命名空间中看两个设备的情况：

```
# ip link show
1: lo: <LOOPBACK,UP,LOWER_UP> mtu 65536 qdisc noqueue state UNKNOWN mode DEFAULT
    Link/loopback: 00:00:00:00:00:00 brd 00:00:00:00:00:00
2: eno16777736: <BROADCAST,MULTICAST,UP,LOWER_UP> mtu 1500 qdisc pfifo_fast state UP mode DEFAULT qlen 1000
    link/ether 00:0c:29:cf:1a:2e brd ff:ff:ff:ff:ff:ff
3: docker0: <NO-CARRIER,BROADCAST,MULTICAST,UP> mtu 1500 qdisc noqueue state UP mode DEFAULT
    link/ether 56:84:7a:fe:97:99 brd ff:ff:ff:ff:ff:ff
20: veth0: <BROADCAST,MULTICAST> mtu 1500 qdisc noop state DOWN mode DEFAULT qlen 1000
    link/ether ea:da:85:a3:75:8a brd ff:ff:ff:ff:ff:ff
```

只剩一个 veth0 设备了，已经看不到另一个设备了，另一个设备已经被转移到另一个网络命名空间中了。

在 netns1 网络命名空间中可以看到 veth1 设备了,符合预期:

```
# ip netns exec netns1 ip link show
1: lo: <LOOPBACK,UP,LOWER_UP> mtu 65536 qdisc noqueue state UNKNOWN mode DEFAULT
    Link/loopback: 00:00:00:00:00:00 brd 00:00:00:00:00:00
19: veth1: <BROADCAST,MULTICAST> mtu 1500 qdisc noop state DOWN mode DEFAULT qlen 1000
link/ether 7e:4a:ae:41:a3:65 brd ff:ff:ff:ff:ff:ff
```

现在看到的结果是,两个不同的命名空间各自有一个 Veth 的"网线头",各显示为一个 Device(在 Docker 的实现里面,它除了将 Veth 放入容器内,还将它的名字改成了 eth0,简直以假乱真,你以为它是一个本地网卡吗)。

现在可以通信了吗?不行,因为它们还没有任何地址,我们现在给它们分配 IP 地址:

```
ip netns exec netns1 ip addr add 10.1.1.1/24 dev veth1
ip addr add 10.1.1.2/24 dev veth0
```

再启动它们:

```
ip netns exec netns1 ip link set dev veth1 up
ip link set dev veth0 up
```

现在两个网络命名空间可以互相通信了:

```
# ping 10.1.1.1
PING 10.1.1.1 (10.1.1.1) 56(84) bytes of data.
64 bytes from 10.1.1.1: icmp_seq=1 ttl=64 time=0.035 ms
64 bytes from 10.1.1.1: icmp_seq=2 ttl=64 time=0.096 ms
^C
--- 10.1.1.1 ping statistics ---
2 packets transmitted, 2 received, 0% packet loss, time 1001ms
rtt min/avg/max/mdev = 0.035/0.065/0.096/0.031 ms

# ip netns exec netns1 ping 10.1.1.2
PING 10.1.1.2 (10.1.1.2) 56(84) bytes of data.
64 bytes from 10.1.1.2: icmp_seq=1 ttl=64 time=0.045 ms
64 bytes from 10.1.1.2: icmp_seq=2 ttl=64 time=0.105 ms
^C
--- 10.1.1.2 ping statistics ---
2 packets transmitted, 2 received, 0% packet loss, time 1000ms
rtt min/avg/max/mdev = 0.045/0.075/0.105/0.030 ms
```

至此，我们就能够理解 Veth 设备对的原理和用法了。在 Docker 内部，Veth 设备对也是连通容器与宿主机的主要网络设备，离开它是不行的。

2. Veth 设备对如何查看对端

我们在操作 Veth 设备对时有一些实用技巧，如下所示。

一旦将 Veth 设备对的对端放入另一个命名空间，在本命名空间中就看不到它了。那么我们怎么知道这个 Veth 设备的对端在哪里呢，也就是说它到底连接到哪个命名空间呢？可以使用 ethtool 工具来查看（当网络命名空间特别多时，这可不是一件很容易的事情）。

首先，在命名空间 netns1 中查询 Veth 设备对端接口在设备列表中的序列号：

```
# ip netns exec netns1 ethtool -S veth1
NIC statistics:
    peer_ifindex: 5
```

得知另一端的接口设备的序列号是 5，我们再到命名空间 netns2 中查看序列号 5 代表什么设备：

```
# ip netns exec netns2 ip link | grep 5      <-- 我们只关注序列号为 5 的设备
veth0
```

好了，我们现在就找到序列号为 5 的设备了，它是 veth0，它的另一端自然就是命名空间 netns1 中的 veth1 了，因为它们互为 peer。

7.2.3 网桥

Linux 可以支持多个不同的网络，它们之间能够相互通信，如何将这些网络连接起来并实现各网络中主机的相互通信呢？可以用网桥。网桥是一个二层的虚拟网络设备，把若干个网络接口"连接"起来，以使得网络接口之间的报文能够互相转发。网桥能够解析收发的报文，读取目标 MAC 地址的信息，和自己记录的 MAC 表结合，来决策报文的转发目标网络接口。为了实现这些功能，网桥会学习源 MAC 地址（二层网桥转发的依据就是 MAC 地址）。在转发报文时，网桥只需要向特定的网口进行转发，来避免不必要的网络交互。如果它遇到一个自己从未学习到的地址，就无法知道这个报文应该向哪个网络接口转发，就将报文广播给所有的网络接口（报文来源的网络接口除外）。

在实际的网络中，网络拓扑不可能永久不变。设备如果被移动到另一个端口上，却没有发送任何数据，网桥设备就无法感知到这个变化，网桥还是向原来的端口转发数据包，在这种情况下数据就会丢失。所以网桥还要对学习到的 MAC 地址表加上超时时间（默认为 5min）。如果网桥收到了对应端口 MAC 地址回发的包，则重置超时时间，否则过了超时时间后，就认为设备已经不在那个端口上了，它就会重新广播发送。

在 Linux 的内部网络栈里实现的网桥设备，作用和上面的描述相同。过去 Linux 主机一般都只有一个网卡，现在多网卡的机器越来越多，而且有很多虚拟的设备存在，所以 Linux 的网桥提供了在这些设备之间互相转发数据的二层设备。

Linux 内核支持网口的桥接（目前只支持以太网接口）。但是与单纯的交换机不同，交换机只是一个二层设备，对于接收到的报文，要么转发，要么丢弃。运行着 Linux 内核的机器本身就是一台主机，有可能是网络报文的目的地，其收到的报文除了转发和丢弃，还可能被送到网络协议栈的上层（网络层），从而被自己（这台主机本身的协议栈）消化，所以我们既可以把网桥看作一个二层设备，也可以把它看作一个三层设备。

1. Linux 网桥的实现

Linux 内核是通过一个虚拟的网桥设备（Net Device）来实现桥接的。这个虚拟设备可以绑定若干个以太网接口设备，从而将它们桥接起来。如图 7.3 所示，这种 Net Device 网桥和普通的设备不同，最明显的一个特性是它还可以有一个 IP 地址。

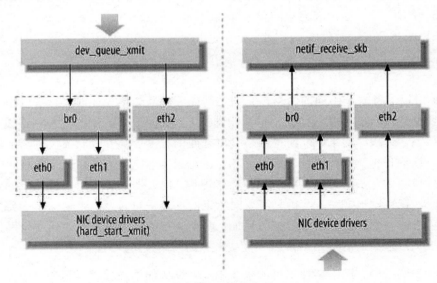

图 7.3　网桥的位置

如图 7.3 所示，网桥设备 br0 绑定了 eth0 和 eth1。对于网络协议栈的上层来说，只看得到 br0 就行。因为桥接是在数据链路层实现的，上层不需要关心桥接的细节，所以协议栈上层需要发送的报文被送到 br0，网桥设备的处理代码判断报文该被转发到 eth0 还是 eth1，或者两者皆转发；反过来，从 eth0 或从 eth1 接收到的报文被提交给网桥的处理代码，在这里会判断报文应该被转发、丢弃还是被提交到协议栈上层。

而有时 eth0、eth1 也可能会作为报文的源地址或目的地址，直接参与报文的发送与接收，从而绕过网桥。

2. 网桥的常用操作命令

Docker 自动完成了对网桥的创建和维护。为了进一步理解网桥，下面举几个常用的网桥操作例子，对网桥进行手工操作：

新增一个网桥设备：

```
#brctl addbr xxxxx
```

之后可以为网桥增加网口，在 Linux 中，一个网口其实就是一个物理网卡。将物理网卡和网桥连接起来：

```
#brctl addif xxxxx ethx
```

网桥的物理网卡作为一个网口，由于在链路层工作，就不再需要 IP 地址了，这样上面的 IP 地址自然失效：

```
#ifconfig ethx 0.0.0.0
```

给网桥配置一个 IP 地址：

```
#ifconfig brxxx xxx.xxx.xxx.xxx
```

这样网桥就有了一个 IP 地址，而连接到上面的网卡就是一个纯链路层设备了。

7.2.4　iptables 和 Netfilter

我们知道，Linux 网络协议栈非常高效，同时比较复杂。如果我们希望在数据的处理过程中对关心的数据进行一些操作，则该怎么做呢？Linux 提供了一套机制来为用户实现自定义的数据包处理过程。

在 Linux 网络协议栈中有一组回调函数挂接点，通过这些挂接点挂接的钩子函数可以在 Linux 网络栈处理数据包的过程中对数据包进行一些操作，例如过滤、修改、丢弃等。整个挂接点技术叫作 Netfilter 和 iptables。

Netfilter 负责在内核中执行各种挂接的规则，运行在内核模式中；而 iptables 是在用户模式下运行的进程，负责协助和维护内核中 Netfilter 的各种规则表。二者互相配合来实现整个 Linux 网络协议栈中灵活的数据包处理机制。

Netfilter 可以挂接的规则点有 5 个，如图 7.4 中的深色椭圆所示。

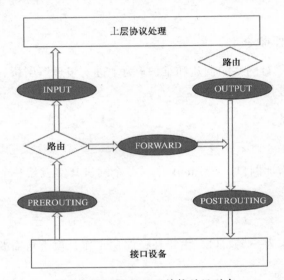

图 7.4　Netfilter 可以挂接的规则点

1. 规则表 Table

这些挂接点能挂接的规则也分不同的类型（也就是规则表 Table），我们可以在不同类型的 Table 中加入我们的规则。目前主要支持的 Table 类型有：RAW、MANGLE、NAT 和 FILTER。

上述 4 个 Table（规则链）的优先级是 RAW 最高，FILTER 最低。

在实际应用中，不同的挂接点需要的规则类型通常不同。例如，在 Input 的挂接点上明显不需要 FILTER 过滤规则，因为根据目标地址已经选择好本机的上层协议栈了，所以无须再挂接 FILTER 过滤规则。目前 Linux 系统支持的不同挂接点能挂接的规则类型如图 7.5 所示。

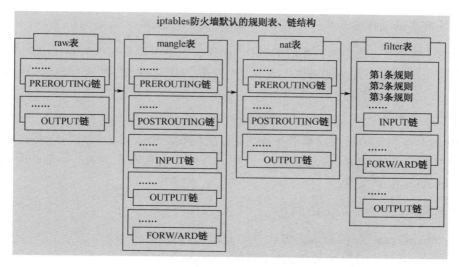

图 7.5 不同的挂接点能挂接的规则类型

当 Linux 协议栈的数据处理运行到挂接点时，它会依次调用挂接点上所有的挂钩函数，直到数据包的处理结果是明确地接受或者拒绝。

2. 处理规则

每个规则的特性都分为以下几部分。

- 表类型（准备干什么事情）。
- 什么挂接点（什么时候起作用）。
- 匹配的参数是什么（针对什么样的数据包）。
- 匹配后有什么动作（匹配后具体的操作是什么）。

前面已经介绍了表类型和挂接点，接下来看看匹配的参数和匹配后的动作。

（1）匹配的参数

匹配的参数用于对数据包或者 TCP 数据连接的状态进行匹配。当有多个条件存在时，它们一起发挥作用，来达到只针对某部分数据进行修改的目的。常见的匹配参数如下。

- 流入、流出的网络接口。
- 来源、目的地址。
- 协议类型。
- 来源、目的端口。

（2）匹配后的动作

一旦有数据匹配，就会执行相应的动作。动作类型既可以是标准的预定义的几个动作，也可以是自定义的模块注册的动作，或者是一个新的规则链，以便更好地组织一组动作。

3. iptables 命令

iptables 命令用于协助用户维护各种规则。我们在使用 Kubernetes、Docker 的过程中，通常都会去查看相关的 Netfilter 配置。这里只介绍如何查看规则表，详细的介绍请参照 Linux 的 iptables 帮助文档。

查看系统中已有规则的方法如下。

- ◎ iptables-save：按照命令的方式打印 iptables 的内容。
- ◎ iptables-vnL：以另一种格式显示 Netfilter 表的内容。

7.2.5 路由

Linux 系统包含一个完整的路由功能。当 IP 层在处理数据发送或者转发时，会使用路由表来决定发往哪里。在通常情况下，如果主机与目的主机直接相连，那么主机可以直接发送 IP 报文到目的主机，这个过程比较简单。例如，通过点对点的链接或网络共享，如果主机与目的主机没有直接相连，那么主机会将 IP 报文发送给默认的路由器，然后由路由器来决定往哪里发送 IP 报文。

路由功能由 IP 层维护的一张路由表来实现。当主机收到数据报文时，它用此表来决策接下来应该做什么操作。当从网络侧接收到数据报文时，IP 层首先会检查报文的 IP 地址是否与主机自身的地址相同。如果数据报文中的 IP 地址是主机自身的地址，那么报文将被发送到传输层相应的协议中。如果报文中的 IP 地址不是主机自身的地址，并且主机配置了路由功能，那么报文将被转发，否则，报文将被丢弃。

路由表中的数据一般是以条目形式存在的。一个典型的路由表条目通常包含以下主要的条目项。

（1）目的 IP 地址：此字段表示目标的 IP 地址。这个 IP 地址可以是某主机的地址，也可以是一个网络地址。如果这个条目包含的是一个主机地址，那么它的主机 ID 将被标记为非零；如果这个条目包含的是一个网络地址，那么它的主机 ID 将被标记为零。

（2）下一个路由器的 IP 地址：这里采用"下一个"的说法，是因为下一个路由器并不总是最终的目的路由器，它很可能是一个中间路由器。条目给出的下一个路由器的地址用来转发在相应接口接收到的 IP 数据报文。

（3）标志：这个字段提供了另一组重要信息，例如，目的 IP 地址是一个主机地址还是一个网络地址。此外，从标志中可以得知下一个路由器是一个真实路由器还是一个直接相连的接口。

（4）网络接口规范：为一些数据报文的网络接口规范，该规范将与报文一起被转发。

在通过路由表转发时，如果任何条目的第 1 个字段完全匹配目的 IP 地址（主机）或部分匹配条目的 IP 地址（网络），它将指示下一个路由器的 IP 地址。这是一个重要的信息，因为这些信息直接告诉主机（具备路由功能的）数据包应该被转发到哪个路由器。而条目中的所有其他字段将提供更多的辅助信息来为路由转发做决定。

如果没有找到一个完全匹配的 IP，就接着搜索相匹配的网络 ID。如果找到，那么该数据报文会被转发到指定的路由器上。可以看出，网络上的所有主机都通过这个路由表中的单个（这个）条目进行管理。

如果上述两个条件都不匹配，那么该数据报文将被转发到一个默认的路由器上。

如果上述步骤都失败，默认路由器也不存在，那么该数据报文最终无法被转发。任何无法投递的数据报文都将产生一个 ICMP 主机不可达或 ICMP 网络不可达的错误，并将此错误返回给生成此数据报文的应用程序。

1. 路由表的创建

Linux 的路由表至少包括两个表(当启用策略路由时,还会有其他表)：一个是 LOCAL，另一个是 MAIN。在 LOCAL 表中会包含所有的本地设备地址。LOCAL 路由表是在配置网络设备地址时自动创建的。LOCAL 表用于供 Linux 协议栈识别本地地址，以及进行本地各个不同网络接口之间的数据转发。

可以通过下面的命令查看 LOCAL 表的内容：

```
# ip route show table local type local
10.1.1.0 dev flannel0  proto kernel  scope host  src 10.1.1.0
127.0.0.0/8 dev lo  proto kernel  scope host  src 127.0.0.1
127.0.0.1 dev lo  proto kernel  scope host  src 127.0.0.1
172.17.42.1 dev docker  proto kernel  scope host  src 172.17.42.1
```

```
192.168.1.128 dev eno16777736 proto kernel scope host src 192.168.1.128
```

MAIN 表用于各类网络 IP 地址的转发。它的建立既可以使用静态配置生成，也可以使用动态路由发现协议生成。动态路由发现协议一般使用组播功能来通过发送路由发现数据，动态地交换和获取网络的路由信息，并更新到路由表中。

Linux 下支持路由发现协议的开源软件有许多，常用的有 Quagga、Zebra 等。7.8 节会介绍如何使用 Quagga 动态容器路由发现的机制来实现 Kubernetes 的网络组网。

2. 路由表的查看

我们可以使用 ip route list 命令查看当前的路由表：

```
# ip route list
192.168.6.0/24 dev eno16777736 proto kernel scope link src 192.168.6.140
metric 1
```

在上面的例子代码中只有一个子网的路由，源地址是 192.168.6.140（本机），目标地址在 192.168.6.0/24 网段的数据包都将通过 eno16777736 接口发送出去。

Netstat-rn 是另一个查看路由表的工具：

```
# netstat -rn
Kernel IP routing table
Destination      Gateway         Genmask         Flags MSS Window  irtt Iface
0.0.0.0          192.168.6.2     0.0.0.0         UG    0 0            0 eth0
192.168.6.0      0.0.0.0         255.255.255.0   U     0 0            0 eth0
```

在它显示的信息中，如果标志是 U，则说明是可达路由；如果标志是 G，则说明这个网络接口连接的是网关，否则说明这个接口直连主机。

7.3 Docker 的网络实现

标准的 Docker 支持以下 4 类网络模式。

◎ host 模式：使用--net=host 指定。
◎ container 模式：使用--net=container:NAME_or_ID 指定。
◎ none 模式：使用--net=none 指定。
◎ bridge 模式：使用--net=bridge 指定，为默认设置。

在 Kubernetes 管理模式下通常只会使用 bridge 模式，所以本节只介绍在 bridge 模式下 Docker 是如何支持网络的。

在 bridge 模式下，Docker Daemon 第 1 次启动时会创建一个虚拟的网桥，默认的名称是 docker0，然后按照 RPC1918 的模型在私有网络空间中给这个网桥分配一个子网。针对由 Docker 创建的每一个容器，都会创建一个虚拟的以太网设备（Veth 设备对），其中一端关联到网桥上，另一端使用 Linux 的网络命名空间技术，映射到容器内的 eth0 设备，然后从网桥的地址段内给 eth0 接口分配一个 IP 地址。

如图 7.6 所示就是 Docker 的默认桥接网络模型。

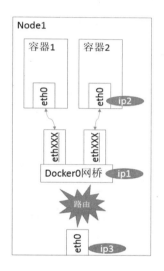

图 7.6　Docker 的默认桥接网络模型

其中 ip1 是网桥的 IP 地址，Docker Daemon 会在几个备选地址段里给它选一个地址，通常是以 172 开头的一个地址。这个地址和主机的 IP 地址是不重叠的。ip2 是 Docker 在启动容器时，在这个地址段选择的一个没有使用的 IP 地址分配给容器。相应的 MAC 地址也根据这个 IP 地址，在 02:42:ac:11:00:00 和 02:42:ac:11:ff:ff 的范围内生成，这样做可以确保不会有 ARP 冲突。

启动后，Docker 还将 Veth 对的名称映射到 eth0 网络接口。ip3 就是主机的网卡地址。

在一般情况下，ip1、ip2 和 ip3 是不同的 IP 段，所以在默认不做任何特殊配置的情况下，在外部是看不到 ip1 和 ip2 的。

这样做的结果就是，在同一台机器内的容器之间可以相互通信，不同主机上的容器不能相互通信，实际上它们甚至有可能在相同的网络地址范围内（不同主机上的 docker0 的地址段可能是一样的）。

为了让它们跨节点互相通信，就必须在主机的地址上分配端口，然后通过这个端口路由或代理到容器上。这种做法显然意味着一定要在容器之间小心谨慎地协调好端口的分配，或者使用动态端口的分配技术。在不同应用之间协调好端口分配是十分困难的事情，特别是集群水平扩展时。而动态的端口分配也会带来高度复杂性，例如：每个应用程序都只能将端口看作一个符号（因为是动态分配的，所以无法提前设置）。而且 API Server 要在分配完后，将动态端口插入配置的合适位置，服务也必须能互相找到对方等。这些都是 Docker 的网络模型在跨主机访问时面临的问题。

1. 查看 Docker 启动后的系统情况

我们已经知道，Docker 网络在 bridge 模式下 Docker Daemon 启动时创建 docker0 网桥，并在网桥使用的网段为容器分配 IP。让我们看看实际的操作。

在刚刚启动 Docker Daemon 并且还没有启动任何容器时，网络协议栈的配置情况如下：

```
# systemctl start docker
# ip addr
1: lo: <LOOPBACK,UP,LOWER_UP> mtu 65536 qdisc noqueue state UNKNOWN
    link/loopback 00:00:00:00:00:00 brd 00:00:00:00:00:00
    inet 127.0.0.1/8 scope host lo
       valid_lft forever preferred_lft forever
    inet6 ::1/128 scope host
       valid_lft forever preferred_lft forever
2: eno16777736: <BROADCAST,MULTICAST,UP,LOWER_UP> mtu 1500 qdisc pfifo_fast state UP qlen 1000
    link/ether 00:0c:29:14:3d:80 brd ff:ff:ff:ff:ff:ff
    inet 192.168.1.133/24 brd 192.168.1.255 scope global eno16777736
       valid_lft forever preferred_lft forever
    inet6 fe80::20c:29ff:fe14:3d80/64 scope link
       valid_lft forever preferred_lft forever
3: docker0: <NO-CARRIER,BROADCAST,MULTICAST,UP> mtu 1500 qdisc noqueue state DOWN
    link/ether 02:42:6e:af:0e:c3 brd ff:ff:ff:ff:ff:ff
    inet 172.17.42.1/24 scope global docker0
       valid_lft forever preferred_lft forever
```

```
# iptables-save
# Generated by iptables-save v1.4.21 on Thu Sep 24 17:11:04 2015
*nat
:PREROUTING ACCEPT [7:878]
:INPUT ACCEPT [7:878]
:OUTPUT ACCEPT [3:536]
:POSTROUTING ACCEPT [3:536]
:DOCKER - [0:0]
-A PREROUTING -m addrtype --dst-type LOCAL -j DOCKER
-A OUTPUT ! -d 127.0.0.0/8 -m addrtype --dst-type LOCAL -j DOCKER
-A POSTROUTING -s 172.17.0.0/16 ! -o docker0 -j MASQUERADE
COMMIT
# Completed on Thu Sep 24 17:11:04 2015
# Generated by iptables-save v1.4.21 on Thu Sep 24 17:11:04 2015
*filter
:INPUT ACCEPT [133:11362]
:FORWARD ACCEPT [0:0]
:OUTPUT ACCEPT [37:5000]
:DOCKER - [0:0]
-A FORWARD -o docker0 -j DOCKER
-A FORWARD -o docker0 -m conntrack --ctstate RELATED,ESTABLISHED -j ACCEPT
-A FORWARD -i docker0 ! -o docker0 -j ACCEPT
-A FORWARD -i docker0 -o docker0 -j ACCEPT
COMMIT
# Completed on Thu Sep 24 17:11:04 2015
```

可以看到，Docker 创建了 docker0 网桥，并添加了 iptables 规则。docker0 网桥和 iptables 规则都处于 root 命名空间中。通过解读这些规则，我们发现，在还没有启动任何容器时，如果启动了 Docker Daemon，那么它已经做好了通信准备。对这些规则的说明如下。

（1）在 NAT 表中有 3 条记录，前两条匹配生效后，都会继续执行 DOCKER 链，而此时 DOCKER 链为空，所以前两条只是做了一个框架，并没有实际效果。

（2）NAT 表第 3 条的含义是，若本地发出的数据包不是发往 docker0 的，即是发往主机之外的设备的，则都需要进行动态地址修改（MASQUERADE），将源地址从容器的地址（172 段）修改为宿主机网卡的 IP 地址，之后就可以发送给外面的网络了。

（3）在 FILTER 表中，第 1 条也是一个框架，因为后继的 DOCKER 链是空的。

（4）在 FILTER 表中，第 3 条是说，docker0 发出的包，如果需要 Forward 到非 docker0

的本地 IP 地址的设备，则是允许的。这样，docker0 设备的包就可以根据路由规则中转到宿主机的网卡设备，从而访问外面的网络。

（5）FILTER 表中，第 4 条是说，docker0 的包还可以被中转给 docker0 本身，即连接在 docker0 网桥上的不同容器之间的通信也是允许的。

（6）FILTER 表中，第 2 条是说，如果接收到的数据包属于以前已经建立好的连接，那么允许直接通过。这样接收到的数据包自然又走回 docker0，并中转到相应的容器。

除了这些 Netfilter 的设置，Linux 的 ip_forward 功能也被 Docker Daemon 打开了：

```
# cat /proc/sys/net/ipv4/ip_forward
1
```

另外，我们可以看到刚刚启动 Docker 后的 Route 表，和启动前没有什么不同：

```
# ip route
default via 192.168.1.2 dev eno16777736  proto static  metric 100
172.17.0.0/16 dev docker  proto kernel  scope link  src 172.17.42.1
192.168.1.0/24 dev eno16777736  proto kernel  scope link  src 192.168.1.132
192.168.1.0/24 dev eno16777736  proto kernel  scope link  src 192.168.1.132  metric 100
```

2. 查看容器启动后的情况（容器无端口映射）

刚才查看了 Docker 服务启动后的网络情况。现在启动一个 Registry 容器（不使用任何端口镜像参数），看一下网络堆栈部分相关的变化：

```
docker run --name register -d registry
# ip addr
1: lo: <LOOPBACK,UP,LOWER_UP> mtu 65536 qdisc noqueue state UNKNOWN
    link/loopback 00:00:00:00:00:00 brd 00:00:00:00:00:00
    inet 127.0.0.1/8 scope host lo
       valid_lft forever preferred_lft forever
    inet6 ::1/128 scope host
       valid_lft forever preferred_lft forever
2: eno16777736: <BROADCAST,MULTICAST,UP,LOWER_UP> mtu 1500 qdisc pfifo_fast state UP qlen 1000
    link/ether 00:0c:29:c8:12:5f brd ff:ff:ff:ff:ff:ff
    inet 192.168.1.132/24 brd 192.168.1.255 scope global eno16777736
       valid_lft forever preferred_lft forever
    inet6 fe80::20c:29ff:fec8:125f/64 scope link
```

```
       valid_lft forever preferred_lft forever
3: docker0: <NO-CARRIER,BROADCAST,MULTICAST,UP> mtu 1500 qdisc noqueue state DOWN
    link/ether 02:42:72:79:b8:88 brd ff:ff:ff:ff:ff:ff
    inet 172.17.42.1/24 scope global docker0
       valid_lft forever preferred_lft forever
    inet6 fe80::42:7aff:fe79:b888/64 scope link
       valid_lft forever preferred_lft forever
13: veth2dc8bbd: <BROADCAST,MULTICAST,UP,LOWER_UP> mtu 1500 qdisc noqueue master docker0 state UP
    link/ether be:d9:19:42:46:18 brd ff:ff:ff:ff:ff:ff
    inet6 fe80::bcd9:19ff:fe42:4618/64 scope link
       valid_lft forever preferred_lft forever

# iptables-save
# Generated by iptables-save v1.4.21 on Thu Sep 24 18:21:04 2015
*nat
:PREROUTING ACCEPT [14:1730]
:INPUT ACCEPT [14:1730]
:OUTPUT ACCEPT [59:4918]
:POSTROUTING ACCEPT [59:4918]
:DOCKER - [0:0]
-A PREROUTING -m addrtype --dst-type LOCAL -j DOCKER
-A OUTPUT ! -d 127.0.0.0/8 -m addrtype --dst-type LOCAL -j DOCKER
-A POSTROUTING -s 172.17.0.0/16 ! -o docker0 -j MASQUERADE
COMMIT
# Completed on Thu Sep 24 18:21:04 2015
# Generated by iptables-save v1.4.21 on Thu Sep 24 18:21:04 2015
*filter
:INPUT ACCEPT [2383:211572]
:FORWARD ACCEPT [0:0]
:OUTPUT ACCEPT [2004:242872]
:DOCKER - [0:0]
-A FORWARD -o docker0 -j DOCKER
-A FORWARD -o docker0 -m conntrack --ctstate RELATED,ESTABLISHED -j ACCEPT
-A FORWARD -i docker0 ! -o docker0 -j ACCEPT
-A FORWARD -i docker0 -o docker0 -j ACCEPT
COMMIT
# Completed on Thu Sep 24 18:21:04 2015

# ip route
default via 192.168.1.2 dev eno16777736 proto static metric 100
```

```
    172.17.0.0/16 dev docker  proto kernel  scope link  src 172.17.42.1
    192.168.1.0/24 dev eno16777736  proto kernel  scope link  src 192.168.1.132
    192.168.1.0/24 dev eno16777736  proto kernel  scope link  src 192.168.1.132
metric 100
```

可以看到如下情况。

（1）宿主机器上的 Netfilter 和路由表都没有变化，说明在不进行端口映射时，Docker 的默认网络是没有特殊处理的。相关的 NAT 和 FILTER 这两个 Netfilter 链还是空的。

（2）宿主机上的 Veth 对已经建立，并连接到容器内。

我们再次进入刚刚启动的容器内，看看网络栈是什么情况。容器内部的 IP 地址和路由如下：

```
# docker exec -ti 24981a750a1a bash
[root@24981a750a1a /]# ip route
default via 172.17.42.1 dev eth0
172.17.0.0/16 dev eth0  proto kernel  scope link  src 172.17.0.10
[root@24981a750a1a /]# ip addr
1: lo: <LOOPBACK,UP,LOWER_UP> mtu 65536 qdisc noqueue state UNKNOWN
    link/loopback 00:00:00:00:00:00 brd 00:00:00:00:00:00
    inet 127.0.0.1/8 scope host lo
       valid_lft forever preferred_lft forever
    inet6 ::1/128 scope host
       valid_lft forever preferred_lft forever
22: eth0: <BROADCAST,MULTICAST,UP,LOWER_UP> mtu 1500 qdisc noqueue state UP
    link/ether 02:42:ac:11:00:0a brd ff:ff:ff:ff:ff:ff
    inet 172.17.0.10/16 scope global eth0
       valid_lft forever preferred_lft forever
    inet6 fe80::42:acff:fe11:a/64 scope link
       valid_lft forever preferred_lft forever
```

可以看到，默认停止的回环设备 lo 已经被启动，外面宿主机连接进来的 Veth 设备也被命名成了 eth0，并且已经配置了地址 172.17.0.10。

路由信息表包含一条到 docker0 的子网路由和一条到 docker0 的默认路由。

3. 查看容器启动后的情况（容器有端口映射）

下面用带端口映射的命令启动 registry：

```
docker run --name register -d -p 1180:5000 registry
```

在启动后查看 iptables 的变化：

```
# iptables-save
# Generated by iptables-save v1.4.21 on Thu Sep 24 18:45:13 2015
*nat
:PREROUTING ACCEPT [2:236]
:INPUT ACCEPT [0:0]
:OUTPUT ACCEPT [0:0]
:POSTROUTING ACCEPT [0:0]
:DOCKER - [0:0]
-A PREROUTING -m addrtype --dst-type LOCAL -j DOCKER
-A OUTPUT ! -d 127.0.0.0/8 -m addrtype --dst-type LOCAL -j DOCKER
-A POSTROUTING -s 172.17.0.0/16 ! -o docker0 -j MASQUERADE
-A POSTROUTING -s 172.17.0.19/32 -d 172.17.0.19/32 -p tcp -m tcp --dport 5000 -j MASQUERADE
-A DOCKER ! -i docker0 -p tcp -m tcp --dport 1180 -j DNAT --to-destination 172.17.0.19:5000
COMMIT
# Completed on Thu Sep 24 18:45:13 2015
# Generated by iptables-save v1.4.21 on Thu Sep 24 18:45:13 2015
*filter
:INPUT ACCEPT [54:4464]
:FORWARD ACCEPT [0:0]
:OUTPUT ACCEPT [41:5576]
:DOCKER - [0:0]
-A FORWARD -o docker0 -j DOCKER
-A FORWARD -o docker0 -m conntrack --ctstate RELATED,ESTABLISHED -j ACCEPT
-A FORWARD -i docker0 ! -o docker0 -j ACCEPT
-A FORWARD -i docker0 -o docker0 -j ACCEPT
-A DOCKER -d 172.17.0.19/32 ! -i docker0 -o docker0 -p tcp -m tcp --dport 5000 -j ACCEPT
COMMIT
# Completed on Thu Sep 24 18:45:13 2015
```

从新增的规则可以看出，Docker 服务在 NAT 和 FILTER 两个表内添加的两个 DOCKER 子链都是给端口映射用的。在本例中我们需要把外面宿主机的 1180 端口映射到容器的 5000 端口。通过前面的分析我们知道，无论是宿主机接收到的还是宿主机本地协议栈发出的，目标地址是本地 IP 地址的包都会经过 NAT 表中的 DOCKER 子链。Docker 为每一个端口映射都在这个链上增加了到实际容器目标地址和目标端口的转换。

经过这个 DNAT 的规则修改后的 IP 包，会重新经过路由模块的判断进行转发。由于目标地址和端口已经是容器的地址和端口，所以数据自然就被转发到docker0上，从而被转发到对应的容器内部。

当然在Forward时，也需要在DOCKER子链中添加一条规则，如果目标端口和地址是指定容器的数据，则允许通过。

在Docker按照端口映射的方式启动容器时，主要的不同就是上述iptables部分。而容器内部的路由和网络设备，都和不做端口映射时一样，没有任何变化。

1. Docker的网络局限

我们从Docker对Linux网络协议栈的操作可以看到，Docker一开始没有考虑到多主机互联的网络解决方案。

Docker一直以来的理念都是"简单为美"，几乎所有尝试Docker的人都被它"用法简单，功能强大"的特性所吸引，这也是Docker迅速走红的一个原因。

我们都知道，虚拟化技术中最为复杂的部分就是虚拟化网络技术，即使是单纯的物理网络部分，也是一个门槛很高的技能领域，通常只被少数网络工程师所掌握，所以我们可以理解结合了物理网络的虚拟网络技术有多难。在Docker之前，所有接触过OpenStack的人都对其网络问题讳莫如深，Docker明智地避开这个"雷区"，让其他专业人员去用现有的虚拟化网络技术解决Docker主机的互联问题，以免让用户觉得Docker太难，从而放弃学习和使用Docker。

Docker成名以后，重新开始重视网络解决方案，收购了一家Docker网络解决方案公司——Socketplane，原因在于这家公司的产品广受好评，但有趣的是Socketplane的方案就是以Open vSwitch为核心的，其还为Open vSwitch提供了Docker镜像，以方便部署程序。之后，Docker开启了一个宏伟的虚拟化网络解决方案——Libnetwork，如图7.7所示是其概念图。

这个概念图没有了IP，也没有了路由，已经颠覆了我们的网络常识，对于不怎么懂网络的大多数人来说，它的确很有诱惑力，未来是否会对虚拟化网络的模型产生深远冲击，我们还不得而知，但它仅仅是Docker官方当前的一次"尝试"。

针对目前Docker的网络实现，Docker使用的Libnetwork组件只是将Docker平台中的网络子系统模块化为一个独立库的简单尝试，离成熟和完善还有一段距离。

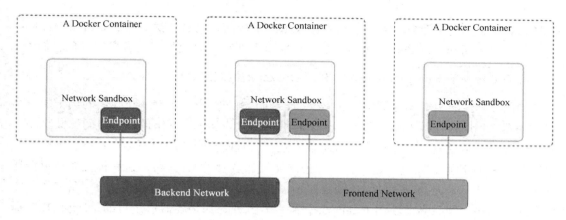

图 7.7　Libnetwork 概念图

7.4　Kubernetes 的网络实现

在实际的业务场景中，业务组件之间的关系十分复杂，特别是随着微服务理念逐步深入人心，应用部署的粒度更加细小和灵活。为了支持业务应用组件的通信，Kubernetes 网络的设计主要致力于解决以下问题。

（1）容器到容器之间的直接通信。

（2）抽象的 Pod 到 Pod 之间的通信。

（3）Pod 到 Service 之间的通信。

（4）集群外部与内部组件之间的通信。

其中第 3 条、第 4 条在之前的章节里都有所讲解，本节对更为基础的第 1 条与第 2 条进行深入分析和讲解。

7.4.1　容器到容器的通信

同一个 Pod 内的容器（Pod 内的容器是不会跨宿主机的）共享同一个网络命名空间，共享同一个 Linux 协议栈。所以对于网络的各类操作，就和它们在同一台机器上一样，它们甚至可以用 localhost 地址访问彼此的端口。

这么做的结果是简单、安全和高效，也能减小将已经存在的程序从物理机或者虚拟机移植到容器下运行的难度。其实，在容器技术出来之前，大家早就积累了如何在一台机器上运行一组应用程序的经验，例如，如何让端口不冲突，以及如何让客户端发现它们等。

我们来看一下 Kubernetes 是如何利用 Docker 的网络模型的。

如图 7.8 中的阴影部分所示，在 Node 上运行着一个 Pod 实例。在我们的例子中，容器就是图 7.8 中的容器 1 和容器 2。容器 1 和容器 2 共享一个网络的命名空间，共享一个命名空间的结果就是它们好像在一台机器上运行，它们打开的端口不会有冲突，可以直接使用 Linux 的本地 IPC 进行通信（例如消息队列或者管道）。其实，这和传统的一组普通程序运行的环境是完全一样的，传统程序不需要针对网络做特别的修改就可以移植了，它们之间的互相访问只需要使用 localhost 就可以。例如，如果容器 2 运行的是 MySQL，那么容器 1 使用 localhost:3306 就能直接访问这个运行在容器 2 上的 MySQL 了。

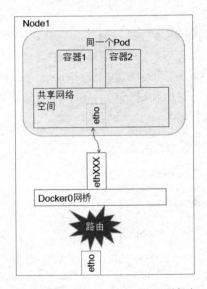

图 7.8 Kubernetes 的 Pod 网络模型

7.4.2 Pod 之间的通信

我们看了同一个 Pod 内的容器之间的通信情况，再看看 Pod 之间的通信情况。

每一个 Pod 都有一个真实的全局 IP 地址，同一个 Node 内的不同 Pod 之间可以直接采用对方 Pod 的 IP 地址通信，而且不需要采用其他发现机制，例如 DNS、Consul 或者 etcd。

Pod 容器既有可能在同一个 Node 上运行，也有可能在不同的 Node 上运行，所以通信也分为两类：同一个 Node 内 Pod 之间的通信和不同 Node 上 Pod 之间的通信。

1. **同一个 Node 内 Pod 之间的通信**

我们看一下同一个 Node 内两个 Pod 之间的关系，如图 7.9 所示。

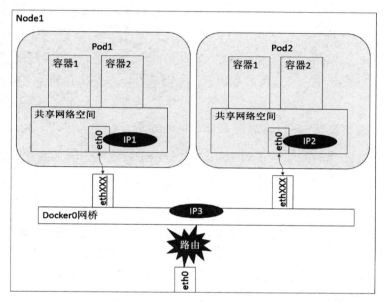

图 7.9 同一个 Node 内两个 Pod 之间的关系

可以看出，Pod1 和 Pod2 都是通过 Veth 连接到同一个 docker0 网桥上的，它们的 IP 地址 IP1、IP2 都是从 docker0 的网段上动态获取的，它们和网桥本身的 IP3 是同一个网段的。

另外，在 Pod1、Pod2 的 Linux 协议栈上，默认路由都是 docker0 的地址，也就是说所有非本地地址的网络数据，都会被默认发送到 docker0 网桥上，由 docker0 网桥直接中转。

综上所述，由于它们都关联在同一个 docker0 网桥上，地址段相同，所以它们之间是能直接通信的。

2. **不同 Node 上 Pod 之间的通信**

Pod 的地址是与 docker0 在同一个网段的，我们知道 docker0 网段与宿主机网卡是两个完全不同的IP网段，并且不同Node之间的通信只能通过宿主机的物理网卡进行，因此要想实现不同Node上Pod容器之间的通信，就必须想办法通过主机的这个IP地址进行寻址和通信。

另一方面，这些动态分配且藏在docker0之后的所谓"私有"IP地址也是可以找到的。Kubernetes会记录所有正在运行的Pod的IP分配信息，并将这些信息保存在etcd中（作为Service的Endpoint）。这些私有IP信息对于Pod到Pod的通信也是十分重要的，因为我们的网络模型要求Pod到Pod使用私有IP进行通信。所以首先要知道这些IP是什么。

之前提到，Kubernetes的网络对Pod的地址是平面的和直达的，所以这些Pod的IP规划也很重要，不能有冲突。只要没有冲突，我们就可以想办法在整个Kubernetes的集群中找到它。

综上所述，要想支持不同Node上Pod之间的通信，就要满足两个条件：

（1）在整个Kubernetes集群中对Pod的IP分配进行规划，不能有冲突；

（2）找到一种办法，将Pod的IP和所在Node的IP关联起来，通过这个关联让Pod可以互相访问。

根据条件1的要求，我们需要在部署Kubernetes时对docker0的IP地址进行规划，保证每个Node上的docker0地址都没有冲突。我们可以在规划后手工配置到每个Node上，或者做一个分配规则，由安装的程序自己去分配占用。例如，Kubernetes的网络增强开源软件Flannel就能够管理资源池的分配。

根据条件2的要求，Pod中的数据在发出时，需要有一个机制能够知道对方Pod的IP地址挂在哪个具体的Node上。也就是说先要找到Node对应宿主机的IP地址，将数据发送到这个宿主机的网卡，然后在宿主机上将相应的数据转发到具体的docker0上。一旦数据到达宿主机Node，则那个Node内部的docker0便知道如何将数据发送到Pod。如图7.10所示。

在图7.10中，IP1对应的是Pod1，IP2对应的是Pod2。Pod1在访问Pod2时，首先要将数据从源Node的eth0发送出去，找到并到达Node2的eth0。即先是从IP3到IP4的递送，之后才是从IP4到IP2的递送。

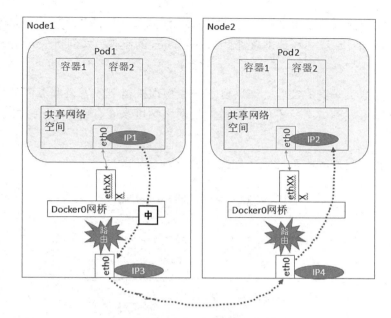

图 7.10 跨 Node 的 Pod 通信

在谷歌的 GCE 环境中，Pod 的 IP 管理（类似 docker0）、分配及它们之间的路由打通都是由 GCE 完成的。Kubernetes 作为主要在 GCE 上面运行的框架，它的设计是假设底层已经具备这些条件，所以它分配完地址并将地址记录下来就完成了它的工作。在实际的 GCE 环境中，GCE 的网络组件会读取这些信息，实现具体的网络打通。

而在实际生产环境中，因为安全、费用、合规等种种原因，Kubernetes 的客户不可能全部使用谷歌的 GCE 环境，所以在实际的私有云环境中，除了需要部署 Kubernetes 和 Docker，还需要额外的网络配置，甚至通过一些软件来实现 Kubernetes 对网络的要求。做到这些后，Pod 和 Pod 之间才能无差别地进行透明通信。

为了达到这个目的，开源界有不少应用增强了 Kubernetes、Docker 的网络，在后面的章节中会介绍几个常用的组件及其组网原理。

7.5 Pod 和 Service 网络实战

Docker 给我们带来了不同的网络模式，Kubernetes 也以一种不同的方式来解决这些网络模式的挑战，但其方式有些难以理解，特别是对于刚开始接触 Kubernetes 的网络的开发者来说。我们在前面学习了 Kubernetes、Docker 的理论,本节将通过一个完整的实验,从部署一个 Pod 开始,一步一步地部署那些 Kubernetes 的组件,来剖析 Kubernetes 在网络层是如何实现及工作的。

这里使用虚拟机来完成实验。如果要部署在物理机器上或者云服务商的环境中,则涉及的网络模型很可能稍微有所不同。不过,从网络角度来看,Kubernetes 的机制是类似且一致的。

好了,来看看我们的实验环境,如图 7.11 所示。

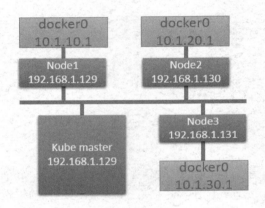

图 7.11 实验环境

Kubernetes 的网络模型要求每个 Node 上的容器都可以相互访问。

默认的 Docker 网络模型提供了一个 IP 地址段是 172.17.0.0/16 的 docker0 网桥。每个容器都会在这个子网内获得 IP 地址,并且将 docker0 网桥的 IP 地址(172.17.42.1)作为其默认网关。需要注意的是,Docker 宿主机外面的网络不需要知道任何关于这个 172.17.0.0/16 的信息或者知道如何连接到其内部,因为 Docker 的宿主机针对容器发出的数据,在物理网卡地址后面都做了 IP 伪装 MASQUERADE(隐含 NAT)。也就是说,在网络上看到的任何容器数据流都来源于那台 Docker 节点的物理 IP 地址。这里所说的网络都指连接这些主机的物理网络。

这个模型便于使用,但是并不完美,需要依赖端口映射的机制。

在 Kubernetes 的网络模型中,每台主机上的 docker0 网桥都是可以被路由到的。也就是说,在部署了一个 Pod 时,在同一个集群内,各主机都可以访问其他主机上的 Pod IP,并不需要在主机上做端口映射。综上所述,我们可以在网络层将 Kubernetes 的节点看作一个路由器。如果将实验环境改画成一个网络图，那么它看起来如图 7.12 所示。

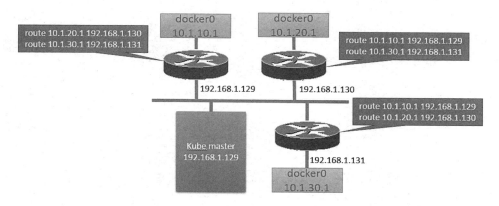

图 7.12　实验环境网络图

为了支持 Kubernetes 网络模型，我们采取了直接路由的方式来实现，在每个 Node 上都配置相应的静态路由项，例如在 192.168.1.129 这个 Node 上配置了两个路由项：

```
#route add -net 10.1.20.0 netmask 255.255.255.0 gw 192.168.130
#route add -net 10.1.30.0 netmask 255.255.255.0 gw 192.168.131
```

这意味着，每一个新部署的容器都将使用这个 Node（docker0 的网桥 IP）作为它的默认网关。而这些 Node（类似路由器）都有其他 docker0 的路由信息，这样它们就能够相互连通了。

接下来通过一些实际的案例，来看看 Kubernetes 在不同的场景下其网络部分到底做了什么。

第 1 步：部署一个 RC/Pod

部署的 RC/Pod 描述文件如下（frontend-controller.yaml）：

```
apiVersion: v1
kind: ReplicationController
metadata:
  name: frontend
  labels:
    name: frontend
spec:
  replicas: 1
```

```yaml
  selector:
    name: frontend
  template:
    metadata:
      labels:
        name: frontend
    spec:
      containers:
      - name: php-redis
        image: kubeguide/guestbook-php-frontend
        env:
        - name: GET_HOSTS_FROM
          value: env
        ports:
        - containerPort: 80
          hostPort: 80
```

为了便于观察,我们假定在一个空的 Kubernetes 集群上运行,提前清理了所有 Replication Controller、Pod 和其他 Service:

```
# kubectl get rc
CONTROLLER    CONTAINER(S)    IMAGE(S)    SELECTOR    REPLICAS
#
# kubectl get services
NAME          LABELS                                          SELECTOR    IP(S)
PORT(S)
kubernetes    component=apiserver,provider=kubernetes  <none>      20.1.0.1    443/TCP
#
# kubectl get pods
NAME       READY      STATUS     RESTARTS    AGE
```

让我们检查一下此时某个 Node 上的网络接口都有哪些。Node1 的状态是:

```
# ifconfig
docker0: flags=4099<UP,BROADCAST,RUNNING,MULTICAST>  mtu 1500
        inet 10.1.10.1  netmask 255.255.255.0  broadcast 10.1.10.255
        inet6 fe80::5484:7aff:fefe:9799  prefixlen 64  scopeid 0x20<link>
        ether 56:84:7a:fe:97:99  txqueuelen 0  (Ethernet)
        RX packets 373245  bytes 170175373 (162.2 MiB)
        RX errors 0  dropped 0  overruns 0  frame 0
        TX packets 353569  bytes 353948005 (337.5 MiB)
        TX errors 0  dropped 0  overruns 0  carrier 0  collisions 0
```

```
eno16777736: flags=4163<UP,BROADCAST,RUNNING,MULTICAST> mtu 1500
        inet 192.168.1.129  netmask 255.255.255.0  broadcast 192.168.1.255
        inet6 fe80::20c:29ff:fe47:6e2c  prefixlen 64  scopeid 0x20<link>
        ether 00:0c:29:47:6e:2c  txqueuelen 1000  (Ethernet)
        RX packets 326552  bytes 286033393 (272.7 MiB)
        RX errors 0  dropped 0  overruns 0  frame 0
        TX packets 219520  bytes 31014871 (29.5 MiB)
        TX errors 0  dropped 0  overruns 0  carrier 0  collisions 0

lo: flags=73<UP,LOOPBACK,RUNNING>  mtu 65536
        inet 127.0.0.1  netmask 255.0.0.0
        inet6 ::1  prefixlen 128  scopeid 0x10<host>
        loop  txqueuelen 0  (Local Loopback)
        RX packets 24095  bytes 2133648 (2.0 MiB)
        RX errors 0  dropped 0  overruns 0  frame 0
        TX packets 24095  bytes 2133648 (2.0 MiB)
        TX errors 0  dropped 0  overruns 0  carrier 0  collisions 0
```

可以看出，有一个docker0网桥和一个本地地址的网络端口。现在部署一下我们在前面准备的RC/Pod配置文件，看看发生了什么：

```
# kubectl create -f frontend-controller.yaml
replicationcontrollers/frontend
#
# kubectl get pods
NAME              READY     STATUS    RESTARTS   AGE    NODE
frontend-4o11g    1/1       Running   0          11s    192.168.1.130
```

可以看到一些有趣的事情。Kubernetes为这个Pod找了一个主机192.168.1.130（Node2）来运行它。另外，这个Pod获得了一个在Node2的docker0网桥上的IP地址。我们登录Node2查看正在运行的容器：

```
# docker ps
  CONTAINER ID       IMAGE                  COMMAND             CREATED          STATUS        PORTS        NAMES
  37b193a4c633       kubeguide/example-guestbook-php-redis      "/bin/sh -c /run.sh"
32 seconds ago       Up 26 seconds          k8s_php-redis.6ad3289e_frontend-n9n1m_
development_813e2dd9-8149-11e5-823b-000c2921ba71_af6dd859
  6d1b99cff4ae       google_containers/pause:latest      "/pause"        35 seconds ago
Up 28 seconds        0.0.0.0:80->80/tcp   k8s_POD.855eeb3d_frontend-4t52y_development_
813e3870-8149-11e5-823b-000c2921ba71_2b66f05e
```

在Node2上现在运行了两个容器，在我们的RC/Pod定义文件中仅仅包含了一个，那么这第2个是从哪里来的呢？第2个看起来运行的是一个叫作google_containers/pause:latest的镜像，而且这个容器已经有端口映射到它上面了，为什么是这样呢？让我们深入容器内部去看一下具体原因。使用Docker的inspect命令来查看容器的详细信息，特别要关注容器的网络模型：

```
# docker inspect 6d1b99cff4ae | grep NetworkMode
      "NetworkMode": "bridge",
# docker inspect 37b193a4c633 | grep NetworkMode
      "NetworkMode": "container:6d1b99cff4ae537689ce87d7528f4ba9dbb40ae711ecc0a5b3f7c39ff5e5e495",
```

有趣的结果是，在查看完每个容器的网络模型后，我们可以看到这样的配置：我们检查的第1个容器是运行了"google_containers/pause:latest"镜像的容器，它使用了Docker默认的网络模型bridge；而我们检查的第2个容器，也就是在RC/Pod中定义运行的php-redis容器，使用了非默认的网络配置和映射容器的模型，指定了映射目标容器为"google_containers/ pause:latest"。

一起来仔细思考这个过程，为什么Kubernetes要这么做呢？

首先，一个Pod内的所有容器都需要共用同一个IP地址，这就意味着一定要使用网络的容器映射模式。然而，为什么不能只启动1个容器，而将第2个容器关联到第1个容器呢？我们认为Kubernetes是从两方面来考虑这个问题的：首先，如果在Pod内有多个容器的话，则可能很难连接这些容器；其次，后面的容器还要依赖第1个被关联的容器，如果第2个容器关联到第1个容器，且第1个容器死掉的话，第2个容器也将死掉。启动一个基础容器，然后将Pod内的所有容器都连接到它上面会更容易一些。因为我们只需要为基础的这个google_containers/pause容器执行端口映射规则，这也简化了端口映射的过程。所以我们启动Pod后的网络模型类似于图7.13。

在这种情况下，实际Pod的IP数据流的网络目标都是这个google_containers/pause容器。图7.13有点儿取巧地显示了是google_containers/pause容器将端口80的流量转发给了相关的容器。而pause容器只是看起来转发了网络流量，但它并没有真的这么做。实际上，应用容器直接监听了这些端口，和google_containers/pause容器共享了同一个网络堆栈。这就是为什么在Pod内部实际容器的端口映射都显示到google_containers/pause容器上了。我们可以通过docker port命令来检验一下：

```
# docker ps
CONTAINER ID        IMAGE
37b193a4c633        kubeguide/example-guestbook-php-redis
6d1b99cff4ae        google_containers/pause:latest
#
# docker port 6d1b99cff4ae
80/tcp -> 0.0.0.0:80
```

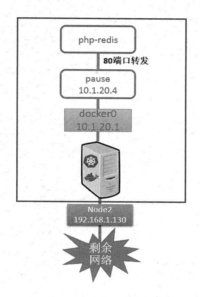

图 7.13　启动 Pod 后的网络模型

综上所述，google_containers/pause 容器实际上只是负责接管这个 Pod 的 Endpoint，并没有做更多的事情。那么 Node 呢？它需要将数据流传给 google_containers/pause 容器吗？我们来检查一下 iptables 的规则，看看有什么发现：

```
# iptables-save
# Generated by iptables-save v1.4.21 on Thu Sep 24 17:15:01 2015
*nat
:PREROUTING ACCEPT [0:0]
:INPUT ACCEPT [0:0]
:OUTPUT ACCEPT [0:0]
:POSTROUTING ACCEPT [0:0]
:DOCKER - [0:0]
:KUBE-NODEPORT-CONTAINER - [0:0]
:KUBE-NODEPORT-HOST - [0:0]
```

```
:KUBE-PORTALS-CONTAINER - [0:0]
:KUBE-PORTALS-HOST - [0:0]
-A PREROUTING -m comment --comment "handle ClusterIPs; NOTE: this must be before
the NodePort rules" -j KUBE-PORTALS-CONTAINER
-A PREROUTING -m addrtype --dst-type LOCAL -j DOCKER
-A PREROUTING -m addrtype --dst-type LOCAL -m comment --comment "handle service
NodePorts; NOTE: this must be the last rule in the chain" -j KUBE-NODEPORT-CONTAINER
-A OUTPUT -m comment --comment "handle ClusterIPs; NOTE: this must be before the
NodePort rules" -j KUBE-PORTALS-HOST
-A OUTPUT ! -d 127.0.0.0/8 -m addrtype --dst-type LOCAL -j DOCKER
-A OUTPUT -m addrtype --dst-type LOCAL -m comment --comment "handle service
NodePorts; NOTE: this must be the last rule in the chain
-A POSTROUTING -s 10.1.20.0/24 ! -o docker0 -j MASQUERADE
-A KUBE-PORTALS-CONTAINER -d 20.1.0.1/32 -p tcp -m comment --comment
"default/kubernetes:" -m tcp --dport 443 -j REDIRECT --to-ports 60339
-A KUBE-PORTALS-HOST -d 20.1.0.1/32 -p tcp -m comment --comment
"default/kubernetes:" -m tcp --dport 443 -j DNAT --to-destination 192.168.1.131:60339
COMMIT
# Completed on Thu Sep 24 17:15:01 2015
# Generated by iptables-save v1.4.21 on Thu Sep 24 17:15:01 2015
*filter
:INPUT ACCEPT [1131:377745]
:FORWARD ACCEPT [0:0]
:OUTPUT ACCEPT [1246:209888]
:DOCKER - [0:0]
-A FORWARD -o docker0 -j DOCKER
-A FORWARD -o docker0 -m conntrack --ctstate RELATED,ESTABLISHED -j ACCEPT
-A FORWARD -i docker0 ! -o docker0 -j ACCEPT
-A FORWARD -i docker0 -o docker0 -j ACCEPT
-A DOCKER -d 172.17.0.19/32 ! -i docker0 -o docker0 -p tcp -m tcp --dport 5000
-j ACCEPT
COMMIT
# Completed on Thu Sep 24 17:15:01 2015
```

上面的这些规则并没有被应用到我们刚刚定义的 Pod 上。当然，Kubernetes 会给每一个 Kubernetes 节点都提供一些默认的服务，上面的规则就是 Kubernetes 的默认服务所需要的。关键是，我们没有看到任何 IP 伪装的规则，并且没有任何指向 Pod 10.1.20.4 内部的端口映射。

第 2 步：发布一个服务

我们已经了解了 Kubernetes 如何处理最基本的元素即 Pod 的连接问题，接下来看一下它是如何处理 Service 的。Service 允许我们在多个 Pod 之间抽象一些服务，而且服务可以通过提供在同一个 Service 的多个 Pod 之间的负载均衡机制来支持水平扩展。我们再次将环境初始化，删除刚刚创建的 RC 或 Pod 来确保集群是空的：

```
# kubectl stop rc frontend
replicationcontroller/frontend
#
# kubectl get rc
CONTROLLER    CONTAINER(S)    IMAGE(S)    SELECTOR    REPLICAS
#
# kubectl get services
NAME          LABELS                                          SELECTOR    IP(S)       PORT(S)
kubernetes    component=apiserver,provider=kubernetes         <none>      20.1.0.1    443/TCP
#
# kubectl get pods
NAME          READY    STATUS    RESTARTS    AGE
```

然后准备一个名为 frontend 的 Service 配置文件：

```
apiVersion: v1
kind: Service
metadata:
  name: frontend
  labels:
    name: frontend
spec:
  ports:
  - port: 80
#   nodePort: 30001
  selector:
    name: frontend
#  type:
#    NodePort
```

接着在 Kubernetes 集群中定义这个服务：

```
# kubectl create -f frontend-service.yaml
services/frontend
```

```
# kubectl get services
NAME          LABELS                                     SELECTOR                 IP(S)           PORT(S)
frontend      name=frontend                              name=frontend            20.1.244.75     80/TCP
kubernetes                component=apiserver,provider=kubernetes    <none>       20.1.0.1        443/TCP
```

在服务正确创建后，可以看到 Kubernetes 集群已经为这个服务分配了一个虚拟 IP 地址 20.1.244.75，这个 IP 地址是在 Kubernetes 的 Portal Network 中分配的。而这个 Portal Network 的地址范围是我们在 Kubmaster 上启动 API 服务进程时，使用 --service-cluster-ip-range=xx 命令行参数指定的：

```
# cat /etc/kubernetes/apiserver
......
# Address range to use for services
KUBE_SERVICE_ADDRESSES="--service-cluster-ip-range=20.1.0.0/16"
......
```

这个 IP 段可以是任何段，只要不和 docker0 或者物理网络的子网冲突就可以。选择任意其他网段的原因是这个网段将不会在物理网络和 docker0 网络上进行路由。这个 Portal Network 针对每一个 Node 都有局部的特殊性，实际上它存在的意义是让容器的流量都指向默认网关（也就是 docker0 网桥）。在继续实验前，先登录到 Node1 上看一下在我们定义服务后发生了什么变化。首先检查一下 iptables 或 Netfilter 的规则：

```
# iptables-save
......
-A KUBE-PORTALS-CONTAINER -d 20.1.244.75/32 -p tcp -m comment --comment "default/frontend:" -m tcp --dport 80 -j REDIRECT --to-ports 59528
-A KUBE-PORTALS-HOST -d 20.1.244.75/32 -p tcp -m comment --comment "default/kubernetes:" -m tcp --dport 80 -j DNAT --to-destination 192.168.1.131:59528
......
```

第 1 行是挂在 PREROUTING 链上的端口重定向规则，所有进入的流量如果满足 20.1.244.75:80，则都会被重定向到端口 33761。第 2 行是挂在 OUTPUT 链上的目标地址 NAT，做了和上述第 1 行规则类似的工作，但针对的是当前主机生成的外出流量。所有主机生成的流量都需要使用这个 DNAT 规则来处理。简而言之，这两个规则使用了不同的方式做了类似的事情，就是将所有从节点生成的发送给 20.1.244.75:80 的流量重定向到本地的 33761 端口。

至此，目标为 Service IP 地址和端口的任何流量都将被重定向到本地的 33761 端口。

这个端口连到哪里去了呢？这就到了 kube-proxy 发挥作用的地方了。这个 kube-proxy 服务给每一个新创建的服务都关联了一个随机的端口号，并且监听那个特定的端口，为服务创建相关的负载均衡对象。在我们的实验中，随机生成的端口刚好是 33761。通过监控 Node1 上的 Kubernetes-Service 的日志，在创建服务时可以看到下面的记录：

```
2612 proxier.go:413] Opened iptables from-containers portal for service "default/frontend:" on TCP 20.1.244.75:80
2612 proxier.go:424] Opened iptables from-host portal for service "default/frontend:" on TCP 20.1.244.75:80
```

现在我们知道，所有流量都被导入 kube-proxy 中了。我们现在需要它完成一些负载均衡的工作，创建 Replication Controller 并观察结果，下面是 Replication Controller 的配置文件：

```
apiVersion: v1
kind: ReplicationController
metadata:
  name: frontend
  labels:
    name: frontend
spec:
  replicas: 3
  selector:
    name: frontend
  template:
    metadata:
      labels:
        name: frontend
    spec:
      containers:
      - name: php-redis
        image: kubeguide/example-guestbook-php-redis
        env:
        - name: GET_HOSTS_FROM
          value: env
        ports:
        - containerPort: 80
#         hostPort: 80
```

在集群发布上述配置文件后，等待并观察，确保所有 Pod 都运行起来了：

```
# kubectl create -f frontend-controller.yaml
replicationcontrollers/frontend
```

```
#
# kubectl get pods -o wide
NAME             READY   STATUS    RESTARTS   AGE   NODE
frontend-64t8q   1/1     Running   0          5s    192.168.1.130
frontend-dzqve   1/1     Running   0          5s    192.168.1.131
frontend-x5dwy   1/1     Running   0          5s    192.168.1.129
```

现在所有的 Pod 都运行起来了，Service 将会把客户端请求负载分发到包含 "name=frontend" 标签的所有 Pod 上。现在的实验环境如图 7.14 所示。

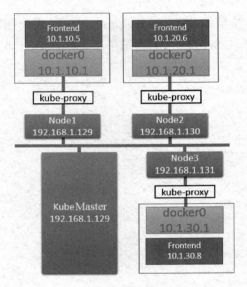

图 7.14　现在的实验环境

Kubernetes 的 kube-proxy 看起来只是一个夹层，但实际上它只是在 Node 上运行的一个服务。上述重定向规则的结果就是针对目标地址为服务 IP 的流量，将 Kubernetes 的 kube-proxy 变成了一个中间的夹层。

为了查看具体的重定向动作，我们会使用 tcpdump 来进行网络抓包操作。

首先，安装 tcpdump：

```
yum -y install tcpdump
```

安装完成后，登录 Node1，运行 tcpdump 命令：

```
tcpdump -nn -q -i eno16777736 port 80
```

需要捕获物理服务器以太网接口的数据包，Node1 机器上的以太网接口名字叫作 eno16777736。

再打开第 1 个窗口运行第 2 个 tcpdump 程序，不过我们需要一些额外的信息去运行它，即挂接在 docker0 桥上的虚拟网卡 Veth 的名称。我们看到只有一个 frontend 容器在 Node1 主机上运行，所以可以使用简单的 "ip addr" 命令来查看唯一的 Veth 网络接口：

```
# ip addr
1: lo: <LOOPBACK,UP,LOWER_UP> mtu 65536 qdisc noqueue state UNKNOWN
    link/loopback 00:00:00:00:00:00 brd 00:00:00:00:00:00
    inet 127.0.0.1/8 scope host lo
       valid_lft forever preferred_lft forever
    inet6 ::1/128 scope host
       valid_lft forever preferred_lft forever
2: eno16777736: <BROADCAST,MULTICAST,UP,LOWER_UP> mtu 1500 qdisc pfifo_fast state UP qlen 1000
    link/ether 00:0c:29:47:6e:2c brd ff:ff:ff:ff:ff:ff
    inet 192.168.1.129/24 brd 192.168.1.255 scope global eno16777736
       valid_lft forever preferred_lft forever
    inet6 fe80::20c:29ff:fe47:6e2c/64 scope link
       valid_lft forever preferred_lft forever
3: docker0: <NO-CARRIER,BROADCAST,MULTICAST,UP> mtu 1500 qdisc noqueue state DOWN
    link/ether 56:84:7a:fe:97:99 brd ff:ff:ff:ff:ff:ff
    inet 10.1.10.1/24 brd 10.1.10.255 scope global docker0
       valid_lft forever preferred_lft forever
    inet6 fe80::5484:7aff:fefe:9799/64 scope link
       valid_lft forever preferred_lft forever
12: veth0558bfa: <BROADCAST,MULTICAST,UP,LOWER_UP> mtu 1500 qdisc noqueue master docker0 state UP
    link/ether 86:82:e5:c8:5a:9a brd ff:ff:ff:ff:ff:ff
    inet6 fe80::8482:e5ff:fec8:5a9a/64 scope link
       valid_lft forever preferred_lft forever
```

复制这个接口的名字，在第 2 个窗口中运行 tcpdump 命令：

```
tcpdump -nn -q -i veth0558bfa host 20.1.244.75
```

同时运行这两个命令，并且将窗口并排放置，以便同时看到两个窗口的输出：

```
# tcpdump -nn -q -i eno16777736 port 80
tcpdump: verbose output suppressed, use -v or -vv for full protocol decode
listening on eno16777736, link-type EN10MB (Ethernet), capture size 65535 bytes
```

```
# tcpdump -nn -q -i veth0558bfa host 20.1.244.75
tcpdump: verbose output suppressed, use -v or -vv for full protocol decode
listening on veth0558bfa, link-type EN10MB (Ethernet), capture size 65535 bytes
```

好了，我们已经在同时捕获两个接口的网络包了。这时再启动第 3 个窗口，运行一个 "docker exec" 命令来连接到我们的 frontend 容器内部（你可以先执行 docker ps 来获得这个容器的 ID）：

```
# docker ps
CONTAINER ID        IMAGE                                          ......
268ccdfb9524        kubeguide/example-guestbook-php-redis          ......
6a519772b27e        google_containers/pause:latest                 ......
```

执行命令进入容器内部：

```
#docker exec -it 268ccdfb9524 bash
# docker exec -it 268ccdfb9524 bash
root@frontend-x5dwy:/#
```

一旦进入运行的容器内部，我们就可以通过 Pod 的 IP 地址来访问服务了。使用 curl 来尝试访问服务：

```
curl 20.1.244.75
```

在使用 curl 访问服务时，将在抓包的两个窗口内看到：

```
20:19:45.208948 IP 192.168.1.129.57452 > 10.1.30.8.8080: tcp 0
20:19:45.209005 IP 10.1.30.8.8080 > 192.168.1.129.57452: tcp 0
20:19:45.209013 IP 192.168.1.129.57452 > 10.1.30.8.8080: tcp 0
20:19:45.209066 IP 10.1.30.8.8080 > 192.168.1.129.57452: tcp 0

20:19:45.209227 IP 10.1.10.5.35225 > 20.1.244.75.80: tcp 0
20:19:45.209234 IP 20.1.244.75.80 > 10.1.10.5.35225: tcp 0
20:19:45.209280 IP 10.1.10.5.35225 > 20.1.244.75.80: tcp 0
20:19:45.209336 IP 20.1.244.75.80 > 10.1.10.5.35225: tcp 0
```

这些信息说明了什么问题呢？让我们在网络图上用实线标出第 1 个窗口中网络抓包信息的含义（物理网卡上的网络流量），并用虚线标出第 2 个窗口中网络抓包信息的含义（docker0 网桥上的网络流量），如图 7.15 所示。

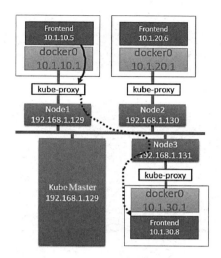

图 7.15 数据流动情况 1

注意，在图 7.15 中，虚线绕过了 Node3 的 kube-proxy，这么做是因为 Node3 上的 kube-proxy 没有参与这次网络交互。换句话说，Node1 的 kube-proxy 服务直接和负载均衡到的 Pod 进行网络交互。

在查看第 2 个捕获包的窗口时，我们能够站在容器的视角看这些流量。首先，容器尝试使用 20.1.244.75:80 打开 TCP 的 Socket 连接。同时，我们可以看到从服务地址 20.1.244.75 返回的数据。从容器的视角来看，整个交互过程都是在服务之间进行的。但是在查看一个捕获包的窗口时（上面的窗口），我们可以看到物理机之间的数据交互，可以看到一个 TCP 连接从 Node1 的物理地址（192.168.1.129）发出，直接连接到运行 Pod 的主机 Node3（192.168.1.131）。总而言之，Kubernetes 的 kube-proxy 作为一个全功能的代理服务器管理了两个独立的 TCP 连接：一个是从容器到 kube-proxy；另一个是从 kube-proxy 到负载均衡的目标 Pod。

如果清理一下捕获的记录，再次运行 curl，则还可以看到网络流量被负载均衡转发到另一个节点 Node2 上了：

```
20:19:45.208948 IP 192.168.1.129.57485 > 10.1.20.6.8080: tcp 0
20:19:45.209005 IP 10.1.20.6.8080 > 192.168.1.129.57485: tcp 0
20:19:45.209013 IP 192.168.1.129.57485 > 10.1.20.6.8080: tcp 0
20:19:45.209066 IP 10.1.20.6.8080 > 192.168.1.129.57485: tcp 0

20:19:45.209227 IP 10.1.10.5.38026 > 20.1.244.75.80: tcp 0
20:19:45.209234 IP 20.1.244.75.80 > 10.1.10.5.38026: tcp 0
```

```
20:19:45.209280 IP 10.1.10.5.38026> 20.1.244.75.80: tcp 0
20:19:45.209336 IP 20.1.244.75.80 > 10.1.10.5.38026: tcp 0
```

这一次，Kubernetes 的 Proxy 将选择运行在 Node2（10.1.20.1）上的 Pod 作为目标地址。网络流动图如图 7.16 所示。

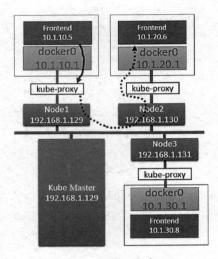

图 7.16　数据流动情况 2

到这里，你肯定已经知道另一个可能的负载均衡的路由结果了。

7.6　CNI 网络模型

随着容器技术在企业生产系统中的逐步落地，用户对容器云的网络特性要求也越来越高。跨主机容器间的网络互通已经成为基本要求，更高的要求包括容器固定 IP 地址、一个容器多个 IP 地址、多个子网隔离、ACL 控制策略、与 SDN 集成等。目前主流的容器网络模型主要有 Docker 公司提出的 Container Network Model（CNM）模型和 CoreOS 公司提出的 Container Network Interface（CNI）模型。

7.6.1　CNM 模型

CNM 模型是由 Docker 公司提出的容器网络模型，现在已经被 Cisco Contiv、Kuryr、Open Virtual Networking（OVN）、Project Calico、VMware、Weave 和 Plumgrid 等项目所

采纳。另外，Weave、Project Calico、Kuryr 和 Plumgrid 等项目也为 CNM 提供了网络插件的具体实现。

CNM 模型主要通过 Network Sandbox、Endpoint 和 Network 这 3 个组件进行实现，如图 7.17 所示。

◎ Network Sandbox：容器内部的网络栈，包括网络接口、路由表、DNS 等配置的管理。Sandbox 可用 Linux 网络命名空间、FreeBSD Jail 等机制进行实现。一个 Sandbox 可以包含多个 Endpoint。

◎ Endpoint：用于将容器内的 Sandbox 与外部网络相连的网络接口。可以使用 veth 对、Open vSwitch 的内部 port 等技术进行实现。一个 Endpoint 仅能够加入一个 Network。

◎ Network：可以直接互连的 Endpoint 的集合。可以通过 Linux 网桥、VLAN 等技术进行实现。一个 Network 包含多个 Endpoint。

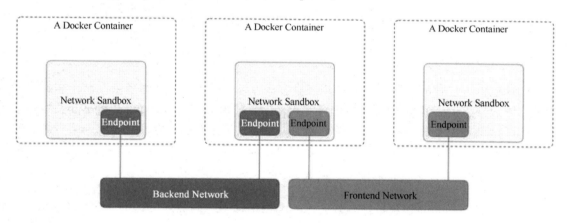

图 7.17　CNM 模型

7.6.2　CNI 模型

CNI 是由 CoreOS 公司提出的另一种容器网络规范，现在已经被 Kubernetes、rkt、Apache Mesos、Cloud Foundry 和 Kurma 等项目采纳。另外，Contiv Networking, Project Calico、Weave、SR-IOV、Cilium、Infoblox、Multus、Romana、Plumgrid 和 Midokura 等项目也为 CNI 提供网络插件的具体实现。图 7.18 描述了容器运行环境与各种网络插件通过 CNI 进行连接的模型。

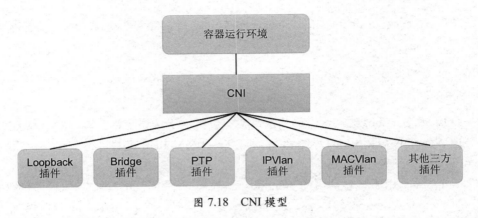

图 7.18 CNI 模型

CNI 定义的是容器运行环境与网络插件之间的简单接口规范，通过一个 JSON Schema 定义 CNI 插件提供的输入和输出参数。一个容器可以通过绑定多个网络插件加入多个网络中。

本节将对 Kubernetes 如何实现 CNI 模型进行详细说明。

1. CNI 规范概述

CNI 提供了一种应用容器的插件化网络解决方案，定义对容器网络进行操作和配置的规范，通过插件的形式对 CNI 接口进行实现。CNI 是由 rkt Networking Proposal 发展而来的，试图提供一种普适的容器网络解决方案。CNI 仅关注在创建容器时分配网络资源，和在销毁容器时删除网络资源，这使得 CNI 规范非常轻巧、易于实现，得到了广泛的支持。

在 CNI 模型中只涉及两个概念：容器和网络。

- 容器（Container）：是拥有独立 Linux 网络命名空间的环境，例如使用 Docker 或 rkt 创建的容器。关键之处是容器需要拥有自己的 Linux 网络命名空间，这是加入网络的必要条件。
- 网络（Network）：表示可以互连的一组实体，这些实体拥有各自独立、唯一的 IP 地址，可以是容器、物理机或者其他网络设备（比如路由器）等。

对容器网络的设置和操作都通过插件（Plugin）进行具体实现，CNI 插件包括两种类型：CNI Plugin 和 IPAM（IP Address Management）Plugin。CNI Plugin 负责为容器配置网络资源，IPAM Plugin 负责对容器的 IP 地址进行分配和管理。IPAM Plugin 作为 CNI Plugin 的一部分，与 CNI Plugin 一起工作。

2. CNI Plugin 插件详解

CNI Plugin 包括 3 个基本接口的定义：添加（ADD）、删除（DELETE）、检查（CHECK）和版本查询（VERSION）。这些接口的具体实现要求插件提供一个可执行的程序，在容器网络添加或删除时进行调用，以完成具体的操作。

（1）添加：将容器添加到某个网络。主要过程为在 Container Runtime 创建容器时，先创建好容器内的网络命名空间（Network Namespace），然后调用 CNI 插件为该 netns 进行网络配置，最后启动容器内的进程。

添加接口的参数如下。

◎ Version：CNI 版本号。
◎ Container ID：容器 ID。
◎ Network namespace path：容器的网络命名空间路径，例如/proc/[pid]/ns/net。
◎ Network configuration：网络配置 JSON 文档，用于描述容器待加入的网络。
◎ Extra arguments：其他参数，提供基于容器的 CNI 插件简单配置机制。
◎ Name of the interface inside the container：容器内的网卡名。

返回的信息如下。

◎ Interfaces list：网卡列表，根据 Plugin 的实现，可能包括 Sandbox Interface 名称、主机 Interface 名称、每个 Interface 的地址等信息。
◎ IPs assigned to the interface：IPv4 或者 IPv6 地址、网关地址、路由信息等。
◎ DNS information：DNS 相关的信息。

（2）删除：容器销毁时将容器从某个网络中删除。

删除接口的参数如下。

◎ Version：CNI 版本号。
◎ Container ID：容器 ID。
◎ Network namespace path：容器的网络命名空间路径，例如/proc/[pid]/ns/net。
◎ Network configuration：网络配置 JSON 文档，用于描述容器待加入的网络。
◎ Extra arguments：其他参数，提供基于容器的 CNI 插件简单配置机制。
◎ Name of the interface inside the container：容器内的网卡名。

（3）检查：检查容器网络是否正确设置。

检查接口的参数如下。

◎ Container ID：容器 ID。
◎ Network namespace path：容器的网络命名空间路径，例如/proc/[pid]/ns/net。
◎ Network configuration：网络配置 JSON 文档，用于描述容器待加入的网络。
◎ Extra arguments：其他参数，提供基于容器的 CNI 插件简单配置机制。
◎ Name of the interface inside the container：容器内的网卡名。

（4）版本查询：查询网络插件支持的 CNI 规范版本号。

无参数，返回值为网络插件支持的 CNI 规范版本号，例如：

```
{
  "cniVersion": "0.4.0", // the version of the CNI spec in use for this output
  "supportedVersions": [ "0.1.0", "0.2.0", "0.3.0", "0.3.1", "0.4.0" ] // the list of CNI spec versions that this plugin supports
}
```

CNI 插件应能够支持通过环境变量和标准输入传入参数。可执行文件通过网络配置参数中的 type 字段标识的文件名在环境变量 CNI_PATH 设定的路径下进行查找。一旦找到，容器运行时将调用该可执行程序，并传入以下环境变量和网络配置参数，供该插件完成容器网络资源和参数的设置。

环境变量参数如下。

◎ CNI_COMMAND：接口方法，包括 ADD、DEL 和 VERSION。
◎ CNI_CONTAINERID：容器 ID。
◎ CNI_NETNS：容器的网络命名空间路径，例如/proc/[pid]/ns/net。
◎ CNI_IFNAME：待设置的网络接口名称。
◎ CNI_ARGS：其他参数，为 key=value 格式，多个参数之间用分号分隔，例如 "FOO=BAR; ABC=123"。
◎ CNI_PATH：可执行文件的查找路径，可以设置多个。

网络配置参数则由一个 JSON 报文组成，以标准输入（stdin）的方式传递给可执行程序。

网络配置参数如下。

◎ cniVersion（string）：CNI 版本号。

◎ name（string）：网络名称，应在一个管理域内唯一。
◎ type（string）：CNI 插件的可执行文件的名称。
◎ args（dictionary）：其他参数。
◎ ipMasq（boolean）：是否设置 IP Masquerade（需插件支持），适用于主机可作为网关的环境中。
◎ ipam：IP 地址管理的相关配置。
  - type（string）：IPAM 可执行的文件名。
◎ dns：DNS 服务的相关配置。
  - nameservers（list of strings）：名字服务器列表，可以使用 IPv4 或 IPv6 地址。
  - domain（string）：本地域名，用于短主机名查询。
  - search（list of strings）：按优先级排序的域名查询列表。
  - options（list of strings）：传递给 resolver 的选项列表。

下面的例子定义了一个名为 dbnet 的网络配置参数，IPAM 使用 host-local 进行设置：

```
{
  "cniVersion": "0.4.0",
  "name": "dbnet",
  "type": "bridge",
  "bridge": "cni0",
  "ipam": {
    "type": "host-local",
    "subnet": "10.1.0.0/16",
    "gateway": "10.1.0.1"
  },
  "dns": {
    "nameservers": [ "10.1.0.1" ]
  }
}
```

3. IPAM Plugin 插件详解

为了减轻 CNI Plugin 对 IP 地址管理的负担，在 CNI 规范中设置了一个新的插件专门用于管理容器的 IP 地址（还包括网关、路由等信息），被称为 IPAM Plugin。通常由 CNI Plugin 在运行时自动调用 IPAM Plugin 完成容器 IP 地址的分配。

IPAM Plugin 负责为容器分配 IP 地址、网关、路由和 DNS，典型的实现包括 host-local 和 dhcp。与 CNI Plugin 类似，IPAM 插件也通过可执行程序完成 IP 地址分配的具体操作。

IPAM 可执行程序也处理传递给 CNI 插件的环境变量和通过标准输入（stdin）传入的网络配置参数。

如果成功完成了容器 IP 地址的分配，则 IPAM 插件应该通过标准输出（stdout）返回以下 JSON 报文：

```
{
  "cniVersion": "0.4.0",
  "ips": [
      {
          "version": "<4-or-6>",
          "address": "<ip-and-prefix-in-CIDR>",
          "gateway": "<ip-address-of-the-gateway>"  (optional)
      },
      ......
  ],
  "routes": [                                        (optional)
      {
          "dst": "<ip-and-prefix-in-cidr>",
          "gw": "<ip-of-next-hop>"                   (optional)
      },
      ......
  ]
  "dns": {
    "nameservers": <list-of-nameservers>             (optional)
    "domain": <name-of-local-domain>                 (optional)
    "search": <list-of-search-domains>               (optional)
    "options": <list-of-options>                     (optional)
  }
}
```

其中包括 ips、routes 和 dns 三段内容。

◎ ips 段：分配给容器的 IP 地址（也可能包括网关）。
◎ routes 段：路由规则记录。
◎ dns 段：DNS 相关的信息。

4. 多网络插件

在很多情况下，一个容器需要连接多个网络，CNI 规范支持为一个容器运行多个 CNI Plugin 来实现这个目标。多个网络插件将按照网络配置列表中的顺序执行，并将前一个网络配置的执行结果传递给后面的网络配置。多网络配置用 JSON 报文进行配置，包括如下信息。

- ◎ cniVersion（string）：CNI 版本号。
- ◎ name（string）：网络名称，应在一个管理域内唯一，将用于下面的所有 Plugin。
- ◎ plugins（list）：网络配置列表。

下面的例子定义了两个网络配置参数，分别作用于两个插件，第 1 个为 bridge，第 2 个为 tuning。CNI 将首先执行第 1 个 bridge 插件设置容器的网络，然后执行第 2 个 tuning 插件：

```
{
  "cniVersion": "0.4.0",
  "name": "dbnet",
  "plugins": [
    {
      "type": "bridge",
      // type (plugin) specific
      "bridge": "cni0",
      // args may be ignored by plugins
      "args": {
        "labels" : {
            "appVersion" : "1.0"
        }
      },
      "ipam": {
        "type": "host-local",
        // ipam specific
        "subnet": "10.1.0.0/16",
        "gateway": "10.1.0.1"
      },
      "dns": {
        "nameservers": [ "10.1.0.1" ]
      }
    },
    {
      "type": "tuning",
      "sysctl": {
        "net.core.somaxconn": "500"
      }
    }
```

    ]
}

在容器运行且执行第 1 个 bridge 插件时,网络配置参数将被设置为:

```
{
  "cniVersion": "0.4.0",
  "name": "dbnet",
  "type": "bridge",
  "bridge": "cni0",
  "args": {
    "labels" : {
        "appVersion" : "1.0"
    }
  },
  "ipam": {
    "type": "host-local",
    // ipam specific
    "subnet": "10.1.0.0/16",
    "gateway": "10.1.0.1"
  },
  "dns": {
    "nameservers": [ "10.1.0.1" ]
  }
}
```

接下来执行第 2 个 tuning 插件,网络配置参数将被设置为:

```
{
  "cniVersion": "0.4.0",
  "name": "dbnet",
  "type": "tuning",
  "sysctl": {
    "net.core.somaxconn": "500"
  },
  "prevResult": {
    "ips": [
        {
          "version": "4",
          "address": "10.0.0.5/32",
          "interface": 2
        }
    ],
```

```
    "interfaces": [
        {
            "name": "cni0",
            "mac": "00:11:22:33:44:55",
        },
        {
            "name": "veth3243",
            "mac": "55:44:33:22:11:11",
        },
        {
            "name": "eth0",
            "mac": "99:88:77:66:55:44",
            "sandbox": "/var/run/netns/blue",
        }
    ],
    "dns": {
      "nameservers": [ "10.1.0.1" ]
    }
  }
}
```

其中，prevResult 字段包含的信息为上一个 bridge 插件执行的结果。

在删除多个 CNI Plugin 时，则以逆序执行删除操作，以上例为例，将先删除 tuning 插件的网络配置，其中 prevResult 字段包含的信息为新增操作（ADD）时补充的信息：

```
{
  "cniVersion": "0.4.0",
  "name": "dbnet",
  "type": "tuning",
  "sysctl": {
    "net.core.somaxconn": "500"
  },
  "prevResult": {
    "ips": [
        {
            "version": "4",
            "address": "10.0.0.5/32",
            "interface": 2
        }
    ],
    "interfaces": [
```

```
        {
            "name": "cni0",
            "mac": "00:11:22:33:44:55",
        },
        {
            "name": "veth3243",
            "mac": "55:44:33:22:11:11",
        },
        {
            "name": "eth0",
            "mac": "99:88:77:66:55:44",
            "sandbox": "/var/run/netns/blue",
        }
    ],
    "dns": {
      "nameservers": [ "10.1.0.1" ]
    }
  }
}
```

然后删除 bridge 插件的网络配置，其中 prevResult 字段包含的信息也是在新增操作（ADD）时补充的信息：

```
{
  "cniVersion": "0.4.0",
  "name": "dbnet",
  "type": "bridge",
  "bridge": "cni0",
  "args": {
    "labels" : {
        "appVersion" : "1.0"
    }
  },
  "ipam": {
    "type": "host-local",
    // ipam specific
    "subnet": "10.1.0.0/16",
    "gateway": "10.1.0.1"
  },
  "dns": {
    "nameservers": [ "10.1.0.1" ]
  },
```

```
    "prevResult": {
     "ips": [
         {
             "version": "4",
             "address": "10.0.0.5/32",
             "interface": 2
         }
     ],
     "interfaces": [
         {
             "name": "cni0",
             "mac": "00:11:22:33:44:55",
         },
         {
             "name": "veth3243",
             "mac": "55:44:33:22:11:11",
         },
         {
             "name": "eth0",
             "mac": "99:88:77:66:55:44",
             "sandbox": "/var/run/netns/blue",
         }
     ],
     "dns": {
       "nameservers": [ "10.1.0.1" ]
     }
   }
 }
```

1. 命令返回信息说明

对于 ADD 或 DELETE 操作，返回码为 0 表示执行成功，非 0 表示失败，并以 JSON 报文的格式通过标准输出（stdout）返回操作的结果。

以 ADD 操作为例，成功将容器添加到网络的结果将返回以下 JSON 报文。其中 ips、routes 和 dns 段的信息应该与 IPAM Plugin（IPAM Plugin 的说明详见下节）返回的结果相同，重要的是 interfaces 段，应通过 CNI Plugin 进行设置并返回：

```
{
 "cniVersion": "0.4.0",
 "interfaces": [                    (this key omitted by IPAM plugins)
```

```
        {
            "name": "<name>",
            "mac": "<MAC address>",    (required if L2 addresses are meaningful)
            "sandbox": "<netns path or hypervisor identifier>" (required for
container/hypervisor interfaces, empty/omitted for host interfaces)
        }
    ],
    "ips": [
        {
            "version": "<4-or-6>",
            "address": "<ip-and-prefix-in-CIDR>",
            "gateway": "<ip-address-of-the-gateway>",  (optional)
            "interface": <numeric index into 'interfaces' list>
        },
        ......
    ],
    "routes": [                                              (optional)
        {
            "dst": "<ip-and-prefix-in-cidr>",
            "gw": "<ip-of-next-hop>"                         (optional)
        },
        ......
    ]
    "dns": {
      "nameservers": <list-of-nameservers>            (optional)
      "domain": <name-of-local-domain>                (optional)
      "search": <list-of-additional-search-domains>   (optional)
      "options": <list-of-options>                    (optional)
    }
}
```

接口调用失败时，返回码不为 0，应通过标准输出返回包含错误信息的如下 JSON 报文：

```
{
  "cniVersion": "0.4.0",
  "code": <numeric-error-code>,
  "msg": <short-error-message>,
  "details": <long-error-message> (optional)
}
```

错误码包括如下内容。

- ◎ CNI 版本不匹配。
- ◎ 在网络配置中存在不支持的字段，详细信息应在 msg 中说明。

7.6.3 在 Kubernetes 中使用网络插件

Kubernetes 目前支持两种网络插件的实现。

- ◎ CNI 插件：根据 CNI 规范实现其接口，以与插件提供者进行对接。
- ◎ kubenet 插件：使用 bridge 和 host-local CNI 插件实现一个基本的 cbr0。

为了在 Kubernetes 集群中使用网络插件，需要在 kubelet 服务的启动参数上设置下面两个参数。

- ◎ --network-plugin-dir：kubelet 启动时扫描网络插件的目录。
- ◎ --network-plugin：网络插件名称，对于 CNI 插件，设置为 cni 即可，无须关注 --network-plugin-dir 的路径。对于 kubenet 插件，设置为 kubenet，目前仅实现了一个简单的 cbr0 Linux 网桥。

在设置 --network-plugin="cni" 时，kubelet 还需设置下面两个参数。

- ◎ --cni-conf-dir：CNI 插件的配置文件目录，默认为 /etc/cni/net.d。该目录下配置文件的内容需要符合 CNI 规范。
- ◎ --cni-bin-dir：CNI 插件的可执行文件目录，默认为 /opt/cni/bin。

目前已有多个开源项目支持以 CNI 网络插件的形式部署到 Kubernetes 集群中，进行 Pod 的网络设置和网络策略的设置，包括 Calico、Canal、Cilium、Contiv、Flannel、Romana、Weave Net 等。

7.7 Kubernetes 网络策略

为了实现细粒度的容器间网络访问隔离策略，Kubernetes 从 1.3 版本开始，由 SIG-Network 小组主导研发了 Network Policy 机制，目前已升级为 networking.k8s.io/v1 稳定版本。Network Policy 的主要功能是对 Pod 间的网络通信进行限制和准入控制，设置方式为将 Pod 的 Label 作为查询条件，设置允许访问或禁止访问的客户端 Pod 列表。目前查询条件可以作用于 Pod 和 Namespace 级别。

为了使用 Network Policy，Kubernetes 引入了一个新的资源对象 NetworkPolicy，供用户设置 Pod 间网络访问的策略。但仅定义一个网络策略是无法完成实际的网络隔离的，还需要一个策略控制器（Policy Controller）进行策略的实现。策略控制器由第三方网络组件提供，目前 Calico、Cilium、Kube-router、Romana、Weave Net 等开源项目均支持网络策略的实现。

Network Policy 的工作原理如图 7.19 所示，policy controller 需要实现一个 API Listener，监听用户设置的 NetworkPolicy 定义，并将网络访问规则通过各 Node 的 Agent 进行实际设置（Agent 则需要通过 CNI 网络插件实现）。

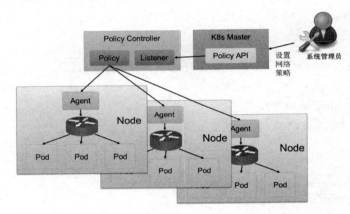

图 7.19　Network Policy 的工作原理

7.7.1　网络策略配置说明

网络策略的设置主要用于对目标 Pod 的网络访问进行限制，在默认情况下对所有 Pod 都是允许访问的，在设置了指向 Pod 的 NetworkPolicy 网络策略之后，到 Pod 的访问才会被限制。

下面通过一个例子对 NetworkPolicy 资源对象的使用进行说明：

```
apiVersion: networking.k8s.io/v1
kind: NetworkPolicy
metadata:
  name: test-network-policy
  namespace: default
spec:
```

```
  podSelector:
    matchLabels:
      role: db
  policyTypes:
  - Ingress
  - Egress
  ingress:
  - from:
    - ipBlock:
        cidr: 172.17.0.0/16
        except:
        - 172.17.1.0/24
    - namespaceSelector:
        matchLabels:
          project: myproject
    - podSelector:
        matchLabels:
          role: frontend
    ports:
    - protocol: TCP
      port: 6379
  egress:
  - to:
    - ipBlock:
        cidr: 10.0.0.0/24
    ports:
    - protocol: TCP
      port: 5978
```

主要的参数说明如下。

◎ podSelector：用于定义该网络策略作用的 Pod 范围，本例的选择条件为包含"role=db"标签的 Pod。

◎ policyTypes：网络策略的类型，包括 ingress 和 egress 两种，用于设置目标 Pod 的入站和出站的网络限制。

◎ ingress：定义允许访问目标 Pod 的入站白名单规则，满足 from 条件的客户端才能访问 ports 定义的目标 Pod 端口号。
- from：对符合条件的客户端 Pod 进行网络放行，规则包括基于客户端 Pod 的 Label、基于客户端 Pod 所在的 Namespace 的 Label 或者客户端的 IP 范围。

- ports：允许访问的目标 Pod 监听的端口号。
◎ egress：定义目标 Pod 允许访问的"出站"白名单规则，目标 Pod 仅允许访问满足 to 条件的服务端 IP 范围和 ports 定义的端口号。
  - to：允许访问的服务端信息，可以基于服务端 Pod 的 Label、基于服务端 Pod 所在的 Namespace 的 Label 或者服务端 IP 范围。
  - ports：允许访问的服务端的端口号。

通过本例的 NetworkPolicy 设置，对目标 Pod 的网络访问的效果如下。

◎ 该网络策略作用于 Namespace "default"中含有"role=db" Label 的全部 Pod。
◎ 允许与目标 Pod 在同一个 Namespace 中的包含"role=frontend"Label 的客户端 Pod 访问目标 Pod。
◎ 允许属于包含"project=myproject"Label 的 Namespace 的客户端 Pod 访问目标 Pod。
◎ 允许从 IP 地址范围"172.17.0.0/16"的客户端 Pod 访问目标 Pod，但是不包括 IP 地址范围"172.17.1.0/24"的客户端。
◎ 允许目标 Pod 访问 IP 地址范围"10.0.0.0/24"并监听 5978 端口的服务。

这里是关于 namespaceSelector 和 podSelector 的说明：在 from 或 to 的配置中，namespaceSelector 和 podSelector 可以单独设置，也可以组合配置。如果仅配置 podSelector，则表示与目标 Pod 属于相同的 Namespace，而组合设置则可以设置 Pod 所属的 Namespace，例如：

```
- from:
  - namespaceSelector:
      matchLabels:
        project: myproject
    podSelector:
      matchLabels:
        role: frontend
```

表示允许访问目标 Pod 的来源客户端 Pod 应具有如下属性：属于有"project=myproject"标签的 Namespace，并且有"role=frontend"标签。

7.7.2 在 Namespace 级别设置默认的网络策略

在 Namespace 级别还可以设置一些默认的全局网络策略，以方便管理员对整个 Namespace 进行统一的网络策略设置。

默认禁止任何客户端访问该 Namespace 中的所有 Pod:

```
apiVersion: networking.k8s.io/v1
kind: NetworkPolicy
metadata:
  name: default-deny
spec:
  podSelector: {}
  policyTypes:
  - Ingress
```

默认允许任何客户端访问该 Namespace 中的所有 Pod:

```
apiVersion: networking.k8s.io/v1
kind: NetworkPolicy
metadata:
  name: allow-all
spec:
  podSelector: {}
  ingress:
  - {}
  policyTypes:
  - Ingress
```

默认禁止该 Namespace 中的所有 Pod 访问外部服务:

```
apiVersion: networking.k8s.io/v1
kind: NetworkPolicy
metadata:
  name: default-deny
spec:
  podSelector: {}
  policyTypes:
  - Egress
```

默认允许该 Namespace 中的所有 Pod 访问外部服务:

```
apiVersion: networking.k8s.io/v1
kind: NetworkPolicy
metadata:
  name: allow-all
spec:
  podSelector: {}
  egress:
```

```
  - {}
  policyTypes:
  - Egress
```

默认禁止任何客户端访问该 Namespace 中的所有 Pod，同时禁止访问外部服务：

```
apiVersion: networking.k8s.io/v1
kind: NetworkPolicy
metadata:
  name: default-deny
spec:
  podSelector: {}
  policyTypes:
  - Ingress
  - Egress
```

7.7.3 NetworkPolicy 的发展

Kubernetes 从 1.12 版本开始，引入了对 SCTP 协议的支持，目前为 Alpha 版本的功能，可以通过打开 --feature-gates=SCTPSupport=true 特性开关启用。开启之后，可以在 NetworkPolicy 资源对象中设置 protocol 字段的值为 SCTP，启用对 SCTP 协议的网络隔离设置。这要求 CNI 插件提供对 SCTP 协议的支持。

7.8 开源的网络组件

Kubernetes 的网络模型假定了所有 Pod 都在一个可以直接连通的扁平网络空间中。这在 GCE 里面是现成的网络模型，Kubernetes 假定这个网络已经存在。而在私有云里搭建 Kubernetes 集群，就不能假定这种网络已经存在了。我们需要自己实现这个网络假设，将不同节点上的 Docker 容器之间的互相访问先打通，然后运行 Kubernetes。

目前已经有多个开源组件支持容器网络模型。本节介绍几个常见的网络组件及其安装配置方法，包括 Flannel、Open vSwitch、直接路由和 Calico。

7.8.1 Flannel

Flannel 之所以可以搭建 Kubernetes 依赖的底层网络，是因为它能实现以下两点。

（1）它能协助 Kubernetes，给每一个 Node 上的 Docker 容器都分配互相不冲突的 IP 地址。

（2）它能在这些 IP 地址之间建立一个覆盖网络（Overlay Network），通过这个覆盖网络，将数据包原封不动地传递到目标容器内。

现在，通过图 7.20 来看看 Flannel 是如何实现这两点的。

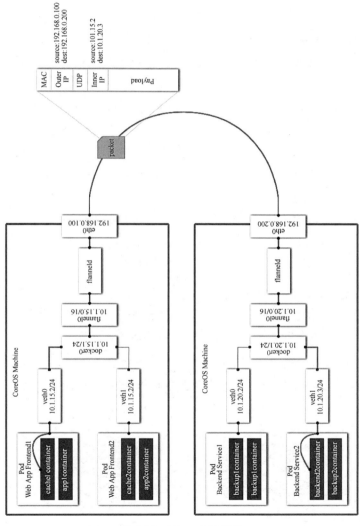

图 7.20　Flannel 的实现

可以看到，Flannel 首先创建了一个名为 flannel0 的网桥，而且这个网桥的一端连接 docker0 网桥，另一端连接一个叫作 flanneld 的服务进程。

flanneld 进程并不简单，它上连 etcd，利用 etcd 来管理可分配的 IP 地址段资源，同时监控 etcd 中每个 Pod 的实际地址，并在内存中建立了一个 Pod 节点路由表；它下连 docker0 和物理网络，使用内存中的 Pod 节点路由表，将 docker0 发给它的数据包包装起来，利用物理网络的连接将数据包投递到目标 flanneld 上，从而完成 Pod 到 Pod 之间的直接地址通信。

Flannel 之间的底层通信协议的可选技术包括 UDP、VxLan、AWS VPC 等多种方式。通过源 flanneld 封包、目标 flanneld 解包，最终 docker0 收到的就是原始的数据，对容器应用来说是透明的，感觉不到中间 Flannel 的存在。

我们看一下 Flannel 是如何做到为不同 Node 上的 Pod 分配的 IP 不产生冲突的。其实想到 Flannel 使用了集中的 etcd 存储就很容易理解了。它每次分配的地址段都在同一个公共区域获取，这样大家自然能够互相协调，不产生冲突了。而且在 Flannel 分配好地址段后，后面的事情是由 Docker 完成的，Flannel 通过修改 Docker 的启动参数将分配给它的地址段传递进去：

```
--bip=172.17.18.1/24
```

通过这些操作，Flannel 就控制了每个 Node 上的 docker0 地址段的地址，也就保障了所有 Pod 的 IP 地址在同一个水平网络中且不产生冲突了。

Flannel 完美地实现了对 Kubernetes 网络的支持，但是它引入了多个网络组件，在网络通信时需要转到 flannel0 网络接口，再转到用户态的 flanneld 程序，到对端后还需要走这个过程的反过程，所以也会引入一些网络的时延损耗。

另外，Flannel 模型默认采用了 UDP 作为底层传输协议，UDP 本身是非可靠协议，虽然两端的 TCP 实现了可靠传输，但在大流量、高并发的应用场景下还需要反复测试，确保没有问题。

Flannel 的安装和配置如下。

1）安装 etcd

由于 Flannel 使用 etcd 作为数据库，所以需要预先安装好 etcd，此处不再赘述。

2）安装 Flannel

需要在每个 Node 上都安装 Flannel。Flannel 软件的下载地址为 https://github.com/coreos/flannel/releases。将下载的压缩包 flannel-<version>-linux-amd64.tar.gz 解压，将二进制文件 flanneld 和 mk-docker-opts.sh 复制到/usr/bin（或其他 PATH 环境变量中的目录）下，即可完成对 Flannel 的安装。

3）配置 Flannel

此处以使用 systemd 系统为例对 flanneld 服务进行配置。编辑服务配置文件/usr/lib/systemd/system/flanneld.service：

```
[Unit]
Description=flanneld overlay address etcd agent
After=network.target
Before=docker.service

[Service]
Type=notify
EnvironmentFile=/etc/sysconfig/flanneld
ExecStart=/usr/bin/flanneld -etcd-endpoints=${FLANNEL_ETCD} $FLANNEL_OPTIONS

[Install]
RequiredBy=docker.service
WantedBy=multi-user.target
```

编辑配置文件/etc/sysconfig/flannel，设置 etcd 的 URL 地址：

```
# flanneld configuration options

# etcd url location.  Point this to the server where etcd runs
FLANNEL_ETCD="http://192.168.18.3:2379"

# etcd config key. This is the configuration key that flannel queries
# For address range assignment
FLANNEL_ETCD_KEY="/coreos.com/network"
```

在启动 flanneld 服务之前，需要在 etcd 中添加一条网络配置记录，这个配置将用于 flanneld 分配给每个 Docker 的虚拟 IP 地址段：

```
# etcdctl set /coreos.com/network/config '{ "Network": "10.1.0.0/16" }'
```

由于 Flannel 将覆盖 docker0 网桥，所以如果 Docker 服务已启动，则需要停止 Docker 服务。

4）启动 flanneld 服务

```
# systemctl restart flanneld
```

5）设置 docker0 网桥的 IP 地址

```
# mk-docker-opts.sh -i
# source /run/flannel/subnet.env
# ifconfig docker0 ${FLANNEL_SUBNET}
```

完成后确认网络接口 docker0 的 IP 地址属于 flannel0 的子网：

```
# ip addr
flannel0: flags=4305<UP,POINTOPOINT,RUNNING,NOARP,MULTICAST> mtu 1472
        inet 10.1.10.0  netmask 255.255.0.0  destination 10.1.10.0
docker0: flags=4163<UP,BROADCAST,RUNNING,MULTICAST> mtu 1500
        inet 10.1.10.1  netmask 255.255.255.0  broadcast 10.1.10.255
```

6）重新启动 Docker 服务

```
# systemctl restart docker
```

至此就完成了 Flannel 覆盖网络的设置。

使用 ping 命令验证各 Node 上 docker0 之间的相互访问。例如在 Node1（docker0 IP=10.1.10.1）机器上 ping Node2 的 docker0（docker0's IP=10.1.30.1），通过 Flannel 能够成功连接其他物理机的 Docker 网络：

```
$ ping 10.1.30.1
PING 10.1.30.1 (10.1.30.1) 56(84) bytes of data.
64 bytes from 10.1.30.1: icmp_seq=1 ttl=62 time=1.15 ms
64 bytes from 10.1.30.1: icmp_seq=2 ttl=62 time=1.16 ms
64 bytes from 10.1.30.1: icmp_seq=3 ttl=62 time=1.57 ms
```

我们也可以在 etcd 中查看 Flannel 设置的 flannel0 地址与物理机 IP 地址的对应规则：

```
# etcdctl ls /coreos.com/network/subnets
/coreos.com/network/subnets/10.1.10.0-24
/coreos.com/network/subnets/10.1.20.0-24
/coreos.com/network/subnets/10.1.30.0-24

# etcdctl get /coreos.com/network/subnets/10.1.10.0-24
```

```
{"PublicIP": "192.168.1.129"}
# etcdctl get /coreos.com/network/subnets/10.1.20.0-24
{"PublicIP": "192.168.1.130"}
# etcdctl get /coreos.com/network/subnets/10.1.30.0-24
{"PublicIP": "192.168.1.131"}
```

7.8.2　Open vSwitch

在了解了 Flannel 后，我们再看看 Open vSwitch 是怎么解决上述两个问题的。

Open vSwitch 是一个开源的虚拟交换机软件，有点儿像 Linux 中的 bridge，但是功能要复杂得多。Open vSwitch 的网桥可以直接建立多种通信通道（隧道），例如 Open vSwitch with GRE/VxLAN。这些通道的建立可以很容易地通过 OVS 的配置命令实现。在 Kubernetes、Docker 场景下，我们主要是建立 L3 到 L3 的隧道。举个例子来看看 Open vSwitch with GRE/VxLAN 的网络架构，如图 7.21 所示。

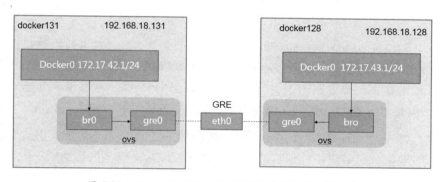

图 7.21　Open vSwitch with GRE/VxLAN 的网络架构

首先，为了避免 Docker 创建的 docker0 地址产生冲突（因为 Docker Daemon 启动且给 docker0 选择子网地址时只有几个备选列表，很容易产生冲突），我们可以将 docker0 网桥删除，手动建立一个 Linux 网桥，然后手动给这个网桥配置 IP 地址范围。

其次，建立 Open vSwitch 的网桥 ovs，使用 ovs-vsctl 命令给 ovs 网桥增加 gre 端口，在添加 gre 端口时要将目标连接的 NodeIP 地址设置为对端的 IP 地址。对每一个对端 IP 地址都需要这么操作（对于大型集群网络，这可是个体力活，要做自动化脚本来完成）。

最后，将 ovs 的网桥作为网络接口，加入 Docker 的网桥上（docker0 或者自己手工建立的新网桥）。

重启 ovs 网桥和 Docker 的网桥，并添加一个 Docker 的地址段到 Docker 网桥的路由规则项，就可以将两个容器的网络连接起来了。

1. 网络通信过程

当容器内的应用访问另一个容器的地址时，数据包会通过容器内的默认路由发送给 docker0 网桥。ovs 的网桥是作为 docker0 网桥的端口存在的，它会将数据发送给 ovs 网桥。ovs 网络已经通过配置建立了和其他 ovs 网桥的 GRE/VxLAN 隧道，自然能将数据送达对端的 Node，并送往 docker0 及 Pod。

通过新增的路由项，Node 本身的应用数据也被路由到 docker0 网桥上，和刚才的通信过程一样，自然也可以访问其他 Node 上的 Pod。

2. OVS with GRE/VxLAN 组网方式的特点

OVS 的优势是，作为开源虚拟交换机软件，它相对成熟和稳定，而且支持各类网络隧道协议，通过了 OpenStack 等项目的考验。

另一方面，在前面介绍 Flannel 时可知，Flannel 除了支持建立覆盖网络，保证 Pod 到 Pod 的无缝通信，还和 Kubernetes、Docker 架构体系紧密结合。Flannel 能够感知 Kubernetes 的 Service，动态维护自己的路由表，还通过 etcd 来协助 Docker 对整个 Kubernetes 集群中 docker0 的子网地址分配。而我们在使用 OVS 时，很多事情就需要手工完成了。

无论是 OVS 还是 Flannel，通过覆盖网络提供的 Pod 到 Pod 通信都会引入一些额外的通信开销，如果是对网络依赖特别重的应用，则需要评估对业务的影响。

Open vSwitch 的安装和配置如下。

以两个 Node 为例，目标网络拓扑如图 7.22 所示。

首先，确保节点 192.168.18.128 的 Docker0 采用了 172.17.43.0/24 网段，而 192.168.18.131 的 Docker0 采用了 172.17.42.0/24 网段，对应的参数为 docker daemon 的启动参数"--bip"设置的值。

第 7 章 网络原理

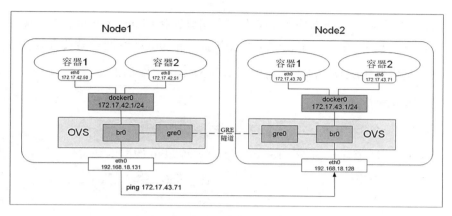

图 7.22 目标网络拓扑

1. 在两个 Node 上安装 ovs

```
# yum install openvswitch-2.4.0-1.x86_64.rpm
```

禁止 selinux，配置后重启 Linux：

```
# vi /etc/selinux/config
SELINUX=disabled
```

查看 Open vSwitch 的服务状态，应该启动 ovsdb-server 与 ovs-vswitchd 两个进程：

```
# service openvswitch status
ovsdb-server is running with pid 2429
ovs-vswitchd is running with pid 2439
```

查看 Open vSwitch 的相关日志，确认没有异常：

```
# more /var/log/messages |grep openv
Nov  2 03:12:52 docker128 openvswitch: Starting ovsdb-server [ OK ]
Nov  2 03:12:52 docker128 openvswitch: Configuring Open vSwitch system IDs
[ OK ]
Nov  2 03:12:52 docker128 kernel: openvswitch: Open vSwitch switching datapath
Nov  2 03:12:52 docker128 openvswitch: Inserting openvswitch module [ OK ]
```

注意，上述操作需要在两个节点机器上分别执行完成。

2. 创建网桥和 GRE 隧道

接下来需要在每个 Node 上都建立 ovs 的网桥 br0，然后在网桥上创建一个 GRE 隧道连接对端网桥,最后把 ovs 的网桥 br0 作为一个端口连接到 docker0 这个 Linux 网桥上(可以认为是交换机互联),这样一来,两个节点机器上的 docker0 网段就能互通了。

下面以节点机器 192.168.18.131 为例,具体的操作步骤如下。

(1) 创建 ovs 网桥:

```
# ovs-vsctl add-br br0
```

(2) 创建 GRE 隧道连接对端,remote_ip 为对端 eth0 的网卡地址:

```
# ovs-vsctl add-port br0 gre1 -- set interface gre1 type=gre option:remote_ip=192.168.18.128
```

(3) 添加 br0 到本地 docker0,使得容器流量通过 OVS 流经 tunnel:

```
# brctl addif docker0 br0
```

(4) 启动 br0 与 docker0 网桥:

```
# ip link set dev br0 up
# ip link set dev docker0 up
```

(5) 添加路由规则。由于 192.168.18.128 与 192.168.18.131 的 docker0 网段分别为 172.17.43.0/24 与 172.17.42.0/24,这两个网段的路由都需要经过本机的 docker0 网桥路由,其中一个 24 网段是通过 OVS 的 GRE 隧道到达对端的,因此需要在每个 Node 上都添加通过 docker0 网桥转发的 172.17.0.0/16 段的路由规则:

```
# ip route add 172.17.0.0/16 dev docker0
```

(6) 清空 Docker 自带的 iptables 规则及 Linux 的规则,后者存在拒绝 icmp 报文通过防火墙的规则:

```
# iptables -t nat -F; iptables -F
```

在 192.168.18.131 上完成上述步骤后,在 192.168.18.128 节点执行同样的操作,注意,GRE 隧道里的 IP 地址要改为对端节点(192.168.18.131)的 IP 地址。

配置完成后,192.168.18.131 的 IP 地址、docker0 的 IP 地址及路由等重要信息显示如下:

```
# ip addr
1: lo: <LOOPBACK,UP,LOWER_UP> mtu 65536 qdisc noqueue state UNKNOWN
    link/loopback 00:00:00:00:00:00 brd 00:00:00:00:00:00
    inet 127.0.0.1/8 scope host lo
```

```
           valid_lft forever preferred_lft forever
   2: eth0: <BROADCAST,MULTICAST,UP,LOWER_UP> mtu 1500 qdisc pfifo_fast state UP
qlen 1000
       link/ether 00:0c:29:55:5e:c3 brd ff:ff:ff:ff:ff:ff
       inet 192.168.18.131/24 brd 192.168.18.255 scope global dynamic eth0
           valid_lft 1369sec preferred_lft 1369sec
   3: ovs-system: <BROADCAST,MULTICAST> mtu 1500 qdisc noop state DOWN
       link/ether a6:15:c3:25:cf:33 brd ff:ff:ff:ff:ff:ff
   4: br0: <BROADCAST,MULTICAST,UP,LOWER_UP> mtu 1500 qdisc noqueue master docker0
state UNKNOWN
       link/ether 92:8d:d0:a4:ca:45 brd ff:ff:ff:ff:ff:ff
   5: docker0: <BROADCAST,MULTICAST,UP,LOWER_UP> mtu 1500 qdisc noqueue state UP
       link/ether 02:42:44:8d:62:11 brd ff:ff:ff:ff:ff:ff
       inet 172.17.42.1/24 scope global docker0
           valid_lft forever preferred_lft forever
```

同样，192.168.18.128 节点的重要信息如下：

```
# ip addr
   1: lo: <LOOPBACK,UP,LOWER_UP> mtu 65536 qdisc noqueue state UNKNOWN
       link/loopback 00:00:00:00:00:00 brd 00:00:00:00:00:00
       inet 127.0.0.1/8 scope host lo
           valid_lft forever preferred_lft forever
   2: eth0: <BROADCAST,MULTICAST,UP,LOWER_UP> mtu 1500 qdisc pfifo_fast state UP
qlen 1000
       link/ether 00:0c:29:e8:02:c7 brd ff:ff:ff:ff:ff:ff
       inet 192.168.18.128/24 brd 192.168.18.255 scope global dynamic eth0
           valid_lft 1356sec preferred_lft 1356sec
   3: ovs-system: <BROADCAST,MULTICAST> mtu 1500 qdisc noop state DOWN
       link/ether fa:6c:89:a2:f2:01 brd ff:ff:ff:ff:ff:ff
   4: br0: <BROADCAST,MULTICAST,UP,LOWER_UP> mtu 1500 qdisc noqueue master docker0
state UNKNOWN
       link/ether ba:89:14:e0:7f:43 brd ff:ff:ff:ff:ff:ff
   5: docker0: <BROADCAST,MULTICAST,UP,LOWER_UP> mtu 1500 qdisc noqueue state UP
       link/ether 02:42:63:a8:14:d5 brd ff:ff:ff:ff:ff:ff
       inet 172.17.43.1/24 scope global docker0
           valid_lft forever preferred_lft forever
```

3. 两个 Node 上容器之间的互通测试

首先，在 192.168.18.128 节点上 ping 192.168.18.131 上的 docker0 地址 172.17.42.1，验证网络的互通性：

```
# ping 172.17.42.1
PING 172.17.42.1 (172.17.42.1) 56(84) bytes of data.
64 bytes from 172.17.42.1: icmp_seq=1 ttl=64 time=1.57 ms
64 bytes from 172.17.42.1: icmp_seq=2 ttl=64 time=0.966 ms
64 bytes from 172.17.42.1: icmp_seq=3 ttl=64 time=1.01 ms
64 bytes from 172.17.42.1: icmp_seq=4 ttl=64 time=1.00 ms
64 bytes from 172.17.42.1: icmp_seq=5 ttl=64 time=1.22 ms
64 bytes from 172.17.42.1: icmp_seq=6 ttl=64 time=0.996 ms
```

下面通过 tshark 抓包工具来分析流量走向。首先，在 192.168.18.128 节点监听在 br0 上是否有 GRE 报文，执行下面的命令，我们发现在 br0 上并没有 GRE 报文：

```
# tshark -i br0 -R ip proto GRE
tshark: -R without -2 is deprecated. For single-pass filtering use -Y.
Running as user "root" and group "root". This could be dangerous.
Capturing on 'br0'
^C
```

在 eth0 上抓包，则发现了 GRE 封装的 ping 包报文通过，说明 GRE 是在物理网络上完成的封包过程：

```
# tshark -i eth0 -R ip proto GRE
tshark: -R without -2 is deprecated. For single-pass filtering use -Y.
Running as user "root" and group "root". This could be dangerous.
Capturing on 'eth0'
    1   0.000000  172.17.43.1 -> 172.17.42.1  ICMP 136 Echo (ping) request
id=0x0970, seq=180/46080, ttl=64
    2   0.000892  172.17.42.1 -> 172.17.43.1  ICMP 136 Echo (ping) reply
id=0x0970, seq=180/46080, ttl=64 (request in 1)
    2 3   1.002014  172.17.43.1 -> 172.17.42.1  ICMP 136 Echo (ping) request
id=0x0970, seq=181/46336, ttl=64
    4   1.002916  172.17.42.1 -> 172.17.43.1  ICMP 136 Echo (ping) reply
id=0x0970, seq=181/46336, ttl=64 (request in 3)
    4 5   2.004101  172.17.43.1 -> 172.17.42.1  ICMP 136 Echo (ping) request
id=0x0970, seq=182/46592, ttl=64
```

至此，基于 OVS 的网络搭建成功，由于 GRE 是点对点的隧道通信方式，所以如果有多个 Node，则需要建立 $N \times (N-1)$ 条 GRE 隧道，即所有 Node 组成一个网状网络，实现了全网互通。

7.8.3 直接路由

我们知道，docker0 网桥上的 IP 地址在 Node 网络上是看不到的。从一个 Node 到一个 Node 内的 docker0 是不通的，因为它不知道某个 IP 地址在哪里。如果能够让这些机器知道对端 docker0 地址在哪里，就可以让这些 docker0 互相通信了。这样，在所有 Node 上运行的 Pod 就都可以互相通信了。

我们可以通过部署 MultiLayer Switch（MLS）来实现这一点，在 MLS 中配置每个 docker0 子网地址到 Node 地址的路由项，通过 MLS 将 docker0 的 IP 寻址定向到对应的 Node 上。

另外，我们可以将这些 docker0 和 Node 的匹配关系配置在 Linux 操作系统的路由项中，这样通信发起的 Node 就能够根据这些路由信息直接找到目标 Pod 所在的 Node，将数据传输过去。如图 7.23 所示。

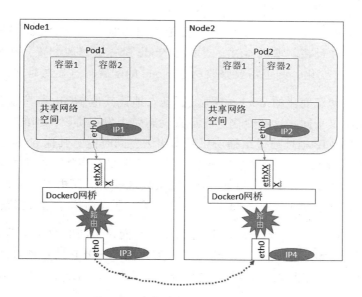

图 7.23 直接路由 Pod 到 Pod 通信

我们在每个 Node 的路由表中增加对方所有 docker0 的路由项。

例如，Pod1 所在 docker0 网桥的 IP 子网是 10.1.10.0，Node 的地址为 192.168.1.128；而 Pod2 所在 docker0 网桥的 IP 子网是 10.1.20.0，Node 的地址为 192.168.1.129。

在 Node1 上用 route add 命令增加一条到 Node2 上 docker0 的静态路由规则：

```
# route add -net 10.1.20.0 netmask 255.255.255.0 gw 192.168.1.129
```

同样，在Node2上增加一条到Node1上docker0的静态路由规则：

```
# route add -net 10.1.10.0 netmask 255.255.255.0 gw 192.168.1.128
```

在Node1上通过ping命令验证到Node2上docker0的网络连通性。这里10.1.20.1为Node2上docker0网桥自身的IP地址：

```
$ ping 10.1.20.1
PING 10.1.20.1 (10.1.20.1) 56(84) bytes of data.
64 bytes from 10.1.20.1: icmp_seq=1 ttl=62 time=1.15 ms
64 bytes from 10.1.20.1: icmp_seq=2 ttl=62 time=1.16 ms
64 bytes from 10.1.20.1: icmp_seq=3 ttl=62 time=1.57 ms
......
```

可以看到，路由转发规则生效，Node1可以直接访问Node2上的docker0网桥，进一步就可以访问属于docker0网段的容器应用了。

在大规模集群中，在每个Node上都需要配置到其他docker0/Node的路由项，这会带来很大的工作量；并且在新增机器时，对所有Node都需要修改配置；在重启机器时，如果docker0的地址有变化，则也需要修改所有Node的配置，这显然是非常复杂的。

为了管理这些动态变化的docker0地址，动态地让其他Node都感知到它，还可以使用动态路由发现协议来同步这些变化。在运行动态路由发现协议代理的Node时，会将本机LOCAL路由表的IP地址通过组播协议发布出去，同时监听其他Node的组播包。通过这样的信息交换，Node上的路由规则就都能够相互学习。当然，路由发现协议本身还是很复杂的，感兴趣的话，可以查阅相关规范。在实现这些动态路由发现协议的开源软件中，常用的有Quagga（http://www.quagga.net）、Zebra等。下面简单介绍直接路由的操作过程。

首先，手工分配Docker bridge的地址，保证它们在不同的网段是不重叠的。建议最好不用Docker Daemon自动创建的docker0（因为我们不需要它的自动管理功能），而是单独建立一个bridge，给它配置规划好的IP地址，然后使用--bridge=XX来指定网桥。

然后，在每个节点上都运行Quagga。

完成这些操作后，我们很快就能得到一个Pod和Pod直接互相访问的环境了。由于路由发现能够被网络上的所有设备接收，所以如果网络上的路由器也能打开RIP协议选项，则能够学习到这些路由信息。通过这些路由器，我们甚至可以在非Node上使用Pod的IP地址直接访问Node上的Pod了。

除了在每台服务器上安装 Quagga 软件并启动，还可以使用 Quagga 容器运行（例如 index.alauda.cn/georce/router）。在每个 Node 上下载该 Docker 镜像：

```
$ docker pull index.alauda.cn/georce/router
```

在运行 Quagga 容器之前，需要确保每个 Node 上 docker0 网桥的子网地址不能重叠，也不能与物理机所在的网络重叠，这需要网络管理员的仔细规划。

下面以 3 个 Node 为例，每个 Node 的 docker0 网桥的地址如下（前提是 Node 物理机的 IP 地址不是 10.1.$X.X$ 地址段）：

```
Node 1: # ifconfig docker0 10.1.10.1/24
Node 2: # ifconfig docker0 10.1.20.1/24
Node 3: # ifconfig docker0 10.1.30.1/24
```

在每个 Node 上启动 Quagga 容器。需要说明的是，Quagga 需要以 --privileged 特权模式运行，并且指定 --net=host，表示直接使用物理机的网络：

```
$ docker run -itd --name=router --privileged --net=host index.alauda.cn/
georce/router
```

启动成功后，各 Node 上的 Quagga 会相互学习来完成到其他机器的 docker0 路由规则的添加。

一段时间后，在 Node1 上使用 route -n 命令来查看路由表，可以看到 Quagga 自动添加了两条到 Node2 和到 Node3 上 docker0 的路由规则：

```
# route -n
Kernel IP routing table
Destination     Gateway         Genmask         Flags Metric Ref    Use Iface
0.0.0.0         192.168.1.128   0.0.0.0         UG    0      0        0 eth0
10.1.10.0       0.0.0.0         255.255.255.0   U     0      0        0 docker0
10.1.20.0       192.168.1.129   255.255.255.0   UG    20     0        0 eth0
10.1.30.0       192.168.1.130   255.255.255.0   UG    20     0        0 eth0
```

在 Node2 上查看路由表，可以看到自动添加了两条到 Node1 和 Node3 上 docker0 的路由规则：

```
# route -n
Kernel IP routing table
Destination     Gateway         Genmask         Flags Metric Ref    Use Iface
0.0.0.0         192.168.1.129   0.0.0.0         UG    0      0        0 eth0
10.1.20.0       0.0.0.0         255.255.255.0   U     0      0        0 docker0
```

10.1.10.0	192.168.1.128	255.255.255.0	UG	20	0	0 eth0
10.1.30.0	192.168.1.130	255.255.255.0	UG	20	0	0 eth0

至此，所有 Node 上的 docker0 就都可以互联互通了。

当然，聪明的你还会有新的疑问：这样做的话，由于每个 Pod 的地址都会被路由发现协议广播出去，会不会存在路由表过大的情况？实际上，路由表通常都会有高速缓存，查找速度会很快，不会对性能产生太大的影响。当然，如果你的集群容量在数千个 Node 以上，则仍然需要测试和评估路由表的效率问题。

7.8.4　Calico 容器网络和网络策略实战

本节以 Calico 为例讲解 Kubernetes 中 CNI 插件和网络策略的原理和应用。

1. Calico 简介

Calico 是一个基于 BGP 的纯三层的网络方案，与 OpenStack、Kubernetes、AWS、GCE 等云平台都能够良好地集成。Calico 在每个计算节点都利用 Linux Kernel 实现了一个高效的 vRouter 来负责数据转发。每个 vRouter 都通过 BGP1 协议把在本节点上运行的容器的路由信息向整个 Calico 网络广播，并自动设置到达其他节点的路由转发规则。Calico 保证所有容器之间的数据流量都是通过 IP 路由的方式完成互联互通的。Calico 节点组网时可以直接利用数据中心的网络结构（L2 或者 L3），不需要额外的 NAT、隧道或者 Overlay Network，没有额外的封包解包，能够节约 CPU 运算，提高网络效率，如图 7.24 所示。

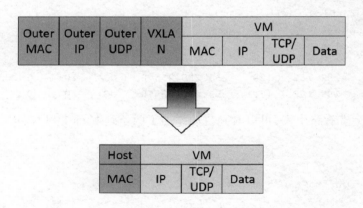

图 7.24　Calico 不使用额外的封包解包

Calico 在小规模集群中可以直接互联，在大规模集群中可以通过额外的 BGP route reflector 来完成，如图 7.25 所示。

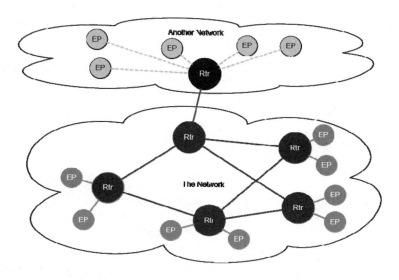

图 7.25　通过 BGP route reflector 连接大规模网络

此外，Calico 基于 iptables 还提供了丰富的网络策略，实现了 Kubernetes 的 Network Policy 策略，提供容器间网络可达性限制的功能。

Calico 的系统架构如图 7.26 所示。

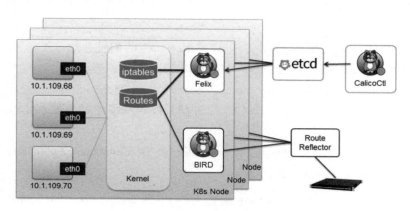

图 7.26　Calico 的系统架构

Calico 的主要组件如下。

- Felix：Calico Agent，运行在每个 Node 上，负责为容器设置网络资源（IP 地址、路由规则、iptables 规则等），保证跨主机容器网络互通。
- etcd：Calico 使用的后端存储。
- BGP Client：负责把 Felix 在各 Node 上设置的路由信息通过 BGP 协议广播到 Calico 网络。
- Route Reflector：通过一个或者多个 BGP Route Reflector 来完成大规模集群的分级路由分发。
- CalicoCtl：Calico 命令行管理工具。

2. 部署 Calico 服务

在 Kubernetes 中部署 Calico 的主要步骤如下。

（1）修改 Kubernetes 服务的启动参数，并重启服务。

- 设置 Master 上 kube-apiserver 服务的启动参数：--allow-privileged=true（因为 calico-node 需要以特权模式运行在各 Node 上）。
- 设置各 Node 上 kubelet 服务的启动参数：--network-plugin=cni（使用 CNI 网络插件）。

本例中的 Kubernetes 集群包括两个 Node：k8s-node-1（IP 地址为 192.168.18.3）和 k8s-node-2（IP 地址为 192.168.18.4）。

（2）创建 Calico 服务，主要包括 calico-node 和 calico policy controller。需要创建的资源对象如下。

- 创建 ConfigMap calico-config，包含 Calico 所需的配置参数。
- 创建 Secret calico-etcd-secrets，用于使用 TLS 方式连接 etcd。
- 在每个 Node 上都运行 calico/node 容器，部署为 DaemonSet。
- 在每个 Node 上都安装 Calico CNI 二进制文件和网络配置参数（由 install-cni 容器完成）。
- 部署一个名为 calico/kube-policy-controller 的 Deployment，以对接 Kubernetes 集群中为 Pod 设置的 Network Policy。

从 Calico 官网下载 Calico 的 YAML 配置文件，下载地址为 http://docs.projectcalico.org/v3.5/getting-started/kubernetes/installation/hosted/calico.yaml，该配置文件包括启动 Calico 所需的全部资源对象的定义，下面对它们逐个进行说明。

(1) Calico 所需的配置，以 ConfigMap 对象进行创建：

```yaml
kind: ConfigMap
apiVersion: v1
metadata:
  name: calico-config
  namespace: kube-system
data:
 # 配置 etcd 的服务 URL
 etcd_endpoints: "http://192.168.18.3:2379"

 # 如果 etcd 启用了 HTTPS 安全认证，则需要配置 etcd 相关证书
 etcd_ca: ""   # "/calico-secrets/etcd-ca"
 etcd_cert: "" # "/calico-secrets/etcd-cert"
 etcd_key: ""  # "/calico-secrets/etcd-key"

 typha_service_name: "none"

 # 设置 Calico 使用的 backend 类型，默认为 bird
 calico_backend: "bird"

 # 设置 MTU 值
 veth_mtu: "1440"

 # 设置 CNI 网络配置文件的内容
 cni_network_config: |-
   {
     "name": "k8s-pod-network",
     "cniVersion": "0.3.0",
     "plugins": [
       {
         "type": "calico",
         "log_level": "info",
         "etcd_endpoints": "__ETCD_ENDPOINTS__",
         "etcd_key_file": "__ETCD_KEY_FILE__",
         "etcd_cert_file": "__ETCD_CERT_FILE__",
         "etcd_ca_cert_file": "__ETCD_CA_CERT_FILE__",
         "mtu": __CNI_MTU__,
         "ipam": {
             "type": "calico-ipam"
         },
```

```
      "policy": {
          "type": "k8s"
      },
      "kubernetes": {
          "kubeconfig": "__KUBECONFIG_FILEPATH__"
      }
    },
    {
      "type": "portmap",
      "snat": true,
      "capabilities": {"portMappings": true}
    }
  ]
}
```

对主要参数说明如下。

◎ etcd_endpoints：Calico 使用 etcd 来保存网络拓扑和状态，该参数指定 etcd 服务的地址。

◎ calico_backend：Calico 的后端，默认为 bird。

◎ cni_network_config：符合 CNI 规范的网络配置。其中 type=calico 表示 kubelet 将从/opt/cni/bin 目录下搜索名为 calico 的可执行文件，并调用它来完成容器网络的设置。ipam 中的 type=calico-ipam 表示 kubelet 将在/opt/cni/bin 目录下搜索名为 calico-ipam 的可执行文件，用于完成容器 IP 地址的分配。

如果 etcd 服务配置了 TLS 安全认证，则还需指定相应的 ca、cert、key 等文件。

（2）访问 etcd 所需的 secret，对于无 TLS 的 etcd 服务，将 data 设置为空即可：

```
apiVersion: v1
kind: Secret
type: Opaque
metadata:
  name: calico-etcd-secrets
  namespace: kube-system
data:
  # 如果配置了 TLS，则需设置相应的证书和密钥文件路径
  # etcd-key: null
  # etcd-cert: null
  # etcd-ca: null
```

（3）calico-node，以 DaemonSet 方式在每个 Node 上都运行一个 calico-node 服务和一个 install-cni 服务：

```yaml
kind: DaemonSet
apiVersion: extensions/v1beta1
metadata:
  name: calico-node
  namespace: kube-system
  labels:
    k8s-app: calico-node
spec:
  selector:
    matchLabels:
      k8s-app: calico-node
  updateStrategy:
    type: RollingUpdate
    rollingUpdate:
      maxUnavailable: 1
  template:
    metadata:
      labels:
        k8s-app: calico-node
      annotations:
        scheduler.alpha.kubernetes.io/critical-pod: ''
    spec:
      nodeSelector:
        beta.kubernetes.io/os: linux
      hostNetwork: true
      tolerations:
        - effect: NoSchedule
          operator: Exists
        - key: CriticalAddonsOnly
          operator: Exists
        - effect: NoExecute
          operator: Exists
      serviceAccountName: calico-node
      terminationGracePeriodSeconds: 0
      initContainers:
        # This container installs the Calico CNI binaries
        # and CNI network config file on each node.
        - name: install-cni
```

```yaml
        image: calico/cni:v3.5.2
        command: ["/install-cni.sh"]
        env:
          - name: CNI_CONF_NAME
            value: "10-calico.conflist"
          - name: CNI_NETWORK_CONFIG
            valueFrom:
              configMapKeyRef:
                name: calico-config
                key: cni_network_config
          - name: ETCD_ENDPOINTS
            valueFrom:
              configMapKeyRef:
                name: calico-config
                key: etcd_endpoints
          - name: CNI_MTU
            valueFrom:
              configMapKeyRef:
                name: calico-config
                key: veth_mtu
          # Prevents the container from sleeping forever.
          - name: SLEEP
            value: "false"
        volumeMounts:
          - mountPath: /host/opt/cni/bin
            name: cni-bin-dir
          - mountPath: /host/etc/cni/net.d
            name: cni-net-dir
          - mountPath: /calico-secrets
            name: etcd-certs
      containers:
        - name: calico-node
          image: calico/node:v3.5.2
          env:
            - name: ETCD_ENDPOINTS
              valueFrom:
                configMapKeyRef:
                  name: calico-config
                  key: etcd_endpoints
            - name: ETCD_CA_CERT_FILE
              valueFrom:
```

```yaml
        configMapKeyRef:
          name: calico-config
          key: etcd_ca
  # Location of the client key for etcd.
  - name: ETCD_KEY_FILE
    valueFrom:
      configMapKeyRef:
        name: calico-config
        key: etcd_key
  - name: ETCD_CERT_FILE
    valueFrom:
      configMapKeyRef:
        name: calico-config
        key: etcd_cert
  - name: CALICO_K8S_NODE_REF
    valueFrom:
      fieldRef:
        fieldPath: spec.nodeName
  # Choose the backend to use.
  - name: CALICO_NETWORKING_BACKEND
    valueFrom:
      configMapKeyRef:
        name: calico-config
        key: calico_backend
  - name: CLUSTER_TYPE
    value: "k8s,bgp"
  - name: IP
    value: "autodetect"
  # 启用 IPIP 模式
  - name: CALICO_IPV4POOL_IPIP
    value: "Always"
  - name: FELIX_IPINIPMTU
    valueFrom:
      configMapKeyRef:
        name: calico-config
        key: veth_mtu
  # 设置容器 IP 网段
  - name: CALICO_IPV4POOL_CIDR
    value: "10.1.0.0/16"
  # 设置 IPIP 模式
  - name: CALICO_IPV4POOL_IPIP
```

```yaml
      value: "always"
    # 查询网卡名的正则表达式
    - name: IP_AUTODETECTION_METHOD
      value: "interface=ens.*"
    - name: IP6_AUTODETECTION_METHOD
      value: "interface=ens.*"
    - name: CALICO_DISABLE_FILE_LOGGING
      value: "true"
    - name: FELIX_DEFAULTENDPOINTTOHOSTACTION
      value: "ACCEPT"
    # 禁用IPv6模式
    - name: FELIX_IPV6SUPPORT
      value: "false"
    # 设置日志级别
    - name: FELIX_LOGSEVERITYSCREEN
      value: "info"
    - name: FELIX_HEALTHENABLED
      value: "true"
  securityContext:
    privileged: true
  resources:
    requests:
      cpu: 250m
  livenessProbe:
    httpGet:
      path: /liveness
      port: 9099
      host: localhost
    periodSeconds: 10
    initialDelaySeconds: 10
    failureThreshold: 6
  readinessProbe:
    exec:
      command:
      - /bin/calico-node
      - -bird-ready
      - -felix-ready
    periodSeconds: 10
  volumeMounts:
  - mountPath: /lib/modules
    name: lib-modules
```

```yaml
        readOnly: true
      - mountPath: /run/xtables.lock
        name: xtables-lock
        readOnly: false
      - mountPath: /var/run/calico
        name: var-run-calico
        readOnly: false
      - mountPath: /var/lib/calico
        name: var-lib-calico
        readOnly: false
      - mountPath: /calico-secrets
        name: etcd-certs
  volumes:
    # Used by calico/node.
    - name: lib-modules
      hostPath:
        path: /lib/modules
    - name: var-run-calico
      hostPath:
        path: /var/run/calico
    - name: var-lib-calico
      hostPath:
        path: /var/lib/calico
    - name: xtables-lock
      hostPath:
        path: /run/xtables.lock
        type: FileOrCreate
    - name: cni-bin-dir
      hostPath:
        path: /opt/cni/bin
    - name: cni-net-dir
      hostPath:
        path: /etc/cni/net.d
    - name: etcd-certs
      secret:
        secretName: calico-etcd-secrets
        defaultMode: 0400
```

在该 Pod 中包括如下两个容器。

◎ install-cni：在 Node 上安装 CNI 二进制文件到/opt/cni/bin 目录下，并安装相应的

网络配置文件到/etc/cni/net.d 目录下,设置为 initContainers 并在运行完成后退出。
- calico-node：Calico 服务程序,用于设置 Pod 的网络资源,保证 Pod 的网络与各 Node 互联互通。它还需要以 hostNetwork 模式运行,直接使用宿主机网络。

calico-node 服务的主要参数如下。

- CALICO_IPV4POOL_CIDR：Calico IPAM 的 IP 地址池,Pod 的 IP 地址将从该池中进行分配。
- CALICO_IPV4POOL_IPIP：是否启用 IPIP 模式。启用 IPIP 模式时,Calico 将在 Node 上创建一个名为 tunl0 的虚拟隧道。
- IP_AUTODETECTION_METHOD：获取 Node IP 地址的方式,默认使用第 1 个网络接口的 IP 地址,对于安装了多块网卡的 Node,可以使用正则表达式选择正确的网卡,例如"interface=ens.*"表示选择名称以 ens 开头的网卡的 IP 地址。
- FELIX_IPV6SUPPORT：是否启用 IPv6。
- FELIX_LOGSEVERITYSCREEN：日志级别。
- securityContext.privileged=true：以特权模式运行。

另外,如果启用 RBAC 权限控制,则可以设置 ServiceAccount。

IP Pool 可以使用两种模式:BGP 或 IPIP。使用 IPIP 模式时,设置 CALICO_IPV4POOL_IPIP="always",不使用 IPIP 模式时,设置 CALICO_IPV4POOL_IPIP="off",此时将使用 BGP 模式。

IPIP 是一种将各 Node 的路由之间做一个 tunnel,再把两个网络连接起来的模式,如图 7.27 所示。启用 IPIP 模式时,Calico 将在各 Node 上创建一个名为 tunl0 的虚拟网络接口。

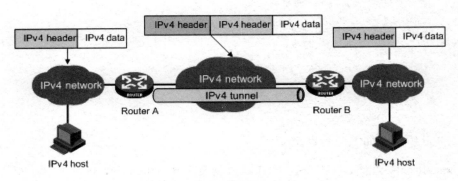

图 7.27　IPIP 模式

BGP 模式则直接使用物理机作为虚拟路由器（vRouter），不再创建额外的 tunnel。

（4）calico-kube-controllers 容器，用于对接 Kubernetes 集群中为 Pod 设置的 Network Policy：

```
apiVersion: extensions/v1beta1
kind: Deployment
metadata:
  name: calico-kube-controllers
  namespace: kube-system
  labels:
    k8s-app: calico-kube-controllers
  annotations:
    scheduler.alpha.kubernetes.io/critical-pod: ''
spec:
  replicas: 1
  strategy:
    type: Recreate
  template:
    metadata:
      name: calico-kube-controllers
      namespace: kube-system
      labels:
        k8s-app: calico-kube-controllers
    spec:
      nodeSelector:
        beta.kubernetes.io/os: linux
      hostNetwork: true
      tolerations:
        - key: CriticalAddonsOnly
          operator: Exists
        - key: node-role.kubernetes.io/master
          effect: NoSchedule
      serviceAccountName: calico-kube-controllers
      containers:
        - name: calico-kube-controllers
          image: calico/kube-controllers:v3.5.2
          env:
            - name: ETCD_ENDPOINTS
              valueFrom:
                configMapKeyRef:
```

```yaml
          name: calico-config
          key: etcd_endpoints
    - name: ETCD_CA_CERT_FILE
      valueFrom:
        configMapKeyRef:
          name: calico-config
          key: etcd_ca
    - name: ETCD_KEY_FILE
      valueFrom:
        configMapKeyRef:
          name: calico-config
          key: etcd_key
    - name: ETCD_CERT_FILE
      valueFrom:
        configMapKeyRef:
          name: calico-config
          key: etcd_cert
    - name: ENABLED_CONTROLLERS
      value: policy,namespace,serviceaccount,workloadendpoint,node
    volumeMounts:
    - mountPath: /calico-secrets
      name: etcd-certs
    readinessProbe:
      exec:
        command:
        - /usr/bin/check-status
        - -r
  volumes:
  - name: etcd-certs
    secret:
      secretName: calico-etcd-secrets
      defaultMode: 0400
```

如果启用 RBAC 权限控制,则可以设置 ServiceAccount。

用户在 Kubernetes 集群中设置了 Pod 的 Network Policy 之后, calico-kube-controllers 就会自动通知各 Node 上的 calico-node 服务,在宿主机上设置相应的 iptables 规则,完成 Pod 间网络访问策略的设置。

修改好相应的参数后,创建 Calico 的各资源对象:

```
# kubectl create -f calico.yaml
```

```
configmap "calico-config" created
secret "calico-etcd-secrets" created
daemonset "calico-node" created
deployment "calico-kube-controllers" created
```

确保 Calico 的各服务正确运行：

```
# kubectl get pods --namespace=kube-system -o wide
NAME                                    READY   STATUS    RESTARTS   AGE   IP             NODE
calico-node-pgwqr                       2/2     Running   0          1m    192.168.18.4   k8s-node-2
calico-node-t3ntq                       2/2     Running   0          1m    192.168.18.3   k8s-node-1
calico-kube-controllers-1838634297-cfddl 1/1    Running   0          2m    192.168.18.3   k8s-node-1
```

calico-node 在正常运行之后，会根据 CNI 规范，在 /etc/cni/net.d/ 目录下生成如下文件和目录，并在 /opt/cni/bin/ 目录下安装二进制文件 calico 和 calico-ipam，供 kubelet 调用。

◎ 10-calico.conf：符合 CNI 规范的网络配置，其中 type=calico 表示该插件的二进制文件名为 calico。
◎ calico-kubeconfig：Calico 所需的 kubeconfig 文件。
◎ calico-tls 目录：以 TLS 方式连接 etcd 的相关文件。

查看 k8s-node-1 服务器的网络接口设置，可以看到一个新的名为 tunl0 的接口，并设置了网络地址为 10.1.109.64/32：

```
# ip addr show
1: lo: <LOOPBACK,UP,LOWER_UP> mtu 65536 qdisc noqueue state UNKNOWN qlen 1
    link/loopback 00:00:00:00:00:00 brd 00:00:00:00:00:00
    inet 127.0.0.1/8 scope host lo
       valid_lft forever preferred_lft forever
    inet6 ::1/128 scope host
       valid_lft forever preferred_lft forever
2: ens33: <BROADCAST,MULTICAST,UP,LOWER_UP> mtu 1500 qdisc pfifo_fast state UP qlen 1000
    link/ether 00:0c:29:1b:c5:fc brd ff:ff:ff:ff:ff:ff
    inet 192.168.18.3/24 brd 192.168.18.255 scope global ens33
       valid_lft forever preferred_lft forever
    inet6 fe80::20c:29ff:fe1b:c5fc/64 scope link
       valid_lft forever preferred_lft forever
3: docker0: <NO-CARRIER,BROADCAST,MULTICAST,UP> mtu 1500 qdisc noqueue state DOWN
    link/ether 02:42:46:ad:a4:38 brd ff:ff:ff:ff:ff:ff
    inet 172.17.1.1/24 scope global docker0
```

```
        valid_lft forever preferred_lft forever
4: tun10@NONE: <NOARP,UP,LOWER_UP> mtu 1440 qdisc noqueue state UNKNOWN qlen 1
    link/ipip 0.0.0.0 brd 0.0.0.0
    inet 10.1.109.64/32 scope global tun10
        valid_lft forever preferred_lft forever
```

查看 k8s-node-2 服务器的网络接口设置，同样可以看到一个新的名为 tun10 的接口，网络地址为 10.1.140.64/32：

```
1: lo: <LOOPBACK,UP,LOWER_UP> mtu 65536 qdisc noqueue state UNKNOWN qlen 1
    link/loopback 00:00:00:00:00:00 brd 00:00:00:00:00:00
    inet 127.0.0.1/8 scope host lo
        valid_lft forever preferred_lft forever
    inet6 ::1/128 scope host
        valid_lft forever preferred_lft forever
2: ens33: <BROADCAST,MULTICAST,UP,LOWER_UP> mtu 1500 qdisc pfifo_fast state UP qlen 1000
    link/ether 00:0c:29:93:71:9e brd ff:ff:ff:ff:ff:ff
    inet 192.168.18.4/24 brd 192.168.18.255 scope global ens33
        valid_lft forever preferred_lft forever
    inet6 fe80::20c:29ff:fe93:719e/64 scope link
        valid_lft forever preferred_lft forever
3: docker0: <NO-CARRIER,BROADCAST,MULTICAST,UP> mtu 1500 qdisc noqueue state DOWN
    link/ether 02:42:d9:08:8e:93 brd ff:ff:ff:ff:ff:ff
    inet 172.17.2.1/24 scope global docker0
        valid_lft forever preferred_lft forever
4: tun10@NONE: <NOARP,UP,LOWER_UP> mtu 1440 qdisc noqueue state UNKNOWN qlen 1
    link/ipip 0.0.0.0 brd 0.0.0.0
    inet 10.1.140.64/32 scope global tun10
        valid_lft forever preferred_lft forever
```

这两个子网都是从 calico-node 设置的 IP 地址池（CALICO_IPV4POOL_CIDR="10.1.0.0/16"）中进行分配的。同时，docker0 对于 Kubernetes 设置 Pod 的 IP 地址将不再起作用。

查看两台主机的路由表。首先，查看 k8s-node-1 服务器的路由表，可以看到一条到 k8s-node-2 的私网 10.1.140.64 的路由转发规则：

```
# ip route
default via 192.168.18.2 dev ens33
blackhole 10.1.109.64/26 proto bird
10.1.140.64/26 via 192.168.18.4 dev tun10 proto bird onlink
```

```
172.17.1.0/24 dev docker0  proto kernel  scope link  src 172.17.1.1
192.168.18.0/24 dev ens33 proto kernel  scope link  src 192.168.18.3  metric 100
```

然后，查看 k8s-node-2 服务器的路由表，可以看到一条到 k8s-node-1 的私网 10.1.109.64/26 的路由转发规则：

```
# ip route
default via 192.168.18.2 dev ens33
blackhole 10.1.140.64/26  proto bird
10.1.109.64/26 via 192.168.18.3 dev tunl0  proto bird onlink
172.17.2.0/24 dev docker0  proto kernel  scope link  src 172.17.2.1
192.168.18.0/24 dev ens33 proto kernel  scope link  src 192.168.18.4  metric 100
```

这样，通过 Calico 就完成了 Node 间容器网络的设置。在后续的 Pod 创建过程中，kubelet 将通过 CNI 接口调用 Calico 进行 Pod 网络的设置，包括 IP 地址、路由规则、iptables 规则等。

如果设置 CALICO_IPV4POOL_IPIP="off"，即不使用 IPIP 模式，则 Calico 将不会创建 tunl0 网络接口，路由规则直接使用物理机网卡作为路由器进行转发。

查看 k8s-node-1 服务器的路由表，可以看到一条到 k8s-node-2 的私网 10.1.140.64 的路由转发规则，将通过本机 ens33 网卡进行转发：

```
# ip route
default via 192.168.18.2 dev ens33
blackhole 10.1.109.64/26  proto bird
10.1.140.64/26 via 192.168.18.4 dev ens33  proto bird
172.17.1.0/24 dev docker0  proto kernel  scope link  src 172.17.1.1
192.168.18.0/24 dev ens33 proto kernel  scope link  src 192.168.18.3  metric 100
```

查看 k8s-node-2 服务器的路由表，可以看到一条到 k8s-node-1 的私网 10.1.109.64/26 的路由转发规则，将通过本机 ens33 网卡进行转发：

```
# ip route
default via 192.168.18.2 dev ens33
blackhole 10.1.140.64/26  proto bird
10.1.109.64/26 via 192.168.18.3 dev ens33  proto bird
172.17.2.0/24 dev docker0  proto kernel  scope link  src 172.17.2.1
192.168.18.0/24 dev ens33 proto kernel  scope link  src 192.168.18.4  metric 100
```

3. Calico 设置容器 IP 地址，跨主机容器网络连通性验证

下面创建几个 Pod，验证 Calico 对它们的网络设置。以第 1 章的 mysql 和 myweb 为例，分别创建 1 个 Pod 和两个 Pod：

```yaml
mysql-rc.yaml
apiVersion: v1
kind: ReplicationController
metadata:
  name: mysql
spec:
  replicas: 1
  selector:
    app: mysql
  template:
    metadata:
      labels:
        app: mysql
    spec:
      containers:
      - name: mysql
        image: mysql
        ports:
        - containerPort: 3306
        env:
        - name: MYSQL_ROOT_PASSWORD
          value: "123456"
```

```yaml
myweb-rc.yaml
apiVersion: v1
kind: ReplicationController
metadata:
  name: myweb
spec:
  replicas: 2
  selector:
    app: myweb
  template:
    metadata:
      labels:
        app: myweb
```

```
    spec:
      containers:
      - name: myweb
        image: kubeguide/tomcat-app:v1
        ports:
        - containerPort: 8080
        env:
        - name: MYSQL_SERVICE_HOST
          value: 'mysql'
        - name: MYSQL_SERVICE_PORT
          value: '3306'
```

```
# kubectl create -f mysql-rc.yaml -f myweb-rc.yaml
replicationcontroller "mysql" created
replicationcontroller "myweb" created
```

查看各 Pod 的 IP 地址，可以看到是通过 Calico 设置的以 10.1 开头的 IP 地址：

```
# kubectl get pod -o wide
NAME             READY   STATUS    RESTARTS   AGE   IP             NODE
mysql-8cztq      1/1     Running   0          2m    10.1.109.71    k8s-node-1
myweb-h4lg3      1/1     Running   0          2m    10.1.109.70    k8s-node-1
myweb-s86sk      1/1     Running   0          2m    10.1.140.66    k8s-node-2
```

进入运行在 k8s-node-2 上的 Pod "myweb-s86sk"：

```
# kubectl exec -ti myweb-s86sk bash
```

在容器内访问运行在 k8s-node-1 上的 Pod "mysql-8cztq" 的 IP 地址 10.1.109.71：

```
root@myweb-s86sk:/usr/local/tomcat# ping 10.1.109.71
PING 10.1.109.71 (10.1.109.71): 56 data bytes
64 bytes from 10.1.109.71: icmp_seq=0 ttl=63 time=0.344 ms
64 bytes from 10.1.109.71: icmp_seq=1 ttl=63 time=0.213 ms
```

在容器内访问物理机 k8s-node-1 的 IP 地址 192.168.18.3：

```
root@myweb-s86sk:/usr/local/tomcat# ping 192.168.18.3
PING 192.168.18.3 (192.168.18.3): 56 data bytes
64 bytes from 192.168.18.3: icmp_seq=0 ttl=64 time=0.327 ms
64 bytes from 192.168.18.3: icmp_seq=1 ttl=64 time=0.182 ms
```

这说明跨主机容器间、容器与宿主机之间的网络都能互联互通了。

查看 k8s-node-2 物理机的网络接口和路由表，可以看到 Calico 为 Pod "myweb-s86sk"

新建了一个网络接口 cali439924adc43，并为其设置了一条路由规则：

```
# ip addr show
1: lo: <LOOPBACK,UP,LOWER_UP> mtu 65536 qdisc noqueue state UNKNOWN qlen 1
......
7: cali439924adc43@if3: <BROADCAST,MULTICAST,UP,LOWER_UP> mtu 1500 qdisc
noqueue state UP
    link/ether e2:e9:9a:55:52:92 brd ff:ff:ff:ff:ff:ff link-netnsid 0
    inet6 fe80::e0e9:9aff:fe55:5292/64 scope link
       valid_lft forever preferred_lft forever

# ip route
default via 192.168.18.2 dev ens33
blackhole 10.1.140.64/26  proto bird
10.1.109.64/26 via 192.168.18.3 dev tunl0  proto bird onlink
10.1.140.66 dev cali439924adc43  scope link
172.17.2.0/24 dev docker0  proto kernel  scope link  src 172.17.2.1
192.168.18.0/24 dev ens33  proto kernel  scope link  src 192.168.18.4  metric 100
```

另外，Calico 为该网络接口 cali439924adc43 设置了一系列 iptables 规则：

```
# iptables -L
......
Chain cali-from-wl-dispatch (2 references)
target     prot opt source               destination
cali-fw-cali439924adc43  all  --  anywhere             anywhere             [goto] /* cali:27N3bvAtjtNgABL_ */
DROP       all  --  anywhere             anywhere             /* cali:tL986QdUS4OiW3mC */ /* Unknown interface */

Chain cali-fw-cali439924adc43 (1 references)
target     prot opt source               destination
ACCEPT     all  --  anywhere             anywhere             /* cali:w_ft-rPVu6fgqGmc */ ctstate RELATED,ESTABLISHED
DROP       all  --  anywhere             anywhere             /* cali:ATcF-FBghYxNthE2 */ ctstate INVALID
MARK       all  --  anywhere             anywhere             /* cali:5mvqaVXl8wQh6vS6 */ MARK and 0xfeffffff
MARK       all  --  anywhere             anywhere             /* cali:nOAdEHYzt1IeVaqu */ /* Start of policies */ MARK and 0xfdffffff

Chain cali-to-wl-dispatch (1 references)
```

```
    target     prot opt source               destination
    cali-tw-cali439924adc43  all  --  anywhere             anywhere             [goto]
/* cali:WibRaHK-UmAeF88Y */

Chain cali-tw-cali439924adc43 (1 references)
    target     prot opt source               destination
    ACCEPT     all  --  anywhere             anywhere             /*
cali:c21cc_VY82hSFHuc */ ctstate RELATED,ESTABLISHED
    DROP       all  --  anywhere             anywhere             /*
cali:6eNswYurPxc_1g2M */ ctstate INVALID
    MARK       all  --  anywhere             anywhere             /*
cali:Y55YBsPr1TihN4NE */ MARK and 0xfeffffff
    MARK       all  --  anywhere             anywhere             /*
cali:hfMD9kYf5exJluSH */ /* Start of policies */ MARK and 0xfdffffff
    ......
```

4. 使用网络策略实现 Pod 间的访问策略

下面以一个提供服务的 Nginx Pod 为例，为两个客户端 Pod 设置不同的网络访问权限，允许包含 Label "role=nginxclient" 的 Pod 访问 Nginx 容器，不包含该 Label 的容器则拒绝访问。为了实现这个需求，需要通过以下步骤完成。

（1）创建 Nginx Pod，并添加 Label "app=nginx"：

```
nginx.yaml
apiVersion: v1
kind: Pod
metadata:
  name: nginx
  labels:
    app: nginx
spec:
  containers:
  - name: nginx
    image: nginx

# kubectl create -f nginx.yaml
pod "nginx" created
```

（2）为 Nginx 设置网络策略，编辑文件 networkpolicy-allow-nginxclient.yaml，内容如下。

```yaml
kind: NetworkPolicy
apiVersion: networking.k8s.io/v1
metadata:
  name: allow-nginxclient
spec:
  podSelector:
    matchLabels:
      app: nginx
  ingress:
  - from:
    - podSelector:
        matchLabels:
          role: nginxclient
    ports:
    - protocol: TCP
      port: 80
```

目标 Pod 应包含 Label "app=nginx"，允许访问的客户端 Pod 包含 Label "role=nginxclient"，并允许客户端访问 mysql 容器的 80 端口。

创建该 NetworkPolicy 资源对象：

```
# kubectl create -f networkpolicy-allow-nginxclient.yaml
networkpolicy "allow-nginxclient" created
```

（3）创建两个客户端 Pod，一个包含 Label "role=nginxclient"，另一个无此 Label。分别进入各 Pod，访问 Nginx 容器，验证网络策略的效果。

```yaml
client1.yaml
apiVersion: v1
kind: Pod
metadata:
  name: client1
  labels:
    role: nginxclient
spec:
  containers:
  - name: client1
    image: busybox
    command: [ "sleep", "3600" ]

client2.yaml
```

```
apiVersion: v1
kind: Pod
metadata:
  name: client2
spec:
  containers:
  - name: client2
    image: busybox
    command: [ "sleep", "3600" ]

# kubectl create -f client1.yaml -f client2.yaml
pod "client1" created
pod "client2" created
```

登录 Pod "client1"：

```
# kubectl exec -ti client1 -- sh
```

尝试连接 Nginx 容器的 80 端口：

```
/ # wget 10.1.109.69
Connecting to 10.1.109.69 (10.1.109.69:80)
index.html           100% |*****************************|   612   0:00:00 ETA
```

成功访问到 Nginx 的服务，说明 NetworkPolicy 生效。

登录 Pod "client2"：

```
# kubectl exec -ti client2 -- sh
```

尝试连接 Nginx 容器的 80 端口：

```
/ # wget --timeout=5 10.1.109.69
Connecting to 10.1.109.69 (10.1.109.69:80)
wget: download timed out
```

访问超时，说明 NetworkPolicy 生效，对没有 Label "role=nginxclient" 的客户端 Pod 拒绝访问。

本例中的网络策略是由 calico-kube-controllers 具体实现的，calico-kube-controllers 持续监听 Kubernetes 中 NetworkPolicy 的定义，与各 Pod 通过 Label 进行关联，将允许访问或拒绝访问的策略通知到各 calico-node 服务，最终 calico-node 完成对 Pod 间网络访问的设置，实现应用的网络隔离。

第 8 章

共享存储原理

8.1 共享存储机制概述

Kubernetes 对于有状态的容器应用或者对数据需要持久化的应用,不仅需要将容器内的目录挂载到宿主机的目录或者 emptyDir 临时存储卷,而且需要更加可靠的存储来保存应用产生的重要数据,以便容器应用在重建之后仍然可以使用之前的数据。不过,存储资源和计算资源(CPU/内存)的管理方式完全不同。为了能够屏蔽底层存储实现的细节,让用户方便使用,同时让管理员方便管理,Kubernetes 从 1.0 版本就引入 PersistentVolume(PV)和 PersistentVolumeClaim(PVC)两个资源对象来实现对存储的管理子系统。

PV 是对底层网络共享存储的抽象,将共享存储定义为一种"资源",比如 Node 也是一种容器应用可以"消费"的资源。PV 由管理员创建和配置,它与共享存储的具体实现直接相关,例如 GlusterFS、iSCSI、RBD 或 GCE 或 AWS 公有云提供的共享存储,通过插件式的机制完成与共享存储的对接,以供应用访问和使用。

PVC 则是用户对存储资源的一个"申请"。就像 Pod "消费" Node 的资源一样,PVC 能够"消费" PV 资源。PVC 可以申请特定的存储空间和访问模式。

使用 PVC "申请"到一定的存储空间仍然不能满足应用对存储设备的各种需求。通常应用程序都会对存储设备的特性和性能有不同的要求,包括读写速度、并发性能、数据冗余等更高的要求,Kubernetes 从 1.4 版本开始引入了一个新的资源对象 StorageClass,用于标记存储资源的特性和性能。到 1.6 版本时,StorageClass 和动态资源供应的机制得到了完善,实现了存储卷的按需创建,在共享存储的自动化管理进程中实现了重要的一步。

通过 StorageClass 的定义,管理员可以将存储资源定义为某种类别(Class),正如存储设备对于自身的配置描述(Profile),例如"快速存储""慢速存储""有数据冗余""无数据冗余"等。用户根据 StorageClass 的描述就能够直观地得知各种存储资源的特性,就可以根据应用对存储资源的需求去申请存储资源了。

Kubernetes 从 1.9 版本开始引入容器存储接口 Container Storage Interface(CSI)机制,目标是在 Kubernetes 和外部存储系统之间建立一套标准的存储管理接口,通过该接口为容器提供存储服务,类似于 CRI(容器运行时接口)和 CNI(容器网络接口)。

下面对 Kubernetes 的 PV、PVC、StorageClass、动态资源供应和 CSI 等共享存储管理机制进行详细说明。

8.2　PV 详解

PV 作为存储资源，主要包括存储能力、访问模式、存储类型、回收策略、后端存储类型等关键信息的设置。下面的例子声明的 PV 具有如下属性：5GiB 存储空间，访问模式为 ReadWriteOnce，存储类型为 slow（要求在系统中已存在名为 slow 的 StorageClass），回收策略为 Recycle，并且后端存储类型为 nfs（设置了 NFS Server 的 IP 地址和路径）：

```
apiVersion: v1
kind: PersistentVolume
metadata:
  name: pv1
spec:
  capacity:
    storage: 5Gi
  accessModes:
    - ReadWriteOnce
  persistentVolumeReclaimPolicy: Recycle
  storageClassName: slow
  nfs:
    path: /tmp
    server: 172.17.0.2
```

Kubernetes 支持的 PV 类型如下。

- AWSElasticBlockStore：AWS 公有云提供的 ElasticBlockStore。
- AzureFile：Azure 公有云提供的 File。
- AzureDisk：Azure 公有云提供的 Disk。
- CephFS：一种开源共享存储系统。
- FC（Fibre Channel）：光纤存储设备。
- FlexVolume：一种插件式的存储机制。
- Flocker：一种开源共享存储系统。
- GCEPersistentDisk：GCE 公有云提供的 PersistentDisk。
- Glusterfs：一种开源共享存储系统。
- HostPath：宿主机目录，仅用于单机测试。
- iSCSI：iSCSI 存储设备。
- Local：本地存储设备，从 Kubernetes 1.7 版本引入，到 1.14 版本时更新为稳定版，目前可以通过指定块（Block）设备提供 Local PV，或通过社区开发的 sig-storage-local-static-provisioner 插件（https://github.com/kubernetes-sigs/sig-storage-local-static-provisioner）来管理 Local PV 的生命周期。

◎ NFS：网络文件系统。
◎ Portworx Volumes：Portworx 提供的存储服务。
◎ Quobyte Volumes：Quobyte 提供的存储服务。
◎ RBD（Ceph Block Device）：Ceph 块存储。
◎ ScaleIO Volumes：DellEMC 的存储设备。
◎ StorageOS：StorageOS 提供的存储服务。
◎ VsphereVolume：VMWare 提供的存储系统。

每种存储类型都有各自的特点，在使用时需要根据它们各自的参数进行设置。

8.2.1 PV 的关键配置参数

1. 存储能力（Capacity）

描述存储设备具备的能力，目前仅支持对存储空间的设置（storage=xx），未来可能加入 IOPS、吞吐率等指标的设置。

2. 存储卷模式（Volume Mode）

Kubernetes 从 1.13 版本开始引入存储卷类型的设置（volumeMode=xxx），可选项包括 Filesystem（文件系统）和 Block（块设备），默认值为 Filesystem。

目前有以下 PV 类型支持块设备类型：

◎ AWSElasticBlockStore
◎ AzureDisk
◎ FC
◎ GCEPersistentDisk
◎ iSCSI
◎ Local volume
◎ RBD（Ceph Block Device）
◎ VsphereVolume（alpha）

下面的例子为使用块设备的 PV 定义：

```
apiVersion: v1
kind: PersistentVolume
metadata:
  name: block-pv
spec:
  capacity:
    storage: 10Gi
  accessModes:
    - ReadWriteOnce
  persistentVolumeReclaimPolicy: Retain
  volumeMode: Block
  fc:
    targetWWNs: ["50060e801049cfd1"]
    lun: 0
    readOnly: false
```

3. 访问模式（Access Modes）

对 PV 进行访问模式的设置，用于描述用户的应用对存储资源的访问权限。访问模式如下。

◎ ReadWriteOnce（RWO）：读写权限，并且只能被单个 Node 挂载。
◎ ReadOnlyMany（ROX）：只读权限，允许被多个 Node 挂载。
◎ ReadWriteMany（RWX）：读写权限，允许被多个 Node 挂载。

某些 PV 可能支持多种访问模式，但 PV 在挂载时只能使用一种访问模式，多种访问模式不能同时生效。

表 8.1 描述了不同的存储提供者支持的访问模式。

表 8.1　不同的存储提供者支持的访问模式

Volume Plugin	ReadWriteOnce	ReadOnlyMany	ReadWriteMany
AWSElasticBlockStore	✓	-	-
AzureFile	✓	✓	✓
AzureDisk	✓	-	-
CephFS	✓	✓	✓
Cinder	✓	-	-

续表

Volume Plugin	ReadWriteOnce	ReadOnlyMany	ReadWriteMany
FC	✓	✓	-
FlexVolume	✓	✓	视驱动而定
Flocker	✓	-	-
GCEPersistentDisk	✓	✓	-
GlusterFS	✓	✓	✓
HostPath	✓	-	-
iSCSI	✓	✓	-
Quobyte	✓	✓	✓
NFS	✓	✓	✓
RBD	✓	✓	-
VsphereVolume	✓	-	-
PortworxVolume	✓	-	✓
ScaleIO	✓	✓	-
StorageOS	✓	-	-

4. 存储类别（Class）

PV 可以设定其存储的类别，通过 storageClassName 参数指定一个 StorageClass 资源对象的名称。具有特定类别的 PV 只能与请求了该类别的 PVC 进行绑定。未设定类别的 PV 则只能与不请求任何类别的 PVC 进行绑定。

5. 回收策略（Reclaim Policy）

通过 PV 定义中的 persistentVolumeReclaimPolicy 字段进行设置，可选项如下。

◎ 保留：保留数据，需要手工处理。
◎ 回收空间：简单清除文件的操作（例如执行 rm -rf /thevolume/* 命令）。
◎ 删除：与 PV 相连的后端存储完成 Volume 的删除操作（如 AWS EBS、GCE PD、Azure Disk、OpenStack Cinder 等设备的内部 Volume 清理）。

目前，只有 NFS 和 HostPath 两种类型的存储支持 Recycle 策略；AWS EBS、GCE PD、

Azure Disk 和 Cinder volumes 支持 Delete 策略。

6. 挂载参数（Mount Options）

在将 PV 挂载到一个 Node 上时，根据后端存储的特点，可能需要设置额外的挂载参数，可以根据 PV 定义中的 mountOptions 字段进行设置。下面的例子为对一个类型为 gcePersistentDisk 的 PV 设置挂载参数：

```yaml
apiVersion: "v1"
kind: "PersistentVolume"
metadata:
  name: gce-disk-1
spec:
  capacity:
    storage: "10Gi"
  accessModes:
    - "ReadWriteOnce"
  mountOptions:
    - hard
    - nolock
    - nfsvers=3
  gcePersistentDisk:
    fsType: "ext4"
    pdName: "gce-disk-1
```

目前，以下 PV 类型支持设置挂载参数：

- AWSElasticBlockStore
- AzureDisk
- AzureFile
- CephFS
- Cinder (OpenStack block storage)
- GCEPersistentDisk
- Glusterfs
- NFS
- Quobyte Volumes
- RBD (Ceph Block Device)
- StorageOS

- VsphereVolume
- iSCSI

7. 节点亲和性（Node Affinity）

PV 可以设置节点亲和性来限制只能通过某些 Node 访问 Volume，可以在 PV 定义中的 nodeAffinity 字段进行设置。使用这些 Volume 的 Pod 将被调度到满足条件的 Node 上。

这个参数仅用于 Local 存储卷，例如：

```
apiVersion: v1
kind: PersistentVolume
metadata:
  name: example-local-pv
spec:
  capacity:
    storage: 5Gi
  accessModes:
  - ReadWriteOnce
  persistentVolumeReclaimPolicy: Delete
  storageClassName: local-storage
  local:
    path: /mnt/disks/ssd1
  nodeAffinity:
    required:
      nodeSelectorTerms:
      - matchExpressions:
        - key: kubernetes.io/hostname
          operator: In
          values:
          - my-node
```

公有云提供的存储卷（如 AWS EBS、GCE PD、Azure Disk 等）都由公有云自动完成节点亲和性设置，无须用户手工设置。

8.2.2 PV 生命周期的各个阶段

某个 PV 在生命周期中可能处于以下 4 个阶段（Phaes）之一。

- Available：可用状态，还未与某个 PVC 绑定。

- Bound：已与某个 PVC 绑定。
- Released：绑定的 PVC 已经删除，资源已释放，但没有被集群回收。
- Failed：自动资源回收失败。

定义了 PV 以后如何使用呢？这时就需要用到 PVC 了。下一节将对 PVC 进行详细说明。

8.3　PVC 详解

PVC 作为用户对存储资源的需求申请，主要包括存储空间请求、访问模式、PV 选择条件和存储类别等信息的设置。下例声明的 PVC 具有如下属性：申请 8GiB 存储空间，访问模式为 ReadWriteOnce，PV 选择条件为包含标签"release=stable"并且包含条件为"environment In [dev]"的标签，存储类别为"slow"（要求在系统中已存在名为 slow 的 StorageClass）：

```yaml
kind: PersistentVolumeClaim
apiVersion: v1
metadata:
  name: myclaim
spec:
  accessModes:
    - ReadWriteOnce
  resources:
    requests:
      storage: 8Gi
  storageClassName: slow
  selector:
    matchLabels:
      release: "stable"
    matchExpressions:
      - {key: environment, operator: In, values: [dev]}
```

PVC 的关键配置参数说明如下。

- 资源请求（Resources）：描述对存储资源的请求，目前仅支持 request.storage 的设置，即存储空间大小。
- 访问模式（Access Modes）：PVC 也可以设置访问模式，用于描述用户应用对存储资源的访问权限。其三种访问模式的设置与 PV 的设置相同。

- 存储卷模式（Volume Modes）：PVC 也可以设置存储卷模式，用于描述希望使用的 PV 存储卷模式，包括文件系统和块设备。
- PV 选择条件（Selector）：通过对 Label Selector 的设置，可使 PVC 对于系统中已存在的各种 PV 进行筛选。系统将根据标签选出合适的 PV 与该 PVC 进行绑定。选择条件可以使用 matchLabels 和 matchExpressions 进行设置，如果两个字段都设置了，则 Selector 的逻辑将是两组条件同时满足才能完成匹配。
- 存储类别（Class）：PVC 在定义时可以设定需要的后端存储的类别（通过 storageClassName 字段指定），以减少对后端存储特性的详细信息的依赖。只有设置了该 Class 的 PV 才能被系统选出，并与该 PVC 进行绑定。

PVC 也可以不设置 Class 需求。如果 storageClassName 字段的值被设置为空（storageClassName=""），则表示该 PVC 不要求特定的 Class，系统将只选择未设定 Class 的 PV 与之匹配和绑定。PVC 也可以完全不设置 storageClassName 字段，此时将根据系统是否启用了名为 DefaultStorageClass 的 admission controller 进行相应的操作。

- 未启用 DefaultStorageClass：等效于 PVC 设置 storageClassName 的值为空（storageClassName=""），即只能选择未设定 Class 的 PV 与之匹配和绑定。
- 启用 DefaultStorageClass：要求集群管理员已定义默认的 StorageClass。如果在系统中不存在默认的 StorageClass，则等效于不启用 DefaultStorageClass 的情况。如果存在默认的 StorageClass，则系统将自动为 PVC 创建一个 PV（使用默认 StorageClass 的后端存储），并将它们进行绑定。集群管理员设置默认 StorageClass 的方法为，在 StorageClass 的定义中加上一个 annotation "storageclass.kubernetes.io/is-default-class= true"。如果管理员将多个 StorageClass 都定义为 default，则由于不唯一，系统将无法为 PVC 创建相应的 PV。

注意，PVC 受限于 Namespace。Pod 在引用 PVC 时同样受 Namespace 的限制，只有相同 Namespace 中的 PVC 才能挂载到 Pod 内。

当 Selector 和 Class 都进行了设置时，系统将选择两个条件同时满足的 PV 与之匹配。

另外，如果资源供应使用的是动态模式，即管理员没有预先定义 PV，仅通过 StorageClass 交给系统自动完成 PV 的动态创建，那么 PVC 再设定 Selector 时，系统将无法为其供应任何存储资源。

在启用动态供应模式的情况下，一旦用户删除了 PVC，与之绑定的 PV 也将根据其默认的回收策略"Delete"被删除。如果需要保留 PV（用户数据），则在动态绑定成功后，用户需要将系统自动生成 PV 的回收策略从"Delete"改成"Retain"。

8.4 PV 和 PVC 的生命周期

我们可以将 PV 看作可用的存储资源，PVC 则是对存储资源的需求，PV 和 PVC 的相互关系遵循如图 8.1 所示的生命周期。

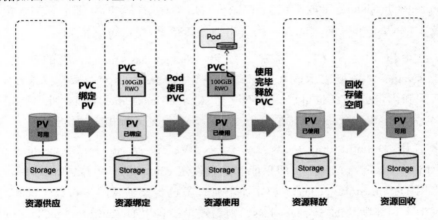

图 8.1　PV 和 PVC 的生命周期

8.4.1　资源供应

Kubernetes 支持两种资源的供应模式：静态模式（Static）和动态模式（Dynamic）。资源供应的结果就是创建好的 PV。

- ◎ **静态模式**：集群管理员手工创建许多 PV，在定义 PV 时需要将后端存储的特性进行设置。
- ◎ **动态模式**：集群管理员无须手工创建 PV，而是通过 StorageClass 的设置对后端存储进行描述，标记为某种类型。此时要求 PVC 对存储的类型进行声明，系统将自动完成 PV 的创建及与 PVC 的绑定。PVC 可以声明 Class 为""，说明该 PVC 禁止使用动态模式。

8.4.2 资源绑定

在用户定义好 PVC 之后，系统将根据 PVC 对存储资源的请求（存储空间和访问模式）在已存在的 PV 中选择一个满足 PVC 要求的 PV，一旦找到，就将该 PV 与用户定义的 PVC 进行绑定，用户的应用就可以使用这个 PVC 了。如果在系统中没有满足 PVC 要求的 PV，PVC 则会无限期处于 Pending 状态，直到等到系统管理员创建了一个符合其要求的 PV。PV 一旦绑定到某个 PVC 上，就会被这个 PVC 独占，不能再与其他 PVC 进行绑定了。在这种情况下，当 PVC 申请的存储空间比 PV 的少时，整个 PV 的空间就都能够为 PVC 所用，可能会造成资源的浪费。如果资源供应使用的是动态模式，则系统在为 PVC 找到合适的 StorageClass 后，将自动创建一个 PV 并完成与 PVC 的绑定。

8.4.3 资源使用

Pod 使用 Volume 的定义，将 PVC 挂载到容器内的某个路径进行使用。Volume 的类型为 persistentVolumeClaim，在后面的示例中再进行详细说明。在容器应用挂载了一个 PVC 后，就能被持续独占使用。不过，多个 Pod 可以挂载同一个 PVC，应用程序需要考虑多个实例共同访问一块存储空间的问题。

8.4.4 资源释放

当用户对存储资源使用完毕后，用户可以删除 PVC，与该 PVC 绑定的 PV 将会被标记为"已释放"，但还不能立刻与其他 PVC 进行绑定。通过之前 PVC 写入的数据可能还被留在存储设备上，只有在清除之后该 PV 才能再次使用。

8.4.5 资源回收

对于 PV，管理员可以设定回收策略，用于设置与之绑定的 PVC 释放资源之后如何处理遗留数据的问题。只有 PV 的存储空间完成回收，才能供新的 PVC 绑定和使用。回收策略详见下节的说明。

下面通过两张图分别对在静态资源供应模式和动态资源供应模式下，PV、PVC、StorageClass 及 Pod 使用 PVC 的原理进行说明。

图 8.2 描述了在静态资源供应模式下，通过 PV 和 PVC 完成绑定，并供 Pod 使用的存储管理机制。

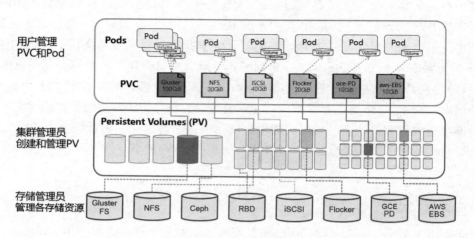

图 8.2　静态资源供应模式下的 PV 和 PVC 原理

图 8.3 描述了在动态资源供应模式下，通过 StorageClass 和 PVC 完成资源动态绑定（系统自动生成 PV），并供 Pod 使用的存储管理机制。

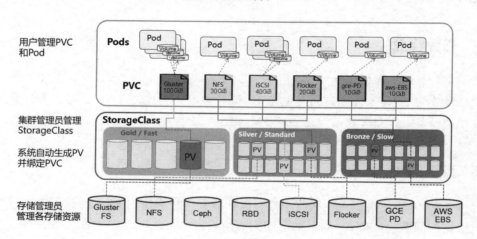

图 8.3　动态资源供应模式下的 StorageClass、PV 和 PVC 原理

接下来看看 StorageClass 的概念和用法。

8.5　StorageClass 详解

StorageClass 作为对存储资源的抽象定义，对用户设置的 PVC 申请屏蔽后端存储的细节，一方面减少了用户对于存储资源细节的关注，另一方面减轻了管理员手工管理 PV 的工作，由系统自动完成 PV 的创建和绑定，实现了动态的资源供应。基于 StorageClass 的动态资源供应模式将逐步成为云平台的标准存储配置模式。

StorageClass 的定义主要包括名称、后端存储的提供者（provisioner）和后端存储的相关参数配置。StorageClass 一旦被创建出来，则将无法修改。如需更改，则只能删除原 StorageClass 的定义重建。下例定义了一个名为 standard 的 StorageClass，提供者为 aws-ebs，其参数设置了一个 type，值为 gp2：

```
kind: StorageClass
apiVersion: storage.k8s.io/v1
metadata:
  name: standard
provisioner: kubernetes.io/aws-ebs
parameters:
  type: gp2
```

8.5.1　StorageClass 的关键配置参数

1. 提供者（Provisioner）

描述存储资源的提供者，也可以看作后端存储驱动。目前 Kubernetes 支持的 Provisioner 都以 "kubernetes.io/" 为开头，用户也可以使用自定义的后端存储提供者。为了符合 StorageClass 的用法，自定义 Provisioner 需要符合存储卷的开发规范，详见 https://github.com/kubernetes/community/blob/master/contributors/design-proposals/volume-provisioning.md 的说明。

2. 参数（Parameters）

后端存储资源提供者的参数设置，不同的 Provisioner 包括不同的参数设置。某些参数可以不显示设定，Provisioner 将使用其默认值。

接下来通过几种常见的 Provisioner 对 StorageClass 的定义进行详细说明。

1）AWS EBS 存储卷

```
kind: StorageClass
apiVersion: storage.k8s.io/v1
metadata:
  name: slow
provisioner: kubernetes.io/aws-ebs
parameters:
  type: io1
  zone: us-east-1d
  iopsPerGB: "10"
```

参数说明如下（详细说明请参考 AWS EBS 文档）。

◎ type：可选项为 io1、gp2、sc1、st1，默认值为 gp2。
◎ zone：AWS zone 的名称。
◎ iopsPerGB：仅用于 io1 类型的 Volume，意为每秒每 GiB 的 I/O 操作数量。
◎ encrypted：是否加密。
◎ kmsKeyId：加密时的 Amazon Resource Name。

2）GCE PD 存储卷

```
kind: StorageClass
apiVersion: storage.k8s.io/v1
metadata:
  name: slow
provisioner: kubernetes.io/gce-pd
parameters:
  type: pd-standard
  zone: us-central1-a
```

参数说明如下（详细说明请参考 GCE 文档）。

◎ type：可选项为 pd-standard、pd-ssd，默认值为 pd-standard。
◎ zone：GCE zone 名称。

3）GlusterFS 存储卷

```
apiVersion: storage.k8s.io/v1
kind: StorageClass
metadata:
```

```
  name: slow
provisioner: kubernetes.io/glusterfs
parameters:
  resturl: "http://127.0.0.1:8081"
  clusterid: "630372ccdc720a92c681fb928f27b53f"
  restauthenabled: "true"
  restuser: "admin"
  secretNamespace: "default"
  secretName: "heketi-secret"
  gidMin: "40000"
  gidMax: "50000"
  volumetype: "replicate:3"
```

参数说明如下（详细说明请参考 GlusterFS 和 Heketi 的文档）。

◎ resturl：Gluster REST 服务（Heketi）的 URL 地址，用于自动完成 GlusterFSvolume 的设置。

◎ restauthenabled：是否对 Gluster REST 服务启用安全机制。

◎ restuser：访问 Gluster REST 服务的用户名。

◎ secretNamespace 和 secretName：保存访问 Gluster REST 服务密码的 Secret 资源对象名。

◎ clusterid：GlusterFS 的 Cluster ID。

◎ gidMin 和 gidMax：StorageClass 的 GID 范围，用于动态资源供应时为 PV 设置的 GID。

◎ volumetype：设置 GlusterFS 的内部 Volume 类型，例如 replicate:3（Replicate 类型，3 份副本）；disperse:4:2（Disperse 类型，数据 4 份，冗余两份；"none"（Distribute 类型）。

4）OpenStack Cinder 存储卷

```
kind: StorageClass
apiVersion: storage.k8s.io/v1
metadata:
  name: gold
provisioner: kubernetes.io/cinder
parameters:
  type: fast
  availability: nova
```

参数说明如下。

◎ type：Cinder 的 VolumeType，默认值为空。
◎ availability：Availability Zone，默认值为空。

其他 Provisioner 的 StorageClass 相关参数设置请参考它们各自的配置手册。

8.5.2 设置默认的 StorageClass

要在系统中设置一个默认的 StorageClass，则首先需要启用名为 DefaultStorageClass 的 admission controller，即在 kube-apiserver 的命令行参数--admission-control 中增加：

```
--admission-control=...,DefaultStorageClass
```

然后，在 StorageClass 的定义中设置一个 annotation：

```
kind: StorageClass
apiVersion: storage.k8s.io/v1
metadata:
  name: gold
  annotations:
    storageclass.beta.kubernetes.io/is-default-class="true"
provisioner: kubernetes.io/gce-pd
parameters:
  type: pd-ssd
```

通过 kubectl create 命令创建成功后，查看 StorageClass 列表，可以看到名为 gold 的 StorageClass 被标记为 default：

```
# kubectl get sc
NAME            TYPE
gold (default)  kubernetes.io/gce-pd
```

8.6 动态存储管理实战：GlusterFS

本节以 GlusterFS 为例，从定义 StorageClass、创建 GlusterFS 和 Heketi 服务、用户申请 PVC 到创建 Pod 使用存储资源，对 StorageClass 和动态资源分配进行详细说明，进一步剖析 Kubernetes 的存储机制。

8.6.1 准备工作

为了能够使用 GlusterFS，首先在计划用于 GlusterFS 的各 Node 上安装 GlusterFS 客户端：

```
# yum install glusterfs glusterfs-fuse
```

GlusterFS 管理服务容器需要以特权模式运行，在 kube-apiserver 的启动参数中增加：

```
--allow-privileged=true
```

给要部署 GlusterFS 管理服务的节点打上 "storagenode=glusterfs" 的标签，是为了将 GlusterFS 容器定向部署到安装了 GlusterFS 的 Node 上：

```
# kubectl label node k8s-node-1 storagenode=glusterfs
# kubectl label node k8s-node-2 storagenode=glusterfs
# kubectl label node k8s-node-3 storagenode=glusterfs
```

8.6.2 创建 GlusterFS 管理服务容器集群

GlusterFS 管理服务容器以 DaemonSet 的方式进行部署，确保在每个 Node 上都运行一个 GlusterFS 管理服务。glusterfs-daemonset.yaml 的内容如下：

```
kind: DaemonSet
apiVersion: extensions/v1beta1
metadata:
  name: glusterfs
  labels:
    glusterfs: daemonset
  annotations:
    description: GlusterFS DaemonSet
    tags: glusterfs
spec:
  template:
    metadata:
      name: glusterfs
      labels:
        glusterfs-node: pod
    spec:
      nodeSelector:
        storagenode: glusterfs
```

```
        hostNetwork: true
        containers:
        - image: gluster/gluster-centos:latest
          name: glusterfs
          volumeMounts:
          - name: glusterfs-heketi
            mountPath: "/var/lib/heketi"
          - name: glusterfs-run
            mountPath: "/run"
          - name: glusterfs-lvm
            mountPath: "/run/lvm"
          - name: glusterfs-etc
            mountPath: "/etc/glusterfs"
          - name: glusterfs-logs
            mountPath: "/var/log/glusterfs"
          - name: glusterfs-config
            mountPath: "/var/lib/glusterd"
          - name: glusterfs-dev
            mountPath: "/dev"
          - name: glusterfs-misc
            mountPath: "/var/lib/misc/glusterfsd"
          - name: glusterfs-cgroup
            mountPath: "/sys/fs/cgroup"
            readOnly: true
          - name: glusterfs-ssl
            mountPath: "/etc/ssl"
            readOnly: true
          securityContext:
            capabilities: {}
            privileged: true
          readinessProbe:
            timeoutSeconds: 3
            initialDelaySeconds: 60
            exec:
              command:
              - "/bin/bash"
              - "-c"
              - systemctl status glusterd.service
          livenessProbe:
            timeoutSeconds: 3
            initialDelaySeconds: 60
```

```yaml
      exec:
        command:
        - "/bin/bash"
        - "-c"
        - systemctl status glusterd.service
    volumes:
    - name: glusterfs-heketi
      hostPath:
        path: "/var/lib/heketi"
    - name: glusterfs-run
    - name: glusterfs-lvm
      hostPath:
        path: "/run/lvm"
    - name: glusterfs-etc
      hostPath:
        path: "/etc/glusterfs"
    - name: glusterfs-logs
      hostPath:
        path: "/var/log/glusterfs"
    - name: glusterfs-config
      hostPath:
        path: "/var/lib/glusterd"
    - name: glusterfs-dev
      hostPath:
        path: "/dev"
    - name: glusterfs-misc
      hostPath:
        path: "/var/lib/misc/glusterfsd"
    - name: glusterfs-cgroup
      hostPath:
        path: "/sys/fs/cgroup"
    - name: glusterfs-ssl
      hostPath:
        path: "/etc/ssl"
```

```
# kubectl create -f glusterfs-daemonset.yaml
daemonset "glusterfs" created
# kubectl get po
NAME                READY   STATUS    RESTARTS   AGE
glusterfs-k2src     1/1     Running   0          1m
glusterfs-q32z2     1/1     Running   0          1m
```

8.6.3 创建 Heketi 服务

Heketi 是一个提供 RESTful API 管理 GlusterFS 卷的框架，并能够在 OpenStack、Kubernetes、OpenShift 等云平台上实现动态存储资源供应，支持 GlusterFS 多集群管理，便于管理员对 GlusterFS 进行操作。图 8.4 简单描述了 Heketi 的作用。

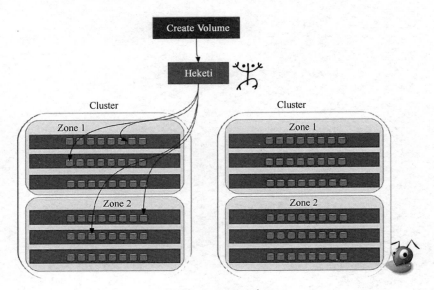

图 8.4　Heketi 的作用

在部署 Heketi 服务之前，需要为它创建一个 ServiceAccount 对象：

```
heketi-service-account.yaml
apiVersion: v1
kind: ServiceAccount
metadata:
  name: heketi-service-account

# kubectl create -f heketi-service-account.yaml
serviceaccount "heketi-service-account" created
```

部署 Heketi 服务：

```
heketi-deployment-svc.yaml
---
kind: Deployment
apiVersion: extensions/v1beta1
```

```yaml
metadata:
  name: deploy-heketi
  labels:
    glusterfs: heketi-deployment
    deploy-heketi: heketi-deployment
  annotations:
    description: Defines how to deploy Heketi
spec:
  replicas: 1
  template:
    metadata:
      name: deploy-heketi
      labels:
        name: deploy-heketi
        glusterfs: heketi-pod
    spec:
      serviceAccountName: heketi-service-account
      containers:
      - image: heketi/heketi
        name: deploy-heketi
        env:
        - name: HEKETI_EXECUTOR
          value: kubernetes
        - name: HEKETI_FSTAB
          value: "/var/lib/heketi/fstab"
        - name: HEKETI_SNAPSHOT_LIMIT
          value: '14'
        - name: HEKETI_KUBE_GLUSTER_DAEMONSET
          value: "y"
        ports:
        - containerPort: 8080
        volumeMounts:
        - name: db
          mountPath: "/var/lib/heketi"
        readinessProbe:
          timeoutSeconds: 3
          initialDelaySeconds: 3
          httpGet:
            path: "/hello"
            port: 8080
        livenessProbe:
```

```
          timeoutSeconds: 3
          initialDelaySeconds: 30
          httpGet:
            path: "/hello"
            port: 8080
      volumes:
      - name: db
        hostPath:
          path: "/heketi-data"
---
kind: Service
apiVersion: v1
metadata:
  name: deploy-heketi
  labels:
    glusterfs: heketi-service
    deploy-heketi: support
  annotations:
    description: Exposes Heketi Service
spec:
  selector:
    name: deploy-heketi
  ports:
  - name: deploy-heketi
    port: 8080
    targetPort: 8080
```

需要注意的是，Heketi 的 DB 数据需要持久化保存，建议使用 hostPath 或其他共享存储进行保存：

```
# kubectl create -f heketi-deployment-svc.yaml
deployment "deploy-heketi" created
service "deploy-heketi" created
```

8.6.4 为 Heketi 设置 GlusterFS 集群

在 Heketi 能够管理 GlusterFS 集群之前，首先要为其设置 GlusterFS 集群的信息。可以用一个 topology.json 配置文件来完成各个 GlusterFS 节点和设备的定义。Heketi 要求在一个 GlusterFS 集群中至少有 3 个节点。在 topology.json 配置文件 hostnames 字段的 manage 上填写主机名，在 storage 上填写 IP 地址，devices 要求为未创建文件系统的裸设备（可以有多块盘），以供 Heketi 自动完成 PV(Physical Volume)、VG(Volume Group)和 LV(Logical Volume)的创建。topology.json 文件的内容如下：

```json
{
  "clusters": [
    {
      "nodes": [
        {
          "node": {
            "hostnames": {
              "manage": [
                "k8s-node-1"
              ],
              "storage": [
                "192.168.18.3"
              ]
            },
            "zone": 1
          },
          "devices": [
            "/dev/sdb"
          ]
        },
        {
          "node": {
            "hostnames": {
              "manage": [
                "k8s-node-2"
              ],
              "storage": [
                "192.168.18.4"
              ]
            },
            "zone": 1
          },
          "devices": [
            "/dev/sdb"
          ]
        },
        {
          "node": {
```

```
              "hostnames": {
                "manage": [
                  "k8s-node-3"
                ],
                "storage": [
                  "192.168.18.5"
                ]
              },
              "zone": 1
            },
            "devices": [
              "/dev/sdb"
            ]
          }
        ]
      }
    ]
}
```

进入 Heketi 容器,使用命令行工具 heketi-cli 完成 GlusterFS 集群的创建:

```
# export HEKETI_CLI_SERVER=http://localhost:8080
# heketi-cli topology load --json=topology.json
Creating cluster ... ID: f643da1cd64691c5705932a46a95d1d5
        Creating node k8s-node-1 ... ID: 883506b091a22bd13f10bc3d0fb51223
                Adding device /dev/sdb ... OK
        Creating node k8s-node-2 ... ID: e64b879689106f82a9c4ac910a865cc8
                Adding device /dev/sdb ... OK
        Creating node k8s-node-3 ... ID: b7783484180f6a592a30baebfb97d9be
                Adding device /dev/sdb ... OK
```

经过这个操作,Heketi 完成了 GlusterFS 集群的创建,同时在 GlusterFS 集群的各个节点的/dev/sdb 盘上成功创建了 PV 和 VG。

查看 Heketi 的 topology 信息,可以看到 Node 和 Device 的详细信息,包括磁盘空间的大小和剩余空间。此时,Volume 和 Brick 还未创建:

```
# heketi-cli topology info
Cluster Id: f643da1cd64691c5705932a46a95d1d5

    Volumes:
```

```
    Nodes:

        Node Id: 883506b091a22bd13f10bc3d0fb51223
        State: online
        Cluster Id: f643da1cd64691c5705932a46a95d1d5
        Zone: 1
        Management Hostname: k8s-node-1
        Storage Hostname: 192.168.18.3
        Devices:
                Id:b474f14b0903ed03ec80d4a989f943f2    Name:/dev/sdb
State:online    Size (GiB):9     Used (GiB):0      Free (GiB):9
                Bricks:

        Node Id: b7783484180f6a592a30baebfb97d9be
        State: online
        Cluster Id: f643da1cd64691c5705932a46a95d1d5
        Zone: 1
        Management Hostname: k8s-node-3
        Storage Hostname: 192.168.18.5
        Devices:
                Id:fac3fa5ac1de3d5bde3aa68f6aa61285    Name:/dev/sdb
State:online    Size (GiB):9     Used (GiB):0      Free (GiB):9
                Bricks:

        Node Id: e64b879689106f82a9c4ac910a865cc8
        State: online
        Cluster Id: f643da1cd64691c5705932a46a95d1d5
        Zone: 1
        Management Hostname: k8s-node-2
        Storage Hostname: 192.168.18.4
        Devices:
                Id:05532e7db723953e8643b64b36aee1d1    Name:/dev/sdb
State:online    Size (GiB):9     Used (GiB):0      Free (GiB):9
                Bricks:
```

8.6.5 定义 StorageClass

准备工作已经就绪，集群管理员现在可以在 Kubernetes 集群中定义一个 StorageClass 了。storageclass-gluster-heketi.yaml 配置文件的内容如下：

```yaml
apiVersion: storage.k8s.io/v1
kind: StorageClass
metadata:
  name: gluster-heketi
provisioner: kubernetes.io/glusterfs
parameters:
  resturl: "http://172.17.2.2:8080"
  restauthenabled: "false"
```

Provisioner 参数必须被设置为 "kubernetes.io/glusterfs"。

resturl 的地址需要被设置为 API Server 所在主机可以访问到的 Heketi 服务的某个地址，可以使用服务 ClusterIP+端口号、容器 IP 地址+端口号，或将服务映射到物理机，使用物理机 IP+NodePort。

创建这个 StorageClass 资源对象：

```
# kubectl create -f storageclass-gluster-heketi.yaml
storageclass "gluster-heketi" created
```

8.6.6 定义 PVC

现在，用户可以申请一个 PVC 了。例如，一个用户申请一个 1GiB 空间的共享存储资源，StorageClass 使用 "gluster-heketi"，未定义任何 Selector，说明使用动态资源供应模式：

```yaml
pvc-gluster-heketi.yaml
kind: PersistentVolumeClaim
apiVersion: v1
metadata:
  name: pvc-gluster-heketi
spec:
  storageClassName: gluster-heketi
  accessModes:
    - ReadWriteOnce
  resources:
    requests:
      storage: 1Gi
# k create -f pvc-gluster-heketi.yaml
persistentvolumeclaim "pvc-gluster-heketi" created
```

PVC 的定义一旦生成，系统便将触发 Heketi 进行相应的操作，主要为在 GlusterFS 集群上创建 brick，再创建并启动一个 Volume。整个过程可以在 Heketi 的日志中查到：

```
......
    [kubeexec] DEBUG 2017/04/26 00:51:30
/src/github.com/heketi/heketi/executors/kubeexec/kubeexec.go:250: Host:
k8s-node-1 Pod: glusterfs-ld7nh Command: gluster --mode=script volume create
vol_87b9314cb76bafacfb7e9cdc04fcaf05 replica 3
192.168.18.3:/var/lib/heketi/mounts/vg_b474f14b0903ed03ec80d4a989f943f2/brick_d0
8520c9ff7b9a0a9165f9815671f2cd/brick
192.168.18.5:/var/lib/heketi/mounts/vg_fac3fa5ac1de3d5bde3aa68f6aa61285/brick_68
18dce118b8a54e9590199d44a3817b/brick
192.168.18.4:/var/lib/heketi/mounts/vg_05532e7db723953e8643b64b36aee1d1/brick_9e
cb8f7fde1ae937011f04401e7c6c56/brick
    Result: volume create: vol_87b9314cb76bafacfb7e9cdc04fcaf05: success: please
start the volume to access data
......
    [kubeexec] DEBUG 2017/04/26 00:51:33
/src/github.com/heketi/heketi/executors/kubeexec/kubeexec.go:250: Host:
k8s-node-1 Pod: glusterfs-ld7nh Command: gluster --mode=script volume start
vol_87b9314cb76bafacfb7e9cdc04fcaf05
    Result: volume start: vol_87b9314cb76bafacfb7e9cdc04fcaf05: success
......
```

查看 PVC 的状态，可见其已经为 Bound（已绑定）：

```
# kubectl get pvc
NAME                  STATUS    VOLUME                                        CAPACITY   ACCESSMODES   STORAGECLASS      AGE
pvc-gluster-heketi    Bound     pvc-783cf949-2a1a-11e7-8717-000c29eaed40      1Gi        RWX           gluster-heketi    6m
```

查看 PV，可见系统自动创建的 PV：

```
# kubectl get pv
NAME                                         CAPACITY   ACCESSMODES   RECLAIMPOLICY   STATUS   CLAIM                        STORAGECLASS      REASON   AGE
pvc-783cf949-2a1a-11e7-8717-000c29eaed40     1Gi        RWX           Delete          Bound    default/pvc-gluster-heketi   gluster-heketi             6m
```

查看该 PV 的详细信息，可以看到其容量、引用的 StorageClass 等信息都已正确设置，状态也为 Bound，回收策略则为默认的 Delete。同时 Gluster 的 Endpoint 和 Path 也由 Heketi 自动完成了设置：

```
# kubectl describe pv pvc-783cf949-2a1a-11e7-8717-000c29eaed40
Name:            pvc-783cf949-2a1a-11e7-8717-000c29eaed40
Labels:          <none>
Annotations:     pv.beta.kubernetes.io/gid=2000
                 pv.kubernetes.io/bound-by-controller=yes
                 pv.kubernetes.io/provisioned-by=kubernetes.io/glusterfs
StorageClass:    gluster-heketi
Status:          Bound
Claim:           default/pvc-gluster-heketi
Reclaim Policy:  Delete
Access Modes:    RWX
Capacity:        1Gi
Message:
Source:
    Type:              Glusterfs (a Glusterfs mount on the host that shares a pod's lifetime)
    EndpointsName:     glusterfs-dynamic-pvc-gluster-heketi
    Path:              vol_87b9314cb76bafacfb7e9cdc04fcaf05
    ReadOnly:          false
Events:          <none>
```

至此，一个可供 Pod 使用的 PVC 就创建成功了。接下来 Pod 就能通过 Volume 的设置将这个 PVC 挂载到容器内部进行使用了。

8.6.7 Pod 使用 PVC 的存储资源

在 Pod 中使用 PVC 定义的存储资源非常容易，只需设置一个 Volume，其类型为 persistentVolumeClaim，即可轻松引用一个 PVC。下例中使用一个 busybox 容器验证对 PVC 的使用，注意 Pod 需要与 PVC 属于同一个 Namespace：

```
pod-use-pvc.yaml
apiVersion: v1
kind: Pod
metadata:
  name: pod-use-pvc
spec:
  containers:
  - name: pod-use-pvc
    image: busybox
    command:
```

```
      - sleep
      - "3600"
    volumeMounts:
    - name: gluster-volume
      mountPath: "/pv-data"
      readOnly: false
  volumes:
  - name: gluster-volume
    persistentVolumeClaim:
      claimName: pvc-gluster-heketi

# kubectl create -f pod-use-pvc.yaml
pod "pod-use-pvc" created
```

进入容器 pod-use-pvc，在/pv-data 目录下创建一些文件：

```
# kubectl exec -ti pod-use-pvc -- /bin/sh
/ # cd /pv-data
/ # touch a
/ # echo "hello" > b
```

可以验证文件 a 和 b 在 GlusterFS 集群中是否正确生成。

至此，使用 Kubernetes 最新的动态存储供应模式，配合 StorageClass 和 Heketi 共同搭建基于 GlusterFS 的共享存储就完成了。有兴趣的读者可以继续尝试 StorageClass 的其他设置，例如调整 GlusterFS 的 Volume 类型、修改 PV 的回收策略等。

在使用动态存储供应模式的情况下，相对于静态模式的优势至少包括如下两点。

（1）管理员无须预先创建大量的 PV 作为存储资源。

（2）用户在申请 PVC 时无法保证容量与预置 PV 的容量完全匹配。从 Kubernetes 1.6 版本开始，建议用户优先考虑使用 StorageClass 的动态存储供应模式进行存储管理。

8.7 CSI 存储机制详解

Kubernetes 从 1.9 版本开始引入容器存储接口 Container Storage Interface（CSI）机制，用于在 Kubernetes 和外部存储系统之间建立一套标准的存储管理接口，通过该接口为容器提供存储服务。CSI 到 Kubernetes 1.10 版本升级为 Beta 版，到 Kubernetes 1.13 版本升级为 GA 版，已逐渐成熟。

8.7.1 CSI 的设计背景

Kubernetes 通过 PV、PVC、Storageclass 已经提供了一种强大的基于插件的存储管理机制,但是各种存储插件提供的存储服务都是基于一种被称为 "in-true"(树内)的方式提供的,这要求存储插件的代码必须被放进 Kubernetes 的主干代码库中才能被 Kubernetes 调用,属于紧耦合的开发模式。这种 "in-tree" 方式会带来一些问题:

◎ 存储插件的代码需要与 Kubernetes 的代码放在同一代码库中,并与 Kubernetes 的二进制文件共同发布;
◎ 存储插件代码的开发者必须遵循 Kubernetes 的代码开发规范;
◎ 存储插件代码的开发者必须遵循 Kubernetes 的发布流程,包括添加对 Kubernetes 存储系统的支持和错误修复;
◎ Kubernetes 社区需要对存储插件的代码进行维护,包括审核、测试等工作;
◎ 存储插件代码中的问题可能会影响 Kubernetes 组件的运行,并且很难排查问题;
◎ 存储插件代码与 Kubernetes 的核心组件(kubelet 和 kube-controller-manager)享有相同的系统特权权限,可能存在可靠性和安全性问题。

Kubernetes 已有的 Flex Volume 插件机制试图通过为外部存储暴露一个基于可执行程序(exec)的 API 来解决这些问题。尽管它允许第三方存储提供商在 Kubernetes 核心代码之外开发存储驱动,但仍然有两个问题没有得到很好的解决:

◎ 部署第三方驱动的可执行文件仍然需要宿主机的 root 权限,存在安全隐患;
◎ 存储插件在执行 mount、attach 这些操作时,通常需要在宿主机上安装一些第三方工具包和依赖库,使得部署过程更加复杂,例如部署 Ceph 时需要安装 rbd 库,部署 GlusterFS 时需要安装 mount.glusterfs 库,等等。

基于以上这些问题和考虑,Kubernetes 逐步推出与容器对接的存储接口标准,存储提供方只需要基于标准接口进行存储插件的实现,就能使用 Kubernetes 的原生存储机制为容器提供存储服务。这套标准被称为 CSI(容器存储接口)。在 CSI 成为 Kubernetes 的存储供应标准之后,存储提供方的代码就能和 Kubernetes 代码彻底解耦,部署也与 Kubernetes 核心组件分离,显然,存储插件的开发由提供方自行维护,就能为 Kubernetes 用户提供更多的存储功能,也更加安全可靠。基于 CSI 的存储插件机制也被称为 "out-of-tree"(树外)的服务提供方式,是未来 Kubernetes 第三方存储插件的标准方案。

8.7.2 CSI 存储插件的关键组件和部署架构

图 8.5 描述了 Kubernetes CSI 存储插件的关键组件和推荐的容器化部署架构。

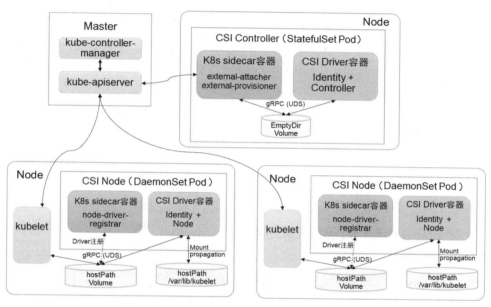

图 8.5 Kubernetes CSI 存储插件的关键组件和推荐的容器化部署架构

其中主要包括两种组件：CSI Controller 和 CSI Node。

1. CSI Controller

CSI Controller 的主要功能是提供存储服务视角对存储资源和存储卷进行管理和操作。在 Kubernetes 中建议将其部署为单实例 Pod，可以使用 StatefulSet 或 Deployment 控制器进行部署，设置副本数量为 1，保证为一种存储插件只运行一个控制器实例。

在这个 Pod 内部署两个容器，如下所述。

（1）与 Master（kube-controller-manager）通信的辅助 sidecar 容器。

在 sidecar 容器内又可以包含 external-attacher 和 external-provisioner 两个容器，它们的功能分别如下。

◎ external-attacher：监控 VolumeAttachment 资源对象的变更，触发针对 CSI 端点的

ControllerPublish 和 ControllerUnpublish 操作。
- ◎ external-provisioner：监控 PersistentVolumeClaim 资源对象的变更，触发针对 CSI 端点的 CreateVolume 和 DeleteVolume 操作。

（2）CSI Driver 存储驱动容器，由第三方存储提供商提供，需要实现上述接口。

这两个容器通过本地 Socket（Unix Domain Socket，UDS），并使用 gPRC 协议进行通信。sidecar 容器通过 Socket 调用 CSI Driver 容器的 CSI 接口，CSI Driver 容器负责具体的存储卷操作。

2. CSI Node

CSI Node 的主要功能是对主机（Node）上的 Volume 进行管理和操作。在 Kubernetes 中建议将其部署为 DaemonSet，在每个 Node 上都运行一个 Pod。

在这个 Pod 中部署以下两个容器：

（1）与 kubelet 通信的辅助 sidecar 容器 node-driver-registrar，主要功能是将存储驱动注册到 kubelet 中；

（2）CSI Driver 存储驱动容器，由第三方存储提供商提供，主要功能是接收 kubelet 的调用，需要实现一系列与 Node 相关的 CSI 接口，例如 NodePublishVolume 接口（用于将 Volume 挂载到容器内的目标路径）、NodeUnpublishVolume 接口（用于从容器中卸载 Volume），等等。

node-driver-registrar 容器与 kubelet 通过 Node 主机的一个 hostPath 目录下的 unix socket 进行通信。CSI Driver 容器与 kubelet 通过 Node 主机的另一个 hostPath 目录下的 unix socket 进行通信，同时需要将 kubelet 的工作目录（默认为/var/lib/kubelet）挂载给 CSI Driver 容器，用于为 Pod 进行 Volume 的管理操作（包括 mount、umount 等）。

8.7.3 CSI 存储插件的使用示例

下面以 csi-hostpath 插件为例，对如何部署 CSI 插件、用户如何使用 CSI 插件提供的存储资源进行详细说明。

（1）设置 Kubernetes 服务启动参数。为 kube-apiserver、kube-controller-manager 和 kubelet 服务的启动参数添加：

```
--feature-gates=VolumeSnapshotDataSource=true,CSINodeInfo=true,CSIDriverRegi
stry=true
```

这 3 个特性开关是 Kubernetes 从 1.12 版本引入的 Alpha 版功能，CSINodeInfo 和 CSIDriverRegistry 需要手工创建其相应的 CRD 资源对象。

Kubernetes 1.10 版本所需的 CSIPersistentVolume 和 MountPropagation 特性开关已经默认启用，KubeletPluginsWatcher 特性开关也在 Kubernetes 1.12 版本中默认启用，无须在命令行参数中指定。

（2）创建 CSINodeInfo 和 CSIDriverRegistry CRD 资源对象。

csidriver.yaml 的内容如下：

```
apiVersion: apiextensions.k8s.io/v1beta1
kind: CustomResourceDefinition
metadata:
  name: csidrivers.csi.storage.k8s.io
  labels:
    addonmanager.kubernetes.io/mode: Reconcile
spec:
  group: csi.storage.k8s.io
  names:
    kind: CSIDriver
    plural: csidrivers
  scope: Cluster
  validation:
   openAPIV3Schema:
     properties:
       spec:
         description: Specification of the CSI Driver.
         properties:
           attachRequired:
             description: Indicates this CSI volume driver requires an attach operation,and that Kubernetes should call attach and wait for any attach operationto complete before proceeding to mount.
             type: boolean
           podInfoOnMountVersion:
             description: Indicates this CSI volume driver requires additional pod
               information (like podName, podUID, etc.) during mount operations.
             type: string
```

```
    version: v1alpha1
```

csinodeinfo.yaml 的内容如下：

```
apiVersion: apiextensions.k8s.io/v1beta1
kind: CustomResourceDefinition
metadata:
  name: csinodeinfos.csi.storage.k8s.io
  labels:
    addonmanager.kubernetes.io/mode: Reconcile
spec:
  group: csi.storage.k8s.io
  names:
    kind: CSINodeInfo
    plural: csinodeinfos
  scope: Cluster
  validation:
    openAPIV3Schema:
      properties:
        spec:
          description: Specification of CSINodeInfo
          properties:
            drivers:
              description: List of CSI drivers running on the node and their specs.
              type: array
              items:
                properties:
                  name:
                    description: The CSI driver that this object refers to.
                    type: string
                  nodeID:
                    description: The node from the driver point of view.
                    type: string
                  topologyKeys:
                    description: List of keys supported by the driver.
                    items:
                      type: string
                    type: array
        status:
          description: Status of CSINodeInfo
          properties:
            drivers:
```

```
                    description: List of CSI drivers running on the node and their
statuses.
                    type: array
                    items:
                      properties:
                        name:
                          description: The CSI driver that this object refers to.
                          type: string
                        available:
                          description: Whether the CSI driver is installed.
                          type: boolean
                        volumePluginMechanism:
                          description: Indicates to external components the required
mechanism
                            to use for any in-tree plugins replaced by this driver.
                          pattern: in-tree|csi
                          type: string
      version: v1alpha1
```

使用 kubectl create 命令完成创建：

```
# kubectl create -f csidriver.yaml
customresourcedefinition.apiextensions.k8s.io/csidrivers.csi.storage.k8s.io created
# kubectl create -f csinodeinfo.yaml
customresourcedefinition.apiextensions.k8s.io/csinodeinfos.csi.storage.k8s.io created
```

（3）创建 csi-hostpath 存储插件相关组件，包括 csi-hostpath-attacher、csi-hostpath-provisioner 和 csi-hostpathplugin（其中包含 csi-node-driver-registrar 和 hostpathplugin）。其中为每个组件都配置了相应的 RBAC 权限控制规则，对于安全访问 Kubernetes 资源对象非常重要。

csi-hostpath-attacher.yaml 的内容如下：

```
# RBAC 相关配置
---
apiVersion: v1
kind: ServiceAccount
metadata:
  name: csi-attacher
  # replace with non-default namespace name
```

```yaml
  namespace: default
---
# Attacher must be able to work with PVs, nodes and VolumeAttachments
kind: ClusterRole
apiVersion: rbac.authorization.k8s.io/v1
metadata:
  name: external-attacher-runner
rules:
  - apiGroups: [""]
    resources: ["persistentvolumes"]
    verbs: ["get", "list", "watch", "update"]
  - apiGroups: [""]
    resources: ["nodes"]
    verbs: ["get", "list", "watch"]
  - apiGroups: ["csi.storage.k8s.io"]
    resources: ["csinodeinfos"]
    verbs: ["get", "list", "watch"]
  - apiGroups: ["storage.k8s.io"]
    resources: ["volumeattachments"]
    verbs: ["get", "list", "watch", "update"]
---
kind: ClusterRoleBinding
apiVersion: rbac.authorization.k8s.io/v1
metadata:
  name: csi-attacher-role
subjects:
  - kind: ServiceAccount
    name: csi-attacher
    # replace with non-default namespace name
    namespace: default
roleRef:
  kind: ClusterRole
  name: external-attacher-runner
  apiGroup: rbac.authorization.k8s.io
---
# Attacher must be able to work with config map in current namespace
# if (and only if) leadership election is enabled
kind: Role
apiVersion: rbac.authorization.k8s.io/v1
metadata:
  # replace with non-default namespace name
```

```yaml
    namespace: default
    name: external-attacher-cfg
rules:
- apiGroups: [""]
  resources: ["configmaps"]
  verbs: ["get", "watch", "list", "delete", "update", "create"]
---
kind: RoleBinding
apiVersion: rbac.authorization.k8s.io/v1
metadata:
  name: csi-attacher-role-cfg
  # replace with non-default namespace name
  namespace: default
subjects:
  - kind: ServiceAccount
    name: csi-attacher
    # replace with non-default namespace name
    namespace: default
roleRef:
  kind: Role
  name: external-attacher-cfg
  apiGroup: rbac.authorization.k8s.io

# Service 和 StatefulSet 的定义
---
kind: Service
apiVersion: v1
metadata:
  name: csi-hostpath-attacher
  labels:
    app: csi-hostpath-attacher
spec:
  selector:
    app: csi-hostpath-attacher
  ports:
    - name: dummy
      port: 12345
---
kind: StatefulSet
apiVersion: apps/v1
metadata:
```

```yaml
    name: csi-hostpath-attacher
spec:
  serviceName: "csi-hostpath-attacher"
  replicas: 1
  selector:
    matchLabels:
      app: csi-hostpath-attacher
  template:
    metadata:
      labels:
        app: csi-hostpath-attacher
    spec:
      serviceAccountName: csi-attacher
      containers:
        - name: csi-attacher
          image: quay.io/k8scsi/csi-attacher:v1.0.1
          imagePullPolicy: IfNotPresent
          args:
            - --v=5
            - --csi-address=$(ADDRESS)
          env:
            - name: ADDRESS
              value: /csi/csi.sock
          volumeMounts:
          - mountPath: /csi
            name: socket-dir
      volumes:
        - hostPath:
            path: /var/lib/kubelet/plugins/csi-hostpath
            type: DirectoryOrCreate
          name: socket-dir
```

csi-hostpath-provisioner.yaml 的内容如下:

```yaml
# RBAC 相关配置
---
apiVersion: v1
kind: ServiceAccount
metadata:
  name: csi-provisioner
  # replace with non-default namespace name
  namespace: default
```

```yaml
---
kind: ClusterRole
apiVersion: rbac.authorization.k8s.io/v1
metadata:
  name: external-provisioner-runner
rules:
  - apiGroups: [""]
    resources: ["secrets"]
    verbs: ["get", "list"]
  - apiGroups: [""]
    resources: ["persistentvolumes"]
    verbs: ["get", "list", "watch", "create", "delete"]
  - apiGroups: [""]
    resources: ["persistentvolumeclaims"]
    verbs: ["get", "list", "watch", "update"]
  - apiGroups: ["storage.k8s.io"]
    resources: ["storageclasses"]
    verbs: ["get", "list", "watch"]
  - apiGroups: [""]
    resources: ["events"]
    verbs: ["list", "watch", "create", "update", "patch"]
  - apiGroups: ["snapshot.storage.k8s.io"]
    resources: ["volumesnapshots"]
    verbs: ["get", "list"]
  - apiGroups: ["snapshot.storage.k8s.io"]
    resources: ["volumesnapshotcontents"]
    verbs: ["get", "list"]
  - apiGroups: ["csi.storage.k8s.io"]
    resources: ["csinodeinfos"]
    verbs: ["get", "list", "watch"]
  - apiGroups: [""]
    resources: ["nodes"]
    verbs: ["get", "list", "watch"]
---
kind: ClusterRoleBinding
apiVersion: rbac.authorization.k8s.io/v1
metadata:
  name: csi-provisioner-role
subjects:
  - kind: ServiceAccount
    name: csi-provisioner
```

```yaml
    # replace with non-default namespace name
    namespace: default
roleRef:
  kind: ClusterRole
  name: external-provisioner-runner
  apiGroup: rbac.authorization.k8s.io
---
# Provisioner must be able to work with endpoints in current namespace
# if (and only if) leadership election is enabled
kind: Role
apiVersion: rbac.authorization.k8s.io/v1
metadata:
  # replace with non-default namespace name
  namespace: default
  name: external-provisioner-cfg
rules:
- apiGroups: [""]
  resources: ["endpoints"]
  verbs: ["get", "watch", "list", "delete", "update", "create"]
---
kind: RoleBinding
apiVersion: rbac.authorization.k8s.io/v1
metadata:
  name: csi-provisioner-role-cfg
  # replace with non-default namespace name
  namespace: default
subjects:
  - kind: ServiceAccount
    name: csi-provisioner
    # replace with non-default namespace name
    namespace: default
roleRef:
  kind: Role
  name: external-provisioner-cfg
  apiGroup: rbac.authorization.k8s.io

# Service 和 StatefulSet 的定义
---
kind: Service
apiVersion: v1
metadata:
```

```yaml
  name: csi-hostpath-provisioner
  labels:
    app: csi-hostpath-provisioner
spec:
  selector:
    app: csi-hostpath-provisioner
  ports:
    - name: dummy
      port: 12345
---
kind: StatefulSet
apiVersion: apps/v1
metadata:
  name: csi-hostpath-provisioner
spec:
  serviceName: "csi-hostpath-provisioner"
  replicas: 1
  selector:
    matchLabels:
      app: csi-hostpath-provisioner
  template:
    metadata:
      labels:
        app: csi-hostpath-provisioner
    spec:
      serviceAccountName: csi-provisioner
      containers:
        - name: csi-provisioner
          image: quay.io/k8scsi/csi-provisioner:v1.0.1
          imagePullPolicy: IfNotPresent
          args:
            - "--provisioner=csi-hostpath"
            - "--csi-address=$(ADDRESS)"
            - "--connection-timeout=15s"
          env:
            - name: ADDRESS
              value: /csi/csi.sock
          volumeMounts:
            - mountPath: /csi
              name: socket-dir
      volumes:
```

```
            - hostPath:
                path: /var/lib/kubelet/plugins/csi-hostpath
                type: DirectoryOrCreate
              name: socket-dir
```

csi-hostpathplugin.yaml 的内容如下：

```
# RBAC 相关配置
---
apiVersion: v1
kind: ServiceAccount
metadata:
  name: csi-node-sa
  # replace with non-default namespace name
  namespace: default

---
kind: ClusterRole
apiVersion: rbac.authorization.k8s.io/v1
metadata:
  name: driver-registrar-runner
rules:
  - apiGroups: [""]
    resources: ["events"]
    verbs: ["get", "list", "watch", "create", "update", "patch"]
  # The following permissions are only needed when running
  # driver-registrar without the --kubelet-registration-path
  # parameter, i.e. when using driver-registrar instead of
  # kubelet to update the csi.volume.kubernetes.io/nodeid
  # annotation. That mode of operation is going to be deprecated
  # and should not be used anymore, but is needed on older
  # Kubernetes versions.
  # - apiGroups: [""]
  #   resources: ["nodes"]
  #   verbs: ["get", "update", "patch"]

---
kind: ClusterRoleBinding
apiVersion: rbac.authorization.k8s.io/v1
metadata:
  name: csi-driver-registrar-role
subjects:
```

```yaml
    - kind: ServiceAccount
      name: csi-node-sa
      # replace with non-default namespace name
      namespace: default
  roleRef:
    kind: ClusterRole
    name: driver-registrar-runner
    apiGroup: rbac.authorization.k8s.io

# DaemonSet 的定义
---
kind: DaemonSet
apiVersion: apps/v1
metadata:
  name: csi-hostpathplugin
spec:
  selector:
    matchLabels:
      app: csi-hostpathplugin
  template:
    metadata:
      labels:
        app: csi-hostpathplugin
    spec:
      serviceAccountName: csi-node-sa
      hostNetwork: true
      containers:
        - name: driver-registrar
          image: quay.io/k8scsi/csi-node-driver-registrar:v1.0.1
          imagePullPolicy: IfNotPresent
          args:
            - --v=5
            - --csi-address=/csi/csi.sock
            - --kubelet-registration-path=/var/lib/kubelet/plugins/csi-hostpath/csi.sock
          env:
            - name: KUBE_NODE_NAME
              valueFrom:
                fieldRef:
                  apiVersion: v1
                  fieldPath: spec.nodeName
```

```yaml
        volumeMounts:
        - mountPath: /csi
          name: socket-dir
        - mountPath: /registration
          name: registration-dir
      - name: hostpath
        image: quay.io/k8scsi/hostpathplugin:v1.0.1
        imagePullPolicy: IfNotPresent
        args:
          - "--v=5"
          - "--endpoint=$(CSI_ENDPOINT)"
          - "--nodeid=$(KUBE_NODE_NAME)"
        env:
          - name: CSI_ENDPOINT
            value: unix:///csi/csi.sock
          - name: KUBE_NODE_NAME
            valueFrom:
              fieldRef:
                apiVersion: v1
                fieldPath: spec.nodeName
        securityContext:
          privileged: true
        volumeMounts:
          - mountPath: /csi
            name: socket-dir
          - mountPath: /var/lib/kubelet/pods
            mountPropagation: Bidirectional
            name: mountpoint-dir
      volumes:
        - hostPath:
            path: /var/lib/kubelet/plugins/csi-hostpath
            type: DirectoryOrCreate
          name: socket-dir
        - hostPath:
            path: /var/lib/kubelet/pods
            type: DirectoryOrCreate
          name: mountpoint-dir
        - hostPath:
            path: /var/lib/kubelet/plugins_registry
            type: Directory
          name: registration-dir
```

使用 kubectl create 命令完成创建：

```
# kubectl create -f csi-hostpath-attacher.yaml
serviceaccount/csi-attacher created
clusterrole.rbac.authorization.k8s.io/external-attacher-runner created
clusterrolebinding.rbac.authorization.k8s.io/csi-attacher-role created
role.rbac.authorization.k8s.io/external-attacher-cfg created
rolebinding.rbac.authorization.k8s.io/csi-attacher-role-cfg created
service/csi-hostpath-attacher created
statefulset.apps/csi-hostpath-attacher created

# kubectl create -f csi-hostpath-provisioner.yaml
serviceaccount/csi-provisioner created
clusterrole.rbac.authorization.k8s.io/external-provisioner-runner created
clusterrolebinding.rbac.authorization.k8s.io/csi-provisioner-role created
role.rbac.authorization.k8s.io/external-provisioner-cfg created
rolebinding.rbac.authorization.k8s.io/csi-provisioner-role-cfg created
service/csi-hostpath-provisioner created
statefulset.apps/csi-hostpath-provisioner created

# kubectl create -f csi-hostpathplugin.yaml
serviceaccount/csi-node-sa created
clusterrole.rbac.authorization.k8s.io/driver-registrar-runner created
clusterrolebinding.rbac.authorization.k8s.io/csi-driver-registrar-role created
daemonset.apps/csi-hostpathplugin created
```

确保 3 个 Pod 都正常运行：

```
# kubectl get pods
NAME                          READY   STATUS    RESTARTS   AGE
csi-hostpath-attacher-0       1/1     Running   1          4m41s
csi-hostpath-provisioner-0    1/1     Running   0          84s
csi-hostpathplugin-t6qzs      2/2     Running   0          39s
```

至此就完成了 CSI 存储插件的部署。

（4）应用容器使用 CSI 存储。应用程序如果希望使用 CSI 存储插件提供的存储服务，则仍然使用 Kubernetes 动态存储管理机制。首先通过创建 StorageClass 和 PVC 为应用容器准备存储资源，然后容器就可以挂载 PVC 到容器内的目录进行使用了。

创建一个 StorageClass,provisioner 为 CSI 存储插件的类型,在本例中为 csi-hostpath:

```
csi-storageclass.yaml
apiVersion: storage.k8s.io/v1
kind: StorageClass
metadata:
  name: csi-hostpath-sc
provisioner: csi-hostpath
reclaimPolicy: Delete
volumeBindingMode: Immediate

# kubectl create -f csi-storageclass.yaml
storageclass.storage.k8s.io/csi-hostpath-sc created
```

创建一个 PVC,引用刚刚创建的 StorageClass,申请存储空间为 1GiB:

```
csi-pvc.yaml
apiVersion: v1
kind: PersistentVolumeClaim
metadata:
  name: csi-pvc
spec:
  accessModes:
  - ReadWriteOnce
  resources:
    requests:
      storage: 1Gi
  storageClassName: csi-hostpath-sc

# kubectl create -f csi-pvc.yaml
persistentvolumeclaim/csi-pvc created
```

查看 PVC 和系统自动创建的 PV,状态为 Bound,说明创建成功:

```
# kubectl get pvc
NAME       STATUS    VOLUME                                     CAPACITY   ACCESS MODES   STORAGECLASS      AGE
csi-pvc    Bound     pvc-f8923093-3e25-11e9-a5fa-000c29069202   1Gi        RWO            csi-hostpath-sc   40s

# kubectl get pv
NAME                                       CAPACITY   ACCESS MODES   RECLAIM POLICY   STATUS   CLAIM              STORAGECLASS      REASON   AGE
```

```
                    pvc-f8923093-3e25-11e9-a5fa-000c29069202     1Gi          RWO            Delete
Bound      default/csi-pvc      csi-hostpath-sc                  42s
```

最后，在应用容器的配置中使用该 PVC：

```
csi-app.yaml
kind: Pod
apiVersion: v1
metadata:
  name: my-csi-app
spec:
  containers:
    - name: my-csi-app
      image: busybox
      imagePullPolicy: IfNotPresent
      command: [ "sleep", "1000000" ]
      volumeMounts:
      - mountPath: "/data"
        name: my-csi-volume
  volumes:
    - name: my-csi-volume
      persistentVolumeClaim:
        claimName: csi-pvc


# k create -f csi-app.yaml
pod/my-csi-app created

# kubectl get pods
NAME                         READY     STATUS      RESTARTS      AGE
my-csi-app                   1/1       Running     0             40s
```

在 Pod 创建成功之后，应用容器中的/data 目录使用的就是 CSI 存储插件提供的存储。

我们通过 kubelet 的日志可以查看到 Volume 挂载的详细过程：

```
   I0304 10:39:27.408018    29488 operation_generator.go:1196] Controller attach
succeeded for volume "pvc-f8923093-3e25-11e9-a5fa-000c29069202" (UniqueName:
"kubernetes.io/csi/csi-hostpath^f89c8e8e-3e25-11e9-8d66-000c29069202") pod
"my-csi-app" (UID: "b624c688-3e26-11e9-a5fa-000c29069202") device path:
"csi-43a8c0897d21520e942e9ceea0b1ddac36c8c462d726780bed5f50841f0b0871"
   I0304 10:39:27.501816    29488 operation_generator.go:501]
MountVolume.WaitForAttach entering for volume
"pvc-f8923093-3e25-11e9-a5fa-000c29069202" (UniqueName:
```

```
"kubernetes.io/csi/csi-hostpath^f89c8e8e-3e25-11e9-8d66-000c29069202") pod
"my-csi-app" (UID: "b624c688-3e26-11e9-a5fa-000c29069202") DevicePath
"csi-43a8c0897d21520e942e9ceea0b1ddac36c8c462d726780bed5f50841f0b0871"
    I0304 10:39:27.504542    29488 operation_generator.go:510]
MountVolume.WaitForAttach succeeded for volume
"pvc-f8923093-3e25-11e9-a5fa-000c29069202" (UniqueName:
"kubernetes.io/csi/csi-hostpath^f89c8e8e-3e25-11e9-8d66-000c29069202") pod
"my-csi-app" (UID: "b624c688-3e26-11e9-a5fa-000c29069202") DevicePath
"csi-43a8c0897d21520e942e9ceea0b1ddac36c8c462d726780bed5f50841f0b0871"
    I0304 10:39:27.506867    29488 csi_attacher.go:360] kubernetes.io/csi:
attacher.MountDevice STAGE_UNSTAGE_VOLUME capability not set. Skipping
MountDevice...
    I0304 10:39:27.506894    29488 operation_generator.go:531]
MountVolume.MountDevice succeeded for volume
"pvc-f8923093-3e25-11e9-a5fa-000c29069202" (UniqueName:
"kubernetes.io/csi/csi-hostpath^f89c8e8e-3e25-11e9-8d66-000c29069202") pod
"my-csi-app" (UID: "b624c688-3e26-11e9-a5fa-000c29069202") device mount path
"/var/lib/kubelet/plugins/kubernetes.io/csi/pv/pvc-f8923093-3e25-11e9-a5fa-000c2
9069202/globalmount"
```

8.7.4 CSI 的发展

目前可用于生产环境的 CSI 插件如表 8.2 所示。

表 8.2 目前可用于生产环境的 CSI 插件里列表

名称	状态/版本号
Alicloud Elastic Block Storage	v1.0.0
Alicloud Elastic File System	v1.0.0
Alicloud OSS	v1.0.0
AWS Elastic Block Storage	v0.2.0
AWS Elastic File System	v0.1.0
AWS FSx for Lustre	v0.1.0
Azure disk	v0.1.0（alpha）
Azure file	v0.1.0（alpha）
CephFS	v1.0.0
Cinder	v0.2.0

续表

名　　称	状态/版本号
Datera	v1.0.0
DigitalOcean Block Storage	v0.4.0
DriveScale	v1.0.0
Ember CSI	v0.2.0（alpha）
GCE Persistent Disk	Beta
Google Cloud Filestore	Alpha
GlusterFS	v1.0.0
Hitachi Vantara	v2.0
Linode Block Storage	v0.0.3
LINSTOR	v0.3.0
MapR	v1.0.0
MooseFS	v0.0.1（alpha）
NetApp	v0.2.0（alpha）
NexentaStor	Beta
Nutanix	beta
OpenSDS	Beta
Portworx	0.3.0
Quobyte	v0.2.0
RBD	v1.0.0
ScaleIO	v0.1.0
StorageOS	v1.0.0
XSKY	Beta
Vault	Alpha
vSphere	v0.1.0
YanRongYun	v1.0.0

实验性的 CSI 插件如表 8.3 所示。

表 8.3 实验性的 CSI 插件列表

名　称	状态/版本号
Flexvolume	示例
HostPath	v0.2.0
In-memory Sample Mock Driver	v0.3.0
NFS	示例
VFS Driver	Released

每种 CSI 存储插件都提供了容器镜像，与 external-attacher、external-provisioner、node-driver-registrar 等 sidecar 辅助容器共同完成存储插件系统的部署，每个插件的部署配置详见官网 https://kubernetes-csi.github.io/docs/drivers.html 中的链接。

Kubernetes 从 1.12 版本开始引入存储卷快照（Volume Snapshots）功能，通过新的 CRD 自定义资源对象 VolumeSnapshotContent、VolumeSnapshot 和 VolumeSnapshotClass 进行管理。VolumeSnapshotContent 定义从当前 PV 创建的快照，类似于一个新的 PV；VolumeSnapshot 定义需要绑定某个快照的请求，类似于 PVC 的定义；VolumeSnapshotClass 用于屏蔽 VolumeSnapshotContent 的细节，类似于 StorageClass 的功能。

下面是一个 VolumeSnapshotContent 的例子：

```
apiVersion: snapshot.storage.k8s.io/v1alpha1
kind: VolumeSnapshotContent
metadata:
  name: new-snapshot-content-test
spec:
  snapshotClassName: csi-hostpath-snapclass
  source:
    name: pvc-test
    kind: PersistentVolumeClaim
  volumeSnapshotSource:
    csiVolumeSnapshotSource:
      creationTime:    1535478900692119403
      driver:          csi-hostpath
      restoreSize:     10Gi
      snapshotHandle:  7bdd0de3-aaeb-11e8-9aae-0242ac110002
```

下面是一个 VolumeSnapshot 的例子：

```yaml
apiVersion: snapshot.storage.k8s.io/v1alpha1
kind: VolumeSnapshot
metadata:
  name: new-snapshot-test
spec:
  snapshotClassName: csi-hostpath-snapclass
  source:
    name: pvc-test
    kind: PersistentVolumeClaim
```

后续要进一步完善的工作如下。

（1）将以下 Alpha 版功能更新到 Beta 版：

◎ Raw Block 类型的 Volumes；
◎ 拓扑感知，Kubernetes 理解和影响 CSI 卷的配置位置（如 zone、region 等）的能力；
◎ 完善基于 CRD 的扩展功能（例如 Skip attach、Pod info on mount 等）。

（2）完善对本地短暂卷（Local Ephemeral Volume）的支持。

（3）将 Kubernetes "in-tree" 存储卷插件迁移到 CSI。

第 9 章

Kubernetes 开发指南

本章将引入 REST 的概念，详细说明 Kubernetes API 的概念和使用方法，并举例说明如何基于 Jersey 和 Fabric8 框架访问 Kubernetes API，深入分析基于这两个框架访问 Kubernetes API 的优缺点，最后对 Kubernetes API 的扩展进行详细说明。下面从 REST 开始说起。

9.1 REST 简述

REST（Representational State Transfer，表述性状态传递）是由 Roy Thomas Fielding 博士在他的论文 *Architectural Styles and the Design of Network-based Software Architectures* 中提出的一个术语。REST 本身只是为分布式超媒体系统设计的一种架构风格，而不是标准。

基于 Web 的架构实际上就是各种规范的集合，比如 HTTP 是一种规范，客户端服务器模式是另一种规范。每当我们在原有规范的基础上增加新的规范时，就会形成新的架构。而 REST 正是这样一种架构，它结合了一系列规范，形成了一种新的基于 Web 的架构风格。

传统的 Web 应用大多是 B/S 架构，涉及如下规范。

（1）客户端-服务器：这种规范的提出，改善了用户接口跨多个平台的可移植性，并且通过简化服务器组件，改善了系统的可伸缩性。最为关键的是通过分离用户接口和数据存储，使得不同的用户终端共享相同的数据成为可能。

（2）无状态性：无状态性是在客户端-服务器规范的基础上添加的又一层规范，它要求通信必须在本质上是无状态的，即从客户端到服务器的每个 request 都必须包含理解该 request 必需的所有信息。这个规范改善了系统的可见性（无状态性使得客户端和服务器端不必保存对方的详细信息，服务器只需要处理当前的 request，而不必了解所有 request 的历史）、可靠性（无状态性减少了服务器从局部错误中恢复的任务量）、可伸缩性（无状态性使得服务器端可以很容易释放资源，因为服务器端不必在多个 request 中保存状态）。同时，这种规范的缺点也是显而易见的，不能将状态数据保存在服务器上，导致增加了在一系列 request 中发送重复数据的开销，严重降低了效率。

（3）缓存：为了改善无状态性带来的网络的低效性，客户端缓存规范出现。缓存规范允许隐式或显式地标记一个 response 中的数据，赋予了客户端缓存 response 数据的功能，这样就可以为以后的 request 共用缓存的数据消除部分或全部交互，提高了网络效率。但是客户端缓存了信息，所以客户端数据与服务器数据不一致的可能性增加，从而降低了可靠性。

B/S 架构的优点是部署非常方便，在用户体验方面却不很理想。为了改善这种状况，REST 规范出现。REST 规范在原有 B/S 架构的基础上增加了三个新规范：统一接口、分层系统和按需代码。

　　（1）统一接口：REST 架构风格的核心特征就是强调组件之间有一个统一的接口，表现为在 REST 世界里，网络上的所有事物都被抽象为资源，REST 通过通用的链接器接口对资源进行操作。这样设计的好处是保证系统提供的服务都是解耦的，可极大简化系统，改善系统的交互性和可重用性。

　　（2）分层系统：分层系统规则的加入提高了各种层次之间的独立性，为整个系统的复杂性设置了边界，通过封装遗留的服务，使新的服务器免受遗留客户端的影响，也提高了系统的可伸缩性。

　　（3）按需代码：REST 允许对客户端的功能进行扩展。比如，通过下载并执行 applet 或脚本形式的代码来扩展客户端的功能。但这在改善系统可扩展性的同时降低了可见性，所以它只是 REST 的一个可选约束。

　　REST 架构是针对 Web 应用而设计的，其目的是为了降低开发的复杂性，提高系统的可伸缩性。REST 提出了如下设计准则。

　　（1）网络上的所有事物都被抽象为资源（Resource）。

　　（2）每个资源都对应唯一的资源标识符（Resource Identifier）。

　　（3）通过通用的连接器接口（Generic Connector Interface）对资源进行操作。

　　（4）对资源的各种操作都不会改变资源标识符。

　　（5）所有操作都是无状态的（Stateless）。

　　REST 中的资源指的不是数据，而是数据和表现形式的组合，比如"最新访问的 10 位会员"和"最活跃的 10 位会员"在数据上可能有重叠或者完全相同，而它们由于表现形式不同，被归为不同的资源，这也就是为什么 REST 的全名是 Representational State Transfer。资源标识符就是 URI（Uniform Resource Identifier），不管是图片、Word 还是视频文件，甚至只是一种虚拟的服务，也不管是 XML、TXT 还是其他文件格式，全部通过 URI 对资源进行唯一标识。

　　REST 是基于 HTTP 的，任何对资源的操作行为都通过 HTTP 来实现。以往的 Web 开发大多数用的是 HTTP 中的 GET 和 POST 方法，很少使用其他方法，这实际上是对 HTTP

的片面理解造成的。HTTP 不仅仅是一个简单的运载数据的协议，还是一个具有丰富内涵的网络软件的协议，它不仅能对互联网资源进行唯一定位，还能告诉我们如何对该资源进行操作。HTTP 把对一个资源的操作限制在 4 种方法（GET、POST、PUT 和 DELETE）中，这正是对资源 CRUD 操作的实现。由于资源和 URI 是一一对应的，在执行这些操作时 URI 没有变化，和以往的 Web 开发有很大的区别，所以极大地简化了 Web 开发，也使得 URI 可以被设计成能更直观地反映资源的结构。这种 URI 的设计被称作 RESTful 的 URI，为开发人员引入了一种新的思维方式：通过 URL 来设计系统结构。当然，这种设计方式对于一些特定情况也是不适用的，也就是说不是所有 URI 都适用于 RESTful。

REST 之所以可以提高系统的可伸缩性，就是因为它要求所有操作都是无状态的。没有了上下文（Context）的约束，做分布式和集群时就更为简单，也可以让系统更为有效地利用缓冲池（Pool），并且由于服务器端不需要记录客户端的一系列访问，也就减少了服务器端的性能损耗。

Kubernetes API 也符合 RESTful 规范，下面对其进行介绍。

9.2 Kubernetes API 详解

9.2.1 Kubernetes API 概述

Kubernetes API 是集群系统中的重要组成部分，Kubernetes 中各种资源（对象）的数据都通过该 API 接口被提交到后端的持久化存储（etcd）中，Kubernetes 集群中的各部件之间通过该 API 接口实现解耦合，同时 Kubernetes 集群中一个重要且便捷的管理工具 kubectl 也是通过访问该 API 接口实现其强大的管理功能的。Kubernetes API 中的资源对象都拥有通用的元数据，资源对象也可能存在嵌套现象，比如在一个 Pod 里面嵌套多个 Container。创建一个 API 对象是指通过 API 调用创建一条有意义的记录，该记录一旦被创建，Kubernetes 就将确保对应的资源对象会被自动创建并托管维护。

在 Kubernetes 系统中，在大多数情况下，API 定义和实现都符合标准的 HTTP REST 格式，比如通过标准的 HTTP 动词（POST、PUT、GET、DELETE）来完成对相关资源对象的查询、创建、修改、删除等操作。但同时，Kubernetes 也为某些非标准的 REST 行为实现了附加的 API 接口，例如 Watch 某个资源的变化、进入容器执行某个操作等。另外，某些 API 接口可能违背严格的 REST 模式，因为接口返回的不是单一的 JSON 对象，而是其他类型的数据，比如 JSON 对象流或非结构化的文本日志数据等。

Kubernetes 开发人员认为，任何成功的系统都会经历一个不断成长和不断适应各种变更的过程。因此，他们期望 Kubernetes API 是不断变更和增长的。同时，他们在设计和开发时，有意识地兼容了已存在的客户需求。通常，我们不希望将新的 API 资源和新的资源域频繁地加入系统，资源或域的删除需要一个严格的审核流程。

Kubernetes API 文档官网为 https://kubernetes.io/docs/reference，可以通过相关链接查看不同版本的 API 文档，例如 Kubernetes 1.14 版本的链接为 https://kubernetes.io/docs/reference/generated/kubernetes-api/v1.14。

在 Kubernetes 1.13 版本及之前的版本中，Master 的 API Server 服务提供了 Swagger 格式的 API 网页。Swagger UI 是一款 REST API 文档在线自动生成和功能测试软件，关于 Swagger 的内容请访问官网 http://swagger.io。我们通过设置 kube-apiserver 服务的启动参数 --enable-swagger-ui=true 来启用 Swagger UI 页面，其访问地址为 http://<master-ip>:<master-port>/swagger-ui/。假设 API Server 启动了 192.168.18.3 服务器上的 8080 端口，则可以通过访问 http://192.168.18.3:8080/swagger-ui/来查看 API 列表，如图 9.1 所示。

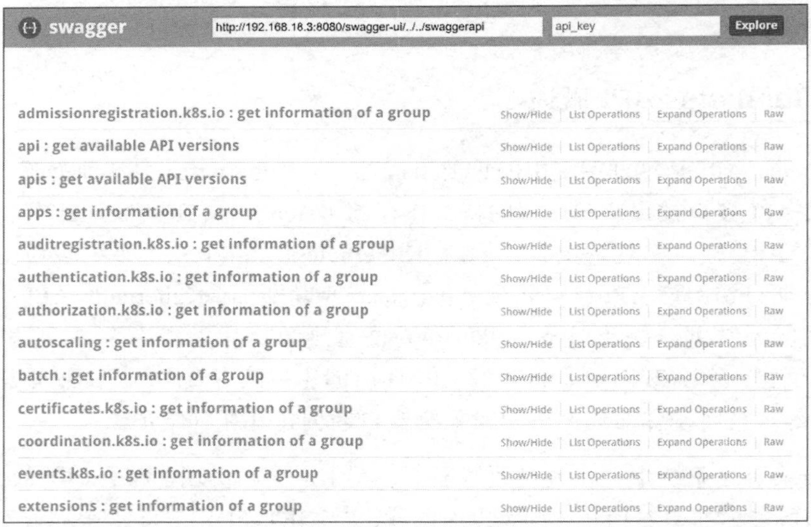

图 9.1　API Server 提供的 swagger-ui 页面

单击 api/v1 可以查看所有 API 的列表，如图 9.2 所示。

图 9.2　查看 API 列表

以创建一个 Pod 为例，找到 Rest API 的访问路径为"/api/v1/namespaces/{namespace}/pods"，如图 9.3 所示。

图 9.3　创建一个 Pod

单击链接展开，即可查看详细的 API 接口说明，如图 9.4 所示。

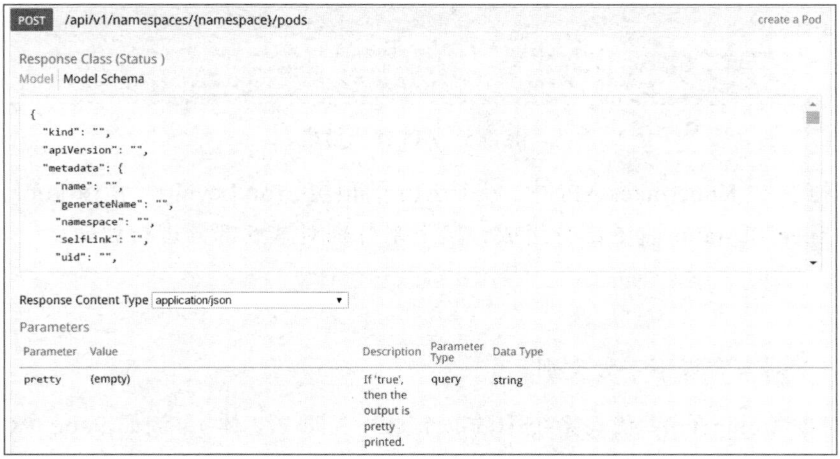

图 9.4　详细的 API 接口说明一

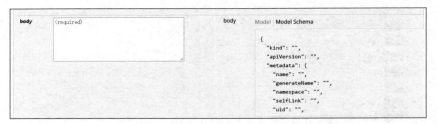

图 9.4　详细的 API 接口说明二

单击 Model 链接，则可以查看文本格式显示的 API 接口描述，如图 9.5 所示。

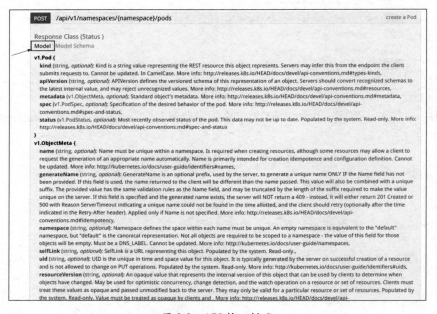

图 9.5　API 接口描述

我们看到，在 Kubernetes API 中，一个 API 的顶层（Top Level）元素由 kind、apiVersion、metadata、spec 和 status 这 5 部分组成，接下来分别对这 5 部分进行说明。

1. kind

kind 表明对象有以下三大类别。

（1）对象（objects）：代表系统中的一个永久资源（实体），例如 Pod、RC、Service、Namespace 及 Node 等。通过操作这些资源的属性，客户端可以对该对象进行创建、修改、删除和获取操作。

（2）列表（list）：一个或多个资源类别的集合。所有列表都通过 items 域获得对象数组，例如 PodLists、ServiceLists、NodeLists。大部分被定义在系统中的对象都有一个返回所有资源集合的端点，以及零到多个返回所有资源集合的子集的端点。某些对象有可能是单例对象（singletons），例如当前用户、系统默认用户等，这些对象没有列表。

（3）简单类别（simple）：该类别包含作用在对象上的特殊行为和非持久实体。该类别限制了使用范围，它有一个通用元数据的有限集合，例如 Binding、Status。

2. apiVersion

apiVersion 表明 API 的版本号，当前版本默认只支持 v1。

3. Metadata

Metadata 是资源对象的元数据定义，是集合类的元素类型，包含一组由不同名称定义的属性。在 Kubernetes 中每个资源对象都必须包含以下 3 种 Metadata。

（1）namespace：对象所属的命名空间，如果不指定，系统则会将对象置于名为 default 的系统命名空间中。

（2）name：对象的名称，在一个命名空间中名称应具备唯一性。

（3）uid：系统为每个对象都生成的唯一 ID，符合 RFC 4122 规范的定义。

此外，每种对象都还应该包含以下几个重要元数据。

（1）labels：用户可定义的"标签"，键和值都为字符串的 map，是对象进行组织和分类的一种手段，通常用于标签选择器，用来匹配目标对象。

（2）annotations：用户可定义的"注解"，键和值都为字符串的 map，被 Kubernetes 内部进程或者某些外部工具使用，用于存储和获取关于该对象的特定元数据。

（3）resourceVersion：用于识别该资源内部版本号的字符串，在用于 Watch 操作时，可以避免在 GET 操作和下一次 Watch 操作之间造成的信息不一致，客户端可以用它来判断资源是否改变。该值应该被客户端看作不透明，且不做任何修改就返回给服务端。客户端不应该假定版本信息具有跨命名空间、跨不同资源类别、跨不同服务器的含义。

（4）creationTimestamp：系统记录创建对象时的时间戳，符合 RFC 3339 规范。

（5）deletionTimestamp：系统记录删除对象时的时间戳，符合 RFC 3339 规范。

（6）selfLink：通过 API 访问资源自身的 URL，例如一个 Pod 的 link 可能是"/api/v1/namespaces/ default/pods/frontend-o8bg4"。

4. spec

spec 是集合类的元素类型，用户对需要管理的对象进行详细描述的主体部分都在 spec 里给出，它会被 Kubernetes 持久化到 etcd 中保存，系统通过 spec 的描述来创建或更新对象，以达到用户期望的对象运行状态。spec 的内容既包括用户提供的配置设置、默认值、属性的初始化值，也包括在对象创建过程中由其他相关组件（例如 schedulers、auto-scalers）创建或修改的对象属性，比如 Pod 的 Service IP 地址。如果 spec 被删除，那么该对象将会从系统中删除。

5. Status

Status 用于记录对象在系统中的当前状态信息，它也是集合类元素类型，status 在一个自动处理的进程中被持久化，可以在流转的过程中生成。如果观察到一个资源丢失了它的状态（Status），则该丢失的状态可能被重新构造。以 Pod 为例，Pod 的 status 信息主要包括 conditions、containerStatuses、hostIP、phase、podIP、startTime 等，其中比较重要的两个状态属性如下。

（1）phase：描述对象所处的生命周期阶段，phase 的典型值是 Pending（创建中）、Running、Active（正在运行中）或 Terminated（已终结），这几种状态对于不同的对象可能有轻微的差别，此外，关于当前 phase 附加的详细说明可能包含在其他域中。

（2）condition：表示条件，由条件类型和状态值组成，目前仅有一种条件类型：Ready，对应的状态值可以为 True、False 或 Unknown。一个对象可以具备多种 condition，而 condition 的状态值也可能不断发生变化，condition 可能附带一些信息，例如最后的探测时间或最后的转变时间。

Kubernetes 从 1.14 版本开始，使用 OpenAPI（https://www.openapis.org）的格式对 API 进行查询，其访问地址为 http://<master-ip>: <master-port>/openapi/v2。例如，使用命令行工具 curl 进行查询：

```
# curl http://192.168.18.3:8080/openapi/v2 | jq
{
```

```json
    "swagger": "2.0",
    "info": {
      "title": "Kubernetes",
      "version": "v1.14.0"
    },
    "paths": {
      "/api/": {
        "get": {
          "description": "get available API versions",
          "consumes": [
            "application/json",
            "application/yaml",
            "application/vnd.kubernetes.protobuf"
          ],
          "produces": [
            "application/json",
            "application/yaml",
            "application/vnd.kubernetes.protobuf"
          ],
          "schemes": [
            "https"
          ],
          "tags": [
            "core"
          ],
          "operationId": "getCoreAPIVersions",
          "responses": {
            "200": {
              "description": "OK",
              "schema": {
                "$ref": "#/definitions/io.k8s.apimachinery.pkg.apis.meta.v1.APIVersions"
              }
            },
            "401": {
              "description": "Unauthorized"
            }
          }
        }
      },
      "/api/v1/": {
```

```
          "get": {
            "description": "get available resources",
            "consumes": [
              "application/json",
              "application/yaml",
              "application/vnd.kubernetes.protobuf"
            ],
            "produces": [
              "application/json",
              "application/yaml",
              "application/vnd.kubernetes.protobuf"
            ],
            "schemes": [
              "https"
            ],
            "tags": [
              "core_v1"
            ],
            "operationId": "getCoreV1APIResources",
            "responses": {
              "200": {
                "description": "OK",
                "schema": {
                  "$ref": "#/definitions/io.k8s.apimachinery.pkg.apis.meta.v1.APIResourceList"
                }
              },
              "401": {
                "description": "Unauthorized"
              }
            }
          }
        },
        "/api/v1/componentstatuses": {
......
```

9.2.2 Kubernetes API 版本的演进策略

为了在兼容旧版本的同时不断升级新的 API，Kubernetes 提供了多版本 API 的支持能力，每个版本的 API 都通过一个版本号路径前缀进行区分，例如/api/v1beta3。在通常情况下，新旧几个不同的 API 版本都能涵盖所有的 Kubernetes 资源对象，在不同的版本之间，这些 API 接口存在一些细微差别。Kubernetes 开发团队基于 API 级别选择版本而不是基于资源和域级别，是为了确保 API 能够清晰、连续地描述一个系统资源和行为的视图，能够控制访问的整个过程和控制实验性 API 的访问。

API 的版本号通常用于描述 API 的成熟阶段，例如：

- v1 表示 GA 稳定版本；
- v1beta3 表示 Beta 版本（预发布版本）；
- v1alpha1 表示 Alpha 版本（实验性的版本）。

当某个 API 的实现达到一个新的 GA 稳定版本时（如 v2），旧的 GA 版本（如 v1）和 Beta 版本（例如 v2beta1）将逐渐被废弃，Kubernetes 建议废弃的时间如下。

- 对于旧的 GA 版本（如 v1），Kubernetes 建议废弃的时间应不少于 12 个月或 3 个大版本 Release 的时间，选择最长的时间。
- 对旧的 Beta 版本（如 v2beta1），Kubernetes 建议废弃的时间应不少于 9 个月或 3 个大版本 Release 的时间，选择最长的时间。
- 对旧的 Alpha 版本，则无须等待，可以直接废弃。

完整的 API 更新和废弃策略请参考官方网站 https://kubernetes.io/docs/reference/using-api/deprecation-policy/的说明。

9.2.3 API Groups（API 组）

为了更容易对 API 进行扩展，Kubernetes 使用 API Groups（API 组）进行标识。API Groups 以 REST URL 中的路径进行定义。当前支持两类 API groups。

- Core Groups（核心组），也可以称之为 Legacy Groups，作为 Kubernetes 最核心的 API，其特点是没有"组"的概念，例如"v1"，在资源对象的定义中表示为"apiVersion: v1"。
- 具有分组信息的 API，以/apis/$GROUP_NAME/$VERSION URL 路径进行标识，在资源对象的定义中表示为 "apiVersion: $GROUP_NAME/$VERSION"，例如："apiVersion: batch/v1" "apiVersion: extensions:v1beta1" "apiVersion: apps/v1beta1" 等，详细的 API 列表请参见官网 https://kubernetes.io/docs/reference，目前根据 Kubernetes 的不同版本有不同的 API 说明页面。

例如，Pod 的 API 说明如图 9.6 所示，由于 Pod 属于核心资源对象，所以不存在某个扩展 API Group，页面显示为 Core，在 Pod 的定义中为 "apiVersion: v1"。

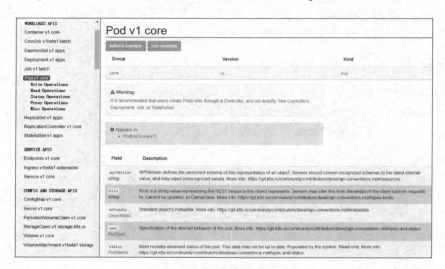

图 9.6　Pod 的 API 说明

StatefulSet 则属于名为 apps 的 API 组，版本号为 v1，在 StatefulSet 的定义中为 "apiVersion: apps/v1"，如图 9.7 所示。

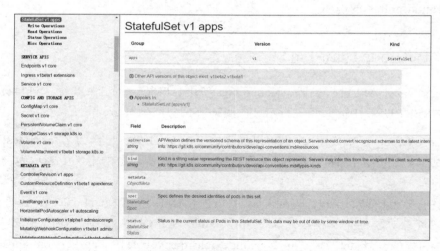

图 9.7　StatefulSet 的 API 说明

如果要启用或禁用特定的 API 组，则需要在 API Server 的启动参数中设置

--runtime-config 进行声明，例如，--runtime-config=batch/v2alpha1 表示启用 API 组 "batch/v2alpha1"；也可以设置--runtime-config=batch/v1=false 表示禁用 API 组 "batch/v1"。多个 API 组的设置以逗号分隔。在当前的 API Server 服务中，DaemonSets、Deployments、HorizontalPodAutoscalers、Ingress、Jobs 和 ReplicaSets 所属的 API 组是默认启用的。

9.2.4 API REST 的方法说明

API 资源使用 REST 模式，对资源对象的操作方法如下。

（1）GET /<资源名的复数格式>：获得某一类型的资源列表，例如 GET /pods 返回一个 Pod 资源列表。

（2）POST /<资源名的复数格式>：创建一个资源，该资源来自用户提供的 JSON 对象。

（3）GET /<资源名复数格式>/<名称>：通过给出的名称获得单个资源，例如 GET /pods/first 返回一个名为 first 的 Pod。

（4）DELETE /<资源名复数格式>/<名称>：通过给出的名称删除单个资源，在删除选项（DeleteOptions）中可以指定优雅删除（Grace Deletion）的时间（GracePeriodSeconds），该选项表明了从服务端接收到删除请求到资源被删除的时间间隔（单位为 s）。不同的类别（Kind）可能为优雅删除时间（Grace Period）声明默认值。用户提交的优雅删除时间将覆盖该默认值，包括值为 0 的优雅删除时间。

（5）PUT /<资源名复数格式>/<名称>：通过给出的资源名和客户端提供的 JSON 对象来更新或创建资源。

（6）PATCH /<资源名复数格式>/<名称>：选择修改资源详细指定的域。

对于 PATCH 操作，目前 Kubernetes API 通过相应的 HTTP 首部 "Content-Type" 对其进行识别。

目前支持以下三种类型的 PATCH 操作。

（1）JSON Patch, Content-Type: application/json-patch+json。在 RFC6902 的定义中，JSON Patch 是执行在资源对象上的一系列操作，例如 {"op": "add", "path": "/a/b/c", "value": ["foo", "bar"]}。详情请查看 RFC6902 说明，网址为 https://tools.ietf.org/html/rfc6902。

（2）Merge Patch, Content-Type: application/merge-json-patch+json。在 RFC7386 的定义

中，Merge Patch 必须包含对一个资源对象的部分描述，这个资源对象的部分描述就是一个 JSON 对象。该 JSON 对象被提交到服务端，并和服务端的当前对象合并，从而创建一个新的对象。详情请查看 RFC73862 说明，网址为 https://tools.ietf.org/html/rfc7386。

（3）Strategic Merge Patch，Content-Type：application/strategic-merge-patch+json。Strategic Merge Patch 是一个定制化的 Merge Patch 实现。接下来将详细讲解 Strategic Merge Patch。

在标准的 JSON Merge Patch 中，JSON 对象总被合并（Merge），但是资源对象中的列表域总被替换，用户通常不希望如此。例如，我们通过下列定义创建一个 Pod 资源对象：

```
spec:
  containers:
    - name: nginx
      image: nginx-1.0
```

接着，我们希望添加一个容器到这个 Pod 中，代码和上传的 JSON 对象如下：

```
PATCH /api/v1/namespaces/default/pods/pod-name
spec:
  containers:
    - name: log-tailer
      image: log-tailer-1.0
```

如果我们使用标准的 Merge Patch，则其中的整个容器列表将被单个"log-tailer"容器所替换，然而我们的目的是使两个容器列表合并。

为了解决这个问题，Strategic Merge Patch 添加元数据到 API 对象中，并通过这些新元数据来决定哪个列表被合并，哪个列表不被合并。当前这些元数据作为结构标签，对于 API 对象自身来说是合法的。对于客户端来说，这些元数据作为 Swagger annotations 也是合法的。在上述例子中向 containers 中添加了 patchStrategy 域，且它的值为 merge，通过添加 patchMergeKey，它的值为 name。也就是说，containers 中的列表将会被合并而不是替换，合并的依据为 name 域的值。

此外，Kubernetes API 添加了资源变动的观察者模式的 API 接口。

- ◎ GET /watch/<资源名复数格式>：随时间变化，不断接收一连串的 JSON 对象，这些 JSON 对象记录了给定资源类别内所有资源对象的变化情况。
- ◎ GET /watch/<资源名复数格式>/<name>：随时间变化，不断接收一连串的 JSON 对象，这些 JSON 对象记录了某个给定资源对象的变化情况。

上述接口改变了返回数据的基本类别，watch 动词返回的是一连串 JSON 对象，而不是单个 JSON 对象。并不是所有对象类别都支持观察者模式的 API 接口，在后续的章节中将会说明哪些资源对象支持这种接口。

另外，Kubernetes 还增加了 HTTP Redirect 与 HTTP Proxy 这两种特殊的 API 接口，前者实现资源重定向访问，后者则实现 HTTP 请求的代理。

9.2.5 API Server 响应说明

API Server 在响应用户请求时附带一个状态码，该状态码符合 HTTP 规范。表 9.1 列出了 API Server 可能返回的状态码。

表 9.1 API Server 可能返回的状态码

状态码	编码	描述
200	OK	表明请求完全成功
201	Created	表明创建类的请求完全成功
204	NoContent	表明请求完全成功，同时 HTTP 响应不包含响应体。在响应 OPTIONS 方法的 HTTP 请求时返回
307	TemporaryRedirect	表明请求资源的地址被改变，建议客户端使用 Location 首部给出的临时 URL 来定位资源
400	BadRequest	表明请求是非法的，建议用户不要重试，修改该请求
401	Unauthorized	表明请求能够到达服务端，且服务端能够理解用户请求，但是拒绝做更多的事情，因为客户端必须提供认证信息。如果客户端提供了认证信息，则返回该状态码，表明服务端指出所提供的认证信息不合适或非法
403	Forbidden	表明请求能够到达服务端，且服务端能够理解用户请求，但是拒绝做更多的事情，因为该请求被设置成拒绝访问。建议用户不要重试，修改该请求
404	NotFound	表明所请求的资源不存在。建议用户不要重试，修改该请求
405	MethodNotAllowed	表明在请求中带有该资源不支持的方法。建议用户不要重试，修改该请求
409	Conflict	表明客户端尝试创建的资源已经存在，或者由于冲突，请求的更新操作不能被完成
422	UnprocessableEntity	表明由于所提供的作为请求部分的数据非法，创建或修改操作不能被完成
429	TooManyRequests	表明超出了客户端访问频率的限制或者服务端接收到多于它能处理的请求。建议客户端读取相应的 Retry-After 首部，然后等待该首部指出的时间后再重试

续表

状态码	编码	描述
500	InternalServerError	表明服务端能被请求访问到,但是不能理解用户的请求;或者在服务端内产生非预期的一个错误,而且该错误无法被认知;或者服务端不能在一个合理的时间内完成处理(这可能是服务器临时负载过重造成的,或和其他服务器通信时的一个临时通信故障造成的)
503	ServiceUnavailable	表明被请求的服务无效。建议用户不要重试修改该请求
504	ServerTimeout	表明请求在给定的时间内无法完成。客户端仅在为请求指定超时(Timeout)参数时得到该响应

在调用 API 接口发生错误时,Kubernetes 将会返回一个状态类别(Status Kind)。下面是两种常见的错误场景。

(1)当一个操作不成功时(例如,当服务端返回一个非 2xx HTTP 状态码时)。

(2)当一个 HTTP DELETE 方法调用失败时。

状态对象被编码成 JSON 格式,同时该 JSON 对象被作为请求的响应体。该状态对象包含人和机器使用的域,在这些域中包含来自 API 的关于失败原因的详细信息。状态对象中的信息补充了对 HTTP 状态码的说明。例如:

```
$ curl -v -k -H "Authorization: Bearer WhCDvq4VPpYhrcfmF6ei7V9qlbqTubUc"
HTTPs://10.240.122.184:443/api/v1/namespaces/default/pods/grafana
> GET /api/v1/namespaces/default/pods/grafana HTTP/1.1
> User-Agent: curl/7.26.0
> Host: 10.240.122.184
> Accept: */*
> Authorization: Bearer WhCDvq4VPpYhrcfmF6ei7V9qlbqTubUc
>
< HTTP/1.1 404 Not Found
< Content-Type: application/json
< Date: Wed, 20 May 2015 18:10:42 GMT
< Content-Length: 232
<
{
  "kind": "Status",
  "apiVersion": "v1",
  "metadata": {},
```

```
"status": "Failure",
"message": "pods \"grafana \"not found",
"reason": "NotFound",
"details": {
 "name": "grafana",
 "kind": "pods"
},
"code": 404
}
```

其中：

- status 域包含两个可能的值：Success 或 Failure。
- message 域包含对错误的描述信息。
- reason 域包含说明该操作失败原因的描述。
- details 可能包含和 reason 域相关的扩展数据。每个 reason 域都可以定义它的扩展的 details 域。该域是可选的，返回数据的格式是不确定的，不同的 reason 类型返回的 details 域的内容不一样。

9.3 使用 Java 程序访问 Kubernetes API

本节介绍如何使用 Java 程序访问 Kubernetes API。在 Kubernetes 官网上列出了多个访问 Kubernetes API 的开源项目，其中有两个是用 Java 语言开发的开源项目，一个是 OSGI，另一个是 Fabric8。在本节所列的两个 Java 开发例子中，一个是基于 Jersey 的，另一个是基于 Fabric8 的。

9.3.1 Jersey

Jersey 是一个 RESTful 请求服务 Java 框架。与 Struts 类似，它可以和 Hibernate、Spring 框架整合。我们不仅能很方便地通过它开发 RESTful Web Service，还可以将它作为客户端方便地访问 RESTful Web Service 服务端。

如果没有一个好的工具包，则很难开发一个能够用不同的媒介（Media）类型无缝地暴露你的数据，以及很好地抽象客户端、服务端通信的底层通信的 RESTful Web Services。为了能够简化使用 Java 开发 RESTful Web Service 及其客户端的流程，业界设计了 JAX-RS API。

Jersey RESTful Web Services 框架是一个开源的高质量的框架，它为用 Java 语言开发 RESTful Web Service 及其客户端而生，支持 JAX-RS APIs。Jersey 不仅支持 JAX-RS APIs，而且在此基础上扩展了 API 接口，这些扩展更加方便并简化了 RESTful Web Services 及其客户端的开发。

由于 Kubernetes API Server 是 RESTful Web Service，因此此处选用 Jersey 框架开发 RESTful Web Service 客户端，用来访问 Kubernetes API。在本例中选用的 Jersey 框架的版本为 1.19，所涉及的 Jar 包如图 9.8 所示。

文件名	日期	类型	大小
commons-codec-1.2.jar	2015/9/13 11:10	Executable Jar File	30 KB
commons-httpclient-3.1.jar	2015/9/13 11:09	Executable Jar File	298 KB
commons-logging-1.0.4.jar	2015/9/13 11:10	Executable Jar File	38 KB
jackson-core-asl-1.9.2.jar	2015/2/11 5:41	Executable Jar File	223 KB
jackson-jaxrs-1.9.2.jar	2015/2/11 5:41	Executable Jar File	18 KB
jackson-mapper-asl-1.9.2.jar	2015/2/11 5:41	Executable Jar File	748 KB
jackson-xc-1.9.2.jar	2015/2/11 5:41	Executable Jar File	27 KB
jersey-apache-client-1.19.jar	2015/2/11 5:41	Executable Jar File	22 KB
jersey-atom-abdera-1.19.jar	2015/2/11 5:41	Executable Jar File	20 KB
jersey-client-1.19.jar	2015/2/11 5:41	Executable Jar File	131 KB
jersey-core-1.19.jar	2015/2/11 5:41	Executable Jar File	427 KB
jersey-guice-1.19.jar	2015/2/11 5:41	Executable Jar File	16 KB
jersey-json-1.19.jar	2015/2/11 5:41	Executable Jar File	162 KB
jersey-multipart-1.19.jar	2015/2/11 5:41	Executable Jar File	53 KB
jersey-server-1.19.jar	2015/2/11 5:41	Executable Jar File	687 KB
jersey-servlet-1.19.jar	2015/2/11 5:41	Executable Jar File	126 KB
jersey-simple-server-1.19.jar	2015/2/11 5:41	Executable Jar File	12 KB
jersey-spring-1.19.jar	2015/2/11 5:41	Executable Jar File	18 KB
jettison-1.1.jar	2015/2/11 5:41	Executable Jar File	67 KB
jsr311-api-1.1.1.jar	2015/2/11 5:41	Executable Jar File	46 KB
oauth-client-1.19.jar	2015/2/11 5:41	Executable Jar File	15 KB
oauth-server-1.19.jar	2015/2/11 5:41	Executable Jar File	30 KB
oauth-signature-1.19.jar	2015/2/11 5:41	Executable Jar File	24 KB

图 9.8 本例所涉及的 Jar 包

对 Kubernetes API 的访问包含如下 3 个方面。

（1）指明访问资源的类型。

（2）访问时的一些选项（参数），比如命名空间、对象的名称、过滤方式（标签和域）、子目录、访问的目标是否是代理和是否用 watch 方式访问等。

（3）访问的方法，比如增、删、改、查。

在使用 Jersey 框架访问 Kubernetes API 之前，我们需要为这 3 个方面定义 3 个对象。第 1 个定义的对象是 ResourceType，它定义了访问资源的类型；第 2 个定义的对象是 Params，它定义了访问 API 时的一些选项，以及通过这些选项如何生成完整的 URL；第 3 个定义的对象是 RestfulClient，它是一个接口，该接口定义了访问 API 的方法（Method）。

（1）ResourceType 是一个 ENUM 类型的对象，定义了 Kubernetes 的各种资源对象类型，代码如下：

```java
package com.hp.k8s.apiclient.imp;
public enum ResourceType {
    NODES("nodes"),
    NAMESPACES("namespaces"),
    SERVICES("services"),
    REPLICATIONCONTROLLERS("replicationcontrollers"),
    PODS("pods"),
    BINDINGS("bindings"),
    ENDPOINTS("endpoints"),
    SERVICEACCOUNTS("serviceaccounts"),
    SECRETS("secrets"),
    EVENTS("events"),
    COMPONENTSTATUSES("componentstatuses"),
    LIMITRANGES("limitranges"),
    RESOURCEQUOTAS("resourcequotas"),
    PODTEMPLATES("podtemplates"),
    PERSISTENTVOLUMECLAIMS("persistentvolumeclaims"); PERSISTENTVOLUMES("persistentvolumes");
    private String type;
    private ResourceType(String type) {
        this.type = type;
    }

    public String getType() {
        return type;
    }
}
```

（2）Params 定义访问 API 时的选项及通过这些选项如何生成完整的 URL，代码如下：

```java
package com.hp.k8s.apiclient.imp;

import java.io.UnsupportedEncodingException;
```

```java
import java.net.URLEncoder;
import java.util.List;
import java.util.Map;
import org.apache.logging.log4j.LogManager;
import org.apache.logging.log4j.Logger;

public class Params {
    private static final Logger LOG = LogManager.getLogger(Params.class.getName());
    private String namespace = null;
    private String name = null;
    private Map<String, String> fields = null;
    private Map<String, String> labels = null;
    private Map<String, String> notLabels = null;
    private Map<String, List<String>> inLabels = null;
    private Map<String, List<String>> notInLabels = null;
    private String json = null;
    private ResourceType resourceType = null;
    private String subPath = null;
    private boolean isVisitProxy = false;
    private boolean isSetWatcher = false;

    // buildPath 方法用于构建访问 URL 的完整路径
    public String buildPath() {
        StringBuilder result = (isVisitProxy ? new StringBuilder("/proxy")
                : (isSetWatcher ? new StringBuilder("/watch") : new StringBuilder("")));
        if (null != namespace)
            result.append("/namespaces/").append(namespace);

        result.append("/").append(resourceType.getType());
        if (null != name)
            result.append("/").append(name);
        if(null!=subPath)
            result.append("/").append(subPath);

        if (null != labels && !labels.isEmpty() || null != notLabels && !notLabels.isEmpty()
                || null != inLabels && inLabels.size() > 0 || null != notInLabels && notInLabels.size() > 0
                || null != fields && fields.size() > 0) {
```

```java
                StringBuilder labelSelectorStr = null;
                StringBuilder fieldSelectorStr = null;
                try {
                    labelSelectorStr = builderLabelSelector();
                    fieldSelectorStr = builderFiledSelector();
                } catch (UnsupportedEncodingException e1) {
                    LOG.error(e1);
                }

                if (labelSelectorStr.length() + fieldSelectorStr.length() > 0)
                    result.append("?");
                if (labelSelectorStr.length() > 0) {
                    result.append("labelSelector=").append(labelSelectorStr.toString());
                    if (fieldSelectorStr.length() > 0) {
                        result.append(",");
                    }
                }
                if (fieldSelectorStr.length() > 0) {
                    result.append("fieldSelector=").append(fieldSelectorStr.toString());
                }
            }
        return result.toString();
    }

    // builderLabelSelector 方法用于构建 Label Selector
    private StringBuilder builderLabelSelector() throws UnsupportedEncodingException {
        StringBuilder result = new StringBuilder();
        if (null != labels) {
            for (String key : labels.keySet()) {
                if (result.length() > 0) {
                    result.append(",");
                }
                result.append(URLEncoder.encode(key + "=" + labels.get(key), "GBK"));
            }
        }

        if (null != notLabels) {
```

```java
            for (String key : labels.keySet()) {
                if (result.length() > 0) {
                    result.append(",");
                }
                result.append(URLEncoder.encode(key + "!=" + labels.get(key),
"GBK"));
            }
        }

        if (null != inLabels) {
            for (String key : inLabels.keySet()) {
                if (result.length() > 0) {
                    result.append(URLEncoder.encode(",","GBK"));
                }
                result.append(URLEncoder.encode(key + " in (" + listToString
(inLabels.get(key), ",") + ")", "GBK"));
            }
        }

        if (null != notInLabels) {
            for (String key : inLabels.keySet()) {
                if (result.length() > 0) {
                    result.append(URLEncoder.encode(",","GBK"));
                }
                result.append(URLEncoder.encode(key + " notin (" + listToString
(inLabels.get(key), ",") + ")", "GBK"));
            }
        }

        LOG.info("label result: " + result);
        return result;
    }

    // builderFieldSelector方法用于构建 Field Selector
    private StringBuilder builderFieldSelector() throws UnsupportedEncoding
Exception {
        StringBuilder result = new StringBuilder();
        if (null != fields) {
            for (String key : fields.keySet()) {
                if (result.length() > 0) {
                    result.append(",");
```

```java
                }
                result.append(URLEncoder.encode(key + "=" + fields.get(key),
"GBK"));
            }
        }
        return result;
    }

    // listToString 方法用于将 List 显示为字符串
    private String listToString(List<String> list, String delim) {
        boolean isFirst = true;
        StringBuilder result = new StringBuilder();
        for (String str : list) {
            if (isFirst) {
                result.append(str);
                isFirst = false;
            } else {
                result.append(delim).append(str);
            }
        }

        return result.toString();
    }

    // 各属性的 get 和 set 方法
    public String getNamespace() {
        return namespace;
    }
    public void setNamespace(String namespace) {
        this.namespace = namespace;
    }
    public String getName() {
        return name;
    }
    public void setName(String name) {
        this.name = name;
    }
    public Map<String, String> getFields() {
        return fields;
    }
    public void setFields(Map<String, String> fields) {
```

```java
        this.fields = fields;
    }
    public Map<String, String> getLabels() {
        return labels;
    }
    public void setLabels(Map<String, String> labels) {
        this.labels = labels;
    }
    public String getJson() {
        return json;
    }
    public void setJson(String json) {
        this.json = json;
    }
    public ResourceType getResourceType() {
        return resourceType;
    }
    public void setResourceType(ResourceType resourceType) {
        this.resourceType = resourceType;
    }
    public String getSubPath() {
        return subPath;
    }
    public void setSubPath(String subPath) {
        this.subPath = subPath;
    }
    public boolean isVisitProxy() {
        return isVisitProxy;
    }
    public void setVisitProxy(boolean isVisitProxy) {
        this.isVisitProxy = isVisitProxy;
    }
    public boolean isSetWatcher() {
        return isSetWatcher;
    }
    public void setSetWatcher(boolean isSetWatcher) {
        this.isSetWatcher = isSetWatcher;
    }
    public Map<String, String> getNotLabels() {
        return notLabels;
    }
```

```java
    public void setNotLabels(Map<String, String> notLabels) {
        this.notLabels = notLabels;
    }
    public Map<String, List<String>> getInLabels() {
        return inLabels;
    }
    public void setInLabels(Map<String, List<String>> inLabels) {
        this.inLabels = inLabels;
    }
    public Map<String, List<String>> getNotInLabels() {
        return notInLabels;
    }
    public void setNotInLabels(Map<String, List<String>> notInLabels) {
        this.notInLabels = notInLabels;
    }
}
```

Params 对象包含的属性说明如表 9.2 所示。

表 9.2　Params 对象包含的属性说明

属　性	说　　明
namespace	String 类型属性，指明资源所在的命名空间，如果没有指定该值，则表明访问所有命名空间下的资源对象
name	String 类型属性，在访问单个资源对象时使用，如果没有指定该值，则表明访问该类资源列表
fields	Map<String, String>类型属性，通过资源对象的域值过滤访问结果
labels	Map<String, String>类型属性，通过指定的标签选择器列表来选择资源对象。选择的资源对象包含在标签列表中所列的标签（即 Map 的 key），且所选资源的标签的 value 和标签列表中的 value 值（即 Map 的 value）相等
notLabels	Map<String, String>类型属性，通过指定的标签选择器列表来选择资源对象。选择的资源对象包含在标签列表中所列的标签（即 Map 的 key），且所选资源的标签的 value 和标签列表中的 value 值（即 Map 的 value）不相等
inLabels	Map<String, List<String>>类型属性，通过指定的标签选择器列表来选择资源对象。Map 对象的 key 值为标签名称，Map 对象的 value 值为该标签可能包含的值
notInLabels	Map<String, List<String>>类型属性，通过指定的标签选择器列表来选择资源对象。Map 对象的 key 值为标签名称，Map 对象的 value 值为列表，表明资源对象包含和 key 值同名的标签，且这些标签的值不在该列表中
json	String 类型属性，在创建或修改资源对象时使用，用于向 API Server 提供资源对象的定义

续表

属性	说明
resourceType	ResourceType 类型属性，用于指明访问资源对象的类型
subPath	String 类型属性，用于指明访问资源的子目录
isVisitProxy	Boolean 类型属性，用于指明是否通过 Proxy 方式访问资源对象
isSetWatcher	Boolean 类型属性，表明是否通过 Watcher 方式访问资源对象

（3）接口对象 RestfulClient 定义了访问 API 接口的所有方法，其代码如下：

```java
package com.hp.k8s.apiclient;
import com.hp.k8s.apiclient.imp.Params;

public interface RestfulClient {
    public String get(Params params);       // 获得单个资源对象
    public String list(Params params);      // 获得资源对象列表
    public String create(Params params);    // 创建资源对象
    public String delete(Params params);    // 删除某个资源对象
    public String update(Params params);    // 部分更新某个资源对象
    public String updateWithMediaType(Params params,String mediaType);
                                            // 通过 mediaType 实现 Merge
    public String replace(Params params);   // 替换某个资源对象
    public String options(Params params);
    public String head(Params params);
}
```

其中，get 和 list 方法对应 Kubernetes API 的 GET 方法；create 方法对应 API 中的 POST 方法；delete 方法对应 API 中的 DELETE 方法；update 方法对应 API 中的 PATCH 方法；replace 方法对应 API 中的 PUT 方法；options 方法对应 API 中的 OPTIONS 方法；head 方法对应 API 中的 HEAD 方法。

该接口基于 Jersey 框架的实现类如下：

```java
package com.hp.k8s.apiclient.imp;

import javax.ws.rs.core.MediaType;
import org.apache.logging.log4j.LogManager;
import org.apache.logging.log4j.Logger;
import com.hp.k8s.apiclient.RestfulClient;
import com.sun.jersey.api.client.Client;
import com.sun.jersey.api.client.WebResource;
```

```java
    import com.sun.jersey.api.client.config.DefaultClientConfig;
    import com.sun.jersey.client.urlconnection.URLConnectionClientHandler;

    public class JerseyRestfulClient implements RestfulClient {
        private static final Logger LOG = LogManager.getLogger(RestfulClient.class.getName());
        private static final String METHOD_PATCH = "PATCH";

        private String _baseUrl = null;
        Client _client = null;

        public JerseyRestfulClient(String baseUrl) {
            DefaultClientConfig config = new DefaultClientConfig();
            config.getProperties().put(URLConnectionClientHandler.PROPERTY_HTTP_URL_CONNECTION_SET_METHOD_WORKAROUND, true);
            _client = Client.create(config);

            this._baseUrl = baseUrl;
        }

        // get 方法对应 Kubernetes API 的 GET 方法
        @Override
        public String get(Params params) {
            WebResource resource = _client.resource(_baseUrl + params.buildPath());
            String response = resource.accept(MediaType.APPLICATION_JSON_TYPE).get(String.class);
            LOG.info("Get one resource:\n" + response);
            return response;
        }

        // list 方法对应 Kubernetes API 的 GET 方法
        @Override
        public String list(Params params) {
            WebResource resource = _client.resource(_baseUrl + params.buildPath());
            LOG.info("URL: " + _baseUrl + params.buildPath());
            String response = resource.accept(MediaType.APPLICATION_JSON_TYPE).get(String.class);
            return response;
        }

        // create 方法对应 Kubernetes API 中的 POST 方法
```

```java
        @Override
        public String create(Params params) {
            WebResource resource = _client.resource(_baseUrl + params.buildPath());
            LOG.info("URL: " + _baseUrl + params.buildPath());
            LOG.info("Create resource: " + params.getJson());
            String response = (null == params.getJson())
                    ?
resource.accept(MediaType.APPLICATION_JSON).post(String.class)
                    : resource.type(MediaType.APPLICATION_JSON).accept(MediaType.
APPLICATION_JSON).post(String.class,params.getJson());
            return response;
        }

        // delete 方法对应 Kubernetes API 中的 DELETE 方法
        @Override
        public String delete(Params params) {
            WebResource resource = _client.resource(_baseUrl + params.buildPath());
            String response = resource.accept(MediaType.APPLICATION_JSON_TYPE).
delete(String.class);
            LOG.info("Detelet resource " + params.getResourceType().getType() + "/"
+ params.getName() + " result:\n" + response);
            return response;
        }

        // update 方法对应 Kubernetes API 中的 PATCH 方法
        @Override
        public String update(Params params) {
            return updateWithMediaType(params, MediaType.APPLICATION_JSON);
        }

        @Override
        public String updateWithMediaType(Params params, String mediaType) {
            WebResource resource = _client.resource(_baseUrl + params.buildPath());
            LOG.info("URL: " + _baseUrl + params.buildPath());
            LOG.info("Patch resource: " + params.getJson());
            String response =
resource.type(mediaType).accept(MediaType.APPLICATION_JSON_TYPE).method(METHOD_
PATCH, String.class,
                    params.getJson());
            LOG.info("Update resource " + params.buildPath() + " result:\n" + response);
            return response;
```

```java
        }

        // replace 方法对应 Kubernetes API 中的 PUT 方法
        @Override
        public String replace(Params params) {
            WebResource resource = _client.resource(_baseUrl + params.buildPath());
            LOG.info("URL: " + _baseUrl + params.buildPath());
            LOG.info("Replace resource: " + params.getJson());
            String response =
resource.type(MediaType.APPLICATION_JSON_TYPE).accept
(MediaType.APPLICATION_JSON_TYPE).put(String.class, params.getJson());
            LOG.info("Replace resource " + params.buildPath() + " result:\n" + response);
            return response;
        }

        // options 方法对应 Kubernetes API 中的 OPTIONS 方法
        @Override
        public String options(Params params) {
            WebResource resource = _client.resource(_baseUrl + params.buildPath());
            String response =
resource.type(MediaType.APPLICATION_JSON_TYPE).accept (MediaType.TEXT_PLAIN_TYPE)
                    .options(String.class);
            LOG.info("Get options for resource " + params.getResourceType().getType() + "/" + params.getName() + " result:\n" + response);
            return response;
        }

        // head 方法对应 Kubernetes API 中的 HEAD 方法
        @Override
        public String head(Params params) {
            WebResource resource = _client.resource(_baseUrl + params.buildPath());
            String response = resource.accept(MediaType.TEXT_PLAIN_TYPE).head().
getResponseStatus().toString();
            LOG.info("Get head for resource " + params.getResourceType().getType() + "/" + params.getName() + " result:\n" + response);
            return response;
        }

        @Override
        public void close() {
```

```
            _client.destroy();
        }
}
```

在该对象中包含如下代码：

```
config.getProperties().put(URLConnectionClientHandler.PROPERTY_HTTP_URL_CONN
ECTION_SET_METHOD_WORKAROUND, true);
```

该段代码的作用是使 Jersey 客户端支持除标准 REST 方法外的方法，比如 PATCH 方法。该段代码能访问除 watcher 外的所有 Kubernetes API 接口，在后续的章节中会举例说明如何访问 Kubernetes API。

9.3.2　Fabric8

Fabric8 包含多款工具包，Kubernetes Client 只是其中之一，也是在 Kubernetes 官网中提到的 Java Client API 之一。本例代码涉及的 Jar 包如图 9.9 所示。

名称	修改日期	类型	大小
dnsjava-2.1.7.jar	2015/8/31 14:23	Executable Jar File	301 KB
fabric8-utils-2.2.22.jar	2015/8/31 14:23	Executable Jar File	134 KB
jackson-annotations-2.6.0.jar	2015/8/31 16:27	Executable Jar File	46 KB
jackson-core-2.6.1.jar	2015/8/31 16:28	Executable Jar File	253 KB
jackson-databind-2.6.1.jar	2015/8/31 15:56	Executable Jar File	1,140 KB
jackson-dataformat-yaml-2.6.1.jar	2015/8/31 15:56	Executable Jar File	313 KB
jackson-module-jaxb-annotations-2.6.0.jar	2015/8/31 16:24	Executable Jar File	32 KB
json-20141113.jar	2015/8/31 14:23	Executable Jar File	64 KB
kubernetes-api-2.2.22.jar	2015/8/31 14:22	Executable Jar File	72 KB
kubernetes-client-1.3.8.jar	2015/8/31 15:37	Executable Jar File	2,262 KB
kubernetes-model-1.0.12.jar	2015/8/31 15:56	Executable Jar File	2,308 KB
log4j-api-2.3.jar	2015/8/31 16:18	Executable Jar File	133 KB
log4j-core-2.3.jar	2015/8/31 15:56	Executable Jar File	808 KB
log4j-slf4j-impl-2.3.jar	2015/8/31 15:56	Executable Jar File	23 KB
oauth-20100527.jar	2015/8/31 15:56	Executable Jar File	44 KB
openshift-client-1.3.2.jar	2015/8/31 14:23	Executable Jar File	24 KB
slf4j-api-1.7.12.jar	2015/8/31 15:56	Executable Jar File	32 KB
sundr-annotations-0.0.25.jar	2015/8/31 15:56	Executable Jar File	146 KB
validation-api-1.1.0.Final.jar	2015/8/31 14:23	Executable Jar File	63 KB

图 9.9　本例代码涉及的 Jar 包

因为该工具包已经对访问 Kubernetes API 客户端做了较好的封装，因此其访问代码比较简单，其具体的访问过程会在后续的章节举例说明。

Fabric 8 的 Kubernetes API 客户端工具包只能访问 Node、Service、Pod、Endpoints、Events、Namespace、PersistenetVolumeclaims、PersistenetVolume、ReplicationController、

ResourceQuota、Secret 和 ServiceAccount 这几种资源类型，不能使用 OPTIONS 和 HEAD 方法访问资源，且不能以代理方式访问资源，但其对以 watcher 方式访问资源做了很好的支持。

9.3.3 使用说明

首先，举例说明对 API 资源的基本访问，也就是对资源的增、删、改、查，以及替换资源的 status。其中会单独对 Node 和 Pod 的特殊接口做举例说明。表 9.3 列出了各资源对象的基本接口。

表 9.3　各资源对象的基本 API 接口

资源类型	方法	URL Path	说明	备注
NODES	GET	/api/v1/nodes	获取 Node 列表	
	POST	/api/v1/nodes	创建一个 Node 对象	
	DELETE	/api/v1/nodes/{name}	删除一个 Node 对象	
	GET	/api/v1/nodes/{name}	获取一个 Node 对象	
	PATCH	/api/v1/nodes/{name}	部分更新一个 Node 对象	
	PUT	/api/v1/nodes/{name}	替换一个 Node 对象	
NAMESPACES	GET	/api/v1/namespaces	获取 Namespace 列表	
	POST	/api/v1/namespaces	创建一个 Namespace 对象	
	DELETE	/api/v1/namespaces/{name}	删除一个 Namespace 对象	
	GET	/api/v1/namespaces/{name}	获取一个 Namespace 对象	
	PATCH	/api/v1/namespaces/{name}	部分更新一个 Namespace 对象	
	PUT	/api/v1/namespaces/{name}	替换一个 Namespace 对象	
	PUT	/api/v1/namespaces/{name}/finalize	替换一个 Namespace 对象的最终方案对象	在 Fabric8 中没有实现
	PUT	/api/v1/namespaces/{name}/status	替换一个 Namespace 对象的状态	在 Fabric8 中没有实现
SERVICES	GET	/api/v1/services	获取 Service 列表	
	POST	/api/v1/services	创建一个 Service 对象	
	GET	/api/v1/namespaces/{namespace}/services	获取某个 Namespace 下的 Service 列表	

续表

资源类型	方法	URL Path	说明	备注
SERVICES	POST	/api/v1/namespaces/{namespace}/services	在某个 Namespace 下创建列表	
	DELETE	/api/v1/namespaces/{namespace}/services/{name}	删除某个 Namespace 下的一个 Service 对象	
	GET	/api/v1/namespaces/{namespace}/services/{name}	获取某个 Namespace 下的一个 Service 对象	
	PATCH	/api/v1/namespaces/{namespace}/services/{name}	部分更新某个 Namespace 下的一个 Service 对象	
	PUT	/api/v1/namespaces/{namespace}/services/{name}	替换某个 Namespace 下的一个 Service 对象	
REPLICATIONCONTROLLERS	GET	/api/v1/replicationcontrollers	获取 RC 列表	
	POST	/api/v1/replicationcontrollers	创建一个 RC 对象	
	GET	/api/v1/namespaces/{namespace}/replicationcontrollers	获取某个 Namespace 下的 RC 列表	
	POST	/api/v1/namespaces/{namespace}/replicationcontrollers	在某个 Namespace 下创建一个 RC 对象	
	DELETE	/api/v1/namespaces/{namespace}/replicationcontrollers/{name}	删除某个 Namespace 下的 RC 对象	
	GET	/api/v1/namespaces/{namespace}/replicationcontrollers/{name}	获取某个 Namespace 下的 RC 对象	
	PATCH	/api/v1/namespaces/{namespace}/replicationcontrollers/{name}	部分更新某个 Namespace 下的 RC 对象	
	PUT	/api/v1/namespaces/{namespace}/replicationcontrollers/{name}	替换某个 Namespace 下的 RC 对象	
PODS	GET	/api/v1/pods	获取一个 Pod 列表	
	POST	/api/v1/pods	创建一个 Pod 对象	
	GET	/api/v1/namespaces/{namespace}/pods	获取某个 Namespace 下的 Pod 列表	
	POST	/api/v1/namespaces/{namespace}/pods	在某个 Namespace 下创建一个 Pod 对象	

续表

资源类型	方法	URL Path	说明	备注
PODS	DELETE	/api/v1/namespaces/{namespace}/pods/{name}	删除某个 Namespace 下的一个 Pod 对象	
	GET	/api/v1/namespaces/{namespace}/pods/{name}	获取某个 Namespace 下的一个 Pod 对象	
	PATCH	/api/v1/namespaces/{namespace}/pods/{name}	部分更新某个 Namespace 下的一个 Pod 对象	
	PUT	/api/v1/namespaces/{namespace}/pods/{name}	替换某个 Namespace 下的一个 Pod 对象	
	PUT	/api/v1/namespaces/{namespace}/pods/{name}/status	替换某个 Namespace 下的一个 Pod 对象状态	在 Fabric8 中没有实现
	POST	/api/v1/namespaces/{namespace}/pods/{name}/binding	创建某个 Namespace 下的一个 Pod 对象的 Binding	在 Fabric8 中没有实现
	GET	/api/v1/namespaces/{namespace}/pods/{name}/exec	连接到某个 Namespace 下的一个 Pod 对象，并执行 exec	在 Fabric8 中没有实现
	POST	/api/v1/namespaces/{namespace}/pods/{name}/exec	连接到某个 Namespace 下的一个 Pod 对象，并执行 exec	在 Fabric8 中没有实现
	GET	/api/v1/namespaces/{namespace}/pods/{name}/log	连接到某个 Namespace 下的一个 Pod 对象，并获取 log 日志信息	在 Fabric8 中没有实现
	GET	/api/v1/namespaces/{namespace}/pods/{name}/portforward	连接到某个 Namespace 下的一个 Pod 对象，并实现端口转发	在 Fabric8 中没有实现
	POST	/api/v1/namespaces/{namespace}/pods/{name}/portforward	连接到某个 Namespace 下的一个 Pod 对象，并实现端口转发	在 Fabric8 中没有实现
BINDINGS	POST	/api/v1/bindings	创建一个 Binding 对象	
	POST	/api/v1/namespaces/{namespace}/bindings	在某个 Namespace 下创建一个 Binding 对象	
ENDPOINTS	GET	/api/v1/endpoints	获取 Endpoint 列表	
	POST	/api/v1/endpoints	创建一个 Endpoint 对象	
	GET	/api/v1/namespaces/{namespace}/endpoints	获取某个 Namespace 下的 Endpoint 对象列表	

续表

资源类型	方法	URL Path	说 明	备 注
ENDPOINTS	POST	/api/v1/namespaces/{namespace}/endpoints	在某个 Namespace 下创建一个 Endpoint 对象	
	DELETE	/api/v1/namespaces/{namespace}/endpoints/{name}	删除某个 Namespace 下的 Endpoint 对象	
	GET	/api/v1/namespaces/{namespace}/endpoints/{name}	获取某个 Namespace 下的 Endpoint 对象	
	PATCH	/api/v1/namespaces/{namespace}/endpoints/{name}	部分更新某个 Namespace 下的 Endpoint 对象	
	PUT	/api/v1/namespaces/{namespace}/endpoints/{name}	替换某个 Namespace 下的 Endpoint 对象	
SERVICEACCOUNTS	GET	/api/v1/serviceaccounts	获取 Serviceaccount 列表	
	POST	/api/v1/serviceaccounts	创建一个 Serviceaccount 对象	
	GET	/api/v1/namespaces/{namespace}/serviceaccounts	获取某个 Namespace 下的 Serviceaccount 对象列表	
	POST	/api/v1/namespaces/{namespace}/serviceaccounts	在某个 Namespace 下创建一个 Serviceaccount 对象	
	DELETE	/api/v1/namespaces/{namespace}/serviceaccounts/{name}	删除某个 Namespace 下的一个 Serviceaccount 对象	
	GET	/api/v1/namespaces/{namespace}/serviceaccounts/{name}	获取某个 Namespace 下的一个 Serviceaccount 对象	
	PATCH	/api/v1/namespaces/{namespace}/serviceaccounts/{name}	部分更新某个 Namespace 下的一个 Serviceaccount 对象	
	PUT	/api/v1/namespaces/{namespace}/serviceaccounts/{name}	替换某个 Namespace 下的一个 Serviceaccount 对象	
SECRETS	GET	/api/v1/secrets	获取 Secret 列表	
	POST	/api/v1/secrets	创建一个 Secret 对象	
	GET	/api/v1/namespaces/{namespace}/secrets	获取某个 Namespace 下的 Secret 列表	

续表

资源类型	方法	URL Path	说明	备注
SECRETS	POST	/api/v1/namespaces/{namespace}/secrets	在某个 Namespace 下创建一个 Secret 对象	
	DELETE	/api/v1/namespaces/{namespace}/secrets/{name}	删除某个 Namespace 下的一个 Secret 对象	
	GET	/api/v1/namespaces/{namespace}/secrets/{name}	获取某个 Namespace 下的一个 Secret 对象	
	PATCH	/api/v1/namespaces/{namespace}/secrets/{name}	部分更新某个 Namespace 下的一个 Secret 对象	
	PUT	/api/v1/namespaces/{namespace}/secrets/{name}	替换某个 Namespace 下的一个 Secret 对象	
EVENTS	GET	/api/v1/events	获取 Event 列表	
	POST	/api/v1/events	创建一个 Event 对象	
	GET	/api/v1/namespaces/{namespace}/events	获取某个 Namespace 下的 Event 列表	
	POST	/api/v1/namespaces/{namespace}/events	在某个 Namespace 下创建一个 Event 对象	
	DELETE	/api/v1/namespaces/{namespace}/events/{name}	删除某个 Namespace 下的一个 Event 对象	
	GET	/api/v1/namespaces/{namespace}/events/{name}	获取某个 Namespace 下的一个 Event 对象	
	PATCH	/api/v1/namespaces/{namespace}/events/{name}	部分更新某个 Namespace 下的一个 Event 对象	
	PUT	/api/v1/namespaces/{namespace}/events/{name}	替换某个 Namespace 下的一个 Event 对象	
COMPONENTSTATUSES	GET	/api/v1/componentstatuses	获取 ComponentStatus 列表	
	GET	/api/v1/namespaces/{namespace}/componentstatuses	获取某个 Namespace 下的 ComponentStatus 列表	
	GET	/api/v1/namespaces/{namespace}/componentstatuses/{name}	获取某个 Namespace 下的一个 ComponentStatus 对象	

续表

资源类型	方法	URL Path	说明	备注
LIMITRANGES	GET	/api/v1/limitranges	获取 LimitRange 列表	
	POST	/api/v1/limitranges	创建一个 LimitRange 对象	
	GET	/api/v1/namespaces/{namespace}/limitranges	获取某个 Namespace 下的 LimitRange 列表	
	POST	/api/v1/namespaces/{namespace}/limitranges	在某个 Namespace 下创建一个 LimitRange 对象	
	DELETE	/api/v1/namespaces/{namespace}/limitranges/{name}	删除某个 Namespace 下的一个 LimitRange 对象	
	GET	/api/v1/namespaces/{namespace}/limitranges/{name}	获取某个 Namespace 下的一个 LimitRange 对象	
	PATCH	/api/v1/namespaces/{namespace}/limitranges/{name}	部分更新某个 Namespace 下的一个 LimitRange 对象	
	PUT	/api/v1/namespaces/{namespace}/limitranges/{name}	替换某个 Namespace 下的一个 LimitRange 对象	
RESOURCEQUOTAS	GET	/api/v1/resourcequotas	获取 ResourceQuota 列表	
	POST	/api/v1/resourcequotas	创建一个 ResourceQuota 对象	
	GET	/api/v1/namespaces/{namespace}/resourcequotas	获取某个 Namespace 下的 ResourceQuota 列表	
	POST	/api/v1/namespaces/{namespace}/resourcequotas	在某个 Namespace 下创建一个 ResourceQuota 对象	
	DELETE	/api/v1/namespaces/{namespace}/resourcequotas/{name}	删除某个 Namespace 下的一个 ResourceQuota 对象	
	GET	/api/v1/namespaces/{namespace}/resourcequotas/{name}	获取某个 Namespace 下的一个 ResourceQuota 对象	
	PATCH	/api/v1/namespaces/{namespace}/resourcequotas/{name}	部分更新某个 Namespace 下的一个 ResourceQuota 对象	

续表

资源类型	方法	URL Path	说明	备注
	PUT	/api/v1/namespaces/{namespace}/resourcequotas/{name}	替换某个 Namespace 下的一个 Resource Quota 对象	
	PUT	/api/v1/namespaces/{namespace}/resourcequotas/{name}/status	替换某个 Namespace 下的一个 Resource Quota 对象状态	在 Fabric8 中没有实现
PODTEMPLATES	GET	/api/v1/podtemplates	获取 PodTemplate 列表	
	POST	/api/v1/podtemplates	创建一个 PodTemplate 对象	
	GET	/api/v1/namespaces/{namespace}/podtemplates	获取某个 Namespace 下的 PodTemplate 列表	
	POST	/api/v1/namespaces/{namespace}/podtemplates	在某个 Namespace 下创建一个 PodTemplate 对象	
	DELETE	/api/v1/namespaces/{namespace}/podtemplates/{name}	删除某个 Namespace 下的一个 PodTemplate 对象	
	GET	/api/v1/namespaces/{namespace}/podtemplates/{name}	获取某个 Namespace 下的一个 PodTemplate 对象	
	PATCH	/api/v1/namespaces/{namespace}/podtemplates/{name}	部分更新某个 Namespace 下的一个 PodTemplate 对象	
	PUT	/api/v1/namespaces/{namespace}/podtemplates/{name}	替换某个 Namespace 下的一个 PodTemplate 对象	
PERSISTENTVOLUMES	GET	/api/v1/persistentvolumes	获取 PersistentVolume 列表	
	POST	/api/v1/persistentvolumes	创建一个 PersistentVolume 对象	
	DELETE	/api/v1/persistentvolumes/{name}	删除一个 PersistentVolume 对象	
	GET	/api/v1/persistentvolumes/{name}	获取一个 PersistentVolume 对象	
	PATCH	/api/v1/persistentvolumes/{name}	部分更新一个 PersistentVolume 对象	
	PUT	/api/v1/persistentvolumes/{name}	替换一个 PersistentVolume 对象	
	PUT	/api/v1/persistentvolumes/{name}/status	替换一个 PersistentVolume 对象状态	在 Fabric8 中没有实现

续表

资源类型	方法	URL Path	说明	备注
PERSISTENTVOLUMECLAIMS	GET	/api/v1/persistentvolumeclaims	获取 PersistentVolumeClaim 列表	
	POST	/api/v1/persistentvolumeclaims	创建一个 PersistentVolumeClaim 对象	
	GET	/api/v1/namespaces/{namespace}/persistentvolumeclaims	获取某个 Namespace 下的 PersistentVolumeClaim 列表	
	POST	/api/v1/namespaces/{namespace}/persistentvolumeclaims	在某个 Namespace 下创建一个 PersistentVolumeClaim 对象	
	DELETE	/api/v1/namespaces/{namespace}/persistentvolumeclaims/{name}	删除某个 Namespace 下的一个 PersistentVolumeClaim 对象	
	GET	/api/v1/namespaces/{namespace}/persistentvolumeclaims/{name}	获取某个 Namespace 下的一个 Persistent VolumeClaim 对象	
	PATCH	/api/v1/namespaces/{namespace}/persistentvolumeclaims/{name}	部分更新某个 Namespace 下的一个 Persistent VolumeClaim 对象	
	PUT	/api/v1/namespaces/{namespace}/persistentvolumeclaims/{name}	替换某个 Namespace 下的一个 Persistent VolumeClaim 对象	
	PUT	/api/v1/namespaces/{namespace}/persistentvolumeclaims/{name}/status	替换某个 Namespace 下的一个 Persistent VolumeClaim 对象状态	在 Fabric8 中没有实现

首先，举例说明如何通过 API 接口来创建资源对象。我们需要创建访问 API Server 的客户端，基于 Jersey 框架的代码如下：

```
RestfulClient _restfulClient = new JerseyRestfulClient("http://192.168.1.128:8080/api/v1");
```

其中，http://192.168.1.128:8080 为 API Server 的地址。基于 Fabric8 框架的代码如下：

```
Config _conf = new Config();
KubernetesClient _kube = new DefaultKubernetesClient("http://192.168.1.128: 8080");
```

分别通过上面的两个客户端创建 Namespace 资源对象，基于 Jersey 框架的代码如下：

```
private void testCreateNamespace() {
    Params params = new Params();
    params.setResourceType(ResourceType.NAMESPACES);
    params.setJson(Utils.getJson("namespace.json"));
```

```
            LOG.info("Result: " + _restfulClient.create(params));
    }
```

其中,"namespace.json"为创建 Namespace 资源对象的 JSON 定义,代码如下:

```
{
  "kind":"Namespace",
  "apiVersion":"v1",
  "metadata":{
    "name": "ns-sample"
  }
}
```

基于 Fabric8 框架的代码如下:

```
    private void testCreateNamespace() {
        Namespace ns = new Namespace();
        ns.setApiVersion(ApiVersion.V_1);
        ns.setKind("Namespace");
        ObjectMeta om = new ObjectMeta();
        om.setName("ns-fabric8");
        ns.setMetadata(om);

        _kube.namespaces().create(ns);

        LOG.info(_kube.namespaces().list().getItems().size());
    }
```

由于 Fabric8 框架对 Kubernetes API 对象做了很好的封装,对其中的大量对象都做了定义,所以用户可以通过其提供的资源对象去定义 Kubernetes API 对象,例如上面例子中的 Namespace 对象。Fabric8 框架中的 kubernetes-model 工具包用于 API 对象的封装。在上面的例子中,通过 Fabric8 框架提供的类创建了一个名为 ns-fabric8 的命名空间对象。

接下来会通过基于 Jeysey 框架的代码创建两个 Pod 资源对象。在两个例子中,一个是在上面创建的"ns-sample" Namespace 中创建 Pod 资源对象,另一个是为后续创建"cluster service"创建的 Pod 资源对象。由于基于 Fabric8 框架创建 Pod 资源对象的方法很简单,因此不再用 Fabric8 框架对上述两个例子做说明。通过基于 Jersey 框架创建这两个 Pod 资源对象的代码如下:

```java
private void testCreatePod() {
    Params params = new Params();
    params.setResourceType(ResourceType.PODS);
    params.setJson(Utils.getJson("podInNs.json"));
    params.setNamespace("ns-sample");
    LOG.info("Result: " + _restfulClient.create(params));

    params.setJson(Utils.getJson("pod4ClusterService.json"));
    LOG.info("Result: " + _restfulClient.create(params));
}
```

其中，podInNs.json 和 pod4ClusterService.json 是创建两个 Pod 资源对象的定义。podInNs.json 文件的内容如下：

```json
{
    "kind":"Pod",
    "apiVersion":"v1",
    "metadata":{
        "name":"pod-sample-in-namespace",
        "namespace": "ns-sample"
    },
    "spec":{
        "containers":[{
            "name":"mycontainer",
            "image":"kubeguide/redis-master"
        }]
    }
}
```

pod4ClusterService.json 文件的内容如下：

```json
{
    "kind":"Pod",
    "apiVersion":"v1",
    "metadata":{
        "name":"pod-sample-4-cluster-service",
        "namespace": "ns-sample",
        "labels":{
            "k8s-cs": "kube-cluster-service",
            "k8s-test": "kube-cluster-test",
            "k8s-sample-app": "kube-service-sample",
```

```
            "kkk": "bbb"
        }
    },
    "spec":{
        "containers":[{
            "name":"mycontainer",
            "image":"kubeguide/redis-master"
        }]
    }
}
```

下面的例子代码用于获取 Pod 资源列表，其中，第 1 部分代码用于获取所有的 Pod 资源对象，第 2、3 部分代码主要用于说明如何使用标签选择 Pod 资源对象，最后一部分代码用于举例说明如何使用 field 选择 Pod 资源对象。代码如下：

```
private void testGetPodList() {
    Params params = new Params();
    params.setResourceType(ResourceType.PODS);
    LOG.info("Result: " + _restfulClient.list(params));

    Map<String, String> labels = new HashMap<String, String>();
    labels.put("k8s-cs", "kube-cluster-service");
    labels.put("k8s-sample-app", "kube-service-sample");
    params.setLabels(labels);
    LOG.info("Result: " + _restfulClient.list(params));
    params.setLabels(null);

    Map<String, List<String>> inLabels = new HashMap<String, List<String>>();
    List list = new ArrayList<String>();
    list.add("kube-cluster-service");
    list.add("kube-cluster");
    inLabels.put("k8s-cs", list);
    params.setInLabels(inLabels);
    LOG.info("Result: " + _restfulClient.list(params));
    params.setInLabels(null);

    Map<String, String> fields = new HashMap<String, String>();
    fields.put("metadata.name", "pod-sample-4-cluster-service");
    params.setNamespace("ns-sample");
    params.setFields(fields);
```

```
    LOG.info("Result: " + _restfulClient.list(params));
}
```

接下来的例子代码用于替换一个 Pod 对象，在通过 Kubernetes API 替换一个 Pod 资源对象时需要注意如下两点。

（1）在替换该资源对象前，先从 API 中获取该资源对象的 JSON 对象，然后在该 JSON 对象的基础上修改需要替换的部分。

（2）在 Kubernetes API 提供的接口中，PUT 方法（replace）只支持替换容器的 image 部分。

代码如下：

```
private void testReplacePod() {
    Params params = new Params();
    params.setNamespace("ns-sample");
    params.setName("pod-sample-in-namespace");
    params.setJson(Utils.getJson("pod4Replace.json"));
    params.setResourceType(ResourceType.PODS);

    LOG.info("Result: " + _restfulClient.replace(params));
}
```

其中，pod4Replace.json 的内容如下：

```
{
  "kind": "Pod",
  "apiVersion": "v1",
  "metadata": {
    "name": "pod-sample-in-namespace",
    "namespace": "ns-sample",
    "selfLink": "/api/v1/namespaces/ns-sample/pods/pod-sample-in-namespace",
    "uid": "084ff63e-59d3-11e5-8035-000c2921ba71",
    "resourceVersion": "45450",
    "creationTimestamp": "2015-09-13T04:51:01Z"
  },
  "spec": {
    "volumes": [
      {
        "name": "default-token-szoje",
        "secret": {
```

```
          "secretName": "default-token-szoje"
        }
      }
    ],
    "containers": [
      {
        "name": "mycontainer",
        "image": "centos",
        "resources": {},
        "volumeMounts": [
          {
            "name": "default-token-szoje",
            "readOnly": true,
            "mountPath": "/var/run/secrets/kubernetes.io/serviceaccount"
          }
        ],
        "terminationMessagePath": "/dev/termination-log",
        "imagePullPolicy": "IfNotPresent"
      }
    ],
    "restartPolicy": "Always",
    "dnsPolicy": "ClusterFirst",
    "serviceAccountName": "default",
    "serviceAccount": "default",
    "nodeName": "192.168.1.129"
  },
  "status": {
    "phase": "Running",
    "conditions": [
      {
        "type": "Ready",
        "status": "True"
      }
    ],
    "hostIP": "192.168.1.129",
    "podIP": "10.1.10.66",
    "startTime": "2015-09-11T15:17:28Z",
    "containerStatuses": [
      {
        "name": "mycontainer",
```

```
        "state": {
         "running": {
           "startedAt": "2015-09-11T15:17:30Z"
          }
        },
        "lastState": {},
        "ready": true,
        "restartCount": 0,
        "image": "kubeguide/redis-master",
        "imageID":
"docker://5630952871a38cddffda9ec611f5978ab0933628fcd54cd7d7677ce6b17de33f",
        "containerID": "docker://7bf0d454c367418348711556e667fd1ef6a04d7153d
24bfcac2e2e06da634a9f"
      }
     ]
   }
 }
```

接下来的两个例子实现了在 9.2.4 节中提到的两种 Merge 方式：Merge Patch 和 Strategic Merge Patch。

Merge Patch 的示例如下：

```
private void testUpdatePod1() {
    Params params = new Params();
    params.setNamespace("ns-sample");
    params.setName("pod-sample-in-namespace");
    params.setJson(Utils.getJson("pod4MergeJsonPatch.json"));
    params.setResourceType(ResourceType.PODS);

    LOG.info("Result: " + _restfulClient.updateWithMediaType(params,
"application/ merge-patch+json"));
}
```

其中，pod4MergeJsonPatch.json 的内容如下：

```
{
    "metadata":{
     "labels":{
       "k8s-cs": "kube-cluster-service",
       "k8s-test": "kube-cluster-test",
       "k8s-sa5555mple-app": "kube-service-sample",
```

```
            "kkk": "bbb4444"
          }
      }
  }
```

Strategic Merge Patch 的示例如下:

```
private void testUpdatePod2() {
    Params params = new Params();
    params.setNamespace("ns-sample");
    params.setName("pod-sample-in-namespace");
    params.setJson(Utils.getJson("pod4StrategicMerge.json"));
    params.setResourceType(ResourceType.PODS);

    LOG.info("Result: " + _restfulClient.updateWithMediaType(params,
"application/strategic-merge-patch+json"));
}
```

其中,pod4StrategicMerge.json 的内容如下:

```
{
    "spec":{
      "containers":[{
          "name":"mycontainer",
          "image":"centos",
          "patchStrategy":"merge",
          "patchMergeKey":"name"
      }]
    }
}
```

接下来实现了修改 Pod 资源对象的状态,代码如下:

```
private void testStatusPod() {
    Params params = new Params();
    params.setNamespace("ns-sample");
    params.setName("pod-sample-in-namespace");
    params.setSubPath("/status");
    params.setJson(Utils.getJson("pod4Status.json"));
    params.setResourceType(ResourceType.PODS);

    _restfulClient.replace(params);
}
```

其中，pod4Status.json 的内容如下：

```json
{
  "kind": "Pod",
  "apiVersion": "v1",
  "metadata": {
    "name": "pod-sample-in-namespace",
    "namespace": "ns-sample",
    "selfLink": "/api/v1/namespaces/ns-sample/pods/pod-sample-in-namespace",
    "uid": "ad1d803f-59ec-11e5-8035-000c2921ba71",
    "resourceVersion": "51640",
    "creationTimestamp": "2015-09-13T07:54:35Z"
  },
  "spec": {
    "volumes": [
      {
        "name": "default-token-szoje",
        "secret": {
          "secretName": "default-token-szoje"
        }
      }
    ],
    "containers": [
      {
        "name": "mycontainer",
        "image": "kubeguide/redis-master",
        "resources": {},
        "volumeMounts": [
          {
            "name": "default-token-szoje",
            "readOnly": true,
            "mountPath": "/var/run/secrets/kubernetes.io/serviceaccount"
          }
        ],
        "terminationMessagePath": "/dev/termination-log",
        "imagePullPolicy": "IfNotPresent"
      }
    ],
    "restartPolicy": "Always",
    "dnsPolicy": "ClusterFirst",
```

```
      "serviceAccountName": "default",
      "serviceAccount": "default",
      "nodeName": "192.168.1.129"
    },
    "status": {
      "phase": "Unknown",
      "conditions": [
        {
          "type": "Ready",
          "status": "false"
        }
      ],
      "hostIP": "192.168.1.129",
      "podIP": "10.1.10.79",
      "startTime": "2015-09-11T18:21:02Z",
      "containerStatuses": [
        {
          "name": "mycontainer",
          "state": {
            "running": {
              "startedAt": "2015-09-11T18:21:03Z"
            }
          },
          "lastState": {},
          "ready": true,
          "restartCount": 0,
          "image": "kubeguide/redis-master",
          "imageID": "docker://5630952871a38cddffda9ec611f5978ab0933628fcd54cd7d7677ce6b17de33f",
          "containerID": "docker://b0e2312643e9a4b59cf1ff5fb7a8468c5777180d5a8ea5f2f0c9dfddcf3f4cd2"
        }
      ]
    }
  }
```

接下来实现了查看 Pod 的 log 日志功能，代码如下：

```
private void testLogPod() {
    Params params = new Params();
    params.setNamespace("ns-sample");
```

```
    params.setName("pod-sample-in-namespace");
    params.setSubPath("/log");
    params.setResourceType(ResourceType.PODS);

    _restfulClient.get(params);
}
```

下面通过 API 访问 Node 的多种接口,代码如下:

```
private void testPoxyNode() {
    Params params = new Params();
    params.setName("192.168.1.129");
    params.setSubPath("pods");
    params.setVisitProxy(true);
    params.setResourceType(ResourceType.NODES);
    _restfulClient.get(params);

    params = new Params();
    params.setName("192.168.1.129");
    params.setSubPath("stats");
    params.setVisitProxy(true);
    params.setResourceType(ResourceType.NODES);
    _restfulClient.get(params);

    params = new Params();
    params.setName("192.168.1.129");
    params.setSubPath("spec");
    params.setVisitProxy(true);
    params.setResourceType(ResourceType.NODES);
    _restfulClient.get(params);

    params = new Params();
    params.setName("192.168.1.129");
    params.setSubPath("run/ns-sample/pod/pod-sample-in-namespace");
    params.setVisitProxy(true);
    params.setResourceType(ResourceType.NODES);
    _restfulClient.get(params);

    params = new Params();
    params.setName("192.168.1.129");
    params.setSubPath("metrics");
```

```
    params.setVisitProxy(true);
    params.setResourceType(ResourceType.NODES);
    _restfulClient.get(params);
}
```

最后，举例说明如何通过 API 删除资源对象 pod，代码如下：

```
private void testDetetePod() {
    Params params = new Params();
    params.setNamespace("ns-sample");
    params.setName("pod-sample-in-namespace");
    params.setResourceType(ResourceType.PODS);
    LOG.info("Result: " + _restfulClient.delete(params));
}
```

通过 API 接口除了能够对资源对象实现前面列出的基本操作，还涉及两类特殊接口，一类是 WATCH，一类是 PROXY。这两类特殊接口所包含的接口如表 9.4 所示。

表 9.4 两类特殊接口所包含的接口

资源类型	类别	方法	URL Path 及其说明
NODES	WATCH	GET	/api/v1/watch/nodes：监听所有节点的变化；
			/api/v1/watch/nodes/{name}：监听单个节点的变化
	PROXY	DELETE	/api/v1/proxy/nodes/{name}/{path:*}：代理 DELETE 请求到节点的某个子目录；
			/api/v1/proxy/nodes/{name}：代理 DELETE 请求到节点
		GET	/api/v1/proxy/nodes/{name}/{path:*}：代理 GET 请求到节点的某个子目录；
			/api/v1/proxy/nodes/{name}：代理 GET 请求到节点
		HEAD	/api/v1/proxy/nodes/{name}/{path:*}：代理 HEAD 请求到节点的某个子目录；
			/api/v1/proxy/nodes/{name}：代理 HEAD 请求到节点
		OPTIONS	/api/v1/proxy/nodes/{name}/{path:*}：代理 OPTIONS 请求到节点的某个子目录；
			/api/v1/proxy/nodes/{name}：代理 OPTIONS 请求到节点
		POST	/api/v1/proxy/nodes/{name}/{path:*}：代理 POST 请求到节点的某个子目录；
			/api/v1/proxy/nodes/{name}：代理 POST 请求到节点
		PUT	/api/v1/proxy/nodes/{name}/{path:*}：代理 PUT 请求到节点的某个子目录；
			/api/v1/proxy/nodes/{name}：代理 PUT 请求到节点

续表

资源类型	类别	方法	URL Path 及其说明
SERVICES	WATCH	GET	/api/v1/watch/services：监听所有 Service 的变化； /api/v1/watch/namespaces/{namespace}/services：监听某个 Namespace 下所有 Service 的变化； /api/v1/watch/namespaces/{namespace}/services/{name}：监听某个 Service 的变化
	PROXY	DELETE	/api/v1/proxy/namespaces/{namespace}/services/{name}/{path:*}：代理 DELETE 请求到 Service 的某个子目录； /api/v1/proxy/namespaces/{namespace}/services/{name}：代理 DELETE 请求到 Service
		GET	/api/v1/proxy/namespaces/{namespace}/services/{name}/{path:*}：代理 GET 请求到 Service 的某个子目录； /api/v1/proxy/namespaces/{namespace}/services/{name}：代理 GET 请求到 Service
		POST	/api/v1/proxy/namespaces/{namespace}/services/{name}/{path:*}：代理 POST 请求到 Service 的某个子目录； /api/v1/proxy/namespaces/{namespace}/services/{name}：代理 POST 请求到 Service
		PUT	/api/v1/proxy/namespaces/{namespace}/services/{name}/{path:*}：代理 PUT 请求到 Service 的某个子目录； /api/v1/proxy/namespaces/{namespace}/services/{name}：代理 PUT 请求到 Service
		HEAD	/api/v1/proxy/namespaces/{namespace}/services/{name}/{path:*}：代理 HEAD 请求到 Service 的某个子目录； /api/v1/proxy/namespaces/{namespace}/services/{name}：代理 HEAD 请求到 Service
		OPTIONS	/api/v1/proxy/namespaces/{namespace}/services/{name}/{path:*}：代理 OPTIONS 请求到 Service 的某个子目录； /api/v1/proxy/namespaces/{namespace}/services/{name}：代理 OPTIONS 请求到 Service

续表

资源类型	类别	方法	URL Path 及其说明
REPLICATIONCONTROLLER	WATCH	GET	/api/v1/watch/replicationcontrollers：监听所有 RC 的变化； /api/v1/watch/namespaces/{namespace}/replicationcontrollers：监听某个 Namespace 下所有 RC 的变化； /api/v1/watch/namespaces/{namespace}/replicationcontrollers/{name}：监听某个 RC 的变化
PODS	WATCH	GET	/api/v1/watch/pods：监听所有 Pod 的变化； /api/v1/watch/namespaces/{namespace}/pods：监听某个 Namespace 下所有 Pod 的变化； /api/v1/watch/namespaces/{namespace}/pods/{name}：监听某个 Pod 的变化
	PROXY	DELETE	/api/v1/namespaces/{namespace}/pods/{name}/proxy/{path:*}：代理 DELETE 请求到 Pod 的某个子目录； /api/v1/namespaces/{namespace}/pods/{name}/proxy：代理 DELETE 请求到 Pod； /api/v1/proxy/namespaces/{namespace}/pods/{name}/{path:*}：代理 DELETE 请求到 Pod 的某个子目录； /api/v1/proxy/namespaces/{namespace}/pods/{name}：代理 DELETE 请求到 Pod
		HEAD	/api/v1/namespaces/{namespace}/pods/{name}/proxy/{path:*}：代理 HEAD 请求到 Pod 的某个子目录； /api/v1/namespaces/{namespace}/pods/{name}/proxy：代理 HEAD 请求到 Pod； /api/v1/proxy/namespaces/{namespace}/pods/{name}/{path:*}：代理 HEAD 请求到 Pod 的某个子目录； /api/v1/proxy/namespaces/{namespace}/pods/{name}：代理 HEAD 请求到 Pod
		OPTIONS	/api/v1/namespaces/{namespace}/pods/{name}/proxy/{path:*}：代理 OPTIONS 请求到 Pod 的某个子目录； /api/v1/namespaces/{namespace}/pods/{name}/proxy：代理 OPTIONS 请求到 Pod； /api/v1/proxy/namespaces/{namespace}/pods/{name}/{path:*}：代理 OPTIONS 请求到 Pod 的某个子目录； /api/v1/proxy/namespaces/{namespace}/pods/{name}：代理 OPTIONS 请求到 Pod

续表

资源类型	类别	方法	URL Path 及其说明
		PUT	/api/v1/namespaces/{namespace}/pods/{name}/proxy/{path:*}：代理 PUT 请求到 Pod 的某个子目录； /api/v1/namespaces/{namespace}/pods/{name}/proxy：代理 PUT 请求到 Pod； /api/v1/proxy/namespaces/{namespace}/pods/{name}/{path:*}：代理 PUT 请求到 Pod 的某个子目录； /api/v1/proxy/namespaces/{namespace}/pods/{name}：代理 PUT 请求到 Pod
		GET	/api/v1/namespaces/{namespace}/pods/{name}/proxy/{path:*}：代理 GET 请求到 Pod 的某个子目录； /api/v1/namespaces/{namespace}/pods/{name}/proxy：代理 GET 请求到 Pod； /api/v1/proxy/namespaces/{namespace}/pods/{name}/{path:*}：代理 GET 请求到 Pod 的某个子目录； /api/v1/proxy/namespaces/{namespace}/pods/{name}：代理 GET 请求到 Pod
		POST	/api/v1/namespaces/{namespace}/pods/{name}/proxy/{path:*}：代理 POST 请求到 Pod 的某个子目录； /api/v1/namespaces/{namespace}/pods/{name}/proxy：代理 POST 请求到 Pod； /api/v1/proxy/namespaces/{namespace}/pods/{name}/{path:*}：代理 POST 请求到 Pod 的某个子目录； /api/v1/proxy/namespaces/{namespace}/pods/{name}：代理 POST 请求到 Pod
ENDPOINTS	WATCH	GET	/api/v1/watch/endpoints：监听所有 Endpoint 的变化； /api/v1/watch/namespaces/{namespace}/endpoints：监听某个 Namespace 下所有 Endpoint 的变化； /api/v1/watch/namespaces/{namespace}/endpoints/{name}：监听某个 Endpoint 的变化
SERVICEACCOUNT	WATCH	GET	/api/v1/watch/serviceaccounts：监听所有 ServiceAccount 的变化； /api/v1/watch/namespaces/{namespace}/serviceaccounts：监听某个 Namespace 下所有 ServiceAccount 的变化； /api/v1/watch/namespaces/{namespace}/serviceaccounts/{name}：监听某个 ServiceAccount 的变化

续表

资源类型	类别	方法	URL Path 及其说明
SECRET	WATCH	GET	/api/v1/watch/secrets：监听所有 Secret 的变化； /api/v1/watch/namespaces/{namespace}/secrets：监听某个 Namespace 下所有 Secret 的变化； /api/v1/watch/namespaces/{namespace}/secrets/{name}：监听某个 Secret 的变化
EVENTS	WATCH	GET	/api/v1/watch/events：监听所有 Event 的变化； /api/v1/watch/namespaces/{namespace}/events：监听某个 Namespace 下所有 Event 的变化； /api/v1/watch/namespaces/{namespace}/events/{name}：监听某个 Event 的变化
LIMITRANGES	WATCH	GET	/api/v1/watch/limitranges：监听所有 Event 的变化； /api/v1/watch/namespaces/{namespace}/limitranges：监听某个 Namespace 下所有 Event 的变化； /api/v1/watch/namespaces/{namespace}/limitranges/{name}：监听某个 Event 的变化
RESOURCEQUOTAS	WATCH	GET	/api/v1/watch/resourcequotas：监听所有 ResourceQuota 的变化； /api/v1/watch/namespaces/{namespace}/resourcequotas：监听某个 Namespace 下所有 ResourceQuota 的变化； /api/v1/watch/namespaces/{namespace}/resourcequotas/{name}：监听某个 ResourceQuota 的变化
PODTEMPLATES	WATCH	GET	/api/v1/watch/podtemplates：监听所有 PodTemplate 的变化； /api/v1/watch/namespaces/{namespace}/podtemplates：监听某个 Namespace 下所有 PodTemplate 的变化； /api/v1/watch/namespaces/{namespace}/podtemplates/{name}：监听某个 PodTemplate 的变化
PERSISTENTVOLUMES	WATCH	GET	/api/v1/watch/persistentvolumes：监听所有 PersistentVolume 的变化； /api/v1/watch/persistentvolumes/{name}：监听某个 PersistentVolume 的变化
PERSISTENTVOLUMECLAIMS	WATCH	GET	/api/v1/watch/persistentvolumeclaims：监听所有 PersistentVolumeClaim 的变化； /api/v1/watch/namespaces/{namespace}/persistentvolumeclaims：监听某个 Namespace 下所有 PersistentVolumeClaim 的变化； /api/v1/watch/namespaces/{namespace}/persistentvolumeclaims/{name}：监听某个 PersistentVolumeClaim 的变化

下面基于 Fabric8 实现对资源对象的监听，代码如下：

```java
private void testWatcher() {
    _kube.pods().watch(new io.fabric8.kubernetes.client.Watcher<Pod>() {
        @Override
        public void eventReceived(Action action, Pod pod) {
            System.out.println(action + ": " + pod);
        }

        @Override
        public void onClose(KubernetesClientException e) {
            System.out.println("Closed: " + e);
        }
    });
}
```

接下来基于 Jersey 框架实现通过 Proxy 方式访问 Pod。由于 API Server 针对 Pod 资源提供了两种 Proxy 访问接口，所以下面分别用两段代码进行示例说明。代码如下：

```java
private void testPoxyPod() {
    // 访问第 1 种 Proxy 接口
    Params params = new Params();
    params.setNamespace("ns-sample");
    params.setName("pod-sample-in-namespace");
    params.setSubPath("/proxy");
    params.setResourceType(ResourceType.PODS);

    _restfulClient.get(params);

    // 访问第 2 种 Proxy 接口
    params = new Params();
    params.setNamespace("ns-sample");
    params.setName("pod-sample-in-namespace");
    params.setVisitProxy(true);
    params.setResourceType(ResourceType.PODS);

    _restfulClient.get(params);
}
```

9.3.4 其他客户端库

为了让开发人员更方便地访问 Kubernetes 的 RESTful API，Kubernetes 社区推出了针对 Go、Python、Java、dotNet、JavaScript 等编程语言的客户端库，这些库由特别兴趣小组（SIG）API Machinary 维护，其官方网站为 https://github.com/kubernetes/community/tree/master/sig-api-machinery。

目前 Kubernetes 官方支持的客户端库如表 9.5 所示。

表 9.5　目前 Kubernetes 官方支持的客户端库

开 发 语 言	客户端库网址
Go	https://github.com/kubernetes/client-go/
Python	https://github.com/kubernetes-client/python/
Java	https://github.com/kubernetes-client/java/
dotNet	https://github.com/kubernetes-client/csharp/
JavaScript	https://github.com/kubernetes-client/javascript

此外，Kubernetes 社区也在开发和维护基于其他开发语言的客户端库，如 9.6 所示。

表 9.6　基于其他开发语言的客户端库

开 发 语 言	客户端库网址
Clojure	github.com/yanatan16/clj-kubernetes-api
Go	github.com/ericchiang/k8s
Java（OSGi）	bitbucket.org/amdatulabs/amdatu-kubernetes
Java（Fabric8、OSGi）	github.com/fabric8io/kubernetes-client
Lisp	github.com/brendandburns/cl-k8s
Lisp	github.com/xh4/cube
Node.js（TypeScript）	github.com/Goyoo/node-k8s-client
Node.js	github.com/tenxcloud/node-kubernetes-client
Node.js	github.com/godaddy/kubernetes-client
Perl	metacpan.org/pod/Net::Kubernetes
PHP	github.com/maclof/kubernetes-client
PHP	github.com/allansun/kubernetes-php-client
Python	github.com/eldarion-gondor/pykube

续表

开 发 语 言	客户端库网址
Python	github.com/mnubo/kubernetes-py
Ruby	github.com/Ch00k/kuber
Ruby	github.com/abonas/kubeclient
Ruby	github.com/kontena/k8s-client
Rust	github.com/ynqa/kubernetes-rust
Scala	github.com/doriordan/skuber
dotNet	github.com/tonnyeremin/kubernetes_gen
DotNet（RestSharp）	github.com/masroorhasan/Kubernetes.DotNet
Elixir	github.com/obmarg/kazan
Haskell	github.com/soundcloud/haskell-kubernetes

9.4 Kubernetes API 的扩展

随着 Kubernetes 的发展，用户对 Kubernetes 的扩展性也提出了越来越高的要求。从 1.7 版本开始，Kubernetes 引入扩展 API 资源的能力，使得开发人员在不修改 Kubernetes 核心代码的前提下可以对 Kubernetes API 进行扩展，仍然使用 Kubernetes 的语法对新增的 API 进行操作，这非常适用于在 Kubernetes 上通过其 API 实现其他功能（例如第三方性能指标采集服务）或者测试实验性新特性（例如外部设备驱动）。

在 Kubernetes 中，所有对象都被抽象定义为某种资源对象，同时系统会为其设置一个 API 入口（API Endpoint），对资源对象的操作（如新增、删除、修改、查看等）都需要通过 Master 的核心组件 API Server 调用资源对象的 API 来完成。与 API Server 的交互可以通过 kubectl 命令行工具或访问其 RESTful API 进行。每个 API 都可以设置多个版本，在不同的 API URL 路径下区分，例如 "/api/v1" 或 "/apis/extensions/v1beta1" 等。使用这种机制后，用户可以很方便地定义这些 API 资源对象(YAML 配置)，并将其提交给 Kubernetes（ 调用 RESTful API ），来完成对容器应用的各种管理工作。

Kubernetes 系统内置的 Pod、RC、Service、ConfigMap、Volume 等资源对象已经能够满足常见的容器应用管理要求，但如果用户希望将其自行开发的第三方系统纳入 Kubernetes，并使用 Kubernetes 的 API 对其自定义的功能或配置进行管理，就需要对 API

进行扩展了。目前 Kubernetes 提供了以下两种机制供用户扩展 API。

（1）使用 CRD 机制：复用 Kubernetes 的 API Server，无须编写额外的 API Server。用户只需要定义 CRD，并且提供一个 CRD 控制器，就能通过 Kubernetes 的 API 管理自定义资源对象了，同时要求用户的 CRD 对象符合 API Server 的管理规范。

（2）使用 API 聚合机制：用户需要编写额外的 API Server，可以对资源进行更细粒度的控制（例如，如何在各 API 版本之间切换），要求用户自行处理对多个 API 版本的支持。

本节主要对 CRD 和 API 聚合这两种 API 扩展机制的概念和用法进行详细说明。

9.4.1 使用 CRD 扩展 API 资源

CRD 是 Kubernetes 从 1.7 版本开始引入的特性，在 Kubernetes 早期版本中被称为 TPR（ThirdPartyResources，第三方资源）。TPR 从 Kubernetes 1.8 版本开始被停用，被 CRD 全面替换。

CRD 本身只是一段声明，用于定义用户自定义的资源对象。但仅有 CRD 的定义并没有实际作用，用户还需要提供管理 CRD 对象的 CRD 控制器（CRD Controller），才能实现对 CRD 对象的管理。CRD 控制器通常可以通过 Go 语言进行开发，并需要遵循 Kubernetes 的控制器开发规范，基于客户端库 client-go 进行开发，需要实现 Informer、ResourceEventHandler、Workqueue 等组件具体的功能处理逻辑，详细的开发过程请参考官方示例（https://github.com/kubernetes/sample-controller）和 client-go 库（https://github.com/kubernetes/sample-controller/blob/master/docs/controller-client-go.md）的详细说明。

1. 创建 CRD 的定义

与其他资源对象一样，对 CRD 的定义也使用 YAML 配置进行声明。以 Istio 系统中的自定义资源 VirtualService 为例，配置文件 crd-virtualservice.yaml 的内容如下：

```
apiVersion: apiextensions.k8s.io/v1beta1
kind: CustomResourceDefinition
metadata:
  name: virtualservices.networking.istio.io
  annotations:
    "helm.sh/hook": crd-install
  labels:
```

```
      app: istio-pilot
spec:
  group: networking.istio.io
  scope: Namespaced
  versions:
  - name: v1alpha3
    served: true
    storage: true
  names:
    kind: VirtualService
    listKind: VirtualServiceList
    singular: virtualservice
    plural: virtualservices
    categories:
    - istio-io
    - networking-istio-io
```

CRD 定义中的关键字段如下。

（1）group：设置 API 所属的组，将其映射为 API URL 中"/apis/"的下一级目录，设置 networking.istio.io 生成的 API URL 路径为"/apis/networking.istio.io"。

（2）scope：该 API 的生效范围，可选项为 Namespaced（由 Namespace 限定）和 Cluster（在集群范围全局生效，不局限于任何 Namespace），默认值为 Namespaced。

（3）versions：设置此 CRD 支持的版本，可以设置多个版本，用列表形式表示。目前还可以设置名为 version 的字段，只能设置一个版本，在将来的 Kubernetes 版本中会被弃用，建议使用 versions 进行设置。如果该 CRD 支持多个版本，则每个版本都会在 API URL "/apis/networking.istio.io" 的下一级进行体现，例如 "/apis/networking.istio.io/v1" 或 "/apis/networking.istio.io/v1alpha3" 等。每个版本都可以设置下列参数。

◎ name：版本的名称，例如 v1、v1alpha3 等。
◎ served：是否启用，在被设置为 true 时表示启用。
◎ storage：是否进行存储，只能有一个版本被设置为 true。

（4）names：CRD 的名称，包括单数、复数、kind、所属组等名称的定义，可以设置如下参数。

- kind：CRD 的资源类型名称，要求以驼峰式命名规范进行命名（单词的首字母都大写），例如 VirtualService。
- listKind：CRD 列表，默认被设置为<kind>List 格式，例如 VirtualServiceList。
- singular：单数形式的名称，要求全部小写，例如 virtualservice。
- plural：复数形式的名称，要求全部小写，例如 virtualservices。
- shortNames：缩写形式的名称，要求全部小写，例如 vs。
- categories：CRD 所属的资源组列表。例如，VirtualService 属于 istio-io 组和 networking-istio-io 组，用户通过查询 istio-io 组和 networking-istio-io 组，也可以查询到该 CRD 实例。

使用 kubectl create 命令完成 CRD 的创建：

```
# kubectl create -f crd-virtualservice.yaml
customresourcedefinition.apiextensions.k8s.io/virtualservices.networking.istio.io created
```

在 CRD 创建成功后，由于本例的 scope 设置了 Namespace 限定，所以可以通过 API Endpoint "/apis/networking.istio.io/v1alpha3/namespaces/<namespace>/virtualservices/" 管理该 CRD 资源。

用户接下来就可以基于该 CRD 的定义创建相应的自定义资源对象了。

2. 基于 CRD 的定义创建自定义资源对象

基于 CRD 的定义，用户可以像创建 Kubernetes 系统内置的资源对象（如 Pod）一样创建 CRD 资源对象。在下面的例子中，virtualservice-helloworld.yaml 定义了一个类型为 VirtualService 的资源对象：

```
apiVersion: networking.istio.io/v1alpha3
kind: VirtualService
metadata:
  name: helloworld
spec:
  hosts:
  - "*"
  gateways:
  - helloworld-gateway
  http:
  - match:
```

```
      - uri:
          exact: /hello
    route:
    - destination:
        host: helloworld
        port:
          number: 5000
```

除了需要设置该 CRD 资源对象的名称,还需要在 spec 段设置相应的参数。在 spec 中可以设置的字段是由 CRD 开发者自定义的,需要根据 CRD 开发者提供的手册进行配置。这些参数通常包含特定的业务含义,由 CRD 控制器进行处理。

使用 kubectl create 命令完成 CRD 资源对象的创建:

```
# kubectl create -f virtualservice-helloworld.yaml
virtualservice.networking.istio.io/helloworld created
```

然后,用户就可以像操作 Kubernetes 内置的资源对象(如 Pod、RC、Service)一样去操作 CRD 资源对象了,包括查看、更新、删除和 watch 等操作。

查看 CRD 资源对象:

```
# kubectl get virtualservice
NAME         AGE
helloworld   1m
```

也可以通过 CRD 所属的 categories 进行查询:

```
# kubectl get istio-io
NAME         AGE
helloworld   1m
# kubectl get networking-istio-io
NAME         AGE
helloworld   1m
```

3. CRD 的高级特性

随着 Kubernetes 的演进,CRD 也在逐步添加一些高级特性和功能,包括 subresources 子资源、校验(Validation)机制、自定义查看 CRD 时需要显示的列,以及 finalizer 预删除钩子。

（1）CRD 的 subresources 子资源

Kubernetes 从 1.11 版本开始，在 CRD 的定义中引入了名为 subresources 的配置，可以设置的选项包括 status 和 scale 两类。

- stcatus：启用 /status 路径，其值来自 CRD 的 .status 字段，要求 CRD 控制器能够设置和更新这个字段的值。
- scale：启用 /scale 路径，支持通过其他 Kubernetes 控制器（如 HorizontalPodAutoscaler 控制器）与 CRD 资源对象实例进行交互。用户通过 kubectl scale 命令也能对该 CRD 资源对象进行扩容或缩容操作，要求 CRD 本身支持以多个副本的形式运行。

下面是一个设置了 subresources 的 CRD 示例：

```
apiVersion: apiextensions.k8s.io/v1beta1
kind: CustomResourceDefinition
metadata:
  name: crontabs.stable.example.com
spec:
  group: stable.example.com
  versions:
  - name: v1
    served: true
    storage: true
  scope: Namespaced
  names:
    plural: crontabs
    singular: crontab
    kind: CronTab
    shortNames:
    - ct
  subresources:
    status: {}
    scale:
      # 定义从 CRD 元数据获取用户期望的副本数量的 JSON 路径
      specReplicasPath: .spec.replicas
      # 定义从 CRD 元数据获取当前运行的副本数量的 JSON 路径
      statusReplicasPath: .status.replicas
      # 定义从 CRD 元数据获取 Label Selector（标签选择器）的 JSON 路径
      labelSelectorPath: .status.labelSelector
```

基于该 CRD 的定义,创建一个自定义资源对象 my-crontab.yaml:

```
apiVersion: "stable.example.com/v1"
kind: CronTab
metadata:
  name: my-new-cron-object
spec:
  cronSpec: "* * * * */5"
  image: my-awesome-cron-image
  replicas: 3
```

之后就能通过 API Endpoint 查看该资源对象的状态了:

/apis/stable.example.com/v1/namespaces/<namespace>/crontabs/status

并查看该资源对象的扩缩容(scale)信息:

/apis/stable.example.com/v1/namespaces/<namespace>/crontabs/scale

用户还可以使用 kubectl scale 命令对 Pod 的副本数量进行调整,例如:

```
# kubectl scale --replicas=5 crontabs/my-new-cron-object
crontabs "my-new-cron-object" scaled
```

(2)CRD 的校验(Validation)机制

Kubernetes 从 1.8 版本开始引入了基于 OpenAPI v3 schema 或 validatingadmissionwebhook 的校验机制,用于校验用户提交的 CRD 资源对象配置是否符合预定义的校验规则。该机制到 Kubernetes 1.13 版本时升级为 Beta 版。要使用该功能,需要为 kube-apiserver 服务开启--feature-gates=CustomResourceValidation=true 特性开关。

下面的例子为 CRD 定义中的两个字段(cronSpec 和 replicas)设置了校验规则:

```
apiVersion: apiextensions.k8s.io/v1beta1
kind: CustomResourceDefinition
metadata:
  name: crontabs.stable.example.com
spec:
  group: stable.example.com
  versions:
  - name: v1
    served: true
    storage: true
```

```
  version: v1
  scope: Namespaced
  names:
    plural: crontabs
    singular: crontab
    kind: CronTab
    shortNames:
    - ct
validation:
  openAPIV3Schema:
    properties:
      spec:
        properties:
          cronSpec:
            type: string
            pattern: '^(\d+|\*)(/\d+)?(\s+(\d+|\*)(/\d+)?){4}$'
          replicas:
            type: integer
            minimum: 1
            maximum: 10
```

校验规则如下。

◎ spec.cronSpec：必须为字符串类型，并且满足正则表达式的格式。
◎ spec.replicas：必须将其设置为 1～10 的整数。

对于不符合要求的 CRD 资源对象定义，系统将拒绝创建。

例如，下面的 my-crontab.yaml 示例违反了 CRD 中 validation 设置的校验规则，即 cronSpec 没有满足正则表达式的格式，replicas 的值大于 10：

```
apiVersion: "stable.example.com/v1"
kind: CronTab
metadata:
  name: my-new-cron-object
spec:
  cronSpec: "* * * *"
  image: my-awesome-cron-image
  replicas: 15
```

创建时，系统将报出 validation 失败的错误信息：

```
# kubectl create -f my-crontab.yaml
The CronTab "my-new-cron-object" is invalid: []: Invalid value:
map[string]interface {}{"apiVersion":"stable.example.com/v1", "kind":"CronTab",
"metadata":map[string]interface {}{"name":"my-new-cron-object",
"namespace":"default", "deletionTimestamp":interface {}(nil),
"deletionGracePeriodSeconds":(*int64)(nil),
"creationTimestamp":"2017-09-05T05:20:07Z",
"uid":"e14d79e7-91f9-11e7-a598-f0761cb232d1", "selfLink":"", "clusterName":""},
"spec":map[string]interface {}{"cronSpec":"* * * *",
"image":"my-awesome-cron-image", "replicas":15}}:
validation failure list:
spec.cronSpec in body should match '^(\d+|\*)(/\d+)?(\s+(\d+|\*)(/\d+)?){4}$'
spec.replicas in body should be less than or equal to 10
```

(3)自定义查看 CRD 时需要显示的列

从 Kubernetes 1.11 版本开始,通过 kubectl get 命令能够显示哪些字段由服务端(API Server)决定,还支持在 CRD 中设置需要在查看(get)时显示的自定义列,在 spec.additionalPrinterColumns 字段设置即可。

在下面的例子中设置了 3 个需要显示的自定义列 Spec、Replicas 和 Age,并在 JSONPath 字段设置了自定义列的数据来源:

```
apiVersion: apiextensions.k8s.io/v1beta1
  kind: CustomResourceDefinition
  metadata:
    name: crontabs.stable.example.com
  spec:
    group: stable.example.com
    version: v1
    scope: Namespaced
    names:
      plural: crontabs
      singular: crontab
      kind: CronTab
      shortNames:
      - ct
    additionalPrinterColumns:
    - name: Spec
      type: string
      description: The cron spec defining the interval a CronJob is run
```

```
      JSONPath: .spec.cronSpec
    - name: Replicas
      type: integer
      description: The number of jobs launched by the CronJob
      JSONPath: .spec.replicas
    - name: Age
      type: date
      JSONPath: .metadata.creationTimestamp
```

通过 kubectl get 命令查看 CronTab 资源对象，会显示出这 3 个自定义的列：

```
# kubectl get crontab my-new-cron-object
NAME                  SPEC        REPLICAS    AGE
my-new-cron-object    * * * * *   1           7s
```

（4）Finalizer（CRD 资源对象的预删除钩子方法）

Finalizer 设置的方法在删除 CRD 资源对象时进行调用，以实现 CRD 资源对象的清理工作。

在下面的例子中为 CRD "CronTab" 设置了一个 finalizer（也可以设置多个），其值为 URL "finalizer.stable.example.com"：

```
apiVersion: "stable.example.com/v1"
kind: CronTab
metadata:
  finalizers:
  - finalizer.stable.example.com
```

在用户发起删除该资源对象的请求时，Kubernetes 不会直接删除这个资源对象，而是在元数据部分设置时间戳 "metadata.deletionTimestamp" 的值，标记为开始删除该 CRD 对象。然后控制器开始执行 finalizer 定义的钩子方法 "finalizer.stable.example.com" 进行清理工作。对于耗时较长的清理操作，还可以设置 metadata.deletionGracePeriodSeconds 超时时间，在超过这个时间后由系统强制终止钩子方法的执行。在控制器执行完钩子方法后，控制器应负责删除相应的 finalizer。当全部 finalizer 都触发控制器执行钩子方法并都被删除之后，Kubernetes 才会最终删除该 CRD 资源对象。

4. 小结

CRD 极大扩展了 Kubernetes 的能力，使用户像操作 Pod 一样操作自定义的各种资源对象。CRD 已经在一些基于 Kubernetes 的第三方开源项目中得到广泛应用，包括 CSI 存储插件、Device Plugin（GPU 驱动程序）、Istio（Service Mesh 管理）等，已经逐渐成为扩展 Kubernetes 能力的标准。

9.4.2　使用 API 聚合机制扩展 API 资源

API 聚合机制是 Kubernetes 1.7 版本引入的特性，能够将用户扩展的 API 注册到 kube-apiserver 上，仍然通过 API Server 的 HTTP URL 对新的 API 进行访问和操作。为了实现这个机制，Kubernetes 在 kube-apiserver 服务中引入了一个 API 聚合层（API Aggregation Layer），用于将扩展 API 的访问请求转发到用户服务的功能。

设计 API 聚合机制的主要目标如下。

- 增加 API 的扩展性：使得开发人员可以编写自己的 API Server 来发布他们的 API，而无须对 Kubernetes 核心代码进行任何修改。
- 无须等待 Kubernetes 核心团队的繁杂审查：允许开发人员将其 API 作为单独的 API Server 发布，使集群管理员不用对 Kubernetes 核心代码进行修改就能使用新的 API，也就无须等待社区繁杂的审查了。
- 支持实验性新特性 API 开发：可以在独立的 API 聚合服务中开发新的 API，不影响系统现有的功能。
- 确保新的 API 遵循 Kubernetes 的规范：如果没有 API 聚合机制，开发人员就可能会被迫推出自己的设计，可能不遵循 Kubernetes 规范。

总的来说，API 聚合机制的目标是提供集中的 API 发现机制和安全的代理功能，将开发人员的新 API 动态地、无缝地注册到 Kubernetes API Server 中进行测试和使用。

下面对 API 聚合机制的使用方式进行详细说明。

1. 在 Master 的 API Server 中启用 API 聚合功能

为了能够将用户自定义的 API 注册到 Master 的 API Server 中，首先需要配置 kube-apiserver 服务的以下启动参数来启用 API 聚合功能。

- --requestheader-client-ca-file=/etc/kubernetes/ssl_keys/ca.crt：客户端 CA 证书。
- --requestheader-allowed-names=：允许访问的客户端 common names 列表，通过 header 中 --requestheader-username-headers 参数指定的字段获取。客户端 common names 的名称需要在 client-ca-file 中进行设置，将其设置为空值时，表示任意客户

端都可访问。
- --requestheader-extra-headers-prefix=X-Remote-Extra-：请求头中需要检查的前缀名。
- --requestheader-group-headers=X-Remote-Group：请求头中需要检查的组名。
- --requestheader-username-headers=X-Remote-User：请求头中需要检查的用户名。
- --proxy-client-cert-file=/etc/kubernetes/ssl_keys/kubelet_client.crt：在请求期间验证 Aggregator 的客户端 CA 证书。
- --proxy-client-key-file=/etc/kubernetes/ssl_keys/kubelet_client.key：在请求期间验证 Aggregator 的客户端私钥。

如果 kube-apiserver 所在的主机上没有运行 kube-proxy，即无法通过服务的 ClusterIP 进行访问，那么还需要设置以下启动参数：

```
--enable-aggregator-routing=true
```

在设置完成重启 kube-apiserver 服务，就启用 API 聚合功能了。

2. 注册自定义 APIService 资源

在启用了 API Server 的 API 聚合功能之后，用户就能将自定义 API 资源注册到 Kubernetes Master 的 API Server 中了。用户只需配置一个 APIService 资源对象，就能进行注册了。APIService 示例的 YAML 配置文件如下：

```yaml
apiVersion: apiregistration.k8s.io/v1beta1
kind: APIService
metadata:
  name: v1beta1.custom.metrics.k8s.io
spec:
  service:
    name: custom-metrics-server
    namespace: custom-metrics
  group: custom.metrics.k8s.io
  version: v1beta1
  insecureSkipTLSVerify: true
  groupPriorityMinimum: 100
  versionPriority: 100
```

在这个 APIService 中设置的 API 组名为 custom.metrics.k8s.io，版本号为 v1beta1，这两个字段将作为 API 路径的子目录注册到 API 路径 "/apis/" 下。注册成功后，就能通过

Master API 路径 "/apis/custom.metrics.k8s.io/v1beta1" 访问自定义的 API Server 了。

在 service 段中通过 name 和 namespace 设置了后端的自定义 API Server，本例中的服务名为 custom-metrics-server，命名空间为 custom-metrics。

通过 kubectl create 命令将这个 APIService 定义发送给 Master，就完成了注册操作。

之后，通过 Master API Server 对 "/apis/custom.metrics.k8s.io/v1beta1" 路径的访问都会被 API 聚合层代理转发到后端服务 custom-metrics-server.custom-metrics.svc 上了。

3. 实现和部署自定义 API Server

仅仅注册 APIService 资源还是不够的，用户对 "/apis/custom.metrics.k8s.io/v1beta1" 路径的访问实际上都被转发给了 custom-metrics-server.custom-metrics.svc 服务。这个服务通常能以普通 Pod 的形式在 Kubernetes 集群中运行。当然，这个服务需要由自定义 API 的开发者提供，并且需要遵循 Kubernetes 的开发规范，详细的开发示例可以参考官方给出的示例（https://github.com/kubernetes/sample-apiserver）。

下面是部署自定义 API Server 的常规操作步骤。

（1）确保 APIService API 已启用，这需要通过 kube-apiserver 的启动参数 --runtime-config 进行设置，默认是启用的。

（2）建议创建一个 RBAC 规则，允许添加 APIService 资源对象，因为 API 扩展对整个 Kubernetes 集群都生效，所以不推荐在生产环境中对 API 扩展进行开发或测试。

（3）创建一个新的 Namespace 用于运行扩展的 API Server。

（4）创建一个 CA 证书用于对自定义 API Server 的 HTTPS 安全访问进行签名。

（5）创建服务端证书和秘钥用于自定义 API Server 的 HTTPS 安全访问。服务端证书应该由上面提及的 CA 证书进行签名，也应该包含含有 DNS 域名格式的 CN 名称。

（6）在新的 Namespace 中使用服务端证书和秘钥创建 Kubernetes Secret 对象。

（7）部署自定义 API Server 实例，通常可以以 Deployment 形式进行部署，并且将之前创建的 Secret 挂载到容器内部。该 Deployment 也应被部署在新的 Namespace 中。

（8）确保自定义的 API Server 通过 Volume 加载了 Secret 中的证书，这将用于后续的 HTTPS 握手校验。

(9)在新的 Namespace 中创建一个 Service Account 对象。

(10)创建一个 ClusterRole 用于对自定义 API 资源进行操作。

(11)使用之前创建的 ServiceAccount 为刚刚创建的 ClusterRole 创建一个 ClusterRolebinding。

(12)使用之前创建的 ServiceAccount 为系统 ClusterRole "system:auth-delegator" 创建一个 ClusterRolebinding,以使其可以将认证决策代理转发给 Kubernetes 核心 API Server。

(13)使用之前创建的 ServiceAccount 为系统 Role "extension-apiserver-authentication-reader" 创建一个 Rolebinding,以允许自定义 API Server 访问名为 "extension-apiserver-authentication" 的系统 ConfigMap。

(14)创建 APIService 资源对象。

(15)访问 APIService 提供的 API URL 路径,验证对资源的访问能否成功。

下面以部署 Metrics Server 为例,说明一个聚合 API 的实现方式。

随着 API 聚合机制的出现,Heapster 也进入弃用阶段,逐渐被 Metrics Server 替代。Metrics Server 通过聚合 API 提供 Pod 和 Node 的资源使用数据,供 HPA 控制器、VPA 控制器及 kubectl top 命令使用。Metrics Server 的源码可以在 GitHub 代码库(https://github.com/kubernetes-incubator/metrics-server)找到,在部署完成后,Metrics Server 将通过 Kubernetes 核心 API Server 的 "/apis/metrics.k8s.io/v1beta1" 路径提供 Pod 和 Node 的监控数据。

首先,部署 Metrics Server 实例,在下面的 YAML 配置中包含一个 ServiceAccount、一个 Deployment 和一个 Service 的定义:

```yaml
---
apiVersion: v1
kind: ServiceAccount
metadata:
  name: metrics-server
  namespace: kube-system
---
apiVersion: extensions/v1beta1
kind: Deployment
metadata:
  name: metrics-server
```

```yaml
  namespace: kube-system
  labels:
    k8s-app: metrics-server
spec:
  selector:
    matchLabels:
      k8s-app: metrics-server
  template:
    metadata:
      name: metrics-server
      labels:
        k8s-app: metrics-server
    spec:
      serviceAccountName: metrics-server
      containers:
      - name: metrics-server
        image: k8s.gcr.io/metrics-server-amd64:v0.3.1
        imagePullPolicy: IfNotPresent
        volumeMounts:
        - name: tmp-dir
          mountPath: /tmp
      volumes:
      - name: tmp-dir
        emptyDir: {}
---
apiVersion: v1
kind: Service
metadata:
  name: metrics-server
  namespace: kube-system
  labels:
    kubernetes.io/name: "Metrics-server"
spec:
  selector:
    k8s-app: metrics-server
  ports:
  - port: 443
    protocol: TCP
    targetPort: 443
```

然后，创建 Metrics Server 所需的 RBAC 权限配置：

```yaml
# 对访问 pods、nodes、nodes/stats 等资源对象进行授权
---
kind: ClusterRole
apiVersion: rbac.authorization.k8s.io/v1
metadata:
  name: system:aggregated-metrics-reader
  labels:
    rbac.authorization.k8s.io/aggregate-to-view: "true"
    rbac.authorization.k8s.io/aggregate-to-edit: "true"
    rbac.authorization.k8s.io/aggregate-to-admin: "true"
rules:
- apiGroups: ["metrics.k8s.io"]
  resources: ["pods"]
  verbs: ["get", "list", "watch"]
---
apiVersion: rbac.authorization.k8s.io/v1
kind: ClusterRole
metadata:
  name: system:metrics-server
rules:
- apiGroups:
  - ""
  resources:
  - pods
  - nodes
  - nodes/stats
  verbs:
  - get
  - list
  - watch
---
apiVersion: rbac.authorization.k8s.io/v1
kind: ClusterRoleBinding
metadata:
  name: system:metrics-server
roleRef:
  apiGroup: rbac.authorization.k8s.io
  kind: ClusterRole
  name: system:metrics-server
subjects:
- kind: ServiceAccount
```

```
  name: metrics-server
  namespace: kube-system

# 定义ClusterRoleBinding，设置为将认证请求转发到Metrics Server上
---
apiVersion: rbac.authorization.k8s.io/v1beta1
kind: ClusterRoleBinding
metadata:
  name: metrics-server:system:auth-delegator
roleRef:
  apiGroup: rbac.authorization.k8s.io
  kind: ClusterRole
  name: system:auth-delegator
subjects:
- kind: ServiceAccount
  name: metrics-server
  namespace: kube-system

# 允许Metrics Server访问系统ConfigMap "extension-apiserver-authentication"
---
apiVersion: rbac.authorization.k8s.io/v1beta1
kind: RoleBinding
metadata:
  name: metrics-server-auth-reader
  namespace: kube-system
roleRef:
  apiGroup: rbac.authorization.k8s.io
  kind: Role
  name: extension-apiserver-authentication-reader
subjects:
- kind: ServiceAccount
  name: metrics-server
  namespace: kube-system
```

最后，定义APIService资源，主要设置自定义API的组（group）、版本号（version）及对应的服务（metrics-server.kube-system）：

```
apiVersion: apiregistration.k8s.io/v1beta1
kind: APIService
metadata:
  name: v1beta1.metrics.k8s.io
spec:
```

```
service:
  name: metrics-server
  namespace: kube-system
group: metrics.k8s.io
version: v1beta1
insecureSkipTLSVerify: true
groupPriorityMinimum: 100
versionPriority: 100
```

在所有资源都成功创建之后,在命名空间 kube-system 中会看到新建的 metrics-server Pod。

通过 Kubernetes Master API Server 的 URL "/apis/metrics.k8s.io/v1beta1" 就能查询到 Metrics Server 提供的 Pod 和 Node 的性能数据了:

```
# curl http://192.168.18.3:8080/apis/metrics.k8s.io/v1beta1/nodes
{
  "kind": "NodeMetricsList",
  "apiVersion": "metrics.k8s.io/v1beta1",
  "metadata": {
    "selfLink": "/apis/metrics.k8s.io/v1beta1/nodes"
  },
  "items": [
    {
      "metadata": {
        "name": "k8s-node-1",
        "selfLink": "/apis/metrics.k8s.io/v1beta1/nodes/k8s-node-1",
        "creationTimestamp": "2019-03-19T00:08:41Z"
      },
      "timestamp": "2019-03-19T00:08:16Z",
      "window": "30s",
      "usage": {
        "cpu": "349414075n",
        "memory": "1182512Ki"
      }
    }
  ]
}

# curl http://192.168.18.3:8080/apis/metrics.k8s.io/v1beta1/pods
{
  "kind": "PodMetricsList",
```

```
        "apiVersion": "metrics.k8s.io/v1beta1",
        "metadata": {
          "selfLink": "/apis/metrics.k8s.io/v1beta1/pods"
        },
        "items": [
          {
            "metadata": {
              "name": "metrics-server-7cb798c45b-4dnmh",
              "namespace": "kube-system",
              "selfLink":
"/apis/metrics.k8s.io/v1beta1/namespaces/kube-system/pods/metrics-server-7cb798c
45b-4dnmh",
              "creationTimestamp": "2019-03-19T00:13:45Z"
            },
            "timestamp": "2019-03-19T00:13:18Z",
            "window": "30s",
            "containers": [
              {
                "name": "metrics-server",
                "usage": {
                  "cpu": "1640261n",
                  "memory": "22240Ki"
                }
              }
            ]
          },
          ......
        ]
      }
```

第 10 章
Kubernetes 集群管理

10.1 Node 的管理

10.1.1 Node 的隔离与恢复

在硬件升级、硬件维护等情况下，我们需要将某些 Node 隔离，使其脱离 Kubernetes 集群的调度范围。Kubernetes 提供了一种机制，既可以将 Node 纳入调度范围，也可以将 Node 脱离调度范围。

创建配置文件 unschedule_node.yaml，在 spec 部分指定 unschedulable 为 true：

```yaml
apiVersion: v1
kind: Node
metadata:
  name: k8s-node-1
  labels:
    kubernetes.io/hostname: k8s-node-1
spec:
  unschedulable: true
```

通过 kubectl replace 命令完成对 Node 状态的修改：

```
$ kubectl replace -f unschedule_node.yaml
node "k8s-node-1" replaced
```

查看 Node 的状态，可以观察到在 Node 的状态中增加了一项 SchedulingDisabled：

```
# kubectl get nodes
NAME            STATUS                      AGE
k8s-node-1      Ready,SchedulingDisabled    1h
```

这样，对于后续创建的 Pod，系统将不会再向该 Node 进行调度。

也可以不使用配置文件，直接使用 kubectl patch 命令完成：

```
$ kubectl patch node k8s-node-1 -p '{"spec":{"unschedulable":true}}'
```

需要注意的是，将某个 Node 脱离调度范围时，在其上运行的 Pod 并不会自动停止，管理员需要手动停止在该 Node 上运行的 Pod。

同样，如果需要将某个 Node 重新纳入集群调度范围，则将 unschedulable 设置为 false，再次执行 kubectl replace 或 kubectl patch 命令就能恢复系统对该 Node 的调度。

另外，使用 kubectl 的子命令 cordon 和 uncordon 也可以实现将 Node 进行隔离调度和恢复调度操作。

例如，使用 kubectl cordon <node_name>对某个 Node 进行隔离调度操作：

```
# kubectl cordon k8s-node-1
node "k8s-node-1" cordoned

# kubectl get nodes
NAME            STATUS                      AGE
k8s-node-1      Ready,SchedulingDisabled    1h
```

使用 kubectl uncordon <node_name>对某个 Node 进行恢复调度操作：

```
# kubectl uncordon k8s-node-1
node "k8s-node-1" uncordoned

# kubectl get nodes
NAME            STATUS      AGE
k8s-node-1      Ready       1h
```

10.1.2 Node 的扩容

在实际生产系统中经常会出现服务器容量不足的情况，这时就需要购买新的服务器，然后将应用系统进行水平扩展来完成对系统的扩容。

在 Kubernetes 集群中，一个新 Node 的加入是非常简单的。在新的 Node 上安装 Docker、kubelet 和 kube-proxy 服务，然后配置 kubelet 和 kube-proxy 的启动参数，将 Master URL 指定为当前 Kubernetes 集群 Master 的地址，最后启动这些服务。通过 kubelet 默认的自动注册机制，新的 Node 将会自动加入现有的 Kubernetes 集群中，如图 10.1 所示。

Kubernetes Master 在接受了新 Node 的注册之后，会自动将其纳入当前集群的调度范围，之后创建容器时，就可以对新的 Node 进行调度了。

通过这种机制，Kubernetes 实现了集群中 Node 的扩容。

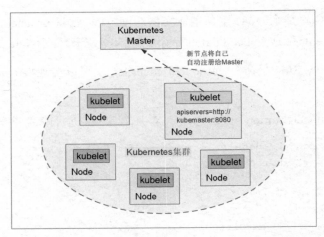

图 10.1　新 Node 自动注册并加入现有的 Kubernetes 集群中

10.2　更新资源对象的 Label

Label 是用户可灵活定义的对象属性，对于正在运行的资源对象，我们随时可以通过 kubectl label 命令进行增加、修改、删除等操作。

例如，要给已创建的 Pod "redis-master-bobr0" 添加一个标签 role=backend：

```
$ kubectl label pod redis-master-bobr0 role=backend
pod "redis-master-bobr0" labeled
```

查看该 Pod 的 Label：

```
$ kubectl get pods -Lrole
NAME                 READY    STATUS     RESTARTS    AGE       ROLE
redis-master-bobr0   1/1      Running    0           3m        backend
```

删除一个 Label 时，只需在命令行最后指定 Label 的 key 名并与一个减号相连即可：

```
$ kubectl label pod redis-master-bobr0 role-
pod "redis-master-bobr0" labeled
```

在修改一个 Label 的值时，需要加上 --overwrite 参数：

```
$ kubectl label pod redis-master-bobr0 role=master --overwrite
pod "redis-master-bobr0" labeled
```

10.3 Namespace：集群环境共享与隔离

在一个组织内部，不同的工作组可以在同一个 Kubernetes 集群中工作，Kubernetes 通过命名空间和 Context 的设置对不同的工作组进行区分，使得它们既可以共享同一个 Kubernetes 集群的服务，也能够互不干扰，如图 10.2 所示。

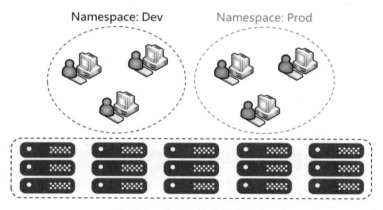

图 10.2 集群环境共享和隔离

假设在我们的组织中有两个工作组：开发组和生产运维组。开发组在 Kubernetes 集群中需要不断创建、修改、删除各种 Pod、RC、Service 等资源对象，以便实现敏捷开发。生产运维组则需要使用严格的权限设置来确保生产系统中的 Pod、RC、Service 处于正常运行状态且不会被误操作。

10.3.1 创建 Namespace

为了在 Kubernetes 集群中实现这两个分组，首先需要创建两个命名空间：

```
namespace-development.yaml:
apiVersion: v1
kind: Namespace
metadata:
  name: development
```

```yaml
namespace-production.yaml:
apiVersion: v1
kind: Namespace
metadata:
  name: production
```

使用 kubectl create 命令完成命名空间的创建：

```
$ kubectl create -f namespace-development.yaml
namespaces/development

$ kubectl create -f namespace-production.yaml
namespaces/production
```

查看系统中的命名空间：

```
$ kubectl get namespaces
NAME            LABELS              STATUS
default         <none>              Active
development     name=development    Active
production      name=production     Active
```

10.3.2 定义 Context（运行环境）

接下来，需要为这两个工作组分别定义一个 Context，即运行环境。这个运行环境将属于某个特定的命名空间。

通过 kubectl config set-context 命令定义 Context，并将 Context 置于之前创建的命名空间中：

```
$ kubectl config set-cluster kubernetes-cluster --server=https://192.168.1.128:8080
$ kubectl config set-context ctx-dev --namespace=development --cluster=kubernetes-cluster --user=dev
$ kubectl config set-context ctx-prod --namespace=production --cluster=kubernetes-cluster --user=prod
```

使用 kubectl config view 命令查看已定义的 Context：

```
$ kubectl config view
apiVersion: v1
clusters:
- cluster:
```

```
    server: http://192.168.1.128:8080
  name: kubernetes-cluster
contexts:
- context:
    cluster: kubernetes-cluster
    namespace: development
  name: ctx-dev
- context:
    cluster: kubernetes-cluster
    namespace: production
  name: ctx-prod
current-context: ctx-dev
kind: Config
preferences: {}
users: []
```

注意,通过 kubectl config 命令在${HOME}/.kube 目录下生成了一个名为 config 的文件,文件的内容为以 kubectl config view 命令查看到的内容。所以,也可以通过手工编辑该文件的方式来设置 Context。

10.3.3　设置工作组在特定 Context 环境下工作

使用 kubectl config use-context <context_name>命令设置当前运行环境。

下面的命令将把当前运行环境设置为 ctx-dev:

```
$ kubectl config use-context ctx-dev
```

运行这个命令后,当前的运行环境被设置为开发组所需的环境。之后的所有操作都将在名为 development 的命名空间中完成。

现在,以 redis-slave RC 为例创建两个 Pod:

```
redis-slave-controller.yaml
apiVersion: v1
kind: ReplicationController
metadata:
  name: redis-slave
  labels:
    name: redis-slave
spec:
```

```
    replicas: 2
    selector:
      name: redis-slave
    template:
      metadata:
        labels:
          name: redis-slave
      spec:
        containers:
        - name: slave
          image: kubeguide/guestbook-redis-slave
          ports:
          - containerPort: 6379

$ kubectl create -f redis-slave-controller.yaml
replicationcontrollers/redis-slave
```

查看创建好的 Pod：

```
$ kubectl get pods
NAME                  READY   STATUS    RESTARTS   AGE
redis-slave-0feq9     1/1     Running   0          6m
redis-slave-6i0g4     1/1     Running   0          6m
```

可以看到容器被正确创建并运行起来了。而且，由于当前运行环境是 ctx-dev，所以不会影响生产运维组的工作。

切换到生产运维组的运行环境：

```
$ kubectl config use-context ctx-prod
```

查看 RC 和 Pod：

```
$ kubectl get rc
CONTROLLER   CONTAINER(S)   IMAGE(S)   SELECTOR   REPLICAS

$ kubectl get pods
NAME      READY    STATUS    RESTARTS   AGE
```

结果为空，说明看不到开发组创建的 RC 和 Pod。

现在也为生产运维组创建两个 redis-slave 的 Pod：

```
$ kubectl create -f redis-slave-controller.yaml
replicationcontrollers/redis-slave
```

查看创建好的 Pod：

```
$ kubectl get pods
NAME                   READY    STATUS     RESTARTS    AGE
redis-slave-a4m7s      1/1      Running    0           12s
redis-slave-xyrkk      1/1      Running    0           12s
```

可以看到容器被正确创建并运行起来了，并且当前运行环境是 ctx-prod，也不会影响开发组的工作。

至此，我们为两个工作组分别设置了两个运行环境，设置好当前运行环境时，各工作组之间的工作将不会相互干扰，并且都能在同一个 Kubernetes 集群中同时工作。

10.4　Kubernetes 资源管理

本节先讲解 Pod 的两个重要参数：CPU Request 与 Memory Request。在大多数情况下，我们在定义 Pod 时并没有定义这两个参数，此时 Kubernetes 会认为该 Pod 所需的资源很少，并可以将其调度到任何可用的 Node 上。这样一来，当集群中的计算资源不很充足时，如果集群中的 Pod 负载突然加大，就会使某个 Node 的资源严重不足。

为了避免系统挂掉，该 Node 会选择 "清理" 某些 Pod 来释放资源，此时每个 Pod 都可能成为牺牲品。但有些 Pod 担负着更重要的职责，比其他 Pod 更重要，比如与数据存储相关的、与登录相关的、与查询余额相关的，即使系统资源严重不足，也需要保障这些 Pod 的存活，Kubernetes 中该保障机制的核心如下。

◎ 通过资源限额来确保不同的 Pod 只能占用指定的资源。
◎ 允许集群的资源被超额分配，以提高集群的资源利用率。
◎ 为 Pod 划分等级，确保不同等级的 Pod 有不同的服务质量（QoS），资源不足时，低等级的 Pod 会被清理，以确保高等级的 Pod 稳定运行。

Kubernetes 集群里的节点提供的资源主要是计算资源，计算资源是可计量的能被申请、分配和使用的基础资源，这使之区别于 API 资源（API Resources，例如 Pod 和 Services 等）。当前 Kubernetes 集群中的计算资源主要包括 CPU、GPU 及 Memory，绝大多数常规应用是用不到 GPU 的，因此这里重点介绍 CPU 与 Memory 的资源管理问题。

CPU 与 Memory 是被 Pod 使用的,因此在配置 Pod 时可以通过参数 CPU Request 及 Memory Request 为其中的每个容器指定所需使用的 CPU 与 Memory 量,Kubernetes 会根据 Request 的值去查找有足够资源的 Node 来调度此 Pod,如果没有,则调度失败。

我们知道,一个程序所使用的 CPU 与 Memory 是一个动态的量,确切地说,是一个范围,跟它的负载密切相关:负载增加时,CPU 和 Memory 的使用量也会增加。因此最准确的说法是,某个进程的 CPU 使用量为 0.1 个 CPU~1 个 CPU,内存占用则为 500MB~1GB。对应到 Kubernetes 的 Pod 容器上,就是下面这 4 个参数:

◎ spec.container[].resources.requests.cpu;
◎ spec.container[].resources.limits.cpu;
◎ spec.container[].resources.requests.memory;
◎ spec.container[].resources.limits.memory。

其中,limits 对应资源量的上限,即最多允许使用这个上限的资源量。由于 CPU 资源是可压缩的,进程无论如何也不可能突破上限,因此设置起来比较容易。对于 Memory 这种不可压缩资源来说,它的 Limit 设置就是一个问题了,如果设置得小了,当进程在业务繁忙期试图请求超过 Limit 限制的 Memory 时,此进程就会被 Kubernetes 杀掉。因此,Memory 的 Request 与 Limit 的值需要结合进程的实际需求谨慎设置。如果不设置 CPU 或 Memory 的 Limit 值,会怎样呢?在这种情况下,该 Pod 的资源使用量有一个弹性范围,我们不用绞尽脑汁去思考这两个 Limit 的合理值,但问题也来了,考虑下面的例子:

Pod A 的 Memory Request 被设置为 1GB,Node A 当时空闲的 Memory 为 1.2GB,符合 Pod A 的需求,因此 Pod A 被调度到 Node A 上。运行 3 天后,Pod A 的访问请求大增,内存需要增加到 1.5GB,此时 Node A 的剩余内存只有 200MB,由于 Pod A 新增的内存已经超出系统资源,所以在这种情况下,Pod A 就会被 Kubernetes 杀掉。

没有设置 Limit 的 Pod,或者只设置了 CPU Limit 或者 Memory Limit 两者之一的 Pod,表面看都是很有弹性的,但实际上,相对于 4 个参数都被设置的 Pod,是处于一种相对不稳定的状态的,它们与 4 个参数都没设置的 Pod 相比,只是稳定一点而已。理解了这一点,就很容易理解 Resource QoS 问题了。

如果我们有成百上千个不同的 Pod,那么先手动设置每个 Pod 的这 4 个参数,再检查并确保这些参数的设置,都是合理的。比如不能出现内存超过 2GB 或者 CPU 占据 2 个核心的 Pod。最后还得手工检查不同租户(Namespace)下的 Pod 的资源使用量是否超过限额。为此,Kubernetes 提供了另外两个相关对象:LimitRange 及 ResourceQuota,前者解决 request 与 limit 参数的默认值和合法取值范围等问题，后者则解决约束租户的资源配额问题。

本章从计算资源管理（Compute Resources）、服务质量管理（QoS）、资源配额管理（LimitRange、ResourceQuota）等方面，对 Kubernetes 集群内的资源管理进行详细说明，并结合实践操作、常见问题分析和一个完整的示例，力求对 Kubernetes 集群资源管理相关的运维工作提供指导。

10.4.1　计算资源管理

1. 详解 Requests 和 Limits 参数

以 CPU 为例，图 10.3 显示了未设置 Limits 和设置了 Requests、Limits 的 CPU 使用率的区别。

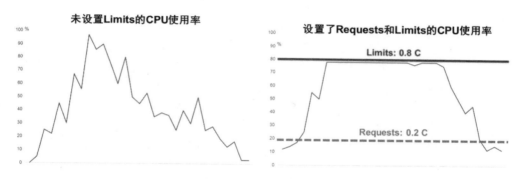

图 10.3　未设置和设置了 Requests 及 Limits 的 CPU 使用率的区别

尽管 Requests 和 Limits 只能被设置到容器上，但是设置 Pod 级别的 Requests 和 Limits 能大大提高管理 Pod 的便利性和灵活性，因此在 Kubernetes 中提供了对 Pod 级别的 Requests 和 Limits 的配置。对于 CPU 和内存而言，Pod 的 Requests 或 Limits 是指该 Pod 中所有容器的 Requests 或 Limits 的总和（对于 Pod 中没有设置 Requests 或 Limits 的容器，该项的值被当作 0 或者按照集群配置的默认值来计算）。下面对 CPU 和内存这两种计算资源的特点进行说明。

1) CPU

CPU 的 Requests 和 Limits 是通过 CPU 数（cpus）来度量的。CPU 的资源值是绝对值，

而不是相对值，比如 0.1CPU 在单核或多核机器上是一样的，都严格等于 0.1 CPU core。

2）Memory

内存的 Requests 和 Limits 计量单位是字节数。使用整数或者定点整数加上国际单位制（International System of Units）来表示内存值。国际单位制包括十进制的 E、P、T、G、M、K、m，或二进制的 Ei、Pi、Ti、Gi、Mi、Ki。KiB 与 MiB 是以二进制表示的字节单位，常见的 KB 与 MB 则是以十进制表示的字节单位，比如：

◎ 1 KB（KiloByte）= 1000 Bytes = 8000 Bits；
◎ 1 KiB（KibiByte）= 2^{10} Bytes = 1024 Bytes = 8192 Bits。

因此，128974848、129e6、129M、123Mi 的内存配置是一样的。

Kubernetes 的计算资源单位是大小写敏感的，因为 m 可以表示千分之一单位（milli unit），而 M 可以表示十进制的 1000，二者的含义不同；同理，小写的 k 不是一个合法的资源单位。

以某个 Pod 中的资源配置为例：

```
apiVersion: v1
kind: Pod
metadata:
  name: frontend
spec:
  containers:
  - name: db
    image: mysql
    resources:
      requests:
        memory: "64Mi"
        cpu: "250m"
      limits:
        memory: "128Mi"
        cpu: "500m"
  - name: wp
    image: wordpress
    resources:
      requests:
        memory: "64Mi"
        cpu: "250m"
```

```
      limits:
        memory: "128Mi"
        cpu: "500m"
```

如上所示,该 Pod 包含两个容器,每个容器配置的 Requests 都是 0.25CPU 和 64MiB(2^{26} Bytes)内存,而配置的 Limits 都是 0.5CPU 和 128MiB(2^{27} Bytes)内存。

这个 Pod 的 Requests 和 Limits 等于 Pod 中所有容器对应配置的总和,所以 Pod 的 Requests 是 0.5CPU 和 128MiB(2^{27} Bytes)内存,Limits 是 1CPU 和 256MiB(2^{28} Bytes)内存。

2. 基于 Requests 和 Limits 的 Pod 调度机制

当一个 Pod 创建成功时,Kubernetes 调度器(Scheduler)会为该 Pod 选择一个节点来执行。对于每种计算资源(CPU 和 Memory)而言,每个节点都有一个能用于运行 Pod 的最大容量值。调度器在调度时,首先要确保调度后该节点上所有 Pod 的 CPU 和内存的 Requests 总和,不超过该节点能提供给 Pod 使用的 CPU 和 Memory 的最大容量值。

例如,某个节点上的 CPU 资源充足,而内存为 4GB,其中 3GB 可以运行 Pod,而某 Pod 的 Memory Requests 为 1GB、Limits 为 2GB,那么在这个节点上最多可以运行 3 个这样的 Pod。

这里需要注意:可能某节点上的实际资源使用量非常低,但是已运行 Pod 配置的 Requests 值的总和非常高,再加上需要调度的 Pod 的 Requests 值,会超过该节点提供给 Pod 的资源容量上限,这时 Kubernetes 仍然不会将 Pod 调度到该节点上。如果 Kubernetes 将 Pod 调度到该节点上,之后该节点上运行的 Pod 又面临服务峰值等情况,就可能导致 Pod 资源短缺。

接着上面的例子,假设该节点已经启动 3 个 Pod 实例,而这 3 个 Pod 的实际内存使用都不足 500MB,那么理论上该节点的可用内存应该大于 1.5GB。但是由于该节点的 Pod Requests 总和已经达到节点的可用内存上限,因此 Kubernetes 不会再将任何 Pod 实例调度到该节点上。

3. Requests 和 Limits 的背后机制

kubelet 在启动 Pod 的某个容器时,会将容器的 Requests 和 Limits 值转化为相应的容器启动参数传递给容器执行器(Docker 或者 rkt)。

如果容器的执行环境是 Docker，那么容器的如下 4 个参数是这样传递给 Docker 的。

1）spec.container[].resources.requests.cpu

这个参数会转化为 core 数（比如配置的 100m 会转化为 0.1），然后乘以 1024，再将这个结果作为 --cpu-shares 参数的值传递给 docker run 命令。在 docker run 命令中，--cpu-share 参数是一个相对权重值（Relative Weight），这个相对权重值会决定 Docker 在资源竞争时分配给容器的资源比例。

举例说明 --cpu-shares 参数在 Docker 中的含义：比如将两个容器的 CPU Requests 分别设置为 1 和 2，那么容器在 docker run 启动时对应的 --cpu-shares 参数值分别为 1024 和 2048，在主机 CPU 资源产生竞争时，Docker 会尝试按照 1∶2 的配比将 CPU 资源分配给这两个容器使用。

这里需要区分清楚的是：这个参数对于 Kubernetes 而言是绝对值，主要用于 Kubernetes 调度和管理；同时 Kubernetes 会将这个参数的值传递给 docker run 的 --cpu-shares 参数。--cpu-shares 参数对于 Docker 而言是相对值，主要用于资源分配比例。

2）spec.container[].resources.limits.cpu

这个参数会转化为 millicore 数（比如配置的 1 被转化为 1000，而配置的 100m 被转化为 100），将此值乘以 100000，再除以 1000，然后将结果值作为 --cpu-quota 参数的值传递给 docker run 命令。docker run 命令中另外一个参数 --cpu-period 默认被设置为 100000，表示 Docker 重新计量和分配 CPU 的使用时间间隔为 100000μs（100ms）。

Docker 的 --cpu-quota 参数和 --cpu-period 参数一起配合完成对容器 CPU 的使用限制：比如 Kubernetes 中配置容器的 CPU Limits 为 0.1，那么计算后 --cpu-quota 为 10000，而 --cpu-period 为 100000，这意味着 Docker 在 100ms 内最多给该容器分配 10ms×core 的计算资源用量，10/100=0.1 core 的结果与 Kubernetes 配置的意义是一致的。

注意：如果 kubelet 的启动参数 --cpu-cfs-quota 被设置为 true，那么 kubelet 会强制要求所有 Pod 都必须配置 CPU Limits（如果 Pod 没有配置，则集群提供了默认配置也可以）。从 Kubernetes 1.2 版本开始，这个 --cpu-cfs-quota 启动参数的默认值就是 true。

3）spec.container[].resources.requests.memory

这个参数值只提供给 Kubernetes 调度器作为调度和管理的依据，不会作为任何参数传递给 Docker。

4）spec.container[].resources.limits.memory

这个参数值会转化为单位为 Bytes 的整数,数值会作为 --memory 参数传递给 docker run 命令。

如果一个容器在运行过程中使用了超出了其内存 Limits 配置的内存限制值,那么它可能会被杀掉,如果这个容器是一个可重启的容器,那么之后它会被 kubelet 重新启动。因此对容器的 Limits 配置需要进行准确测试和评估。

与内存 Limits 不同的是,CPU 在容器技术中属于可压缩资源,因此对 CPU 的 Limits 配置一般不会因为偶然超标使用而导致容器被系统杀掉。

4. 计算资源使用情况监控

Pod 的资源用量会作为 Pod 的状态信息一同上报给 Master。如果在集群中配置了 Heapster 来监控集群的性能数据,那么还可以从 Heapster 中查看 Pod 的资源用量信息。

5. 计算资源相关常见问题分析

(1) Pod 状态为 Pending,错误信息为 FailedScheduling。如果 Kubernetes 调度器在集群中找不到合适的节点来运行 Pod,那么这个 Pod 会一直处于未调度状态,直到调度器找到合适的节点为止。每次调度器尝试调度失败时,Kubernetes 都会产生一个事件,我们可以通过下面这种方式来查看事件的信息:

```
$ kubectl describe pod frontend | grep -A 3 Events
Events:
  FirstSeen  LastSeen  Count  From         Subobject  PathReason         Message
  36s        5s        6      {scheduler}             FailedScheduling   Failed for
reason PodExceedsFreeCPU and possibly others
```

在上面这个例子中,名为 frontend 的 Pod 由于节点的 CPU 资源不足而调度失败(PodExceedsFreeCPU),同样,如果内存不足,则也可能导致调度失败(PodExceedsFreeMemory)。

如果一个或者多个 Pod 调度失败且有这类错误,那么可以尝试以下几种解决方法。

- ◎ 添加更多的节点到集群中。
- ◎ 停止一些不必要的运行中的 Pod,释放资源。
- ◎ 检查 Pod 的配置,错误的配置可能导致该 Pod 永远无法被调度执行。比如整个集群中所有节点都只有 1 CPU,而 Pod 配置的 CPU Requests 为 2,该 Pod 就不会被调度执行。

我们可以使用 kubectl describe nodes 命令来查看集群中节点的计算资源容量和已使用量：

```
$ kubectl describe nodes k8s-node-1
Name:             k8s-node-1
...
Capacity:
 cpu:          1
 memory: 464Mi
 pods:         40
Allocated resources (total requests):
 cpu:          910m
 memory: 2370Mi
 pods:         4
...
Pods:                   (4 in total)
  Namespace            Name                                     CPU(milliCPU)   Memory(bytes)
  frontend             webserver-ffj8j                          500 (50% of total)    2097152000 (50% of total)
  kube-system          fluentd-cloud-logging-k8s-node-1         100 (10% of total)    209715200 (5% of total)
  kube-system          kube-dns-v8-qopgw                        310 (31% of total)    178257920 (4% of total)
  TotalResourceLimits:
   CPU(milliCPU):       910 (91% of total)
   Memory(bytes):       2485125120 (59% of total)
...
```

超过可用资源容量上限（Capacity）和已分配资源量（Allocated resources）差额的 Pod 无法运行在该 Node 上。在这个例子中，如果一个 Pod 的 Requests 超过 90 millicpus 或者超过 1341MiB 内存，就无法运行在这个节点上。在 10.4.4 节中，我们还可以配置针对一组 Pod 的 Requests 和 Limits 总量的限制，这种限制可以作用于命名空间，通过这种方式可以防止一个命名空间下的用户将所有资源据为己有。

（2）容器被强行终止（Terminated）。如果容器使用的资源超过了它配置的 Limits，那么该容器可能会被强制终止。我们可以通过 kubectl describe pod 命令来确认容器是否因为这个原因被终止：

```
$ kubectl describe pod simmemleak-hra99
Name:                       simmemleak-hra99
Namespace:                  default
Image(s):                   saadali/simmemleak
Node:                       192.168.18.3
Labels:                     name=simmemleak
Status:                     Running
Reason:
Message:
IP:                         172.17.1.3
Replication Controllers:    simmemleak (1/1 replicas created)
Containers:
  simmemleak:
    Image:  saadali/simmemleak
    Limits:
      cpu:                  100m
      memory:               50Mi
    State:                  Running
      Started:              Tue, 07 Jul 2015 12:54:41 -0700
    Last Termination State: Terminated
      Exit Code:            1
      Started:              Fri, 07 Jul 2015 12:54:30 -0700
      Finished:             Fri, 07 Jul 2015 12:54:33 -0700
    Ready:                  False
    Restart Count:          5
Conditions:
  Type      Status
  Ready     False
Events:
  FirstSeen                         LastSeen                         Count   From
SubobjectPath                   Reason          Message
  Tue, 07 Jul 2015 12:53:51 -0700   Tue, 07 Jul 2015 12:53:51 -0700   1
{scheduler }                                                              scheduled
Successfully assigned simmemleak-hra99 to kubernetes-node-tf0f
  Tue, 07 Jul 2015 12:53:51 -0700   Tue, 07 Jul 2015 12:53:51 -0700   1       {kubelet
kubernetes-node-tf0f}       implicitly required container POD   pulled       Pod
container image "gcr.io/google_containers/pause:0.8.0" already present on machine
  Tue, 07 Jul 2015 12:53:51 -0700   Tue, 07 Jul 2015 12:53:51 -0700   1       {kubelet
kubernetes-node-tf0f}       implicitly required container POD   created      Created
with docker id 6a41280f516d
  Tue, 07 Jul 2015 12:53:51 -0700   Tue, 07 Jul 2015 12:53:51 -0700   1       {kubelet
```

```
kubernetes-node-tf0f}         implicitly required container POD      started       Started
with docker id 6a41280f516d
       Tue, 07 Jul 2015 12:53:51 -0700    Tue, 07 Jul 2015 12:53:51 -0700  1           {kubelet
kubernetes-node-tf0f}         spec.containers{simmemleak}             created       Created
with docker id 87348f12526a
```

Restart Count: 5 说明这个名为 simmemleak 的容器被强制终止并重启了 5 次。

我们可以在使用 kubectl get pod 命令时添加 -o go-template=... 格式参数来读取已终止容器之前的状态信息：

```
    $ kubectl get pod -o
go-template='{{range.status.containerStatuses}}{{"Container Name:
"}}{{.name}}{{"\r\nLastState: "}}{{.lastState}}{{end}}' simmemleak-60xbc
    Container Name: simmemleak
    LastState: map[terminated:map[exitCode:137 reason:OOM Killed
startedAt:2015-07-07T20:58:43Z finishedAt:2015-07-07T20:58:43Z
containerID:docker://0e4095bba1feccdfe7ef9fb6ebffe972b4b14285d5acdec6f0d3ae8a22f
ad8b2]]
```

可以看到这个容器因为 reason:OOM Killed 而被强制终止，说明这个容器的内存超过了限制（Out of Memory）。

6. 对大内存页（Huge Page）资源的支持

在计算机发展的早期阶段，程序员是直接对内存物理地址编程的，并且需要自己管理内存，很容易由于内存地址错误导致操作系统崩溃，而且出现了一些恶意程序对操作系统进行破坏。后来人们将硬件和软件（操作系统）相结合，推出了虚拟地址的概念，同时推出了内存页的概念，以及 CPU 的逻辑内存地址与物理内存（条）地址的映射关系。

在现代操作系统中，内存是以 Page（页，有时也可以称之为 Block）为单位进行管理的，而不以字节为单位，包括内存的分配和回收都基于 Page。典型的 Page 大小为 4KB，因此用户进程申请 1MB 内存就需要操作系统分配 256 个 Page，而 1GB 内存对应 26 万多个 Page！

为了实现快速内存寻址，CPU 内部以硬件方式实现了一个高性能的内存地址映射的缓存表——TLB（Translation Lookaside Buffer），用来保存逻辑内存地址与物理内存的对应关系。若目标地址的内存页物理地址不在 TLB 的缓存中或者 TLB 中的缓存记录失效，CPU 就需要切换到低速的、以软件方式实现的内存地址映射表进行内存寻址，这将大大降低

CPU 的运算速度。针对缓存条目有限的 TLB 缓存表，提高 TLB 效率的最佳办法就是将内存页增大，这样一来，一个进程所需的内存页数量会相应减少很多。如果把内存页从默认的 4KB 改为 2MB，那么 1GB 内存就只对应 512 个内存页了，TLB 的缓存命中率会大大增加。这是不是意味着我们可以任意指定内存页的大小，比如 1314MB 的内存页？答案是否定的，因为这是由 CPU 来决定的，比如常见的 Intel X86 处理器可以支持的大内存页通常是 2MB，个别型号的高端处理器则支持 1GB 的大内存页。

在 Linux 平台下，对于那些需要大量内存（1GB 以上内存）的程序来说，大内存页的优势是很明显的，因为 Huge Page 大大提升了 TLB 的缓存命中率，又因为 Linux 对 Huge Page 提供了更为简单、便捷的操作接口，所以可以把它当作文件来进行读写操作。Linux 使用 Huge Page 文件系统 hugetlbfs 支持巨页，这种方式更为灵活，我们可以设置 Huge Page 的大小，比如 1GB、2GB 甚至 2.5GB，然后设置有多少物理内存用于分配 Huge Page，这样就设置了一些预先分配好的 Huge Page。可以将 hugetlbfs 文件系统挂载在 /mnt/huge 目录下，通过执行下面的指令完成设置：

```
# mkdir /mnt/huge
# mount -t hugetlbfs nodev /mnt/huge
```

在设置完成后，用户进程就可以使用 mmap 映射 Huge Page 目标文件来使用大内存页了，Intel DPDK 便采用了这种做法，测试表明应用使用大内存页比使用 4KB 的内存页性能提高了 10%～15%。

Kubernetes 1.14 版本对 Linux Huge Page 的支持正式更新为 GA 稳定版。我们可以将 Huge Page 理解为一种特殊的计算资源：拥有大内存页的资源。而拥有 Huge Page 资源的 Node 也与拥有 GPU 资源的 Node 一样，属于一种新的可调度资源节点（Schedulable Resource Node）。

Huge Page 也支持 ResourceQuota 来实现配额限制，类似 CPU 或者 Memory，但不同于 CPU 或者内存，Huge Page 资源属于不可超限使用的资源，拥有 Huge Page 能力的 Node 会将自身支持的 Huge Page 的能力信息自动上报给 Kubernetes Master。

为此，Kubernetes 引入了一个新的资源类型 hugepages-<size>，来表示大内存页这种特殊的资源，比如 hugepages-2Mi 表示 2MiB 规格的大内存页资源。一个能提供 2MiB 规格 Huge Page 的 Node，会上报自己拥有 Hugepages-2Mi 的大内存页资源属性，供需要这种规格的大内存页资源的 Pod 使用，而需要 Huge Page 资源的 Pod 只要给出相关的 Huge Page 的声明，就可以被正确调度到匹配的目标 Node 上了。相关例子如下：

```yaml
apiVersion: v1
kind: Pod
metadata:
  generateName: hugepages-volume-
spec:
  containers:
  - image: fedora:latest
    command:
    - sleep
    - inf
    name: example
    volumeMounts:
    - mountPath: /hugepages
      name: hugepage
    resources:
      limits:
        hugepages-2Mi: 100Mi
        memory: 100Mi
      requests:
        memory: 100Mi
  volumes:
  - name: hugepage
    emptyDir:
      medium: HugePages
```

在上面的定义中有以下几个关键点:

◎ Huge Page 需要被映射到 Pod 的文件系统中;
◎ Huge Page 申请的 request 与 limit 必须相同,即申请固定大小的 Huge Page,不能是可变的;
◎ 在目前的版本中,Huge Page 属于 Pod 级别的资源,未来计划成为 Container 级别的资源,即实现更细粒度的资源管理;
◎ 存储卷 emptyDir(挂载到容器内的 /hugepages 目录)的后台是由 Huge Page 支持的,因此应用不能使用超过 request 声明的内存大小。

在 Kubernetes 未来的版本中,计划继续实现下面的一些高级特性:

◎ 支持容器级别的 Huge Page 的隔离能力;
◎ 支持 NUMA 亲和能力以提升服务的质量;
◎ 支持 LimitRange 配置 Huge Page 资源限额。

10.4.2 资源配置范围管理（LimitRange）

在默认情况下，Kubernetes 不会对 Pod 加上 CPU 和内存限制，这意味着 Kubernetes 系统中任何 Pod 都可以使用其所在节点的所有可用的 CPU 和内存。通过配置 Pod 的计算资源 Requests 和 Limits，我们可以限制 Pod 的资源使用，但对于 Kubernetes 集群管理员而言，配置每一个 Pod 的 Requests 和 Limits 是烦琐的，而且很受限制。更多时候，我们需要对集群内 Requests 和 Limits 的配置做一个全局限制。常见的配置场景如下。

- 集群中的每个节点都有 2GB 内存，集群管理员不希望任何 Pod 申请超过 2GB 的内存：因为在整个集群中都没有任何节点能满足超过 2GB 内存的请求。如果某个 Pod 的内存配置超过 2GB，那么该 Pod 将永远都无法被调度到任何节点上执行。为了防止这种情况的发生，集群管理员希望能在系统管理功能中设置禁止 Pod 申请超过 2GB 内存。
- 集群由同一个组织中的两个团队共享，分别运行生产环境和开发环境。生产环境最多可以使用 8GB 内存，而开发环境最多可以使用 512MB 内存。集群管理员希望通过为这两个环境创建不同的命名空间，并为每个命名空间设置不同的限制来满足这个需求。
- 用户创建 Pod 时使用的资源可能会刚好比整个机器资源的上限稍小，而恰好剩下的资源大小非常尴尬：不足以运行其他任务但整个集群加起来又非常浪费。因此，集群管理员希望设置每个 Pod 都必须至少使用集群平均资源值（CPU 和内存）的 20%，这样集群能够提供更好的资源一致性的调度，从而减少了资源浪费。

针对这些需求，Kubernetes 提供了 LimitRange 机制对 Pod 和容器的 Requests 和 Limits 配置进一步做出限制。在下面的示例中，将说明如何将 LimitsRange 应用到一个 Kubernetes 的命名空间中，然后说明 LimitRange 的几种限制方式，比如最大及最小范围、Requests 和 Limits 的默认值、Limits 与 Requests 的最大比例上限等。

下面通过 LimitRange 的设置和应用对其进行说明。

1. 创建一个 Namespace

创建一个名为 limit-example 的 Namespace：

```
$ kubectl create namespace limit-example
namespace "limit-example" created
```

2. 为 Namespace 设置 LimitRange

为 Namespace "limit-example" 创建一个简单的 LimitRange。创建 limits.yaml 配置文件，内容如下：

```yaml
apiVersion: v1
kind: LimitRange
metadata:
  name: mylimits
spec:
  limits:
  - max:
      cpu: "4"
      memory: 2Gi
    min:
      cpu: 200m
      memory: 6Mi
    maxLimitRequestRatio:
      cpu: 3
      memory: 2
    type: Pod
  - default:
      cpu: 300m
      memory: 200Mi
    defaultRequest:
      cpu: 200m
      memory: 100Mi
    max:
      cpu: "2"
      memory: 1Gi
    min:
      cpu: 100m
      memory: 3Mi
    maxLimitRequestRatio:
      cpu: 5
      memory: 4
    type: Container
```

创建该 LimitRange：

```
$ kubectl create -f limits.yaml --namespace=limit-example
limitrange "mylimits" created
```

查看 namespace limit-example 中的 LimitRange：

```
$ kubectl describe limits mylimits --namespace=limit-example
Name:       mylimits
Namespace:  limit-example
Type         Resource    Min    Max    Default Request    Default Limit    Max Limit/Request Ratio
----         --------    ---    ---    ---------------    -------------    -----------------------
Pod          cpu         200m   4      -                  -                3
Pod          memory      6Mi    2Gi    -                  -                2
Container    cpu         100m   2      200m               300m             5
Container    memory      3Mi    1Gi    100Mi              200Mi            4
```

下面解释 LimitRange 中各项配置的意义和特点。

（1）不论是 CPU 还是内存，在 LimitRange 中，Pod 和 Container 都可以设置 Min、Max 和 Max Limit/Requests Ratio 参数。Container 还可以设置 Default Request 和 Default Limit 参数，而 Pod 不能设置 Default Request 和 Default Limit 参数。

（2）对 Pod 和 Container 的参数解释如下。

◎ Container 的 Min（上面的 100m 和 3Mi）是 Pod 中所有容器的 Requests 值下限；Container 的 Max（上面的 2 和 1Gi）是 Pod 中所有容器的 Limits 值上限；Container 的 Default Request（上面的 200m 和 100Mi）是 Pod 中所有未指定 Requests 值的容器的默认 Requests 值；Container 的 Default Limit（上面的 300m 和 200Mi）是 Pod 中所有未指定 Limits 值的容器的默认 Limits 值。对于同一资源类型，这 4 个参数必须满足以下关系：Min ≤ Default Request ≤ Default Limit ≤ Max。

◎ Pod 的 Min（上面的 200m 和 6Mi）是 Pod 中所有容器的 Requests 值的总和下限；Pod 的 Max（上面的 4 和 2Gi）是 Pod 中所有容器的 Limits 值的总和上限。当容器未指定 Requests 值或者 Limits 值时，将使用 Container 的 Default Request 值或者 Default Limit 值。

◎ Container 的 Max Limit/Requests Ratio（上面的 5 和 4）限制了 Pod 中所有容器的 Limits 值与 Requests 值的比例上限；而 Pod 的 Max Limit/Requests Ratio（上面的 3 和 2）限制了 Pod 中所有容器的 Limits 值总和与 Requests 值总和的比例上限。

（3）如果设置了 Container 的 Max，那么对于该类资源而言，整个集群中的所有容器都必须设置 Limits，否则无法成功创建。Pod 内的容器未配置 Limits 时，将使用 Default Limit

的值（本例中的 300m CPU 和 200MiB 内存），如果也未配置 Default，则无法成功创建。

（4）如果设置了 Container 的 Min，那么对于该类资源而言，整个集群中的所有容器都必须设置 Requests。如果创建 Pod 的容器时未配置该类资源的 Requests，那么在创建过程中会报验证错误。Pod 里容器的 Requests 在未配置时，可以使用默认值 defaultRequest（本例中的 200m CPU 和 100MiB 内存）；如果未配置而又没有使用默认值 defaultRequest，那么会默认等于该容器的 Limits；如果此时 Limits 也未定义，就会报错。

（5）对于任意一个 Pod 而言，该 Pod 中所有容器的 Requests 总和必须大于或等于 6MiB，而且所有容器的 Limits 总和必须小于或等于 1GiB；同样，所有容器的 CPU Requests 总和必须大于或等于 200m，而且所有容器的 CPU Limits 总和必须小于或等于 2。

（6）Pod 里任何容器的 Limits 与 Requests 的比例都不能超过 Container 的 Max Limit/Requests Ratio；Pod 里所有容器的 Limits 总和与 Requests 的总和的比例不能超过 Pod 的 Max Limit/Requests Ratio。

3. 创建 Pod 时触发 LimitRange 限制

最后，让我们看看 LimitRange 生效时对容器的资源限制效果。

命名空间中 LimitRange 只会在 Pod 创建或者更新时执行检查。如果手动修改 LimitRange 为一个新的值，那么这个新的值不会去检查或限制之前已经在该命名空间中创建好的 Pod。

如果在创建 Pod 时配置的资源值（CPU 或者内存）超过了 LimitRange 的限制，那么该创建过程会报错，在错误信息中会说明详细的错误原因。

下面通过创建一个单容器 Pod 来展示默认限制是如何被配置到 Pod 上的：

```
$ kubectl run nginx --image=nginx --replicas=1 --namespace=limit-example
deployment "nginx" created
```

查看已创建的 Pod：

```
$ kubectl get pods --namespace=limit-example
NAME                      READY   STATUS    RESTARTS   AGE
nginx-2040093540-s8vzu    1/1     Running   0          11s
```

查看该 Pod 的 resources 相关信息：

```
$ kubectl get pods nginx-2040093540-s8vzu --namespace=limit-example -o yaml |
grep resources -C 8
    resourceVersion: "57"
    selfLink: /api/v1/namespaces/limit-example/pods/nginx-2040093540-ivimu
    uid: 67b20741-f53b-11e5-b066-64510658e388
spec:
  containers:
  - image: nginx
    imagePullPolicy: Always
    name: nginx
    resources:
      limits:
        cpu: 300m
        memory: 200Mi
      requests:
        cpu: 200m
        memory: 100Mi
    terminationMessagePath: /dev/termination-log
    volumeMounts:
```

由于该 Pod 未配置资源 Requests 和 Limits，所以使用了 namespace limit-example 中的默认 CPU 和内存定义的 Requests 和 Limits 值。

下面创建一个超出资源限制的 Pod（使用 3 CPU）：

```
invalid-pod.yaml:
apiVersion: v1
kind: Pod
metadata:
  name: invalid-pod
spec:
  containers:
  - name: kubernetes-serve-hostname
    image: gcr.io/google_containers/serve_hostname
    resources:
      limits:
        cpu: "3"
        memory: 100Mi
```

创建该 Pod，可以看到系统报错了，并且提供的错误原因为超过资源限制：

```
$ kubectl create -f invalid-pod.yaml --namespace=limit-example
Error from server: error when creating "invalid-pod.yaml": Pod "invalid-pod" is
```

```
forbidden: [Maximum cpu usage per Pod is 2, but limit is 3., Maximum cpu usage per
Container is 2, but limit is 3.]
```

接下来的例子展示了 LimitRange 对 maxLimitRequestRatio 的限制过程：

```
limit-test-nginx.yaml:
apiVersion: v1
kind: Pod
metadata:
  name: limit-test-nginx
  labels:
    name: limit-test-nginx
spec:
  containers:
  - name: limit-test-nginx
    image: nginx
    resources:
      limits:
        cpu: "1"
        memory: 512Mi
      requests:
        cpu: "0.8"
        memory: 250Mi
```

由于 limit-test-nginx 这个 Pod 的全部内存 Limits 总和与 Requests 总和的比例为 512：250，大于在 LimitRange 中定义的 Pod 的最大比率 2（maxLimitRequestRatio.memory=2），因此创建失败：

```
$ kubectl create -f limit-test-nginx.yaml --namespace=limit-example
Error from server: error when creating "limit-test-nginx.yaml": pods
"limit-test-nginx" is forbidden: [memory max limit to request ratio per Pod is 2,
but provided ratio is 2.048000.]
```

下面的例子为满足 LimitRange 限制的 Pod：

```
valid-pod.yaml:
apiVersion: v1
kind: Pod
metadata:
  name: valid-pod
  labels:
    name: valid-pod
spec:
```

```
    containers:
    - name: kubernetes-serve-hostname
      image: gcr.io/google_containers/serve_hostname
      resources:
        limits:
          cpu: "1"
          memory: 512Mi
```

创建 Pod 将会成功：

```
$ kubectl create -f valid-pod.yaml --namespace=limit-example
pod "valid-pod" created
```

查看该 Pod 的资源信息：

```
$ kubectl get pods valid-pod --namespace=limit-example -o yaml | grep -C 6 resources
  uid: 3b1bfd7a-f53c-11e5-b066-64510658e388
spec:
  containers:
  - image: gcr.io/google_containers/serve_hostname
    imagePullPolicy: Always
    name: kubernetes-serve-hostname
    resources:
      limits:
        cpu: "1"
        memory: 512Mi
      requests:
        cpu: "1"
        memory: 512Mi
```

可以看到该 Pod 配置了明确的 Limits 和 Requests，因此该 Pod 不会使用在 namespace limit-example 中定义的 default 和 defaultRequest。

需要注意的是，CPU Limits 强制配置这个选项在 Kubernetes 集群中默认是开启的；除非集群管理员在部署 kubelet 时，通过设置参数--cpu-cfs-quota=false 来关闭该限制：

```
$ kubelet --help
Usage of kubelet
....
--cpu-cfs-quota[=true]: Enable CPU CFS quota enforcement for containers that specify CPU limits
$ kubelet --cpu-cfs-quota=false ...
```

如果集群管理员希望对整个集群中容器或者 Pod 配置的 Requests 和 Limits 做限制，那么可以通过配置 Kubernetes 命名空间中的 LimitRange 来达到该目的。在 Kubernetes 集群中，如果 Pod 没有显式定义 Limits 和 Requests，那么 Kubernetes 系统会将该 Pod 所在的命名空间中定义的 LimitRange 的 default 和 defaultRequests 配置到该 Pod 上。

10.4.3 资源服务质量管理（Resource QoS）

本节对 Kubernetes 如何根据 Pod 的 Requests 和 Limits 配置来实现针对 Pod 的不同级别的资源服务质量控制（QoS）进行说明。

在 Kubernetes 的资源 QoS 体系中，需要保证高可靠性的 Pod 可以申请可靠资源，而一些不需要高可靠性的 Pod 可以申请可靠性较低或者不可靠的资源。在 10.4.1 节中讲到了容器的资源配置分为 Requests 和 Limits，其中 Requests 是 Kubernetes 调度时能为容器提供的完全可保障的资源量（最低保障），而 Limits 是系统允许容器运行时可能使用的资源量的上限（最高上限）。Pod 级别的资源配置是通过计算 Pod 内所有容器的资源配置的总和得出来的。

Kubernetes 中 Pod 的 Requests 和 Limits 资源配置有如下特点。

（1）如果 Pod 配置的 Requests 值等于 Limits 值，那么该 Pod 可以获得的资源是完全可靠的。

（2）如果 Pod 的 Requests 值小于 Limits 值，那么该 Pod 获得的资源可分成两部分：

◎ 完全可靠的资源，资源量的大小等于 Requests 值；
◎ 不可靠的资源，资源量最大等于 Limits 与 Requests 的差额，这份不可靠的资源能够申请到多少，取决于当时主机上容器可用资源的余量。

通过这种机制，Kubernetes 可以实现节点资源的超售（Over Subscription），比如在 CPU 完全充足的情况下，某机器共有 32GiB 内存可提供给容器使用，容器配置为 Requests 值 1GiB，Limits 值为 2GiB，那么在该机器上最多可以同时运行 32 个容器，每个容器最多可以使用 2GiB 内存，如果这些容器的内存使用峰值能错开，那么所有容器都可以正常运行。

超售机制能有效提高资源的利用率，同时不会影响容器申请的完全可靠资源的可靠性。

1. Requests 和 Limits 对不同计算资源类型的限制机制

根据前面的内容可知，容器的资源配置满足以下两个条件：

◎ Requests<=节点可用资源；
◎ Requests<=Limits。

Kubernetes 根据 Pod 配置的 Requests 值来调度 Pod，Pod 在成功调度之后会得到 Requests 值定义的资源来运行；而如果 Pod 所在机器上的资源有空余，则 Pod 可以申请更多的资源，最多不能超过 Limits 的值。下面看一下 Requests 和 Limits 针对不同计算资源类型的限制机制的差异。这种差异主要取决于计算资源类型是可压缩资源还是不可压缩资源。

1）可压缩资源

◎ Kubernetes 目前支持的可压缩资源是 CPU。
◎ Pod 可以得到 Pod 的 Requests 配置的 CPU 使用量，而能否使用超过 Requests 值的部分取决于系统的负载和调度。不过由于目前 Kubernetes 和 Docker 的 CPU 隔离机制都是在容器级别隔离的，所以 Pod 级别的资源配置并不能完全得到保障；Pod 级别的 cgroups 正在紧锣密鼓地开发中，如果将来引入，就可以确保 Pod 级别的资源配置准确运行。
◎ 空闲 CPU 资源按照容器 Requests 值的比例分配。举例说明：容器 A 的 CPU 配置为 Requests 1 Limits 10，容器 B 的 CPU 配置为 Request 2 Limits 8，A 和 B 同时运行在一个节点上，初始状态下容器的可用 CPU 为 3cores，那么 A 和 B 恰好得到在它们的 Requests 中定义的 CPU 用量，即 1CPU 和 2CPU。如果 A 和 B 都需要更多的 CPU 资源，而恰好此时系统的其他任务释放出 1.5CPU，那么这 1.5CPU 将按照 A 和 B 的 Requests 值的比例 1：2 分配给 A 和 B，即最终 A 可使用 1.5CPU，B 可使用 3CPU。
◎ 如果 Pod 使用了超过在 Limits 10 中配置的 CPU 用量，那么 cgroups 会对 Pod 中的容器的 CPU 使用进行限流（Throttled）；如果 Pod 没有配置 Limits 10，那么 Pod 会尝试抢占所有空闲的 CPU 资源（Kubernetes 从 1.2 版本开始默认开启 --cpu-cfs-quota，因此在默认情况下必须配置 Limits）。

2）不可压缩资源

◎ Kubernetes 目前支持的不可压缩资源是内存。
◎ Pod 可以得到在 Requests 中配置的内存。如果 Pod 使用的内存量小于它的 Requests

的配置，那么这个 Pod 可以正常运行（除非出现操作系统级别的内存不足等严重问题）；如果 Pod 使用的内存量超过了它的 Requests 的配置，那么这个 Pod 有可能被 Kubernetes 杀掉：比如 Pod A 使用了超过 Requests 而不到 Limits 的内存量，此时同一机器上另外一个 Pod B 之前只使用了远少于自己的 Requests 值的内存，此时程序压力增大，Pod B 向系统申请的总量不超过自己的 Requests 值的内存，那么 Kubernetes 可能会直接杀掉 Pod A；另外一种情况是 Pod A 使用了超过 Requests 而不到 Limits 的内存量，此时 Kubernetes 将一个新的 Pod 调度到这台机器上，新的 Pod 需要使用内存，而只有 Pod A 使用了超过了自己的 Requests 值的内存，那么 Kubernetes 也可能会杀掉 Pod A 来释放内存资源。
- 如果 Pod 使用的内存量超过了它的 Limits 设置，那么操作系统内核会杀掉 Pod 所有容器的所有进程中使用内存最多的一个，直到内存不超过 Limits 为止。

2. 对调度策略的影响

- Kubernetes 的 kube-scheduler 通过计算 Pod 中所有容器的 Requests 的总和来决定对 Pod 的调度。
- 不管是 CPU 还是内存，Kubernetes 调度器和 kubelet 都会确保节点上所有 Pod 的 Requests 的总和不会超过在该节点上可分配给容器使用的资源容量上限。

3. 服务质量等级（QoS Classes）

在一个超用（Over Committed，容器 Limits 总和大于系统容量上限）系统中，由于容器负载的波动可能导致操作系统的资源不足，最终可能导致部分容器被杀掉。在这种情况下，我们当然会希望优先杀掉那些不太重要的容器，那么如何衡量重要程度呢？Kubernetes 将容器划分成 3 个 QoS 等级：Guaranteed（完全可靠的）、Burstable（弹性波动、较可靠的）和 BestEffort（尽力而为、不太可靠的），这三种优先级依次递减，如图 10.4 所示。

图 10.4　QoS 等级和优先级的关系

从理论上来说，QoS 级别应该作为一个单独的参数来提供 API，并由用户对 Pod 进行配置，这种配置应该与 Requests 和 Limits 无关。但在当前版本的 Kubernetes 的设计中，为了简化模式及避免引入太多的复杂性，QoS 级别直接由 Requests 和 Limits 来定义。在 Kubernetes 中容器的 QoS 级别等于容器所在 Pod 的 QoS 级别，而 Kubernetes 的资源配置定义了 Pod 的三种 QoS 级别，如下所述。

1）Guaranteed

如果 Pod 中的所有容器对所有资源类型都定义了 Limits 和 Requests，并且所有容器的 Limits 值都和 Requests 值全部相等（且都不为 0），那么该 Pod 的 QoS 级别就是 Guaranteed。注意：在这种情况下，容器可以不定义 Requests，因为 Requests 值在未定义时默认等于 Limits。

在下面这两个例子中定义的 Pod QoS 级别就是 Guaranteed：

```
containers:
  name: foo
    resources:
      limits:
        cpu: 10m
        memory: 1Gi
  name: bar
    resources:
      limits:
        cpu: 100m
        memory: 100Mi
```

在上面的例子中未定义 Requests 值，所以其默认等于 Limits 值。而在下面这个例子中定义的 Requests 和 Limits 的值完全相同：

```
containers:
  name: foo
    resources:
      limits:
        cpu: 10m
        memory: 1Gi
      requests:
        cpu: 10m
        memory: 1Gi
  name: bar
    resources:
```

```
        limits:
            cpu: 100m
            memory: 100Mi
        requests:
            cpu: 10m
            memory: 1Gi
```

2）BestEffort

如果 Pod 中所有容器都未定义资源配置（Requests 和 Limits 都未定义），那么该 Pod 的 QoS 级别就是 BestEffort。

例如下面这个 Pod 定义：

```
containers:
   name: foo
      resources:
   name: bar
      resources:
```

3）Burstable

当一个 Pod 既不为 Guaranteed 级别，也不为 BestEffort 级别时，该 Pod 的 QoS 级别就是 Burstable。Burstable 级别的 Pod 包括两种情况。第 1 种情况：Pod 中的一部分容器在一种或多种资源类型的资源配置中定义了 Requests 值和 Limits 值（都不为 0），且 Requests 值小于 Limits 值；第 2 种情况：Pod 中的一部分容器未定义资源配置（Requests 和 Limits 都未定义）。注意：在容器未定义 Limits 时，Limits 值默认等于节点资源容量的上限。

下面几个例子中的 Pod 的 QoS 等级都是 Burstable。

（1）容器 foo 的 CPU Requests 不等于 Limits：

```
containers:
   name: foo
      resources:
         limits:
            cpu: 10m
            memory: 1Gi
         requests:
            cpu: 5m
            memory: 1Gi
   name: bar
      resources:
```

```
        limits:
            cpu: 10m
            memory: 1Gi
        requests:
            cpu: 10m
            memory: 1Gi
```

（2）容器 bar 未定义资源配置而容器 foo 定义了资源配置：

```
containers:
  name: foo
    resources:
        limits:
            cpu: 10m
            memory: 1Gi
        requests:
            cpu: 10m
            memory: 1Gi
  name: bar
```

（3）容器 foo 未定义 CPU，而容器 bar 未定义内存：

```
containers:
  name: foo
    resources:
        limits:
            memory: 1Gi
  name: bar
    resources:
        limits:
            cpu: 100m
```

（4）容器 bar 未定义资源配置，而容器 foo 未定义 Limits 值：

```
containers:
  name: foo
    resources:
        requests:
            cpu: 10m
            memory: 1Gi
  name: bar
```

4）Kubernetes QoS 的工作特点

Pod 的 CPU Requests 无法得到满足（比如节点的系统级任务占用过多的 CPU 导致无法分配足够的 CPU 给容器使用）时，容器得到的 CPU 会被压缩限流。

由于内存是不可压缩的资源，所以针对内存资源紧缺的情况，会按照以下逻辑进行处理。

（1）BestEffort Pod 的优先级最低，在这类 Pod 中运行的进程会在系统内存紧缺时被第一优先杀掉。当然，从另外一个角度来看，BestEffort Pod 由于没有设置资源 Limits，所以在资源充足时，它们可以充分使用所有的闲置资源。

（2）Burstable Pod 的优先级居中，这类 Pod 初始时会分配较少的可靠资源，但可以按需申请更多的资源。当然，如果整个系统内存紧缺，又没有 BestEffort 容器可以被杀掉以释放资源，那么这类 Pod 中的进程可能会被杀掉。

（3）Guaranteed Pod 的优先级最高，而且一般情况下这类 Pod 只要不超过其资源 Limits 的限制就不会被杀掉。当然，如果整个系统内存紧缺，又没有其他更低优先级的容器可以被杀掉以释放资源，那么这类 Pod 中的进程也可能会被杀掉。

5）OOM 计分系统

OOM（Out Of Memory）计分规则包括如下内容。

- OOM 计分的计算方法为：计算进程使用内存在系统中占的百分比，取其中不含百分号的数值，再乘以 10 的结果，这个结果是进程 OOM 的基础分；将进程 OOM 基础分的分值再加上这个进程的 OOM 分数调整值 OOM_SCORE_ADJ 的值，作为进程 OOM 的最终分值（除 root 启动的进程外）。在系统发生 OOM 时，OOM Killer 会优先杀掉 OOM 计分更高的进程。
- 进程的 OOM 计分的基本分数值范围是 0～1000，如果 A 进程的调整值 OOM_SCORE_ADJ 减去 B 进程的调整值的结果大于 1000，那么 A 进程的 OOM 计分最终值必然大于 B 进程，会优先杀掉 A 进程。
- 不论调整 OOM_SCORE_ADJ 值为多少，任何进程的最终分值范围也是 0～1000。

在 Kubernetes，不同 QoS 的 OOM 计分调整值规则如表 10.1 所示。

表 10.1　不同 QoS 的 OOM 计分调整值

QoS 等级	oom_score_adj
Guaranteed	-998
BestEffort	1000
Burstable	min(max(2, 1000 - (1000 * memoryRequestBytes) / machineMemoryCapacityBytes), 999)

其中：

- BestEffort Pod 设置 OOM_SCORE_ADJ 调整值为 1000，因此 BestEffort Pod 中容器里所有进程的 OOM 最终分肯定是 1000。
- Guaranteed Pod 设置 OOM_SCORE_ADJ 调整值为-998，因此 Guaranteed Pod 中容器里所有进程的 OOM 最终分一般是 0 或者 1（因为基础分不可能是 1000）。
- Burstable Pod 规则分情况说明：如果 Burstable Pod 的内存 Requests 超过了系统可用内存的 99.8%，那么这个 Pod 的 OOM_SCORE_ADJ 调整值固定为 2；否则，设置 OOM_SCORE_ADJ 调整值为 1000 − 10×(% of memory requested)；如果内存 Requests 为 0，那么 OOM_SCORE_ADJ 调整值固定为 999。这样的规则能确保 OOM_SCORE_ADJ 调整值的范围为 2～999，而 Burstable Pod 中所有进程的 OOM 最终分数范围为 2～1000。Burstable Pod 进程的 OOM 最终分数始终大于 Guaranteed Pod 的进程得分，因此它们会被优先杀掉。如果一个 Burstable Pod 使用的内存比它的内存 Requests 少，那么可以肯定的是它的所有进程的 OOM 最终分数会小于 1000，此时能确保它的优先级高于 BestEffort Pod。如果在一个 Burstable Pod 的某个容器中某个进程使用的内存比容器的 Requests 值高，那么这个进程的 OOM 最终分数会是 1000，否则它的 OOM 最终分会小于 1000。假设在下面的容器中有一个占用内存非常大的进程，那么当一个使用内存超过其 Requests 的 Burstable Pod 与另外一个使用内存少于其 Requests 的 Burstable Pod 发生内存竞争冲突时，前者的进程会被系统杀掉。如果在一个 Burstable Pod 内部有多个进程的多个容器发生内存竞争冲突，那么此时 OOM 评分只能作为参考，不能保证完全按照资源配置的定义来执行 OOM Kill。

OOM 还有一些特殊的计分规则，如下所述。

- kubelet 进程和 Docker 进程的调整值 OOM_SCORE_ADJ 为-998。
- 如果配置进程调整值 OOM_SCORE_ADJ 为-999,那么这类进程不会被 OOM Killer 杀掉。

6）QoS 的演进

目前 Kubernetes 基于 QoS 的超用机制日趋完善，但还有一些问题需要解决。

（1）内存 Swap 的支持。当前的 QoS 策略都是假定主机不启用内存 Swap。如果主机启用了 Swap，那么上面的 QoS 策略可能会失效。举例说明：两个 Guaranteed Pod 都刚好达到了内存 Limits，那么由于内存 Swap 机制，它们还可以继续申请使用更多的内存。如果 Swap 空间不足，最终这两个 Pod 中的进程就可能会被杀掉。由于 Kubernetes 和 Docker 尚不支持内存 Swap 空间的隔离机制，所以这一功能暂时还未实现。

（2）更丰富的 QoS 策略。当前的 QoS 策略都是基于 Pod 的资源配置（Requests 和 Limits）来定义的，而资源配置本身又承担着对 Pod 资源管理和限制的功能。两种不同维度的功能使用同一个参数来配置，可能会导致某些复杂需求无法满足，比如当前 Kubernetes 无法支持弹性的、高优先级的 Pod。自定义 QoS 优先级能提供更大的灵活性，完美地实现各类需求，但同时会引入更高的复杂性，而且过于灵活的设置会给予用户过高的权限，对系统管理也提出了更大的挑战。

10.4.4 资源配额管理（Resource Quotas）

如果一个 Kubernetes 集群被多个用户或者多个团队共享，就需要考虑资源公平使用的问题，因为某个用户可能会使用超过基于公平原则分配给其的资源量。

Resource Quotas 就是解决这个问题的工具。通过 ResourceQuota 对象，我们可以定义资源配额，这个资源配额可以为每个命名空间都提供一个总体的资源使用的限制：它可以限制命名空间中某种类型的对象的总数目上限，也可以设置命名空间中 Pod 可以使用的计算资源的总上限。

典型的资源配额使用方式如下。

- ◎ 不同的团队工作在不同的命名空间下，目前这是非约束性的，在未来的版本中可能会通过 ACL（Access Control List，访问控制列表）来实现强制性约束。
- ◎ 集群管理员为集群中的每个命名空间都创建一个或者多个资源配额项。
- ◎ 当用户在命名空间中使用资源（创建 Pod 或者 Service 等）时，Kubernetes 的配额系统会统计、监控和检查资源用量，以确保使用的资源用量没有超过资源配额的配置。
- ◎ 如果在创建或者更新应用时资源使用超过了某项资源配额的限制，那么创建或者更新的请求会报错（HTTP 403 Forbidden），并给出详细的出错原因说明。
◎ 如果命名空间中的计算资源（CPU 和内存）的资源配额启用，那么用户必须为相应的资源类型设置 Requests 或 Limits；否则配额系统可能会直接拒绝 Pod 的创建。这里可以使用 LimitRange 机制来为没有配置资源的 Pod 提供默认资源配置。

下面的例子展示了一个非常适合使用资源配额来做资源控制管理的场景。

◎ 集群共有 32GB 内存和 16 CPU，两个小组。A 小组使用 20GB 内存和 10 CPU，B 小组使用 10GB 内存和 2 CPU，剩下的 2GB 内存和 2 CPU 作为预留。
◎ 在名为 testing 的命名空间中，限制使用 1 CPU 和 1GB 内存；在名为 production 的命名空间中，资源使用不受限制。

在使用资源配额时，需要注意以下两点。

◎ 如果集群中总的可用资源小于各命名空间中资源配额的总和，那么可能会导致资源竞争。资源竞争时，Kubernetes 系统会遵循先到先得的原则。
◎ 不管是资源竞争还是配额的修改，都不会影响已经创建的资源使用对象。

1. 在 Master 中开启资源配额选型

资源配额可以通过在 kube-apiserver 的 --admission-control 参数值中添加 ResourceQuota 参数进行开启。如果在某个命名空间的定义中存在 ResourceQuota，那么对于该命名空间而言，资源配额就是开启的。一个命名空间可以有多个 ResourceQuota 配置项。

1）计算资源配额（Compute Resource Quota）

资源配额可以限制一个命名空间中所有 Pod 的计算资源的总和。目前支持的计算资源类型如表 10.2 所示。

表 10.2 ResourceQuota 支持限制的计算资源类型

资源名称	说明
Cpu	所有非终止状态的 Pod，CPU Requests 的总和不能超过该值
limits.cpu	所有非终止状态的 Pod，CPU Limits 的总和不能超过该值
limits.memory	所有非终止状态的 Pod，内存 Limits 的总和不能超过该值
Memory	所有非终止状态的 Pod，内存 Requests 的总和不能超过该值
requests.cpu	所有非终止状态的 Pod，CPU Requests 的总和不能超过该值
requests.memory	所有非终止状态的 Pod，内存 Requests 的总和不能超过该值

2）存储资源配额（Volume Count Quota）

可以在给定的命名空间中限制所使用的存储资源（Storage Resources）的总量，目前支持的存储资源名称如表 10.3 所示。

表 10.3　ResourceQuota 支持限制的计算资源类型

资源名称	说明
requests.storage	所有 PVC，存储请求总量不能超过此值
Persistentvolumeclaims	在该命名空间中能存在的持久卷的总数上限
.storageclass.storage.k8s.io/requests.storage	和该存储类关联的所有 PVC，存储请求总和不能超过此值
.storageclass.storage.k8s.io/persistentvolumeclaims	和该存储类关联的所有 PVC 的总数

3）对象数量配额（Object Count Quota）

指定类型的对象数量可以被限制。表 10.4 列出了 ResourceQuota 支持限制的对象类型。

表 10.4　ResourceQuota 支持限制的对象类型

资源名称	说明
Configmaps	在该命名空间中能存在的 ConfigMap 的总数上限
Pods	在该命名空间中能存在的非终止状态 Pod 的总数上限。Pod 终止状态等价于 Pod 的 status.phase in (Failed, Succeeded) = true
Replicationcontrollers	在该命名空间中能存在的 RC 的总数上限
Resourcequotas	在该命名空间中能存在的资源配额项的总数上限
Services	在该命名空间中能存在的 Service 的总数上限
services.loadbalancers	在该命名空间中能存在的负载均衡的总数上限
services.nodeports	在该命名空间中能存在的 NodePort 的总数上限
Secrets	在该命名空间中能存在的 Secret 的总数上限

例如，我们可以通过资源配额来限制在命名空间中能创建的 Pod 的最大数量。这种设置可以防止某些用户大量创建 Pod 而迅速耗尽整个集群的 Pod IP 和计算资源。

2. 配额的作用域（Quota Scopes）

每项资源配额都可以单独配置一组作用域，配置了作用域的资源配额只会对符合其作用域的资源使用情况进行计量和限制，作用域范围内超过了资源配额的请求都会报验证错误。表 10.5 列出了 ResourceQuota 的 4 种作用域。

表 10.5　ResourceQuota 的 4 种作用域

作用域	说明
Terminating	匹配所有 spec.activeDeadlineSeconds 不小于 0 的 Pod
NotTerminating	匹配所有 spec.activeDeadlineSeconds 是 nil 的 Pod
BestEffort	匹配所有 QoS 是 BestEffort 的 Pod
NotBestEffort	匹配所有 QoS 不是 BestEffort 的 Pod

其中，BestEffort 作用域可以限定资源配额来追踪 pods 资源的使用，Terminating、NotTerminating 和 NotBestEffort 这三种作用域可以限定资源配额来追踪以下资源的使用。

- cpu
- limits.cpu
- limits.memory
- memory
- pods
- requests.cpu
- requests.memory

3. 在资源配额（ResourceQuota）中设置 Requests 和 Limits

资源配额也可以设置 Requests 和 Limits。

如果在资源配额中指定了 requests.cpu 或 requests.memory，那么它会强制要求每个容器都配置自己的 CPU Requests 或 CPU Limits（可使用 LimitRange 提供的默认值）。

同理，如果在资源配额中指定了 limits.cpu 或 limits.memory，那么它也会强制要求每个容器都配置自己的内存 Requests 或内存 Limits（可使用 LimitRange 提供的默认值）。

4. 资源配额的定义

下面通过几个例子对资源配额进行设置和应用。

与 LimitRange 相似，ResourceQuota 也被设置在 Namespace 中。创建名为 myspace 的 Namespace：

```
$ kubectl create namespace myspace
namespace "myspace" created
```

创建 ResourceQuota 配置文件 compute-resources.yaml，用于设置计算资源的配额：

```yaml
apiVersion: v1
kind: ResourceQuota
metadata:
  name: compute-resources
spec:
  hard:
    pods: "4"
    requests.cpu: "1"
    requests.memory: 1Gi
    limits.cpu: "2"
    limits.memory: 2Gi
```

创建该项资源配额：

```
$ kubectl create -f compute-resources.yaml --namespace=myspace
resourcequota "compute-resources" created
```

创建另一个名为 object-counts.yaml 的文件，用于设置对象数量的配额：

```yaml
apiVersion: v1
kind: ResourceQuota
metadata:
  name: object-counts
spec:
  hard:
    configmaps: "10"
    persistentvolumeclaims: "4"
    replicationcontrollers: "20"
    secrets: "10"
    services: "10"
    services.loadbalancers: "2"
```

创建该 ResourceQuota：

```
$ kubectl create -f object-counts.yaml --namespace=myspace
resourcequota "object-counts" created
```

查看各 ResourceQuota 的详细信息：

```
$ kubectl describe quota compute-resources --namespace=myspace
Name:           compute-resources
Namespace:      myspace
Resource        Used Hard
```

```
--------                          ----    ----
limits.cpu                        0       2
limits.memory                     0       2Gi
pods                              0       4
requests.cpu                      0       1
requests.memory                   0       1Gi

$ kubectl describe quota object-counts --namespace=myspace
Name:                    object-counts
Namespace:               myspace
Resource                 Used    Hard
--------                 ----    ----
configmaps               0       10
persistentvolumeclaims   0       4
replicationcontrollers   0       20
secrets                  1       10
services                 0       10
services.loadbalancers   0       2
```

2. 资源配额与集群资源总量的关系

资源配额与集群资源总量是完全独立的。资源配额是通过绝对的单位来配置的，这也就意味着如果在集群中新添加了节点，那么资源配额不会自动更新，而该资源配额所对应的命名空间中的对象也不能自动增加资源上限。

在某些情况下，我们可能希望资源配额支持更复杂的策略，如下所述。

◎ 对于不同的租户，按照比例划分整个集群的资源。
◎ 允许每个租户都能按照需要来提高资源用量，但是有一个较宽容的限制，以防止意外的资源耗尽情况发生。
◎ 探测某个命名空间的需求，添加物理节点并扩大资源配额值。

这些策略可以通过将资源配额作为一个控制模块、手动编写一个控制器来监控资源使用情况，并调整命名空间上的资源配额来实现。

资源配额将整个集群中的资源总量做了一个静态划分，但它并没有对集群中的节点做任何限制：不同命名空间中的 Pod 仍然可以运行在同一个节点上。

10.4.5　ResourceQuota 和 LimitRange 实践

根据前面对资源管理的介绍，这里将通过一个完整的例子来说明如何通过资源配额和资源配置范围的配合来控制一个命名空间的资源使用。

集群管理员根据集群用户的数量来调整集群配置，以达到这个目的：能控制特定命名空间中的资源使用量，最终实现集群的公平使用和成本控制。

需要实现的功能如下。

◎ 限制运行状态的 Pod 的计算资源用量。
◎ 限制持久存储卷的数量以控制对存储的访问。
◎ 限制负载均衡器的数量以控制成本。
◎ 防止滥用网络端口这类稀缺资源。
◎ 提供默认的计算资源 Requests 以便于系统做出更优化的调度。

1. 创建命名空间

创建名为 quota-example 的命名空间，namespace.yaml 文件的内容如下：

```
apiVersion: v1
kind: Namespace
metadata:
  name: quota-example

$ kubectl create -f namespace.yaml
namespace "quota-example" created
```

查看命名空间：

```
$ kubectl get namespaces
NAME            STATUS    AGE
default         Active    2m
kube-system     Active    2m
quota-example   Active    39s
```

2. 设置限定对象数目的资源配额

通过设置限定对象的数量的资源配额，可以控制以下资源的数量：

◎ 持久存储卷；

◎ 负载均衡器；
◎ NodePort。

创建名为 object-counts 的 ResourceQuota：

```yaml
object-counts.yaml:
apiVersion: v1
kind: ResourceQuota
metadata:
  name: object-counts
spec:
  hard:
    persistentvolumeclaims: "2"
    services.loadbalancers: "2"
    services.nodeports: "0"
```

```
$ kubectl create -f object-counts.yaml --namespace=quota-example
resourcequota "object-counts" created
```

配额系统会检测到资源项配额的创建，并且会统计和限制该命名空间中的资源消耗。

查看该配额是否生效：

```
$ kubectl describe quota object-counts --namespace=quota-example
Name:                   object-counts
Namespace:              quota-example
Resource                Used    Hard
--------                ----    ----
persistentvolumeclaims  0       2
services.loadbalancers  0       2
services.nodeports      0       0
```

至此，配额系统会自动阻止那些使资源用量超过资源配额限定值的请求。

3. 设置限定计算资源的资源配额

下面再创建一项限定计算资源的资源配额，以限制该命名空间中的计算资源的使用总量。

创建名为 compute-resources 的 ResourceQuota：

```yaml
apiVersion: v1
kind: ResourceQuota
```

```
metadata:
  name: compute-resources
spec:
  hard:
    pods: "4"
    requests.cpu: "1"
    requests.memory: 1Gi
    limits.cpu: "2"
    limits.memory: 2Gi

$ kubectl create -f compute-resources.yaml --namespace=quota-example
resourcequota "compute-resources" created
```

查看该配额是否生效：

```
$ kubectl describe quota compute-resources --namespace=quota-example
Name:               compute-resources
Namespace:          quota-example
Resource            Used  Hard
--------            ----  ----
limits.cpu          0     2
limits.memory       0     2Gi
pods                0     4
requests.cpu        0     1
requests.memory     0     1Gi
```

配额系统会自动防止在该命名空间中同时拥有超过 4 个非"终止态"的 Pod。此外，由于该项资源配额限制了 CPU 和内存的 Limits 和 Requests 的总量，因此会强制要求该命名空间下的所有容器都显式定义 CPU 和内存的 Limits 和 Requests（可使用默认值，Requests 默认等于 Limits）。

4. 配置默认 Requests 和 Limits

在命名空间已经配置了限定计算资源的资源配额的情况下，如果尝试在该命名空间下创建一个不指定 Requests 和 Limits 的 Pod，那么 Pod 的创建可能会失败。下面是一个失败的例子。

创建一个 Nginx 的 Deployment：

```
$ kubectl run nginx --image=nginx --replicas=1 --namespace=quota-example
deployment "nginx" created
```

查看创建的 Pod，会发现 Pod 没有创建成功：

```
$ kubectl get pods --namespace=quota-example
```

再查看一下 Deployment 的详细信息：

```
$ kubectl describe deployment nginx --namespace=quota-example
Name:                   nginx
Namespace:              quota-example
CreationTimestamp:      Mon, 06 Jun 2016 16:11:37 -0400
Labels:                 run=nginx
Selector:               run=nginx
Replicas:               0 updated | 1 total | 0 available | 1 unavailable
StrategyType:           RollingUpdate
MinReadySeconds:        0
RollingUpdateStrategy:  1 max unavailable, 1 max surge
OldReplicaSets:         <none>
NewReplicaSet:          nginx-3137573019 (0/1 replicas created)
......
```

该 Deployment 尝试创建一个 Pod，但是失败了，查看其中 ReplicaSet 的详细信息：

```
$ kubectl describe rs nginx-3137573019 --namespace=quota-example
Name:           nginx-3137573019
Namespace:      quota-example
Image(s):       nginx
Selector:       pod-template-hash=3137573019,run=nginx
Labels:         pod-template-hash=3137573019
                run=nginx
Replicas:       0 current / 1 desired
Pods Status:    0 Running / 0 Waiting / 0 Succeeded / 0 Failed
No volumes.
Events:
  FirstSeen  LastSeen  Count  From                     SubobjectPath  Type      Reason        Message
  ---------  --------  -----  ----                     -------------  ----      ------        -------
  4m         7s        11     {replicaset-controller}                 Warning   FailedCreate  Error creating: pods "nginx-3137573019-" is forbidden: Failed quota: compute-resources: must specify limits.cpu,limits.memory,requests.cpu,requests.memory
```

可以看到 Pod 创建失败的原因:Master 拒绝这个 ReplicaSet 创建 Pod,因为在这个 Pod 中没有指定 CPU 和内存的 Requests 和 Limits。

为了避免这种失败,我们可以使用 LimitRange 来为这个命名空间下的所有 Pod 都提供一个资源配置的默认值。下面的例子展示了如何为这个命名空间添加一个指定默认资源配置的 LimitRange。

创建一个名为 limits 的 LimitRange:

```
limits.yaml:
apiVersion: v1
kind: LimitRange
metadata:
  name: limits
spec:
  limits:
  - default:
      cpu: 200m
      memory: 512Mi
    defaultRequest:
      cpu: 100m
      memory: 256Mi
    type: Container

$ kubectl create -f limits.yaml --namespace=quota-example
limitrange "limits" created

$ kubectl describe limits limits --namespace=quota-example
Name:           limits
Namespace:      quota-example
Type        Resource  Min  Max  Default Request  Default Limit  Max Limit/Request Ratio
----        --------  ---  ---  ---------------  -------------  -----------------------
Container   memory    -    -    256Mi            512Mi          -
Container   cpu       -    -    100m             200m           -
```

在 LimitRange 创建成功后,用户在该命名空间下创建未指定资源限制的 Pod 的请求时,系统会自动为该 Pod 设置默认的资源限制。

例如,每个新建的未指定资源限制的 Pod 都等价于使用下面的资源限制:

```
$ kubectl run nginx \
  --image=nginx \
  --replicas=1 \
  --requests=cpu=100m,memory=256Mi \
  --limits=cpu=200m,memory=512Mi \
  --namespace=quota-example
```

至此，我们已经为该命名空间配置好默认的计算资源了，我们的 ReplicaSet 应该能够创建 Pod 了。查看一下，创建 Pod 成功：

```
$ kubectl get pods --namespace=quota-example
NAME                    READY   STATUS    RESTARTS   AGE
nginx-3137573019-fvrig  1/1     Running   0          6m
```

接下来可以随时查看资源配额的使用情况：

```
$ kubectl describe quota --namespace=quota-example
Name:           compute-resources
Namespace:      quota-example
Resource        Used    Hard
--------        ----    ----
limits.cpu      200m    2
limits.memory   512Mi   2Gi
pods            1       4
requests.cpu    100m    1
requests.memory 256Mi   1Gi


Name:           object-counts
Namespace:      quota-example
Resource        Used    Hard
--------        ----    ----
persistentvolumeclaims  0   2
services.loadbalancers  0   2
services.nodeports      0   0
```

可以看到每个 Pod 在创建时都会消耗指定的资源量，而这些使用量都会被 Kubernetes 准确跟踪、监控和管理。

5. 指定资源配额的作用域

假设我们并不想为某个命名空间配置默认的计算资源配额，而是希望限定在命名空间内运行的 QoS 为 BestEffort 的 Pod 总数,例如让集群中的部分资源运行 QoS 为非 BestEffort 的服务,而让闲置的资源运行 QoS 为 BestEffort 的服务,即可避免集群的所有资源仅被大量的 BestEffort Pod 耗尽。这可以通过创建两个资源配额来实现,如下所述。

创建一个名为 quota-scopes 的命名空间:

```
$ kubectl create namespace quota-scopes
namespace "quota-scopes" created
```

创建一个名为 best-effort 的 ResourceQuota,指定 Scope 为 BestEffort:

```
apiVersion: v1
kind: ResourceQuota
metadata:
  name: best-effort
spec:
  hard:
    pods: "10"
  scopes:
  - BestEffort

$ kubectl create -f best-effort.yaml --namespace=quota-scopes
resourcequota "best-effort" created
```

再创建一个名为 not-best-effort 的 ResourceQuota,指定 Scope 为 NotBestEffort:

```
apiVersion: v1
kind: ResourceQuota
metadata:
  name: not-best-effort
spec:
  hard:
    pods: "4"
    requests.cpu: "1"
    requests.memory: 1Gi
    limits.cpu: "2"
    limits.memory: 2Gi
  scopes:
  - NotBestEffort

$ kubectl create -f not-best-effort.yaml --namespace=quota-scopes
resourcequota "not-best-effort" created
```

查看创建成功的 ResourceQuota：

```
$ kubectl describe quota --namespace=quota-scopes
Name:        best-effort
Namespace:   quota-scopes
Scopes:      BestEffort
 * Matches all pods that have best effort quality of service.
Resource   Used Hard
--------   ---- ----
pods       0    10

Name:        not-best-effort
Namespace:   quota-scopes
Scopes:      NotBestEffort
 * Matches all pods that do not have best effort quality of service.
Resource         Used Hard
--------         ---- ----
limits.cpu       0    2
limits.memory    0    2Gi
pods             0    4
requests.cpu     0    1
requests.memory  0    1Gi
```

之后，没有配置 Requests 的 Pod 将会被名为 best-effort 的 ResourceQuota 限制；而配置了 Requests 的 Pod 会被名为 not-best-effort 的 ResourceQuota 限制。

创建两个 Deployment：

```
$ kubectl run best-effort-nginx --image=nginx --replicas=8
--namespace=quota-scopes
  deployment "best-effort-nginx" created

$ kubectl run not-best-effort-nginx \
  --image=nginx \
  --replicas=2 \
  --requests=cpu=100m,memory=256Mi \
  --limits=cpu=200m,memory=512Mi \
  --namespace=quota-scopes
  deployment "not-best-effort-nginx" created
```

名为 best-effort-nginx 的 Deployment 因为没有配置 Requests 和 Limits，所以它的 QoS 级别为 BestEffort，因此它的创建过程由 best-effort 资源配额项来限制，而 not-best-effort

资源配额项不会对它进行限制。best-effort 资源配额项没有限制 Requests 和 Limits，因此 best-effort-nginx Deployment 可以成功创建 8 个 Pod。

名为 not-best-effort-nginx 的 Deployment 因为配置了 Requests 和 Limits，且二者不相等，所以它的 QoS 级别为 Burstable，因此它的创建过程由 not-best-effort 资源配额项限制，而 best-effort 资源配额项不会对它进行限制。not-best-effort 资源配额项限制了 Pod 的 Requests 和 Limits 的总上限，not-best-effort-nginx Deployment 并没有超过这个上限，所以可以成功创建两个 Pod。

查看已经创建的 Pod：

```
$ kubectl get pods --namespace=quota-scopes
NAME                                        READY   STATUS    RESTARTS   AGE
best-effort-nginx-3488455095-2qb41          1/1     Running   0          51s
best-effort-nginx-3488455095-3go7n          1/1     Running   0          51s
best-effort-nginx-3488455095-9o2xg          1/1     Running   0          51s
best-effort-nginx-3488455095-eyg40          1/1     Running   0          51s
best-effort-nginx-3488455095-gcs3v          1/1     Running   0          51s
best-effort-nginx-3488455095-rq8p1          1/1     Running   0          51s
best-effort-nginx-3488455095-udhhd          1/1     Running   0          51s
best-effort-nginx-3488455095-zmk12          1/1     Running   0          51s
not-best-effort-nginx-2204666826-7sl61      1/1     Running   0          23s
not-best-effort-nginx-2204666826-ke746      1/1     Running   0          23s
```

可以看到 10 个 Pod 都创建成功。

再看一下两个资源配额项的使用情况：

```
$ kubectl describe quota --namespace=quota-scopes
Name:          best-effort
Namespace:     quota-scopes
Scopes:        BestEffort
 * Matches all pods that have best effort quality of service.
Resource       Used  Hard
--------       ----  ----
pods           8     10


Name:          not-best-effort
Namespace:     quota-scopes
Scopes:        NotBestEffort
 * Matches all pods that do not have best effort quality of service.
```

```
Resource              Used   Hard
--------              ----   ----
limits.cpu            400m   2
limits.memory         1Gi    2Gi
pods                  2      4
requests.cpu          200m   1
requests.memory       512Mi  1Gi
```

可以看到 best-effort 资源配额项已经统计了在 best-effort-nginx Deployment 中创建的 8 个 Pod 的资源使用信息，not-best-effort 资源配额项也已经统计了在 not-best-effort-nginx Deployment 中创建的两个 Pod 的资源使用信息。

通过这个例子我们可以看到：资源配额的作用域（Scopes）提供了一种将资源集合分割的机制，可以使集群管理员更加方便地监控和限制不同类型的对象对各类资源的使用情况，同时为资源分配和限制提供更大的灵活度和便利性。

10.4.6 资源管理总结

Kubernetes 中资源管理的基础是容器和 Pod 的资源配置（Requests 和 Limits）。容器的资源配置指定了容器请求的资源和容器能使用的资源上限，Pod 的资源配置则是 Pod 中所有容器的资源配置总和上限。

通过资源配额机制，我们可以对命名空间中所有 Pod 使用资源的总量进行限制，也可以对这个命名空间中指定类型的对象的数量进行限制。使用作用域可以让资源配额只对符合特定范围的对象加以限制，因此作用域机制可以使资源配额的策略更加丰富、灵活。

如果需要对用户的 Pod 或容器的资源配置做更多的限制，则可以使用资源配置范围（LimitRange）来达到这个目的。LimitRange 可以有效地限制 Pod 和容器的资源配置的最大、最小范围，也可以限制 Pod 和容器的 Limits 与 Requests 的最大比例上限，此外 LimitRange 还可以为 Pod 中的容器提供默认的资源配置。

Kubernetes 基于 Pod 的资源配置实现了资源服务质量（QoS）。不同 QoS 级别的 Pod 在系统中拥有不同的优先级：高优先级的 Pod 有更高的可靠性，可以用于运行可靠性要求较高的服务；低优先级的 Pod 可以实现集群资源的超售，可以有效地提高集群资源利用率。

上面的多种机制共同组成了当前版本 Kubernetes 的资源管理体系。这个资源管理体系可以满足大部分的资源管理需求。同时，Kubernetes 的资源管理体系仍然在不停地发展和进化，对于目前无法满足的更复杂、更个性化的需求，我们可以继续关注 Kubernetes 未来的发展和变化。

10.5　资源紧缺时的 Pod 驱逐机制

如何在系统硬件资源紧缺的情况下保证 Node 的稳定性，是 kubelet 需要解决的一个重要问题。尤其对于内存和磁盘这种不可压缩的资源，紧缺就意味着不稳定。下面对驱逐的策略、信号、阈值、监控频率和驱逐操作进行详细说明。

10.5.1　驱逐策略

kubelet 持续监控主机的资源使用情况，并尽量防止计算资源被耗尽。一旦出现资源紧缺的迹象，kubelet 就会主动终止一个或多个 Pod 的运行，以回收紧缺的资源。当一个 Pod 被终止时，其中的容器会全部停止，Pod 的状态会被设置为 Failed。

10.5.2　驱逐信号

在表 10.6 中提到了一些信号，kubelet 能够利用这些信号作为决策依据来触发驱逐行为。其中，描述列中的内容来自 kubelet Summary API；每个信号都支持整数值或者百分比的表示方法，百分比的分母部分就是各个信号相关资源的总量。

表 10.6　驱逐信号及其描述

驱逐信号	描述
memory.available	memory.available := node.status.capacity[memory] - node.stats.memory.workingSet
nodefs.available	nodefs.available := node.stats.fs.available
nodefs.inodesFree	nodefs.inodesFree := node.stats.fs.inodesFree
imagefs.available	imagefs.available := node.stats.runtime.imagefs.available
imagefs.inodesFree	imagefs.inodesFree := node.stats.runtime.imagefs.inodesFree

memory.available 的值取自 cgroupfs，而不是 free -m 命令，这是因为 free -m 不支持在容器内工作。如果用户使用了 node allocatable 功能，则除了节点自身的内存需要判断，还需要利用 cgroup 根据用户 Pod 部分的情况进行判断。下面的脚本展示了 kubelet 计算 memory.available 的过程：

```bash
#!/bin/bash
#!/usr/bin/env bash

# This script reproduces what the kubelet does
# to calculate memory.available relative to root cgroup.

# current memory usage
memory_capacity_in_kb=$(cat /proc/meminfo | grep MemTotal | awk '{print $2}')
memory_capacity_in_bytes=$((memory_capacity_in_kb * 1024))
memory_usage_in_bytes=$(cat /sys/fs/cgroup/memory/memory.usage_in_bytes)
memory_total_inactive_file=$(cat /sys/fs/cgroup/memory/memory.stat | grep total_inactive_file | awk '{print $2}')

memory_working_set=$memory_usage_in_bytes
if [ "$memory_working_set" -lt "$memory_total_inactive_file" ];
then
    memory_working_set=0
else
    memory_working_set=$((memory_usage_in_bytes - memory_total_inactive_file))
fi

memory_available_in_bytes=$((memory_capacity_in_bytes - memory_working_set))
memory_available_in_kb=$((memory_available_in_bytes / 1024))
memory_available_in_mb=$((memory_available_in_kb / 1024))

echo "memory.capacity_in_bytes $memory_capacity_in_bytes"
echo "memory.usage_in_bytes $memory_usage_in_bytes"
echo "memory.total_inactive_file $memory_total_inactive_file"
echo "memory.working_set $memory_working_set"
echo "memory.available_in_bytes $memory_available_in_bytes"
echo "memory.available_in_kb $memory_available_in_kb"
echo "memory.available_in_mb $memory_available_in_mb"
```

kubelet 假设 inactive_file（不活跃 LRU 列表中的 file-backed 内存，以字节为单位）在紧缺情况下可以回收，因此对其进行了排除。

kubelet 支持以下两种文件系统。

（1）nodefs：保存 kubelet 的卷和守护进程日志等。

（2）imagefs：在容器运行时保存镜像及可写入层。

kubelet 使用 cAdvisor 自动监控这些文件系统。kubelet 不关注其他文件系统，不支持所有其他类型的配置，例如保存在独立文件系统中的卷和日志。

磁盘压力相关的资源回收机制正在逐渐被驱逐策略接管，未来会停止对现有垃圾收集方式的支持。

10.5.3 驱逐阈值

kubelet 可以定义驱逐阈值，一旦超出阈值，就会触发 kubelet 进行资源回收操作。

阈值的定义方式为：

```
<eviction-signal> <operator> <quantity>
```

其中：

- 在表 10.6 中列出了驱逐信号的名称。
- 当前仅支持一个 operator（运算符）：<（小于）。
- quantity 需要符合 Kubernetes 的数量表达方式，也可以用以%结尾的百分比表示。

例如，如果一个节点有 10GiB 内存，我们希望在可用内存不足 1GiB 时进行驱逐，就可以用下面任一方式来定义驱逐阈值。

- memory.available<10%。
- memory.available<1GiB。

驱逐阈值又可以通过软阈值和硬阈值两种方式进行设置。

1. 驱逐软阈值

驱逐软阈值由一个驱逐阈值和一个管理员设定的宽限期共同定义。当系统资源消耗达到软阈值时，在这一状况的持续时间达到宽限期之前，kubelet 不会触发驱逐动作。如果没有定义宽限期，则 kubelet 会拒绝启动。

另外，可以定义终止 Pod 的宽限期。如果定义了这一宽限期，那么 kubelet 会使用 pod.Spec.TerminationGracePeriodSeconds 和最大宽限期这两个值之间较小的数值进行宽限，如果没有指定，则 kubelet 会立即杀掉 Pod。

软阈值的定义包括以下几个参数。

- ◎ --eviction-soft：描述驱逐阈值（例如 memory.available<1.5GiB），如果满足这一条件的持续时间超过宽限期，就会触发对 Pod 的驱逐动作。
- ◎ --eviction-soft-grace-period：驱逐宽限期（例如 memory.available=1m30s），用于定义达到软阈值之后持续时间超过多久才进行驱逐。
- ◎ --eviction-max-pod-grace-period：在达到软阈值后，终止 Pod 的最大宽限时间（单位为 s）。

2. 驱逐硬阈值

硬阈值没有宽限期，如果达到了硬阈值，则 kubelet 会立即杀掉 Pod 并进行资源回收。

硬阈值的定义包括参数--eviction-hard：驱逐硬阈值，一旦达到阈值，就会触发对 Pod 的驱逐操作。

kubelet 的默认硬阈值定义如下：

```
--eviction-hard=memory.available<100Mi
```

10.5.4 驱逐监控频率

kubelet 的--housekeeping-interval 参数定义了一个时间间隔，kubelet 每隔一个这样的时间间隔就会对驱逐阈值进行评估。

10.5.5 节点的状况

kubelet 会将一个或多个驱逐信号与节点的状况对应起来。

无论触发了硬阈值还是软阈值，kubelet 都会认为当前节点的压力太大，如表 10.7 所示为节点状况与驱逐信号的对应关系。

表 10.7 节点状况与驱逐信号的对应关系

节点状况	驱逐信号	描述
MemoryPressure	memory.available	节点的可用内存达到了驱逐阈值
DiskPressure	nodefs.available, nodefs.inodesFree, imagefs.available, imagefs.inodesFree	节点的 root 文件系统或者镜像文件系统的可用空间达到了驱逐阈值

kubelet 会持续向 Master 报告节点状态的更新过程，这一频率由参数--node-status-update- frequency 指定，默认为 10s。

10.5.6 节点状况的抖动

如果一个节点的状况在软阈值的上下抖动，但是又没有超过宽限期，则会导致该节点的相应状态在 True 和 False 之间不断变换，可能会对调度的决策过程产生负面影响。

要防止这种状况，可以使用参数--eviction-pressure-transition-period（在脱离压力状态前需要等待的时间，默认值为 5m0s），为 kubelet 设置在脱离压力状态之前需要等待的时间。

这样一来，kubelet 在把压力状态设置为 False 之前，会确认在检测周期之内该节点没有达到驱逐阈值。

10.5.7 回收 Node 级别的资源

如果达到了驱逐阈值，并且也过了宽限期，kubelet 就会回收超出限量的资源，直到驱逐信号量回到阈值以内。

kubelet 在驱逐用户 Pod 之前，会尝试回收 Node 级别的资源。在观测到磁盘压力的情况下，基于服务器是否为容器运行时定义了独立的 imagefs，会导致不同的资源回收过程。

1. 有 Imagefs 的情况

（1）如果 nodefs 文件系统达到了驱逐阈值，则 kubelet 会删掉死掉的 Pod、容器来清理空间。

（2）如果 imagefs 文件系统达到了驱逐阈值，则 kubelet 会删掉所有无用的镜像来清理空间。

2. 没有 Imagefs 的情况

如果 nodefs 文件系统达到了驱逐阈值，则 kubelet 会按照下面的顺序来清理空间。

（1）删除死掉的 Pod、容器。

（2）删除所有无用的镜像。

10.5.8 驱逐用户的 Pod

如果 kubelet 无法通过节点级别的资源回收获取足够的资源，就会驱逐用户的 Pod。

kubelet 会按照下面的标准对 Pod 的驱逐行为进行判断。

◎ Pod 要求的服务质量。
◎ 根据 Pod 调度请求的被耗尽资源的消耗量。

接下来，kubelet 按照下面的顺序驱逐 Pod。

◎ BestEffort：紧缺资源消耗最多的 Pod 最先被驱逐。
◎ Burstable：根据相对请求来判断，紧缺资源消耗最多的 Pod 最先被驱逐，如果没有 Pod 超出它们的请求，则策略会瞄准紧缺资源消耗量最大的 Pod。
◎ Guaranteed：根据相对请求来判断，紧缺资源消耗最多的 Pod 最先被驱逐，如果没有 Pod 超出它们的请求，策略会瞄准紧缺资源消耗量最大的 Pod。

Guaranteed Pod 永远不会因为其他 Pod 的资源消费被驱逐。如果系统进程（例如 kubelet、docker、journald 等）消耗了超出 system-reserved 或者 kube-reserved 的资源，而在这一节点上只运行了 Guaranteed Pod，那么为了保证节点的稳定性并降低异常消耗对其他 Guaranteed Pod 的影响，必须选择一个 Guaranteed Pod 进行驱逐。

本地磁盘是一种 BestEffort 资源。如有必要，kubelet 会在 DiskPressure 的情况下，对 Pod 进行驱逐以回收磁盘资源。kubelet 会按照 QoS 进行评估。如果 kubelet 判定缺乏 inode 资源，就会通过驱逐最低 QoS 的 Pod 的方式来回收 inodes。如果 kubelet 判定缺乏磁盘空间，就会在相同 QoS 的 Pod 中，选择消耗最多磁盘空间的 Pod 进行驱逐。下面针对有 Imagefs 和没有 Imagefs 的两种情况，说明 kubelet 在驱逐 Pod 时选择 Pod 的排序算法，然后按顺序对 Pod 进行驱逐。

1. 有 Imagefs 的情况

如果 nodefs 触发了驱逐，则 kubelet 会根据 nodefs 的使用情况（以 Pod 中所有容器的本地卷和日志所占的空间进行计算）对 Pod 进行排序。

如果 imagefs 触发了驱逐，则 kubelet 会根据 Pod 中所有容器消耗的可写入层的使用空间进行排序。

2. 没有 Imagefs 的情况

如果 nodefs 触发了驱逐，则 kubelet 会对各个 Pod 中所有容器的总体磁盘消耗（以本地卷+日志+所有容器的写入层所占的空间进行计算）进行排序。

10.5.9 资源最少回收量

在某些场景下，驱逐 Pod 可能只回收了很少的资源，这就导致了 kubelet 反复触发驱逐阈值。另外，回收磁盘这样的资源，是需要消耗时间的。

要缓和这种状况，kubelet 可以对每种资源都定义 minimum-reclaim。kubelet 一旦监测到了资源压力，就会试着回收不少于 minimum-reclaim 的资源数量，使得资源消耗量回到期望的范围。

例如，可以配置 --eviction-minimum-reclaim 如下：

```
--eviction-hard=memory.available<500Mi,nodefs.available<1Gi,imagefs.available<100Gi
--eviction-minimum-reclaim="memory.available=0Mi,nodefs.available=500Mi,imagefs.available=2Gi"`
```

这样配置的效果如下。

- ◎ 当 memory.available 超过阈值触发了驱逐操作时，kubelet 会启动资源回收，并保证 memory.available 至少有 500MiB。
- ◎ 如果是 nodefs.available 超过阈值并触发了驱逐操作，则 kubelet 会恢复 nodefs.available 到至少 1.5GiB。
- ◎ 对于 imagefs.available 超过阈值并触发了驱逐操作的情况，kubelet 会保证 imagefs.available 恢复到最少 102GiB。

在默认情况下，所有资源的 eviction-minimum-reclaim 都为 0。

10.5.10 节点资源紧缺情况下的系统行为

1. 调度器的行为

在节点资源紧缺的情况下，节点会向 Master 报告这一状况。在 Master 上运行的调度器（Scheduler）以此为信号，不再继续向该节点调度新的 Pod。如表 10.8 所示为节点状况与调度行为的对应关系。

表 10.8　节点状况与调度行为的对应关系

节 点 状 况	调 度 行 为
MemoryPressure	不再调度新的 BestEffort Pod 到这个节点
DiskPressure	不再向这一节点调度 Pod

2. Node 的 OOM 行为

如果节点在 kubelet 能够回收内存之前遭遇了系统的 OOM（内存不足），节点则依赖 oom_killer 的设置进行响应（OOM 评分系统详见 10.4 节的描述）。

kubelet 根据 Pod 的 QoS 为每个容器都设置了一个 oom_score_adj 值，如表 10.9 所示。

表 10.9　kubelet 根据 Pod 的 QoS 为每个容器设置了一个 oom_score_adj 值

QoS 等级	oom_score_adj
Guaranteed	-998
BestEffort	1000
Burstable	min(max(2, 1000 - (1000 * memoryRequestBytes) / machineMemoryCapacityBytes), 999)

如果 kubelet 无法在系统 OOM 之前回收足够的内存，则 oom_killer 会根据内存使用比率来计算 oom_score，将得出的结果和 oom_score_adj 相加，得分最高的 Pod 首先被驱逐。

这个策略的思路是，QoS 最低且相对于调度的 Request 来说消耗最多内存的 Pod 会首先被驱逐，来保障内存的回收。

与 Pod 驱逐不同，如果一个 Pod 的容器被 OOM 杀掉，则是可能被 kubelet 根据 RestartPolicy 重启的。

3. 对 DaemonSet 类型的 Pod 驱逐的考虑

通过 DaemonSet 创建的 Pod 具有在节点上自动重启的特性，因此我们不希望 kubelet 驱逐这种 Pod；然而 kubelet 目前并没有能力分辨 DaemonSet 的 Pod，所以无法单独为其制定驱逐策略，所以强烈建议不要在 DaemonSet 中创建 BestEffort 类型的 Pod，避免产生驱逐方面的问题。

10.5.11 可调度的资源和驱逐策略实践

假设一个集群的资源管理需求如下。

- 节点内存容量：10GiB。
- 保留10%的内存给系统守护进程（内核、kubelet等）。
- 在内存使用率达到95%时驱逐Pod，以此降低系统压力并防止系统OOM。

为了满足这些需求，kubelet应该设置如下参数：

```
--eviction-hard=memory.available<500Mi
--system-reserved=memory=1.5Gi
```

在这个配置方式中隐式包含这样一个设置：系统预留内存也包括资源驱逐阈值。

如果内存占用超出这一设置，则要么是Pod占用了超过其Request的内存，要么就是系统使用了超过500MiB内存。

在这种设置下，节点一旦开始接近内存压力，调度器就不会向该节点部署Pod，并且假定这些Pod使用的资源数量少于其请求的资源数量。

10.5.12 现阶段的问题

1. kubelet无法及时观测到内存压力

kubelet目前从cAdvisor定时获取内存使用状况的统计情况。如果内存使用在这个时间段内发生了快速增长，且kubelet无法观察到MemoryPressure，则可能会触发OOMKiller。Kubernetes正在尝试将这一过程集成到memcg通知API中来减少这一延迟，而不是让内核首先发现这一情况。

对用户来说，一个较为可靠的处理方式就是设置驱逐阈值为大约75%，这样就降低了发生OOM的几率，提高了驱逐的标准，有助于集群状态的平衡。

2. kubelet可能会错误地驱逐更多的Pod

这也是状态搜集存在时间差导致的。未来可能会通过按需获取根容器的统计信息来减少计算偏差（https://github.com/google/cadvisor/issues/1247）。

10.6 Pod Disruption Budget（主动驱逐保护）

在 Kubernetes 集群运行的过程中，许多管理操作都可能对 Pod 进行主动驱逐，"主动"一词意味着这一操作可以安全地延迟一段时间，目前主要针对以下两种场景。

◎ 节点的维护或升级时（kubectl drain）。
◎ 对应用的自动缩容操作（autoscaling down）。

未来，rescheduler 也可能执行这个操作。

作为对比，由于节点不可用（Not Ready）导致的 Pod 驱逐就不能被称为主动了。

对于主动驱逐的场景来说，应用如果能够保持存活的 Pod 的数量，则会非常有用。通过使用 PodDisruptionBudget，应用可以保证那些会主动移除 Pod 的集群操作永远不会在同一时间停掉太多 Pod，从而导致服务中断或者服务降级等。例如，在对某些 Node 进行维护时，系统应该保证应用以不低于一定数量的 Pod 保障服务的正常运行。kubectl drain 操作将遵循 PodDisruptionBudget 的设定，如果在该节点上运行了属于同一服务的多个 Pod，则为了保证最少存活数量，系统将确保每终止一个 Pod 后，一定会在另一台健康的 Node 上启动新的 Pod，再继续终止下一个 Pod。

PodDisruptionBudget 资源对象在 Kubernetes 1.5 版本时升级为 Beta 版本，用于指定一个 Pod 集合在一段时间内存活的最小实例数量或者百分比。一个 PodDisruptionBudget 作用于一组被同一个控制器管理的 Pod，例如 ReplicaSet 或 RC。与通常的 Pod 删除不同，驱逐 Pod 的控制器将使用/eviction 接口对 Pod 进行驱逐，如果这一主动驱逐行为违反了 PodDisruptionBudget 的约定，就会被 API Server 拒绝。

PodDisruptionBudget 本身无法真正保障指定数量或百分比的 Pod 的存活。例如，在一个节点中包含了目标 Pod 中的一个，如果节点故障，就会导致 Pod 数量少于 minAvailable。PodDisruptionBudget 对象的保护作用仅仅针对主动驱逐的场景，而非所有场景。

对 PodDisruptionBudget 的定义包括如下两部分。

◎ Label Selector：用于筛选被管理的 Pod。
◎ minAvailable：指定驱逐过程需要保障的最少 Pod 数量。minAvailable 可以是一个数字，也可以是一个百分比，例如 100%就表示不允许进行主动驱逐。

PodDisruptionBudget 示例如下。

（1）首先创建一个 Deployment，Pod 数量为 3 个：

```
apiVersion: extensions/v1beta1
kind: Deployment
metadata:
  name: nginx
  labels:
    name: nginx
spec:
  replicas: 3
  selector:
    matchLabels:
      name: nginx
  template:
    metadata:
      labels:
        name: nginx
    spec:
      containers:
      - name: nginx
        image: nginx
        ports:
        - containerPort: 80
          protocol: TCP
```

创建后通过 kubectl get pods 命令查看 Pod 的创建情况：

```
NAME                        READY   STATUS    RESTARTS   AGE
nginx-1968750913-0k01k      1/1     Running   0          13m
nginx-1968750913-1dpcn      1/1     Running   0          19m
nginx-1968750913-n326r      1/1     Running   0          13m
```

（2）接下来创建一个 PodDisruptionBudget 对象：

```
apiVersion: policy/v1beta1
kind: PodDisruptionBudget
metadata:
  name: nginx
spec:
  minAvailable: 3
  selector:
    matchLabels:
      name: nginx
```

PodDisruptionBudget 使用的是和 Deployment 一样的 Label Selector，并且设置存活 Pod 的数量不得少于 3 个。

（3）主动驱逐验证。对 Pod 的主动驱逐操作将通过驱逐 API（/eviction）来完成。可以将这个 API 看作受策略控制的对 Pod 的 DELETE 操作。要实现一次主动驱逐（更准确的说法是创建一个 eviction），则需要 POST 一个 JSON 请求，以 eviction.json 文件格式表示，内容如下：

```
{
  "apiVersion": "policy/v1beta1",
  "kind": "Eviction",
  "metadata": {
    "name": "nginx-1968750913-0k01k",
    "namespace": "default"
  }
}
```

用 curl 命令执行 eviction 操作：

```
$ curl -v -H 'Content-type: application/json' http://<k8s_master>/api/v1/namespaces/default/pods/nginx-1968750913-0k01k/eviction -d @eviction.json
```

由于 PodDisruptionBudget 设置存活的 Pod 数量不能少于 3 个，因此驱逐操作会失败，在返回的错误信息中会包含如下内容：

```
"message": "Cannot evict pod as it would violate the pod's disruption budget."
```

使用 kubectl get pods 查看 Pod 列表，会看到 Pod 的数量和名称都没有发生变化。

最后用 kubectl delete pdb nginx 命令删除 pdb 对象。

再次执行上文中的 curl 指令，会执行成功。通过 kubectl get pods 命令查看 Pod 列表，会发现 Pod 的数量虽然没有发生变化，但是指定的 Pod 已经消失，取而代之的是一个新的 Pod。

10.7 Kubernetes 集群的高可用部署方案

Kubernetes 作为容器应用的管理平台，通过对 Pod 的运行状况进行监控，并且根据主机或容器失效的状态将新的 Pod 调度到其他 Node 上，实现了应用层的高可用性。针对

Kubernetes 集群，高可用性还应包含以下两个层面的考虑：etcd 数据存储的高可用性和 Kubernetes Master 组件的高可用性。

下面通过两种方式对 Kubernetes 的高可用部署方案进行说明，一种是通过手工方式对每个组件进行配置和部署；另一种是通过 kubeadm 工具进行部署。

10.7.1 手工方式的高可用部署方案

1. etcd 的高可用部署

etcd 在整个 Kubernetes 集群中处于中心数据库的地位，为保证 Kubernetes 集群的高可用性，首先需要保证数据库不是单一故障点。一方面，etcd 需要以集群的方式进行部署，以实现 etcd 数据存储的冗余、备份与高可用；另一方面，etcd 存储的数据本身也应考虑使用可靠的存储设备。

etcd 集群的部署可以使用静态配置，也可以通过 etcd 提供的 REST API 在运行时动态添加、修改或删除集群中的成员。本节将对 etcd 集群的静态配置进行说明。关于动态修改的操作方法请参考 etcd 官方文档的说明。

首先，规划一个至少 3 台服务器（节点）的 etcd 集群，在每台服务器上都安装 etcd。

部署一个由 3 台服务器组成的 etcd 集群，其配置如表 10.10 所示，其集群部署实例如图 10.5 所示。

表 10.10　etcd 集群的配置

etcd 实例的名称	IP 地址
etcd1	10.0.0.1
etcd2	10.0.0.2
etcd3	10.0.0.3

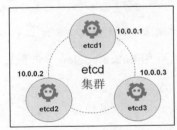

图 10.5　etcd 集群部署实例

然后修改每台服务器上 etcd 的配置文件/etc/etcd/etcd.conf。

以 etcd1 为创建集群的实例，需要将其 ETCD_INITIAL_CLUSTER_STATE 设置为 new。etcd1 的完整配置如下：

```
# [member]
```

```
    ETCD_NAME=etcd1                    # etcd 实例的名称
    ETCD_DATA_DIR="/var/lib/etcd"      # etcd 数据保存目录
    ETCD_LISTEN_CLIENT_URLS="http://10.0.0.1:2379,http://127.0.0.1:2379"    # 供外
部客户端使用的 URL
    ETCD_ADVERTISE_CLIENT_URLS="http://10.0.0.1:2379,http://127.0.0.1:2379"    #
广播给外部客户端使用的 URL
    #[cluster]
    ETCD_LISTEN_PEER_URLS="http://10.0.0.1:2380"    # 集群内部通信使用的 URL
    ETCD_INITIAL_ADVERTISE_PEER_URLS="http://10.0.0.1:2380"    # 广播给集群内其他成员
访问的 URL
    ETCD_INITIAL_CLUSTER="etcd1=http://10.0.0.1:2380,etcd2=http://10.0.0.2:2380,
etcd3=http://10.0.0.3:2380"    # 初始集群成员列表
    ETCD_INITIAL_CLUSTER_STATE="new"    # 初始集群状态，new 为新建集群
    ETCD_INITIAL_CLUSTER_TOKEN="etcd-cluster"    # 集群名称
```

启动 etcd1 服务器上的 etcd 服务：

```
$ systemctl restart etcd
```

启动完成后，就创建了一个名为 etcd-cluster 的集群。

etcd2 和 etcd3 为加入 etcd-cluster 集群的实例，需要将其 ETCD_INITIAL_CLUSTER_STATE 设置为 new。etcd2 的完整配置如下（etcd3 的配置略）：

```
# [member]
    ETCD_NAME=etcd2                    # etcd 实例名称
    ETCD_DATA_DIR="/var/lib/etcd"      # etcd 数据保存目录
    ETCD_LISTEN_CLIENT_URLS="http://10.0.0.2:2379,http://127.0.0.1:2379"    # 供外
部客户端使用的 URL
    ETCD_ADVERTISE_CLIENT_URLS="http://10.0.0.2:2379,http://127.0.0.1:2379"    #
广播给外部客户端使用的 URL
    #[cluster]
    ETCD_LISTEN_PEER_URLS="http://10.0.0.2:2380"    # 集群内部通信使用的 URL
    ETCD_INITIAL_ADVERTISE_PEER_URLS="http://10.0.0.2:2380"    # 广播给集群内其他成员
使用的 URL
    ETCD_INITIAL_CLUSTER="etcd1=http://10.0.0.1:2380,etcd2=http://10.0.0.2:2380,
etcd3=http://10.0.0.3:2380"    # 初始集群成员列表
    ETCD_INITIAL_CLUSTER_STATE="new"    # 初始集群状态，new 为新建集群
    ETCD_INITIAL_CLUSTER_TOKEN="etcd-cluster"    # 集群名称
```

启动 etcd2 和 etcd3 服务器上的 etcd 服务：

```
$ systemctl restart etcd
```

启动完成后，在任意 etcd 节点执行 etcdctl cluster-health 命令来查询集群的运行状态：

```
$ etcdctl cluster-health
cluster is healthy
member ce2a822cea30bfca is healthy
member acda82ba1cf790fc is healthy
member eba209cd0012cd2 is healthy
```

在任意 etcd 节点执行 etcdctl member list 命令来查询集群的成员列表：

```
$ etcdctl member list
ce2a822cea30bfca: name=default peerURLs=http://10.0.0.1:2380,http://127.0.0.1:7001 clientURLs=http://10.0.0.1:2379,http://127.0.0.1:2379
acda82ba1cf790fc: name=default peerURLs=http://10.0.0.2:2380,http://127.0.0.1:7001 clientURLs=http://10.0.0.2:2379,http://127.0.0.1:2379
eba209cd40012cd2: name=default peerURLs=http://10.0.0.3:2380,http://127.0.0.1:7001 clientURLs=http://10.0.0.3:2379,http://127.0.0.1:2379
```

至此，一个 etcd 集群就创建成功了。

以 kube-apiserver 为例，将访问 etcd 集群的参数设置为：

```
--etcd-servers=http://10.0.0.1:2379,http://10.0.0.2:2379,http://10.0.0.3:2379
```

在 etcd 集群成功启动之后，如果需要对集群成员进行修改，则请参考官方文档的详细说明：https://github.com/coreos/etcd/blob/master/Documentation/runtime-configuration.md#cluster-reconfiguration-operations。

对于在 etcd 中需要保存的数据的可靠性，可以考虑使用 RAID 磁盘阵列、高性能存储设备、共享存储文件系统，或者使用云服务商提供的存储系统等来实现。

2. Master 的高可用部署

在 Kubernetes 系统中，Master 服务扮演着总控中心的角色，主要的三个服务 kube-apiserver、kube-controller-mansger 和 kube-scheduler 通过不断与工作节点上的 kubelet 和 kube-proxy 进行通信来维护整个集群的健康工作状态。如果 Master 的服务无法访问某个 Node，则会将该 Node 标记为不可用，不再向其调度新建的 Pod。但对 Master 自身则需要进行额外监控，使 Master 不成为集群的单故障点，所以对 Master 服务也需要进行高可用部署。

以 Master 的 kube-apiserver、kube-controller-mansger 和 kube-scheduler 三个服务作为

一个部署单元，类似于 etcd 集群的典型部署配置。使用至少三台服务器安装 Master 服务，并且需要保证任何时候总有一套 Master 能够正常工作。图 10.6 展示了一种典型的部署方式。

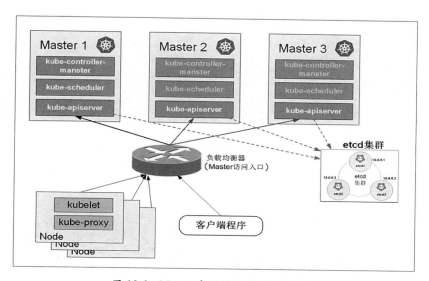

图 10.6　Master 高可用的典型部署方式

Kubernetes 建议 Master 的 3 个组件都以容器的形式启动，启动它们的基础工具是 kubelet，所以它们都将以 Static Pod 的形式启动并由 kubelet 监控和自动重启。kubelet 本身的高可用则通过操作系统来完成，例如使用 Linux 的 Systemd 系统进行管理。

注意，如果之前已运行过这 3 个进程，则需要先停止它们，然后启动 kubelet 服务，这 3 个主进程都将通过 kubelet 以容器的形式启动和运行。

接下来分别对 kube-apiserver 和 kube-controller-manager、kube-scheduler 的高可用部署进行说明。

1）kube-apiserver 的高可用部署

根据第 2 章的介绍，为 kube-apiserver 预先创建所有需要的 CA 证书和基本鉴权文件等内容，然后在每台服务器上都创建其日志文件：

```
# touch /var/log/kube-apiserver.log
```

假设 kubelet 的启动参数指定 --config=/etc/kubernetes/manifests，即 Static Pod 定义文件所在的目录，接下来就可以创建 kube-apiserver.yaml 配置文件用于启动 kube-apiserver 了：

```yaml
kube-apiserver.yaml
apiVersion: v1
kind: Pod
metadata:
  name: kube-apiserver
spec:
  hostNetwork: true
  containers:
  - name: kube-apiserver
    image: gcr.io/google_containers/kube-apiserver:9680e782e08a1a1c94c656190011bd02
    command:
    - /bin/sh
    - -c
    - /usr/local/bin/kube-apiserver --etcd-servers=http://127.0.0.1:2379
    --admission-control=NamespaceLifecycle,LimitRanger,SecurityContextDeny,ServiceAccount,ResourceQuota
      --service-cluster-ip-range=169.169.0.0/16 --v=2
      --allow-privileged=False 1>>/var/log/kube-apiserver.log 2>&1
    ports:
    - containerPort: 443
      hostPort: 443
      name: https
    - containerPort: 7080
      hostPort: 7080
      name: http
    - containerPort: 8080
      hostPort: 8080
      name: local
    volumeMounts:
    - mountPath: /srv/kubernetes
      name: srvkube
      readOnly: true
    - mountPath: /var/log/kube-apiserver.log
      name: logfile
    - mountPath: /etc/ssl
      name: etcssl
      readOnly: true
    - mountPath: /usr/share/ssl
      name: usrsharessl
      readOnly: true
```

```yaml
    - mountPath: /var/ssl
      name: varssl
      readOnly: true
    - mountPath: /usr/ssl
      name: usrssl
      readOnly: true
    - mountPath: /usr/lib/ssl
      name: usrlibssl
      readOnly: true
    - mountPath: /usr/local/openssl
      name: usrlocalopenssl
      readOnly: true
    - mountPath: /etc/openssl
      name: etcopenssl
      readOnly: true
    - mountPath: /etc/pki/tls
      name: etcpkitls
      readOnly: true
  volumes:
  - hostPath:
      path: /srv/kubernetes
    name: srvkube
  - hostPath:
      path: /var/log/kube-apiserver.log
    name: logfile
  - hostPath:
      path: /etc/ssl
    name: etcssl
  - hostPath:
      path: /usr/share/ssl
    name: usrsharessl
  - hostPath:
      path: /var/ssl
    name: varssl
  - hostPath:
      path: /usr/ssl
    name: usrssl
  - hostPath:
      path: /usr/lib/ssl
    name: usrlibssl
  - hostPath:
```

```
        path: /usr/local/openssl
      name: usrlocalopenssl
    - hostPath:
        path: /etc/openssl
      name: etcopenssl
    - hostPath:
        path: /etc/pki/tls
      name: etcpkitls
```

其中：

- kube-apiserver 需要使用 hostNetwork 模式，即直接使用宿主机网络，以使客户端能够通过物理机访问其 API。
- 镜像的 tag 来源于 Kubernetes 发布包中的 kube-apiserver.docker_tag 文件：kubernetes/server/kubernetes-server-linux-amd64/server/bin/kube-apiserver.docker_tag。
- --etcd-servers：指定 etcd 服务的 URL 地址。
- 再加上其他必要的启动参数，包括--admission-control、--service-cluster-ip-range、CA 证书相关配置等内容。
- 端口号的设置都配置了 hostPort，将容器内的端口号直接映射为宿主机的端口号。

将 kube-apiserver.yaml 文件复制到 kubelet 监控的/etc/kubernetes/manifests 目录下，kubelet 将会自动创建在 YAML 文件中定义的 kube-apiserver 的 Pod。

接下来在另外两台服务器上重复该操作，使得在每台服务器上都启动一个 kube-apiserver 的 Pod。

2）为 kube-apiserver 配置负载均衡器

至此，我们启动了三个 kube-apiserver 实例，这三个 kube-apiserver 都可以正常工作，我们需要通过统一的、可靠的、允许部分 Master 发生故障的方式来访问它们，这可以通过部署一个负载均衡器来实现。

在不同的平台下，负载均衡的实现方式不同：在一些公用云如 GCE、AWS、阿里云上都有现成的实现方案；对于本地集群，我们可以选择硬件或者软件来实现负载均衡，比如 Kubernetes 社区推荐的方案 HAProxy 和 Keepalived，其中 HAProxy 负责负载均衡，而 Keepalived 负责对 HAProxy 进行监控和故障切换。

在完成 API Server 的负载均衡配置之后，对其访问还需要注意以下内容。

- 如果Master开启了安全认证机制，那么需要确保在CA证书中包含负载均衡服务节点的IP。
- 对于外部的访问，比如通过kubectl访问API Server，需要配置为访问API Server对应的负载均衡器的IP地址。

3）kube-controller-manager和kube-scheduler的高可用配置

不同于API Server，Master中另外两个核心组件kube-controller-manager和kube-scheduler会修改集群的状态信息，因此对于kube-controller-manager和kube-scheduler而言，高可用不仅意味着需要启动多个实例，还需要这些个实例能实现选举并选举出leader，以保证同一时间只有一个实例可以对集群状态信息进行读写，避免出现同步问题和一致性问题。Kubernetes对于这种选举机制的实现是采用租赁锁（lease-lock）来实现的，我们可以通过在kube-controller- manager和kube-scheduler的每个实例的启动参数中设置--leader-elect=true，来保证同一时间只会运行一个可修改集群信息的实例。

Scheduler和Controller Manager高可用的具体实现方式如下。

首先，在每个Master上都创建相应的日志文件：

```
# touch /var/log/kube-scheduler.log
# touch /var/log/kube-controller-manager.log
```

然后，创建kube-controller-manager和kube-scheduler的YAML配置文件。

kube-controller-manager的YAML配置文件如下：

```
kube-controller-manager.yaml:
apiVersion: v1
kind: Pod
metadata:
  name: kube-controller-manager
spec:
  hostNetwork: true
  containers:
  - name: kube-controller-manager
    image: gcr.io/google_containers/kube-controller-manager:fda24638d51a48baa13c35337fcd4793
    command:
    - /bin/sh
    - -c
    - /usr/local/bin/kube-controller-manager --master=127.0.0.1:8080
```

```
        --v=2 --leader-elect=true 1>>/var/log/kube-controller-manager.log 2>&1
  livenessProbe:
    httpGet:
      path: /healthz
      port: 10252
    initialDelaySeconds: 15
    timeoutSeconds: 1
  volumeMounts:
  - mountPath: /srv/kubernetes
    name: srvkube
    readOnly: true
  - mountPath: /var/log/kube-controller-manager.log
    name: logfile
  - mountPath: /etc/ssl
    name: etcssl
    readOnly: true
  - mountPath: /usr/share/ssl
    name: usrsharessl
    readOnly: true
  - mountPath: /var/ssl
    name: varssl
    readOnly: true
  - mountPath: /usr/ssl
    name: usrssl
    readOnly: true
  - mountPath: /usr/lib/ssl
    name: usrlibssl
    readOnly: true
  - mountPath: /usr/local/openssl
    name: usrlocalopenssl
    readOnly: true
  - mountPath: /etc/openssl
    name: etcopenssl
    readOnly: true
  - mountPath: /etc/pki/tls
    name: etcpkitls
    readOnly: true
volumes:
- hostPath:
    path: /srv/kubernetes
  name: srvkube
```

```
      - hostPath:
          path: /var/log/kube-controller-manager.log
        name: logfile
      - hostPath:
          path: /etc/ssl
        name: etcssl
      - hostPath:
          path: /usr/share/ssl
        name: usrsharessl
      - hostPath:
          path: /var/ssl
        name: varssl
      - hostPath:
          path: /usr/ssl
        name: usrssl
      - hostPath:
          path: /usr/lib/ssl
        name: usrlibssl
      - hostPath:
          path: /usr/local/openssl
        name: usrlocalopenssl
      - hostPath:
          path: /etc/openssl
        name: etcopenssl
      - hostPath:
          path: /etc/pki/tls
        name: etcpkitls
```

其中：

◎ kube-controller-manager 需要使用 hostNetwork 模式，即直接使用宿主机网络。

◎ 镜像的 tag 来源于 kubernetes 发布包中的 kube-controller-manager.docker_tag 文件：kubernetes/server/kubernetes-server-linux-amd64/server/bin/kube-controller-manager.docker_tag。

◎ --master：指定 kube-apiserver 服务的 URL 地址。

◎ --leader-elect=true：使用 leader 选举机制。

kube-scheduler 的 YAML 配置文件如下：

```
kube-scheduler.yaml:
apiVersion: v1
```

```yaml
  kind: Pod
  metadata:
    name: kube-scheduler
  spec:
    hostNetwork: true
    containers:
    - name: kube-scheduler
      image: gcr.io/google_containers/kube-scheduler:34d0b8f8b31e27937327961528739bc9
      command:
      - /bin/sh
      - -c
      - /usr/local/bin/kube-scheduler --master=127.0.0.1:8080 --v=2 --leader-elect=true 1>>/var/log/kube-scheduler.log 2>&1
      livenessProbe:
        httpGet:
          path: /healthz
          port: 10251
        initialDelaySeconds: 15
        timeoutSeconds: 1
      volumeMounts:
      - mountPath: /var/log/kube-scheduler.log
        name: logfile
      - mountPath: /var/run/secrets/kubernetes.io/serviceaccount
        name: default-token-s8ejd
        readOnly: true
    volumes:
    - hostPath:
        path: /var/log/kube-scheduler.log
      name: logfile
```

其中：

◎ kube-scheduler 需要使用 hostNetwork 模式，即直接使用宿主机网络。

◎ 镜像的 tag 来源于 kubernetes 发布包中的 kube-scheduler.docker_tag 文件：kubernetes/server/kubernetes-server-linux-amd64/server/bin/kube-scheduler.docker_tag。

◎ --master：指定 kube-apiserver 服务的 URL 地址。

◎ --leader-elect=true：使用 leader 选举机制。

将这两个 YAML 文件复制到 kubelet 监控的 /etc/kubernetes/manifests 目录下，kubelet 会自动创建在 YAML 文件中定义的 kube-controller-manager 和 kube-scheduler 的 Pod。

至此，我们完成了Kubernetes Master组件的高可用配置。

10.7.2 使用kubeadm的高可用部署方案

kubeadm提供了对Master的高可用部署方案，到Kubernetes 1.13版本时Kubeadm达到GA稳定阶段，这表明kubeadm不仅能够快速部署一个符合一致性要求的Kubernetes集群，更具备足够的弹性，能够支持多种实际生产需求。在Kubernetes 1.14版本中又加入了方便证书传递的--experimental-upload-certs参数，减少了安装过程中的大量证书复制工作。

下面以Kubeadm 1.14版本为例，介绍在CentOS中基于kubeadm的高可用集群部署过程。

kubeadm提供了两种不同的高可用方案。

（1）堆叠方案：etcd服务和控制平面被部署在同样的节点中，对基础设施的要求较低，对故障的应对能力也较低，如图10.7所示。

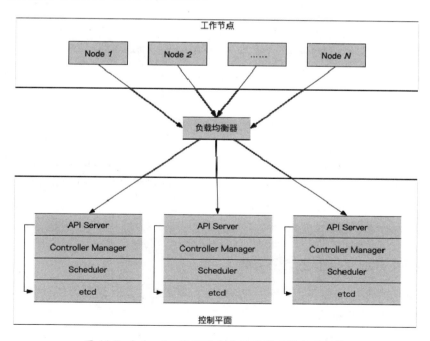

图10.7 kubeadm使用堆叠方案的高可用部署架构

（2）外置 etcd 方案：etcd 和控制平面被分离，需要更多的硬件，也有更好的保障能力，如图 10.8 所示。

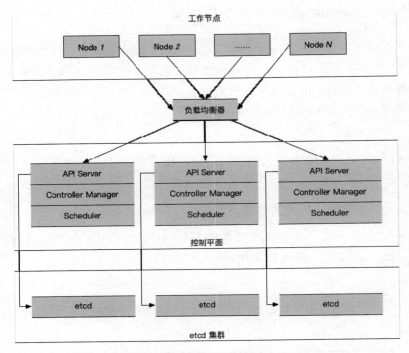

图 10.8　kubeadm 使用外置 etcd 方案的高可用部署架构

1. 准备工作

两种部署方案都有同样的准备工作要做。除了需要符合 kubeadm 部署的基本要求，因为在下面的部署过程中需要使用 SCP 进行传输，所以还需要操作机具备 SSH 到所有节点的连接能力。为了保证高可用性，Maste 服务器至少需要有三台；外置 etcd 的方案则需要有三台以上的 etcd 服务器。

2. 为 API Server 提供负载均衡服务

从前面所示的架构图中可以看到，所有节点都需要通过负载均衡器和 API Server 进行通信。负载均衡器有非常多的方案，需要根据实际情况进行选择。

下面简单介绍一个基于 HAProxy 的简易方案。

在 CentOS 下使用 yum 安装 HAProxy：

```
# yum install -y haproxy
Downloading packages:
......
Running transaction check
haproxy.x86_64 0:1.5.18-8.el7
```

接下来对 HAProxy 进行配置，使之为三台 API Server 提供负载均衡服务。编辑 /etc/haproxy.cfg，输入如下内容：

```
listen stats      0.0.0.0:12345
       mode          http
       log           global
       maxconn 10
       stats enable
       stats hide-version
       stats refresh 30s
       stats show-node
       stats auth admin:p@ssw0rd
       stats uri  /stats
frontend kube-api-https
    bind 0.0.0.0:12567
    mode tcp
    default_backend kube-api-server
backend kube-api-server
    balance roundrobin
    mode tcp
    server kubenode1 10.211.55.30:6443 check
    server kubenode2 10.211.55.31:6443 check
    server kubenode3 10.211.55.32:6443 check
```

以上代码是非常简化的版本，假设三台 Master 的地址为 10.211.55.30-32。我们定义了一个代理端口，用 TCP 转发的方式在 12567 端口对工作节点提供 API Server 服务，并且在 12345 端口上开放了 HAProxy 的状态服务，用于观察服务状态。

运行如下命令，启动服务并将其设置为自动启动：

```
# systemctl enable haproxy
# systemctl start haproxy
```

用浏览器打开 http://10.211.55.30:12345/stats，查看 HAProxy 的状态页面，大致会看到如图 10.9 所示的结果。倒数第 2 行至倒数第 4 行代表我们的三个尚未启动的 API Server 都处于 DOWN 状态，这很正常，我们还没开始进行服务器设置。

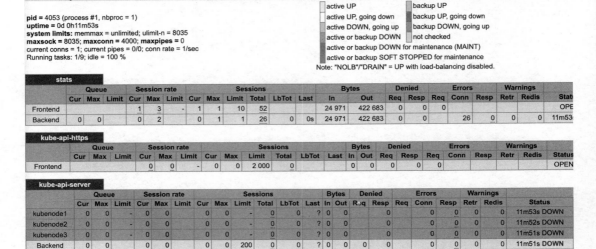

图 10.9　HAProxy 页面显示 3 个 API Server 均为 DOWN 状态

3. 启动第 1 组控制平面

使用如下代码创建一个 YAML 文件，将其保存为 kubeadm-config.yaml，用于为 kubeadm 的安装提供配置：

```
apiVersion: kubeadm.k8s.io/v1beta1
kind: ClusterConfiguration
kubernetesVersion: stable
controlPlaneEndpoint: "10.211.55.30:12567"
```

然后启动 kubeadm 的初始化过程：

```
# kubeadm init --config=kubeadm-config.yaml --experimental-upload-certs
......
You can now join any number of the control-plane node running the following command
```

```
on each as root:

    kubeadm join 10.211.55.30:12567 --token w2q4bd.tfe9027rhslua757 \
        --discovery-token-ca-cert-hash
sha256:58da1a080ea16e30c55f155d094fbf29b3741ee3c123edb169764a203a356be9 \
        --experimental-control-plane --certificate-key
0b606621ec79b3b7a23413d1f098149cb328a44f12ac47ef0dd124fcc7a00f79

    Please note that the certificate-key gives access to cluster sensitive data, keep
it secret!
    As a safeguard, uploaded-certs will be deleted in two hours; If necessary, you
can use
    "kubeadm init phase upload-certs --experimental-upload-certs" to reload certs
afterward.

    Then you can join any number of worker nodes by running the following on each
as root:

    kubeadm join 10.211.55.30:12567 --token w2q4bd.tfe9027rhslua757 \
        --discovery-token-ca-cert-hash
sha256:58da1a080ea16e30c55f155d094fbf29b3741ee3c123edb169764a203a356be9
```

这个初始化过程和之前的情况有所不同，加入了一个新的参数：--experimental-upload-certs。这个参数专门用于高可用部署，可以将需要在不同的控制平面之间传递的证书文件上传到集群中，以 Secret 形式保存起来，并且使用 Token 进行加密。值得注意的是，这个 Secret 会在两个小时后过期，如果过期就需要使用 kubeadm init phase upload-certs --experimental-upload-certs 命令重新生成。

在输出内容中包含加入新的控制平面的命令、和工作节点一致的参数 --toke 和 --discovery-token-ca-cert-hash，以及 --experimental-control-plane 及 --certificate-key 两个参数，用于获取和使用前面生成的 Secret。

打开新生成的 kubeconfig 文件，会发现其中的 server 字段使用了我们的负载均衡地址：

```
# cat ~/.kube/config | grep server
    server: https://10.211.55.30:12567
```

再次打开我们的 HAProxy 状态页面，界面如图 10.10 所示。

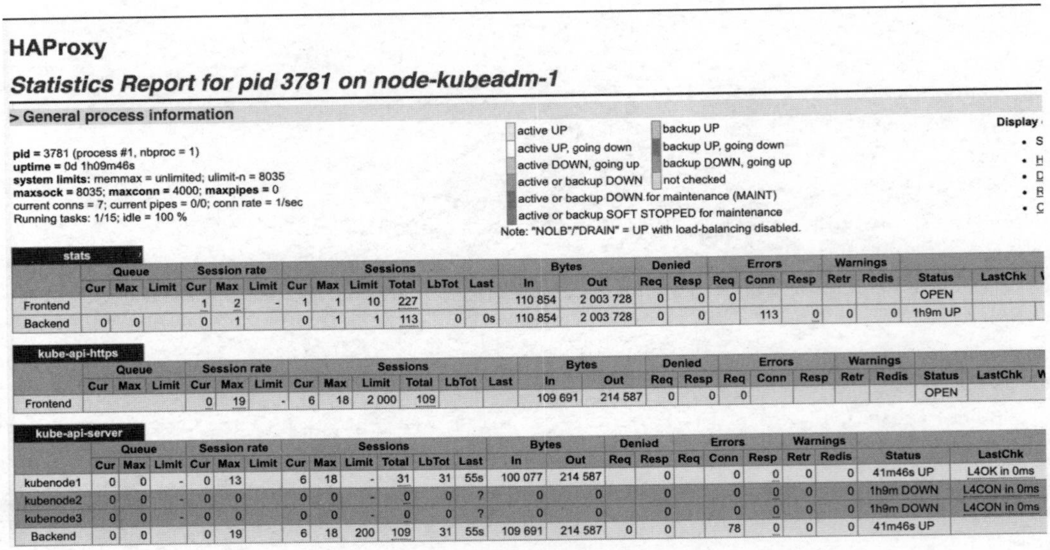

```
HAProxy
Statistics Report for pid 3781 on node-kubeadm-1
```

图10.10　HAProxy 页面显示 1 个 API Server 为 UP 状态

可以看到，kube-api-server 组已经可用。

4. 启用网络

这里以 Weave Net 为例，启用容器网络：

```
# kubectl apply -f "https://cloud.weave.works/k8s/net?k8s-version=$(kubectl version | base64 | tr -d '\n')"
```

在网络启动成功后，我们可以运行如下命令检查当前运行的 Pod：

```
# kubectl get po,no --all-namespaces
NAMESPACE     NAME                                             READY   STATUS    RESTARTS   AGE
kube-system   pod/coredns-fb8b8dccf-4qs4m                      1/1     Running   0          47m
kube-system   pod/coredns-fb8b8dccf-svzrx                      1/1     Running   0          47m
kube-system   pod/etcd-node-kubeadm-1                          1/1     Running   0          46m
kube-system   pod/kube-apiserver-node-kubeadm-1                1/1     Running   0          46m
kube-system   pod/kube-controller-manager-node-kubeadm-1       1/1     Running   0          46m
kube-system   pod/kube-proxy-p5bvv                             1/1     Running   0          47m
kube-system   pod/kube-scheduler-node-kubeadm-1                1/1     Running   0          46m
kube-system   pod/weave-net-wtzt7                              2/2     Running   0          3m49s

NAMESPACE     NAME                   STATUS   ROLES    AGE   VERSION
              node/node-kubeadm-1    Ready    master   47m   v1.14.0
```

可以看到，Pod 都已经开始运行，并且 Master 已经启动。

5. 加入新的控制平面

根据安装第 1 组控制平面时的提示，登录第 2 个控制平面所在的服务器，用下列命令部署新的控制平面：

```
# kubeadm join 10.211.55.30:12567 \
--token w2q4bd.tfe9027rhslua757 \
    --discovery-token-ca-cert-hash
sha256:58da1a080ea16e30c55f155d094fbf29b3741ee3c123edb169764a203a356be9 \
    --experimental-control-plane \
    --certificate-key
0b606621ec79b3b7a23413d1f098149cb328a44f12ac47ef0dd124fcc7a00f79
...
[upload-config] storing the configuration used in ConfigMap "kubeadm-config" in the "kube-system" Namespace
[mark-control-plane] Marking the node node-kubeadm-2 as control-plane by adding the label "node-role.kubernetes.io/master=''"
[mark-control-plane] Marking the node node-kubeadm-2 as control-plane by adding the taints [node-role.kubernetes.io/master:NoSchedule]

This node has joined the cluster and a new control plane instance was created:

* Certificate signing request was sent to apiserver and approval was received.
* The Kubelet was informed of the new secure connection details.
* Control plane (master) label and taint were applied to the new node.
* The Kubernetes control plane instances scaled up.
* A new etcd member was added to the local/stacked etcd cluster
```

可以看到提示加入新的控制平面，此时回到 10.211.55.30，再次使用 kubectl 获取节点列表：

```
# kubectl get nodes
NAME             STATUS   ROLES    AGE   VERSION
node-kubeadm-1   Ready    master   63m   v1.14.0
node-kubeadm-2   Ready    master   97s   v1.14.0
```

可以看到，新的控制平面已经加入。

同样，进入第 3 个节点，再次加入第 3 组控制平面。

如此一来,我们就有了3个控制平面,通过HAProxy提供的12456端口提供服务:

```
# kubectl get nodes
NAME              STATUS   ROLES    AGE     VERSION
node-kubeadm-1    Ready    master   70m     v1.14.0
node-kubeadm-2    Ready    master   8m33s   v1.14.0
node-kubeadm-3    Ready    master   32s     v1.14.0
```

再看HAProxy的状态页面,如图10.11所示。

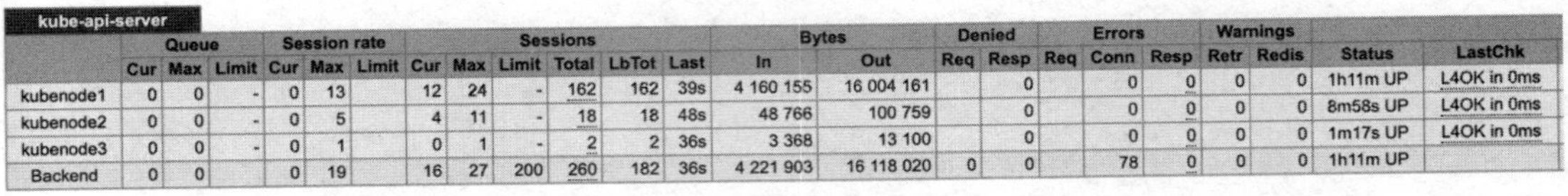

kube-api-server	Queue			Session rate			Sessions					Bytes		Denied		Errors		Warnings		Status	LastChk
	Cur	Max	Limit	Cur	Max	Limit	Cur	Max	Limit	Total	LbTot	Last	In	Out	Req	Resp	Req	Conn	Resp	Retr	Redis		
kubenode1	0	0	-	0	13	-	12	24	-	162	162	39s	4 160 155	16 004 161	0	0	0	0	0	0	0	1h11m UP	L4OK in 0ms
kubenode2	0	0	-	0	5	-	4	11	-	18	18	48s	48 766	100 759	0	0	0	0	0	0	0	8m58s UP	L4OK in 0ms
kubenode3	0	0	-	0	1	-	0	1	-	2	2	36s	3 368	13 100	0	0	0	0	0	0	0	1m17s UP	L4OK in 0ms
Backend	0	0		0	19		16	27	200	260	182	36s	4 221 903	16 118 020	0		78	0	0			1h11m UP	

图10.11 HAProxy页面显示3个API Server均为UP状态

三组控制平面都已成功启动。

6. 外置 etcd 方案的差异

前面几节介绍了堆叠方案的 kubeadm 高可用部署步骤。外置 etcd 的方案与堆叠方案差异不大,进行如下设置即可。

(1)需要为 kubeadm 设置一组高可用的 etcd 集群。

(2)将访问 etcd 所需的证书复制到控制平面所在的服务器上。

(3)在创建 kubeadm-config.yaml 时需要加入 etcd 的访问信息,例如:

```
apiVersion: kubeadm.k8s.io/v1beta1
kind: ClusterConfiguration
kubernetesVersion: stable
controlPlaneEndpoint: "LOAD_BALANCER_DNS:LOAD_BALANCER_PORT"
etcd:
    external:
        endpoints:
        - https://ETCD_0_IP:2379
        - https://ETCD_1_IP:2379
        - https://ETCD_2_IP:2379
        caFile: /etc/kubernetes/pki/etcd/ca.crt
        certFile: /etc/kubernetes/pki/apiserver-etcd-client.crt
        keyFile: /etc/kubernetes/pki/apiserver-etcd-client.key
```

简单来说,将 etcd 所需的证书文件以配置文件的方式指派给控制平面即可。

7. 补充说明

在上述内容中以 Kubernetes 1.14 为例简单讲解了使用 kubeadm 部署高可用 Kubernetes 集群的过程。在完成控制平面的部署之后,就可以在其中加入工作节点了。

这是一个非常简化的过程,HAProxy 成为新的单点,可能导致整体发生故障,因此在实际工作中需要根据生产环境的具体情况,换用硬件负载均衡器方案,或者使用软件进行 VIP、DNS 等相关高可用保障,从而消除对单一 HAProxy 的依赖。

10.8 Kubernetes 集群监控

Kubernetes 的早期版本依靠 Heapster 来实现完整的性能数据采集和监控功能,Kubernetes 从 1.8 版本开始,性能数据开始以 Metrics API 的方式提供标准化接口,并且从 1.10 版本开始将 Heapster 替换为 Metrics Server。在 Kubernetes 新的监控体系中,Metrics Server 用于提供核心指标(Core Metrics),包括 Node、Pod 的 CPU 和内存使用指标。对其他自定义指标(Custom Metrics)的监控则由 Prometheus 等组件来完成。

10.8.1 通过 Metrics Server 监控 Pod 和 Node 的 CPU 和内存资源使用数据

Metrics Server 在部署完成后,将通过 Kubernetes 核心 API Server 的 "/apis/metrics.k8s.io/v1beta1" 路径提供 Pod 和 Node 的监控数据。Metrics Server 源代码和部署配置可以在 GitHub 代码库(https://github.com/kubernetes-incubator/metrics-server)找到。

首先,部署 Metrics Server 实例,在下面的 YAML 配置中包含 ServiceAccount、Deployment 和 Service 的定义:

```
---
apiVersion: v1
kind: ServiceAccount
metadata:
  name: metrics-server
  namespace: kube-system
---
```

```yaml
apiVersion: extensions/v1beta1
kind: Deployment
metadata:
  name: metrics-server
  namespace: kube-system
  labels:
    k8s-app: metrics-server
spec:
  selector:
    matchLabels:
      k8s-app: metrics-server
  template:
    metadata:
      name: metrics-server
      labels:
        k8s-app: metrics-server
    spec:
      serviceAccountName: metrics-server
      containers:
      - name: metrics-server
        image: k8s.gcr.io/metrics-server-amd64:v0.3.1
        imagePullPolicy: IfNotPresent
        volumeMounts:
        - name: tmp-dir
          mountPath: /tmp
      volumes:
      - name: tmp-dir
        emptyDir: {}
---
apiVersion: v1
kind: Service
metadata:
  name: metrics-server
  namespace: kube-system
  labels:
    kubernetes.io/name: "Metrics-server"
spec:
  selector:
    k8s-app: metrics-server
  ports:
  - port: 443
```

```
      protocol: TCP
      targetPort: 443
```

然后，创建 metrics-server 所需的 RBAC 权限配置，此处略。

最后，创建 APIService 资源，将监控数据通过 "/apis/metrics.k8s.io/v1beta1" 路径提供：

```
apiVersion: apiregistration.k8s.io/v1beta1
kind: APIService
metadata:
  name: v1beta1.metrics.k8s.io
spec:
  service:
    name: metrics-server
    namespace: kube-system
  group: metrics.k8s.io
  version: v1beta1
  insecureSkipTLSVerify: true
  groupPriorityMinimum: 100
  versionPriority: 100
```

在部署完成后确保 metrics-server 的 Pod 启动成功：

```
# kubectl -n kube-system get pod -l k8s-app=metrics-server
NAME                                READY   STATUS    RESTARTS   AGE
metrics-server-7cb798c45b-4dnmh     1/1     Running   0          5m
```

使用 kubectl top nodes 和 kubectl top pods 命令监控 CPU 和内存资源的使用情况：

```
# kubectl top nodes
NAME         CPU(cores)   CPU%   MEMORY(bytes)   MEMORY%
k8s-node-1   319m         7%     1167Mi          67%

# kubectl top pods --all-namespaces
NAMESPACE     NAME                                  CPU(cores)   MEMORY(bytes)
kube-system   coredns-767997f5b5-sfz2w              6m           36Mi
kube-system   metrics-server-7cb798c45b-4dnmh       3m           22Mi
......
```

Metrics Server 提供的数据也可以供 HPA 控制器使用，以实现基于 CPU 使用率或内存使用值的 Pod 自动扩缩容功能。

10.8.2 Prometheus+Grafana 集群性能监控平台搭建

Prometheus 是由 SoundCloud 公司开发的开源监控系统,是继 Kubernetes 之后 CNCF 第 2 个毕业的项目,在容器和微服务领域得到了广泛应用。Prometheus 的主要特点如下。

◎ 使用指标名称及键值对标识的多维度数据模型。
◎ 采用灵活的查询语言 PromQL。
◎ 不依赖分布式存储,为自治的单节点服务。
◎ 使用 HTTP 完成对监控数据的拉取。
◎ 支持通过网关推送时序数据。
◎ 支持多种图形和 Dashboard 的展示,例如 Grafana。

Prometheus 生态系统由各种组件组成,用于功能的扩充。

◎ Prometheus Server:负责监控数据采集和时序数据存储,并提供数据查询功能。
◎ 客户端 SDK:对接 Prometheus 的开发工具包。
◎ Push Gateway:推送数据的网关组件。
◎ 第三方 Exporter:各种外部指标收集系统,其数据可以被 Prometheus 采集。
◎ AlertManager:告警管理器。
◎ 其他辅助支持工具。

Prometheus 的核心组件 Prometheus Server 的主要功能包括:从 Kubernetes Master 获取需要监控的资源或服务信息;从各种 Exporter 抓取(Pull)指标数据,然后将指标数据保存在时序数据库(TSDB)中;向其他系统提供 HTTP API 进行查询;提供基于 PromQL 语言的数据查询;可以将告警数据推送(Push)给 AlertManager,等等。

Prometheus 的系统架构如图 10.12 所示。

在部署 Prometheus 时可以直接基于官方提供的镜像进行,也可以通过 Operator 模式进行。本文以直接部署为例,Operator 模式的部署案例可以参考 3.12.2 节的示例。

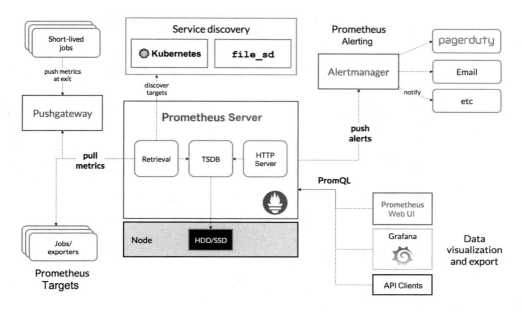

图 10.12　Prometheus 系统架构图

下面对部署 Prometheus 服务的过程进行说明。

首先，创建一个 ConfigMap 用于保存 Prometheus 的主配置文件 prometheus.yml，其中可以配置需要监控的 Kubernetes 集群的资源对象或服务（如 Service、Pod、Node 等）：

```yaml
apiVersion: v1
kind: ConfigMap
metadata:
  name: prometheus-config
  namespace: kube-system
  labels:
    kubernetes.io/cluster-service: "true"
    addonmanager.kubernetes.io/mode: EnsureExists
data:
  prometheus.yml: |
    global:
      scrape_interval: 30s
    scrape_configs:
    - job_name: prometheus
      static_configs:
      - targets:
        - localhost:9090
```

```yaml
    - job_name: kubernetes-apiservers
      kubernetes_sd_configs:
      - role: endpoints
      relabel_configs:
      - action: keep
        regex: default;kubernetes;https
        source_labels:
        - __meta_kubernetes_namespace
        - __meta_kubernetes_service_name
        - __meta_kubernetes_endpoint_port_name
      scheme: https
      tls_config:
        ca_file: /var/run/secrets/kubernetes.io/serviceaccount/ca.crt
        insecure_skip_verify: true
      bearer_token_file: /var/run/secrets/kubernetes.io/serviceaccount/token
    - job_name: kubernetes-nodes-kubelet
    ......
    - job_name: kubernetes-service-endpoints
    ......
    - job_name: kubernetes-services
    ......
    - job_name: kubernetes-pods
    ......
```

接下来部署 Prometheus Deployment 和 Service：

```yaml
---
apiVersion: extensions/v1beta1
kind: Deployment
metadata:
  name: prometheus
  namespace: kube-system
  labels:
    k8s-app: prometheus
    kubernetes.io/cluster-service: "true"
    addonmanager.kubernetes.io/mode: Reconcile
spec:
  replicas: 1
  selector:
    matchLabels:
      k8s-app: prometheus
  template:
```

```yaml
metadata:
  labels:
    k8s-app: prometheus
  annotations:
    scheduler.alpha.kubernetes.io/critical-pod: ''
spec:
  priorityClassName: system-cluster-critical
  initContainers:
  - name: "init-chown-data"
    image: "busybox:latest"
    imagePullPolicy: "IfNotPresent"
    command: ["chown", "-R", "65534:65534", "/data"]
    volumeMounts:
    - name: storage-volume
      mountPath: /data
      subPath: ""
  containers:
    - name: prometheus-server-configmap-reload
      image: "jimmidyson/configmap-reload:v0.1"
      imagePullPolicy: "IfNotPresent"
      args:
      - --volume-dir=/etc/config
      - --webhook-url=http://localhost:9090/-/reload
      volumeMounts:
      - name: config-volume
        mountPath: /etc/config
        readOnly: true
    - name: prometheus-server
      image: "prom/prometheus:v2.8.0"
      imagePullPolicy: "IfNotPresent"
      args:
      - **--config.file=/etc/config/prometheus.yml**
      - **--storage.tsdb.path=/data**
      - --web.console.libraries=/etc/prometheus/console_libraries
      - --web.console.templates=/etc/prometheus/consoles
      - --web.enable-lifecycle
      ports:
      - containerPort: 9090
      readinessProbe:
        httpGet:
          path: /-/ready
```

```yaml
        port: 9090
      initialDelaySeconds: 30
      timeoutSeconds: 30
    livenessProbe:
      httpGet:
        path: /-/healthy
        port: 9090
      initialDelaySeconds: 30
      timeoutSeconds: 30
    volumeMounts:
    - name: config-volume
      mountPath: /etc/config
    - name: storage-volume
      mountPath: /data
      subPath: ""
  terminationGracePeriodSeconds: 300
  volumes:
  - name: config-volume
    configMap:
      name: prometheus-config
  - name: storage-volume
    hostPath:
      path: /prometheus-data
      type: Directory
# Prometheus 的关键启动参数包括：
# --config.file，指定配置文件 prometheus.yml 的路径
# --storage.tsdb.path，指定数据的存储路径

---
kind: Service
apiVersion: v1
metadata:
  name: prometheus
  namespace: kube-system
  labels:
    kubernetes.io/name: "Prometheus"
    kubernetes.io/cluster-service: "true"
    addonmanager.kubernetes.io/mode: Reconcile
spec:
  type: NodePort
  ports:
```

```
    - name: http
      port: 9090
      nodePort: 9090
      protocol: TCP
      targetPort: 9090
  selector:
    k8s-app: prometheus
```

在部署完成后,确保 Prometheus 运行成功:

```
# kubectl -n kube-system get pods -l k8s-app=prometheus
NAME                             READY    STATUS    RESTARTS    AGE
prometheus-5fbb5ddd4f-2wrxj      2/2      Running   0           3m32s
```

Prometheus 提供了一个简单的 Web 页面用于查看已采集的监控数据,上面的 Service 定义了 NodePort 为 9090,我们可以通过访问 Node 的 9090 端口访问这个页面,如图 10.13 所示。

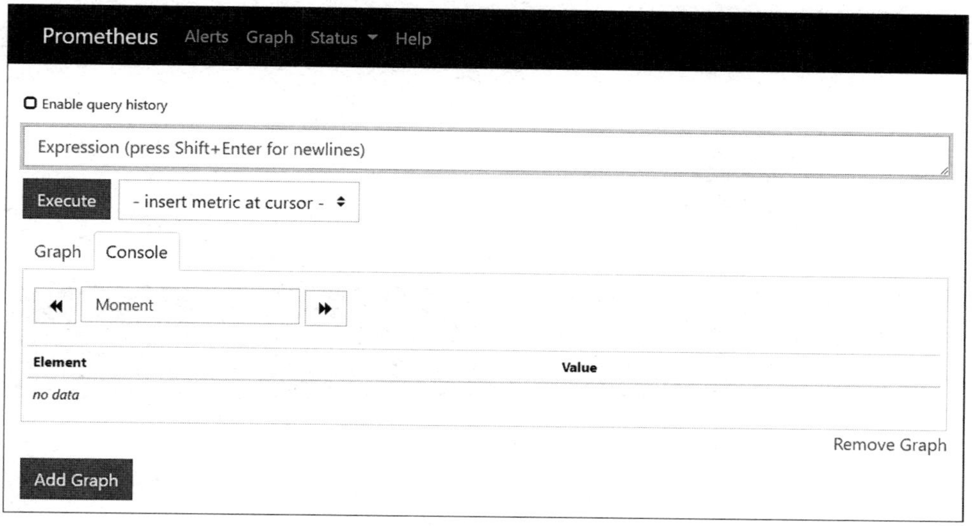

图 10.13　Prometheus 提供的 Web 页面

在 Prometheus 提供的 Web 页面上,可以输入 PromQL 查询语句对指标数据进行查询,也可以选择一个指标进行查看,例如选择 container_network_receive_bytes_total 指标查看容器的网络接收字节数,如图 10.14 所示。

单击 Graph 标签,可以查看该指标的时序图,如图 10.15 所示。

图 10.14　在 Prometheus 页面查询指标的值

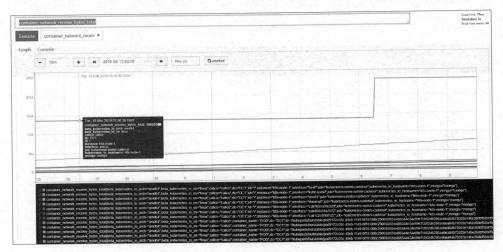

图 10.15　在 Prometheus 页面查询指标的时序图

接下来可以针对各种系统和服务部署各种 Exporter 进行指标数据的采集。目前 Prometheus 支持多种开源软件的 Exporter，包括数据库、硬件系统、消息系统、存储系统、HTTP 服务器、日志服务，等等，可以从 Prometheus 的官网 https://prometheus.io/docs/instrumenting/exporters/ 获取各种 Exporter 的信息。

下面以官方维护的 node_exporter 为例进行部署。node_exporter 主要用于采集主机相关的性能指标数据，其官网为 https://github.com/prometheus/node_exporter。node_exporter 的 YAML 配置文件如下：

```yaml
---
apiVersion: extensions/v1beta1
kind: DaemonSet
metadata:
  name: node-exporter
  namespace: kube-system
  labels:
    k8s-app: node-exporter
    kubernetes.io/cluster-service: "true"
    addonmanager.kubernetes.io/mode: Reconcile
    version: v0.17.0
spec:
  updateStrategy:
    type: OnDelete
  template:
    metadata:
      labels:
        k8s-app: node-exporter
        version: v0.17.0
      annotations:
        scheduler.alpha.kubernetes.io/critical-pod: ''
    spec:
      priorityClassName: system-node-critical
      containers:
        - name: prometheus-node-exporter
          image: "prom/node-exporter:v0.17.0"
          imagePullPolicy: "IfNotPresent"
          args:
            - --path.procfs=/host/proc
            - --path.sysfs=/host/sys
          ports:
            - name: metrics
              containerPort: 9100
              hostPort: 9100
          volumeMounts:
            - name: proc
              mountPath: /host/proc
              readOnly: true
            - name: sys
              mountPath: /host/sys
              readOnly: true
```

```yaml
      resources:
        limits:
          cpu: 1
          memory: 512Mi
        requests:
          cpu: 100m
          memory: 50Mi
      hostNetwork: true
      hostPID: true
      volumes:
        - name: proc
          hostPath:
            path: /proc
        - name: sys
          hostPath:
            path: /sys
# node-exporter 将读取宿主机上/proc 和/sys 目录下的内容，获取主机级别的性能指标数据

---
apiVersion: v1
kind: Service
metadata:
  name: node-exporter
  namespace: kube-system
  annotations:
    prometheus.io/scrape: "true"
  labels:
    kubernetes.io/cluster-service: "true"
    addonmanager.kubernetes.io/mode: Reconcile
    kubernetes.io/name: "NodeExporter"
spec:
  clusterIP: None
  ports:
    - name: metrics
      port: 9100
      protocol: TCP
      targetPort: 9100
  selector:
    k8s-app: node-exporter
```

在部署完成后，在每个 Node 上都运行了一个 node-exporter Pod：

```
# kubectl -n kube-system get pods -l k8s-app: node-exporter
NAME                         READY   STATUS    RESTARTS   AGE
node-exporter-2x4fq          1/1     Running   0          15m
node-exporter-saz2w          1/1     Running   0          15m
node-exporter-kr8wc          1/1     Running   0          15m
......
```

从 Prometheus 的 Web 页面就可以查看 node-exporter 采集的 Node 指标数据了,如图 10.16 所示。

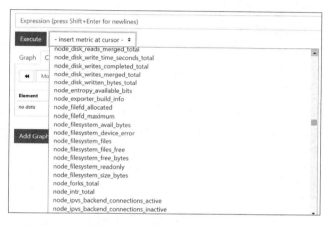

图 10.16 在 Prometheus 的 Web 页面查询指标的时序图

最后,部署 Grafana 用于展示专业的监控页面,其 YAML 配置文件如下:

```
---
kind: Deployment
apiVersion: extensions/v1beta1
metadata:
  name: grafana
  namespace: kube-system
  labels:
    k8s-app: grafana
    kubernetes.io/cluster-service: "true"
    addonmanager.kubernetes.io/mode: Reconcile
spec:
  replicas: 1
  selector:
    matchLabels:
      k8s-app: grafana
```

```yaml
  template:
    metadata:
      labels:
        k8s-app: grafana
      annotations:
        scheduler.alpha.kubernetes.io/critical-pod: ''
    spec:
      priorityClassName: system-cluster-critical
      tolerations:
      - key: node-role.kubernetes.io/master
        effect: NoSchedule
      - key: "CriticalAddonsOnly"
        operator: "Exists"
      containers:
      - name: grafana
        image: grafana/grafana:6.0.1
        imagePullPolicy: IfNotPresent
        resources:
          limits:
            cpu: 1
            memory: 1Gi
          requests:
            cpu: 100m
            memory: 100Mi
        env:
        - name: GF_AUTH_BASIC_ENABLED
          value: "false"
        - name: GF_AUTH_ANONYMOUS_ENABLED
          value: "true"
        - name: GF_AUTH_ANONYMOUS_ORG_ROLE
          value: Admin
        - name: GF_SERVER_ROOT_URL
          value: /api/v1/namespaces/kube-system/services/grafana/proxy/
        ports:
        - name: ui
          containerPort: 3000

---
apiVersion: v1
kind: Service
metadata:
```

```
    name: grafana
    namespace: kube-system
    labels:
      kubernetes.io/cluster-service: "true"
      addonmanager.kubernetes.io/mode: Reconcile
      kubernetes.io/name: "Grafana"
spec:
  ports:
    - port: 80
      protocol: TCP
      targetPort: ui
  selector:
    k8s-app: grafana
```

部署完成后，通过 Kubernetes Master 的 URL 访问 Grafana 页面，例如 http://192.168.18.3:8080/api/v1/namespaces/kube-system/services/grafana/proxy。

在 Grafana 的设置页面添加类型为 Prometheus 的数据源，输入 Prometheus 服务的 URL（如 http://prometheus:9090）进行保存，如图 10.17 所示。

图 10.17 Grafana 配置数据源页面

在 Grafana 的 Dashboard 控制面板导入预置的 Dashboard 模板，以显示各种监控图表。Grafana 官网（https://grafana.com/dashboards）提供了许多针对 Kubernetes 集群监控的 Dashboard 模板，可以下载、导入并使用。图 10.18 显示了一个可以监控集群 CPU、内存、文件系统、网络吞吐率的 Dashboard。

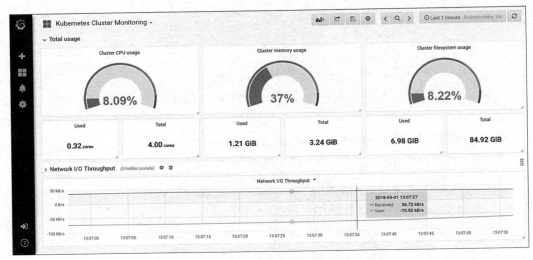

图 10.18　Grafana 监控页面

至此，基于 Prometheus+Grafana 的 Kubernetes 集群监控系统就搭建完成了。

10.9　集群统一日志管理

在 Kubernetes 集群环境中，一个完整的应用或服务都会涉及为数众多的组件运行，各组件所在的 Node 及实例数量都是可变的。日志子系统如果不做集中化管理，则会给系统的运维支撑造成很大的困难，因此有必要在集群层面对日志进行统一收集和检索等工作。

在容器中输出到控制台的日志，都会以 *-json.log 的命名方式保存在/var/lib/docker/containers/目录下，这就为日志采集和后续处理奠定了基础。

Kubernetes 推荐采用 Fluentd+Elasticsearch+Kibana 完成对系统和容器日志的采集、查询和展现工作。

部署统一的日志管理系统，需要以下两个前提条件。

◎ API Server 正确配置了 CA 证书。
◎ DNS 服务启动、运行。

10.9.1 系统部署架构

该系统的逻辑架构如图 10.19 所示。

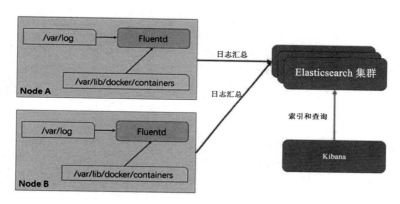

图 10.19　Fluentd+Elasticsearch+Kibana 系统的逻辑架构图

在各 Node 上都运行了一个 Fluentd 容器，采集本节点 /var/log 和 /var/lib/docker/containers 两个目录下的日志进程，将其汇总到 Elasticsearch 集群，最终通过 Kibana 完成和用户的交互工作。

这里有一个特殊的需求：Fluentd 必须在每个 Node 上运行。为了满足这一需求，我们通过以下几种方式部署 Fluentd。

◎ 直接在 Node 主机上部署 Fluentd。
◎ 利用 kubelet 的 --config 参数，为每个 Node 都加载 Fluentd Pod。
◎ 利用 DaemonSet 让 Fluentd Pod 在每个 Node 上运行。

10.9.2　创建 Elasticsearch RC 和 Service

Elasticsearch 的 RC 和 Service 定义如下：

```
elasticsearch-rc-svc.yml
---
```

```yaml
apiVersion: v1
kind: ReplicationController
metadata:
  name: elasticsearch-logging-v1
  namespace: kube-system
  labels:
    k8s-app: elasticsearch-logging
    version: v1
    kubernetes.io/cluster-service: "true"
spec:
  replicas: 2
  selector:
    k8s-app: elasticsearch-logging
    version: v1
  template:
    metadata:
      labels:
        k8s-app: elasticsearch-logging
        version: v1
        kubernetes.io/cluster-service: "true"
    spec:
      containers:
      - image: gcr.io/google_containers/elasticsearch:1.8
        name: elasticsearch-logging
        resources:
          # keep request = limit to keep this container in guaranteed class
          limits:
            cpu: 100m
          requests:
            cpu: 100m
        ports:
        - containerPort: 9200
          name: db
          protocol: TCP
        - containerPort: 9300
          name: transport
          protocol: TCP
        volumeMounts:
        - name: es-persistent-storage
          mountPath: /data
      volumes:
```

```yaml
      - name: es-persistent-storage
        emptyDir: {}
---
apiVersion: v1
kind: Service
metadata:
  name: elasticsearch-logging
  namespace: kube-system
  labels:
    k8s-app: elasticsearch-logging
    kubernetes.io/cluster-service: "true"
    kubernetes.io/name: "Elasticsearch"
spec:
  ports:
  - port: 9200
    protocol: TCP
    targetPort: db
  selector:
    k8s-app: elasticsearch-logging
```

执行 kubectl create -f elastic-search.yml 命令完成创建。

在命令成功执行后,首先验证 Pod 的运行情况。通过 kubectl get pods --namespaces=kube-system 获取运行中的 Pod:

```
# kubectl get pods --namespaces=kube-system
NAMESPACE     NAME                                  READY   STATUS    RESTARTS   AGE
kube-system   elasticsearch-logging-v1-59qvp        1/1     Running   0          18h
kube-system   elasticsearch-logging-v1-xnv14        1/1     Running   0          18h
```

接下来通过 Elasticsearch 页面验证其功能。

首先,执行# kubectl cluster-info 命令获取 Elasticsearch 服务的地址:

```
# kubectl cluster-info
Elasticsearch is running at
http://192.168.18.3:8080/api/v1/proxy/namespaces/kube-system/services/elasticsearch-logging
```

然后,使用# kubectl proxy 命令对 API Server 进行代理,在成功执行后输出如下内容:

```
# kubectl proxy
Starting to serve on 127.0.0.1:8001
```

这样就可以在浏览器上访问 URL 地址 http://192.168.18.3:8001/api/v1/proxy/namespaces/kube-system/services/elasticsearch-logging，来验证 Elasticsearch 的运行情况了，返回的内容是一个 JSON 文档：

```
{
"status": 200,
"name": "Emplate",
"cluster_name": "kubernetes-logging",
"version": {
    "number": "1.5.2",
    "build_hash": "62ff9868b4c8a0c45860bebb259e21980778ab1c",
    "build_timestamp": "2015-04-27T09:21:06Z",
    "build_snapshot": false,
    "lucene_version": "4.10.4"
},
"tagline": "You Know, for Search"
}
```

10.9.3 在每个 Node 上启动 Fluentd

Fluentd 的 DaemonSet 定义如下：

```
fluentd-ds.yml
---
apiVersion: extensions/v1beta1
kind: DaemonSet
metadata:
  name: fluentd-cloud-logging
  namespace: kube-system
  labels:
    k8s-app: fluentd-cloud-logging
spec:
  template:
    metadata:
      namespace: kube-system
      labels:
        k8s-app: fluentd-cloud-logging
    spec:
      containers:
      - name: fluentd-cloud-logging
```

```
    image: gcr.io/google_containers/fluentd-elasticsearch:1.17
    resources:
      limits:
        cpu: 100m
        memory: 200Mi
    env:
    - name: FLUENTD_ARGS
      value: -q
    volumeMounts:
    - name: varlog
      mountPath: /var/log
      readOnly: false
    - name: containers
      mountPath: /var/lib/docker/containers
      readOnly: false
    volumes:
    - name: containers
      hostPath:
        path: /var/lib/docker/containers
    - name: varlog
      hostPath:
        path: /var/log
```

通过 kubectl create 命令创建 Fluentd 容器：

```
# kubectl create -f fluentd-ds.yml
```

查看创建的结果：

```
# kubectl get daemonset
NAME                    DESIRED   CURRENT   NODE-SELECTOR   AGE
fluentd-cloud-logging   3         3         <none>          1h

# kubectl get pods
NAMESPACE   NAME                          READY   STATUS    RESTARTS   AGE
            fluentd-cloud-logging-7tw9z   1/1     Running   0          18h
            fluentd-cloud-logging-aqdn1   1/1     Running   0          18h
            fluentd-cloud-logging-o4usx   1/1     Running   0          18h
```

结果显示 Fluentd DaemonSet 正常运行，还启动了 3 个 Pod，与集群中的 Node 数量一致。

接下来使用 # kubectl logs fluentd-cloud-logging-7tw9z 命令查看 Pod 的日志，在

Elasticsearch 正常工作的情况下，我们会看到类似下面这样的日志内容：

```
# kubectl logs fluentd-cloud-logging-7tw9z
Connection opened to Elasticsearch cluster =>
{:host=>"elasticsearch-logging", :port=>9200, :scheme=>"http"}
```

说明 Fluentd 与 Elasticsearch 已经正确建立了连接。

10.9.4　运行 Kibana

至此已经运行了 Elasticsearch 和 Fluentd，数据的采集和汇聚已经完成，接下来使用 Kibana 展示和操作数据。

Kibana 的 RC 和 Service 定义如下：

```
kibana-rc-svc.yml
---
apiVersion: v1
kind: ReplicationController
metadata:
  name: kibana-logging-v1
  namespace: kube-system
  labels:
    k8s-app: kibana-logging
    version: v1
    kubernetes.io/cluster-service: "true"
spec:
  replicas: 1
  selector:
    k8s-app: kibana-logging
    version: v1
  template:
    metadata:
      labels:
        k8s-app: kibana-logging
        version: v1
        kubernetes.io/cluster-service: "true"
    spec:
      containers:
      - name: kibana-logging
        image: gcr.io/google_containers/kibana:1.3
```

```yaml
        resources:
          # keep request = limit to keep this container in guaranteed class
          limits:
            cpu: 100m
          requests:
            cpu: 100m
        env:
          - name: "ELASTICSEARCH_URL"
            value: "http://elasticsearch-logging:9200"
        ports:
        - containerPort: 5601
          name: ui
          protocol: TCP
---
apiVersion: v1
kind: Service
metadata:
  name: kibana-logging
  namespace: kube-system
  labels:
    k8s-app: kibana-logging
    kubernetes.io/cluster-service: "true"
    kubernetes.io/name: "Kibana"
spec:
  ports:
  - port: 5601
    protocol: TCP
    targetPort: ui
  selector:
    k8s-app: kibana-logging
```

通过 kubectl create -f kibana-rc-svc.yml 命令创建 Kibana 的 RC 和 Service：

```
# kubectl create -f kibana-rc-svc.yml
replicationcontroller "kibana-logging-v1" created
service "kibana-logging" created
```

查看 Kibana 的运行情况：

```
# kubectl get pods
NAMESPACE   NAME                      READY   STATUS    RESTARTS   AGE
default     kibana-logging-v1-o1akg   1/1     Running   0          1h
```

```
# kubectl get svc
NAME              CLUSTER-IP         EXTERNAL-IP    PORT(S)      AGE
kibana-logging    169.169.195.177    <none>         5601/TCP     1h

# kubectl get rc
NAME                 DESIRED    CURRENT    AGE
kibana-logging-v1    1          1          1h
```

结果表明运行均已成功。通过 kubectl cluster-info 命令获取 Kibana 服务的 URL 地址：

```
# kubectl cluster-info
Kibana is running at
http://127.0.0.1:8080/api/v1/proxy/namespaces/kube-system/
services/kibana-logging
```

同样通过 kubectl proxy 命令启动代理，在出现"Starting to serve on 127.0.0.1:8001"字样之后，用浏览器访问 URL 地址 http://192.168.18.3:8001/api/v1/proxy/namespaces/kube-system/services/kibana-logging 即可访问 Kibana 页面。

第 1 次进入页面时需要进行一些设置，如图 10.20 所示，选择所需选项后单击 create 按钮。

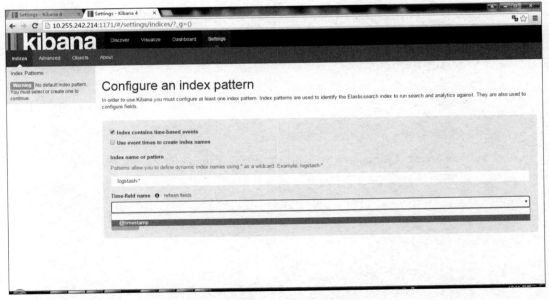

图 10.20　Kibana 创建索引页面

然后单击 discover 按钮，就可以正常查询日志了，如图 10.21 所示。

图 10.21　Kibana 查询日志页面

在搜索栏输入"error"关键字，可以搜索出包含该关键字的日志记录，如图 10.22 所示。

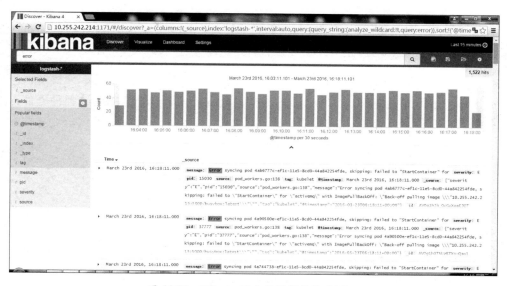

图 10.22　Kibana 日志关键字的搜索页面

同时，通过左边菜单中 Fields 相关的内容对查询的内容进行限定，如图 10.23 所示。

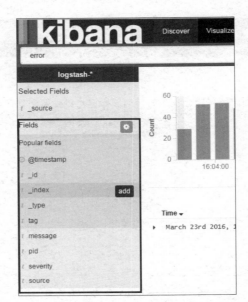

图 10.23　对查询的内容进行限定

至此，Kubernetes 集群范围内的统一日志收集和查询系统就搭建完成了。

10.10　Kubernetes 的审计机制

Kubernetes 为了加强对集群操作的安全监管，从 1.4 版本开始引入审计机制，主要体现为审计日志（Audit Log）。审计日志按照时间顺序记录了与安全相关的各种事件，这些事件有助于系统管理员快速、集中了解以下问题：

◎ 发生了什么事情？
◎ 作用于什么对象？
◎ 在什么时间发生？
◎ 谁（从哪儿）触发的？
◎ 在哪儿观察到的？
◎ 活动的后继处理行为是怎样的？

下面是两条 Pod 操作的审计日志示例。

第 1 条：

```
2017-03-21T03:57:09.106841886-04:00 AUDIT:
id="c939d2a7-1c37-4ef1-b2f7-4ba9b1e43b53" ip="127.0.0.1" method="GET" user="admin"
groups="\"system:masters\",\"system:authenticated\"" as="<self>"
asgroups="<lookup>" namespace="default" uri="/api/v1/namespaces/default/pods"
```

第 2 条：

```
2017-03-21T03:57:09.108403639-04:00 AUDIT:
id="c939d2a7-1c37-4ef1-b2f7-4ba9b1e43b53" response="200"
```

API Server 把客户端的请求（Request）的处理流程视为一个"链条"，这个链条上的每个"节点"就是一个状态（Stage），从开始到结束的所有 Request Stage 如下。

- RequestReceived：在 Audit Handler 收到请求后生成的状态。
- ResponseStarted：响应的 Header 已经发送但 Body 还没有发送的状态，仅对长期运行的请求（Long-running Requests）有效，例如 Watch。
- ResponseComplete：Body 已经发送完成。
- Panic：严重错误（Panic）发生时的状态。

Kubernets 从 1.7 版本开始引入高级审计特性（AdvancedAuditing），可以自定义审计策略（选择记录哪些事件）和审计存储后端（日志和 Webhook）等，开启方法为增加 kube-apiserver 的启动参数 --feature-gates=AdvancedAuditing=true。注意：在开启 AdvancedAuditing 后，日志的格式有一些修改，例如新增了上述 Stage 信息；从 Kubernets 1.8 版本开始，该参数默认为 true。

如图 10.24 所示，kube-apiserver 在收到一个请求后（如创建 Pod 的请求），会根据 Audit Policy（审计策略）对此请求做出相应的处理。

我们可以将 Audit Policy 视作一组规则，这组规则定义了有哪些事件及数据需要记录（审计）。当一个事件被处理时，规则列表会依次尝试匹配该事件，第 1 个匹配的规则会决定审计日志的级别（Audit Level），目前定义的几种级别如下（按级别从低到高排列）。

- None：不生成审计日志。
- Metadata：只记录 Request 请求的元数据如 requesting user、timestamp、resource、verb 等，但不记录请求及响应的具体内容。
- Request：记录 Request 请求的元数据及请求的具体内容。
- RequestResponse：记录事件的元数据，以及请求与应答的具体内容。

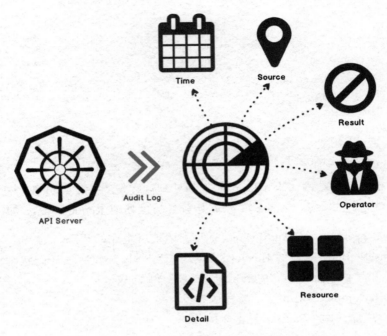

图 10.24　基于审计策略记录审计日志

None 以上的级别会生成相应的审计日志并将审计日志输出到后端，当前的后端实现如下。

（1）Log backend：以本地日志文件记录保存，为 JSON 日志格式，我们需要对 API Server 的启动命令设置下列参数。

◎ --audit-log-path：指定日志文件的保存路径。
◎ --audit-log-maxage：设定审计日志文件保留的最大天数。
◎ --audit-log-maxbackup：设定审计日志文件最多保留多少个。
◎ --audit-log-maxsize：设定审计日志文件的单个大小，单位为 MB，默认为 100MB。

审计日志文件以 audit-log-maxsize 设置的大小为单位，在写满后，kube-apiserver 将以时间戳重命名原文件，然后继续写入 audit-log-path 指定的审计日志文件；audit-log-maxbackup 和 audit-log-maxage 参数则用于 kube-apiserver 自动删除旧的审计日志文件。

（2）Webhook backend：回调外部接口进行通知，审计日志以 JSON 格式发送（POST 方式）给 Webhook Server，支持 batch 和 blocking 这两种通知模式，相关配置参数如下。

◎ --audit-webhook-config-file:指定 Webhook backend 的配置文件。
◎ --audit-webhook-mode:确定采用哪种模式回调通知。
◎ --audit-webhook-initial-backoff:指定回调失败后第 1 次重试的等待时间,后继重试等待时间则呈指数级递增。

Webhook backend 的配置文件采用了 kubeconfig 格式,主要内容包括远程审计服务的地址和相关鉴权参数,配置示例如下:

```
clusters:
  - name: name-of-remote-audit-service
    cluster:
      certificate-authority: /path/to/ca.pem      # 远程审计服务的 CA 证书
      server: https://audit.example.com/audit     # 远程审计服务 URL,必须是 HTTPS
                                                  # API server 的 Webhook 配置
users:
  - name: name-of-api-server
    user:
      client-certificate: /path/to/cert.pem       # Webhook 插件使用的证书文件
      client-key: /path/to/key.pem                # 与证书匹配的私钥文件
current-context: webhook
contexts:
- context:
    cluster: name-of-remote-audit-service
    user: name-of-api-sever
  name: webhook
```

--audit-webhook-mode 则包括以下选项。

◎ batch:批量模式,缓存事件并以异步批量方式通知,是默认的工作模式。
◎ blocking:阻塞模式,事件按顺序逐个处理,这种模式会阻塞 API Server 的响应,可能导致性能问题。
◎ blocking-strict:与阻塞模式类似,不同的是当一个 Request 在 RequestReceived 阶段发生审计失败时,整个 Request 请求会被认为失败。

(3)Batching Dynamic backend:一种动态配置的 Webhook backend,是通过 AuditSink API 动态配置的,在 Kubernetes 1.13 版本中引入。

需要注意的是,开启审计功能会增加 API Server 的内存消耗量,因为此时需要额外的内存来存储每个请求的审计上下文数据,而增加的内存量与审计功能的配置有关,比如更详细的审计日志所需的内存更多。我们可以通过 kube-apiserver 中的 --audit-policy-file 参数指定一个 Audit Policy 文件名来开启 API Server 的审计功能。如下 Audit Policy 文件可作参考：

```yaml
apiVersion: audit.k8s.io/v1
kind: Policy
# 对于 RequestReceived 状态的请求不做审计日志记录
omitStages:
  - "RequestReceived"
rules:
  # 记录对 Pod 请求的审计日志，输出级别为 RequestResponse
  - level: RequestResponse
    resources:
    - group: ""   #core API group
      resources: ["pods"]
  #记录对 pods/log 与 pods/ status 请求的审计日志，输出级别为 Metadata
  - level: Metadata
    resources:
    - group: ""
      resources: ["pods/log", "pods/status"]
  # 记录对核心 API 与扩展 API 的所有请求，输出级别为 Request
  - level: Request
    resources:
    - group: "" # core API group
    - group: "extensions" # Version of group should NOT be included
```

通常审计日志可以以本地日志文件方式保存，然后使用 Fluentd 作为 Agent 采集该日志并存储到 Elasticsearch，用 Kibana 等 UI 界面对日志进行展示和查询。

10.11 使用 Web UI（Dashboard）管理集群

Kubernetes 的 Web UI 网页管理工具 kubernetes-dashboard 可提供部署应用、资源对象管理、容器日志查询、系统监控等常用的集群管理功能。为了在页面上显示系统资源的使用情况，要求部署 Metrics Server，详见 10.8 节的说明。

可通过 https://rawgit.com/kubernetes/dashboard/ master/src/deploy/kubernetes-dashboard.yaml 页面下载部署 kubernetes-dashboard 的 YAML 配置文件。该配置文件的内容如下，其中包含 Deployment 和 Service 的定义：

```yaml
kind: Deployment
```

```yaml
apiVersion: extensions/v1beta1
metadata:
  labels:
    app: kubernetes-dashboard
  name: kubernetes-dashboard
  namespace: kube-system
spec:
  replicas: 1
  revisionHistoryLimit: 10
  selector:
    matchLabels:
      app: kubernetes-dashboard
  template:
    metadata:
      labels:
        app: kubernetes-dashboard
      annotations:
        scheduler.alpha.kubernetes.io/tolerations: |
          [
            {
              "key": "dedicated",
              "operator": "Equal",
              "value": "master",
              "effect": "NoSchedule"
            }
          ]
    spec:
      containers:
      - name: kubernetes-dashboard
        image: gcr.io/google_containers/kubernetes-dashboard-amd64:v1.6.0
        imagePullPolicy: Always
        ports:
        - containerPort: 9090
          protocol: TCP
        args:
        livenessProbe:
          httpGet:
            path: /
            port: 9090
          initialDelaySeconds: 30
          timeoutSeconds: 30
```

```yaml
---
kind: Service
apiVersion: v1
metadata:
  labels:
    app: kubernetes-dashboard
  name: kubernetes-dashboard
  namespace: kube-system
spec:
  type: NodePort
  ports:
  - port: 80
    targetPort: 9090
    nodePort: 9090
  selector:
    app: kubernetes-dashboard
```

这里，Service 设置了 NodePort 映射到物理机的端口号，用于客户端浏览器访问。

使用 kubectl create 命令进行部署：

```
# kubectl create -f kubernetes-dashboard.yaml
deployment "kubernetes-dashboard" created
service "kubernetes-dashboard" created
```

打开浏览器，输入某 Node 的 IP 和 9090 端口号，例如 http://192.168.18.3:9090，就能访问 Dashboard 的页面了，如图 10.25 所示。

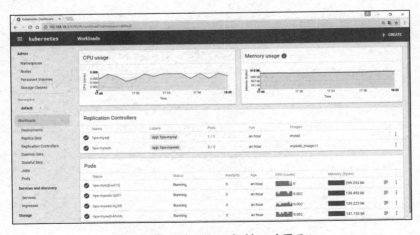

图 10.25　kubernetes-dashboard 页面

主页会默认显示 RC 和 Pod 列表，并显示各 Pod 的 CPU 和内存的性能指标。

单击右上角的"+ CREATE"按钮，将跳转到部署应用的页面。在这个页面可以通过设置相关参数或者直接通过 YAML 或 JSON 文件部署应用，如图 10.26 所示。

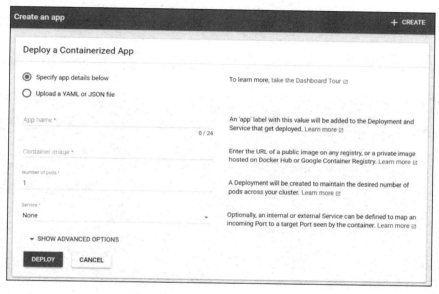

图 10.26　部署应用的页面

通过左侧的菜单，可以查看 Admin、Workloads、Service、Storage、Config 等各类资源对象的列表和详细信息。

例如，查看 Service 列表，如图 10.27 所示。

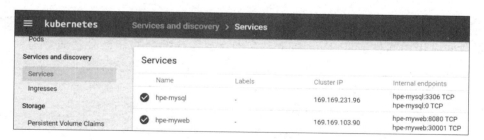

图 10.27　Service 列表

在 Pod 列表中，在各 Pod 右侧可以查看容器应用的日志，如图 10.28 所示。

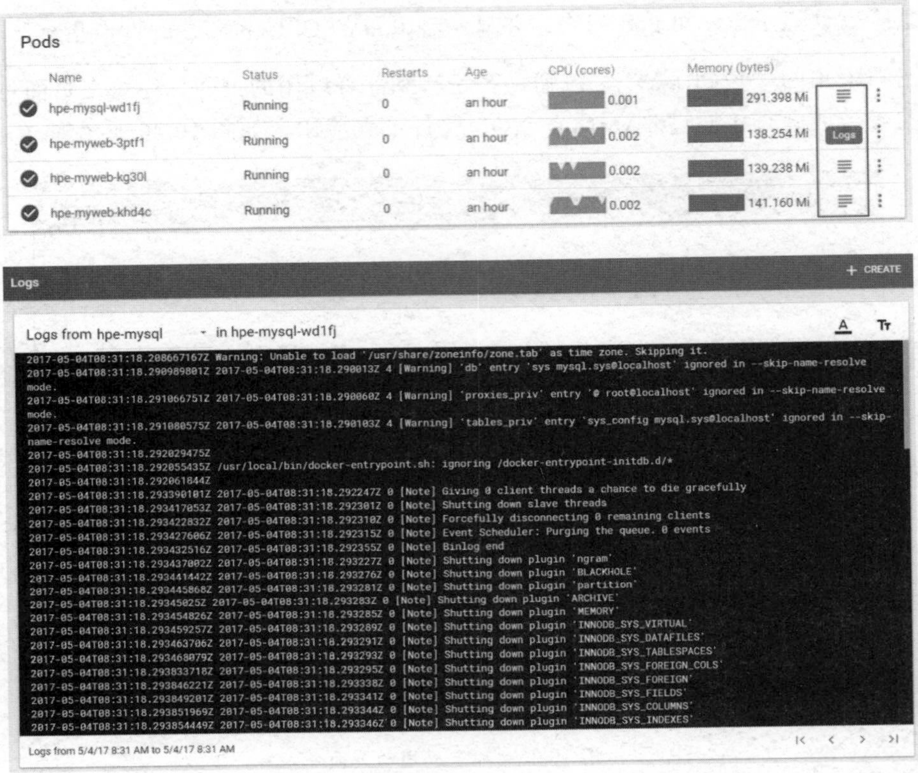

图 10.28　查看容器应用的日志

10.12　Helm：Kubernetes 应用包管理工具

随着容器技术逐渐被企业接受，在 Kubernetes 上已经能便捷地部署简单的应用了。但对于复杂的应用中间件，在 Kubernetes 上进行容器化部署并非易事，通常需要先研究 Docker 镜像的运行需求、环境变量等内容，并能为这些容器定制存储、网络等设置，最后设计和编写 Deployment、ConfigMap、Service 及 Ingress 等相关 YAML 配置文件，再提交给 Kubernetes 部署。这些复杂的过程将逐步被 Helm 应用包管理工具实现。

10.12.1　Helm 概述

Helm 是一个由 CNCF 孵化和管理的项目，用于对需要在 Kubernetes 上部署的复杂应用进行定义、安装和更新。Helm 以 Chart 的方式对应用软件进行描述，可以方便地创建、版本化、共享和发布复杂的应用软件。

10.12.2　Helm 的主要概念

Helm 的主要概念如下。

- Chart：一个 Helm 包，其中包含运行一个应用所需要的工具和资源定义，还可能包含 Kubernetes 集群中的服务定义，类似于 Homebrew 中的 formula、APT 中的 dpkg 或者 Yum 中的 RPM 文件。
- Release：在 Kubernetes 集群上运行的一个 Chart 实例。在同一个集群上，一个 Chart 可以被安装多次。例如有一个 MySQL Chart，如果想在服务器上运行两个 MySQL 数据库，就可以基于这个 Chart 安装两次。每次安装都会生成新的 Release，会有独立的 Release 名称。
- Repository：用于存放和共享 Chart 仓库。

简单来说，Helm 整个系统的主要任务就是，在仓库中查找需要的 Chart，然后将 Chart 以 Release 的形式安装到 Kubernetes 集群中。

10.12.3　安装 Helm

Helm 由 HelmClient 和 TillerServer 两个组件组成，下面讲解这两个组件。

1. HelmClient

HelmClient 是一个客户端，拥有对 Repository、Chart、Release 等对象的管理能力，可以通过二进制文件或脚本进行安装。

通过二进制文件安装时，从 https://github.com/kubernetes/helm/releases 下载二进制文件，将其解压并复制到执行目录即可。

通过脚本安装的代码如下：

```
# curl https://raw.githubusercontent.com/kubernetes/helm/master/scripts/get | bash
```

2. TillerServer

TillerServer 负责客户端指令和 Kubernetes 集群之间的交互，根据 Chart 定义，生成和管理各种 Kubernetes 的资源对象。

对 TillerServer 的安装可以使用 helm init 命令进行（官方推荐），这一命令会在 kubectl 当前 context 指定集群内的 kube-system 命名空间创建一个 Deployment 和一个 Service，运行 TillerServer 服务。

在 Deployment 中使用的镜像是 gcr.io/kubernetes-helm/tiller:v[helm-version]，可以通过 helm version 命令获得其 Helm 版本。如果无法连接互联网获取该镜像，则可以先通过一台能够联网的服务器下载这个镜像并保存到镜像私库，利用 helm init 子命令的 --tiller-image 参数来指定私库中的镜像执行初始化过程。

安装结束之后，用 helm version 命令验证安装情况，一切正常的话，会分别显示 Tiller 和 Helm 的版本信息。一个常见的问题是，Tiller 部分显示一个错误信息 "uid : unable to do port forwarding: socat not found"，这是因为所在节点没有 socat，无法进行端口转发，这时在主机上安装 socat 软件即可。

对 TillerServer 的安装还可以在本地进行，在服务器本地直接运行 Tiller，这种安装方式需要让 Helm 指定要连接的服务地址，有以下两种方法。

◎ 使用 --host 指定 Tiller 运行的监听地址。
◎ 设置 HELM_HOST 环境变量。

需要注意的是，Tiller 仍会使用 kubectl 配置中的 context 连接 Kubernetes 集群。

10.12.4 Helm 的常见用法

下面介绍 Helm 的常见用法，包括搜索 Chart、安装 Chart、自定义 Chart 配置、更新或回滚 Release、删除 Release、创建自定义 Chart、搭建私有仓库等。

1. helm search：搜索可用的 Chart

Helm 在初始化完成之后，默认配置为使用官方的 Kubernetes Chart 仓库。官方仓库包含大量的经过组织和持续维护的 Chart，这个仓库通常被命名为 stable。

使用 helm search 命令查找可安装的 Chart：

```
$ helm search
NAME                VERSION   DESCRIPTION
local/gitlab        0.1.3     A Helm chart for Kubernetes
local/grafana       0.1.0
local/influxdb      0.1.3     Fast, reliable, scalable, and easy to use open-...
local/jenkins       0.1.8     A Helm chart for Kubernetes
......
```

在没有过滤的情况下,通过 helm search 命令会显示所有可用的 Chart,可以使用参数进行过滤:

```
$ helm search mysql
NAME              VERSION  DESCRIPTION
local/mysql       0.2.6    Fast, reliable, scalable, and easy to use open-...
stable/mysql      0.2.6    Fast, reliable, scalable, and easy to use open-...
stable/percona    0.1.0    free, fully compatible, enhanced, open source d...
stable/mariadb    0.6.0    Fast, reliable, scalable, and easy to use open-...
```

为什么在列表中会有 MariaDB?因为在 MariaDB 的描述信息中包含了 mysql 关键字。可以通过 helm inspect <chart_name> 命令查看 Chart 的详细信息:

```
$ helm inspect stable/mariadb
description: Fast, reliable, scalable, and easy to use open-source relational database
    system. MariaDB Server is intended for mission-critical, heavy-load production systems
    as well as for embedding into mass-deployed software.
engine: gotpl
home: https://mariadb.org
icon: https://bitnami.com/assets/stacks/mariadb/img/mariadb-stack-220x234.png
keywords:
- mariadb
- mysql
- database
- sql
......
```

在找到需要安装的 Chart 之后,就可以进行安装了。

2. helm install:安装 Chart

使用 helm install 命令安装应用,最简单的参数是 Chart 的名称。以 MariaDB 为例:

```
$ helm install stable/mariadb
NAME:   womping-bobcat
LAST DEPLOYED: Wed May 31 21:13:58 2017
NAMESPACE: default
STATUS: DEPLOYED

RESOURCES:
==> v1/Service
NAME                      CLUSTER-IP     EXTERNAL-IP  PORT(S)    AGE
womping-bobcat-mariadb    10.31.255.15   <none>       3306/TCP   1s

==> v1beta1/Deployment
NAME                      DESIRED  CURRENT  UP-TO-DATE  AVAILABLE  AGE
womping-bobcat-mariadb    1        1        1           0          1s

==> v1/Secret
NAME                      TYPE     DATA  AGE
womping-bobcat-mariadb    Opaque   2     1s

==> v1/ConfigMap
NAME                      DATA  AGE
womping-bobcat-mariadb    1     1s

==> v1/PersistentVolumeClaim
NAME                      STATUS  VOLUME  CAPACITY  ACCESSMODES  STORAGECLASS
......
```

至此，MariaDB 就安装完成了。可以看到系统创建了一个新的名为 womping-bobcat 的 Release 对象，可以在 helm install 命令中使用 --name 参数修改其名称。

在安装过程中，Helm 客户端会输出一些有用的信息，例如 Release 的状态及额外的配置步骤等。

Helm 不会等待所有创建过程的完成，这是因为有些 Chart 的 Docker 镜像较大，会消耗很长的时间进行下载和创建。

在 helm install 命令的执行过程中，可以使用 helm status 命令跟踪 Release 的状态：

```
$ helm status womping-bobcat                                        [21:28:26]
LAST DEPLOYED: Wed May 31 21:13:58 2017
NAMESPACE: default
STATUS: DEPLOYED
```

```
RESOURCES:
==> v1/Secret
NAME                        TYPE      DATA    AGE
womping-bobcat-mariadb      Opaque    2       14m

==> v1/ConfigMap
NAME                        DATA    AGE
womping-bobcat-mariadb      1       14m
......
```

在成功安装 Chart 后，系统会在 kube-system 命名空间内创建一个 ConfigMap 用于保存 Release 对象的数据：

```
$ kubectl get configmap
NAMESPACE   NAME                        DATA    AGE
default     womping-bobcat-mariadb      1       18m
```

3. 自定义 Chart 的配置

前面介绍的安装过程使用的是 Chart 的默认配置。然而在很多情况下，我们希望使用自定义的配置部署应用。

首先，通过 helm inspect 命令查看 Chart 的可配置内容：

```
$ helm inspect stable/mariadb
......
## Bitnami MariaDB image version
## ref: https://hub.docker.com/r/bitnami/mariadb/tags/
##
## Default: none
image: bitnami/mariadb:10.1.22-r1

## Specify an imagePullPolicy (Required)
## It's recommended to change this to 'Always' if the image tag is 'latest'
## ref: http://kubernetes.io/docs/user-guide/images/#updating-images
imagePullPolicy: IfNotPresent

## Specify password for root user
## ref:
https://github.com/bitnami/bitnami-docker-mariadb/blob/master/README.md#setting-the-root-password-on-first-run
##
# mariadbRootPassword:
```

```
    ## Create a database user
    ## ref:
https://github.com/bitnami/bitnami-docker-mariadb/blob/master/README.md#creating
-a-database-user-on-first-run
    ##
    # mariadbUser:
    # mariadbPassword:
    ......
```

用户可以编写一个 YAML 配置文件来覆盖上面这些设置，然后利用这一文件来给安装过程提供配置。例如，我们可以自定义额外的两个配置文件 config.yaml 和 config2.yaml，用于在执行 helm install 命令安装 MariaDB 之后，在 MariaDB 启动时自动创建名为 firstdb 和 seconddb 的数据库，并且设置 root 用户的密码：

```
$ echo 'mariadbDatabase: firstdb' > config.yaml
$ echo 'mariadbRootPassword: abcdefgh' > config2.yaml
$ echo 'mariadbDatabase: seconddb' >> config2.yaml
$ helm install stable/mariadb -f config.yaml -f config2.yaml
```

在安装完成之后，可以登录 MariaDB Pod 查看数据库是否创建成功。

自定义 Chart 的配置有两种方法。

◎ --values 或者-f：使用 YAML 配置文件进行参数配置，可以设置多个文件，最后一个优先生效。多个文件中的重复 value 会进行覆盖操作，不同的 value 会叠加生效。上面的例子使用的就是这种方式。
◎ --set：在命令行中直接设置参数。

如果同时使用两个参数，则--set 会以高优先级合并到--values 中。

10.12.5 --set 的格式和限制

--set 参数可以使用 0 个或多个名称/值的组合，最简单的方式是--set name=value，YAML 配置文件中的等效描述是：

```
name: value
```

多个值可以使用逗号进行分隔，例如--set a=b,c=d 的 YAML 配置等效于下面的描述：

```
a: b
c: d
```

还可以用来表达多层结构的变量--set outer.inner=value：

```
outer:
  inner: value
```

大括号（{}）可以用来表达列表数据，例如--set name={a,b,c}会被翻译成：

```
name:
  - a
  - b
  - c
```

有时需要在--set 时使用一些特殊字符，这里可以使用斜线进行转义，比如--set name=value1\,value2。类似地，可以对点符号"."进行转义，这样 Chart 使用 toYaml 函数解析注解、标签或者 node selector 时就很方便了，例如：--set nodeSelector."kubernetes\.io/role"=master。

尽管如此，--set 语法的表达能力依然无法和 YAML 语言相提并论，尤其是在处理集合时。目前没有方法能够实现"把列表中第 3 个元素设置为 XXX"这样的语法。

10.12.6　更多的安装方法

使用 helm install 命令时，可以通过多种安装源进行安装。

- ◎ 上面用到的 Chart 仓库。
- ◎ 本地的 Chart 压缩包（helm install foo-0.1.1.tgz）。
- ◎ 一个 Chart 目录（helm install path/to/foo）。
- ◎ 一个完整的 URL（helm install https://example.com/charts/foo-1.2.3.tgz）。

10.12.7　helm upgrade 和 helm rollback：应用的更新或回滚

当一个 Chart 发布新版本或者需要修改一个 Release 的配置时，就需要使用 helm upgrade 命令了。

helm upgrade 命令会利用用户提供的更新信息来对 Release 进行更新。因为 Kubernetes Chart 可能会有很大的规模或者相对复杂的关系，所以 Helm 会尝试进行最小影响范围的更新，只更新相对于上一个 Release 来说发生变化的内容。

例如，我们要更新一个 Release 的资源限制，则可以创建 config3.yaml 配置文件，内容如下：

```
resources:
  requests:
    memory: 256Mi
    cpu: 500m
```

使用upgrade命令完成更新：

```
$ helm upgrade -f config3.yaml nomadic-terrier stable/mariadb
```

看到更新提示之后，我们可以用Helm的list指令查看Release的信息，会发现revision一列发生了变化。接下来通过kubectl的get pods，deploy指令可以看到Pod已经更新；通过kubectl describe deploy指令还会看到Deployment的更新过程和一系列的ScalingReplicaSet事件。

如果对更新后的Release不满意，则可以使用helm rollback命令对Release进行回滚，例如：

```
$ helm rollback nomadic-terrier 2
```

这个命令将把名为nomadic-terrier的Release回滚到版本2。

在执行命令之后，同样可以使用前面提到的几个查询指令，会看到类似的结果。

最后，可以使用helm history <release_name>命令查看一个Release的变更历史。

10.12.8　helm install/upgrade/rollback命令的常用参数

Helm有很多参数可以帮助用户指导命令的行为。本节介绍一些常用参数，用户可以使用helm <command> help命令获取所有参数的列表。

◎ --timeout：等待Kubernetes命令完成的时间，单位是s，默认值为300，也就是5min。
◎ --wait：等待Pod，直到其状态变为ready，PVC才完成绑定。Deployment完成其最低就绪要求的Pod创建，并且服务有了IP地址，才认为Release创建成功。这一等待过程会一直持续到超过--timeout，超时后这一Release被标记为FAILED（注意：当Deployment的replicas被设置为1，同时滚动更新策略的maxUnavailable不为0时，--wait才会因为最小就绪Pod数量达成而返回ready状态）。
◎ --no-hooks：该命令会跳过Hook执行。
◎ --recreate-pods：会引起所有Pod的重建（Deployment所属的Pod除外）。

10.12.9　helm delete：删除一个 Release

通过 helm delete 命令可以删除一个 Release，例如通过 helm delete happy-panda 命令可以从集群中删除名为 happy-panda 的 Release。

可以通过 helm list 命令列出在集群中部署的 Release。如果给 list 加上 --deleted 参数，则会列出所有删除的 Release；--all 参数会列出所有的 Release，包含删除的、现存的及失败的 Release。

正因为 Helm 会保存所有被删除 Release 的信息，所以 Release 的名字是不可复用的（如果坚持复用，则可以使用 --replace 参数，这一操作不建议在生产环境中使用），这样被删除的 Release 也可以被回滚，甚至重新激活。

10.12.10　helm repo：仓库的使用

我们在前面使用的 Chart 来自 stable 仓库。我们也可以对 Helm 进行配置，让其使用其他仓库。Helm 在 helm repo 命令中提供了很多仓库相关的工具。

◎ helm repo list：列出所有仓库。
◎ helm repo add：添加仓库，例如从 repo_url 中添加名为 dev 的仓库 helm repo add dev http://<repo_url>/dev-charts。
◎ helm repo update：更新仓库中的 Chart 信息。

10.12.11　自定义 Chart

用户可以将自己的应用定义为 Chart 并进行打包部署，本节对其进行简单介绍，详细的开发指南参见 https://docs.helm.sh/developing-charts。

自定义 Chart 时需要使用符合 Helm 规范的一组目录和配置文件来完成。

10.12.12　对 Chart 目录结构和配置文件的说明

Chart 是一个包含一系列文件的目录。目录的名称就是 Chart 的名称（不包含版本信息），例如一个 WordPress 的 Chart 就被会存储在 wordpress 目录下。

该目录的文件结构如下：

```
wordpress/
  Chart.yaml              # 用于描述 Chart 信息的 YAML 文件
  LICENSE                 # 可选：Chart 的许可信息
  README.md               # 可选：README 文件
  values.yaml             # 默认的配置值
  charts/                 # 可选：包含该 Chart 所依赖的其他 Chart，依赖管理推荐采用
                          # requirements.yaml 文件来进行
  templates/              # 可选：结合 values.yaml，能够生成 Kubernetes 的 manifest 文件
  templates/NOTES.txt     # 可选：文本文件，用法描述
```

charts/子目录和 requirements.yaml 的区别在于，前者需要提供整个 Chart 文件，后者仅需要注明依赖 Chart 的仓库信息，例如一个 requirements.yaml 可以被定义为：

```
dependencies:
  - name: apache
    version: 1.2.3
    repository: http://example.com/charts
  - name: mysql
    version: 3.2.1
    repository: http://another.example.com/charts
```

10.12.13　对 Chart.yaml 文件的说明

Chart.yaml 文件（注意首字母大写）是个必要文件，包含如下内容。

- name：Chart 的名称，必选。
- version：SemVer 2 规范的版本号，必选。
- description：项目的描述，可选。
- keywords：一个用于描述项目的关键字列表，可选。
- home：项目的主页，可选。
- sources：一个 URL 列表，说明项目的源代码位置，可选。
- maintainers：维护者列表，可选。
- name：管理员的名字，必选。
- email：管理员的邮件，必选。
- engine：模板引擎的名称，默认是 gotpl，可选。
- icon：一个指向 svg 或 png 图像的 URL，作为 Chart 的图标，可选。

- appVersion：在 Chart 中包含的应用的版本，无须遵循 SemVer 规范，可选。
- deprecated：布尔值，该 Chart 是否标注为 "弃用"，可选。
- tillerVersion：可选，该 Chart 所需的 Tiller 版本。取值应该是一个 SemVer 的范围，例如 ">2.0.0"。

10.12.14 快速制作自定义的 Chart

同其他软件开发过程一样，快速制作一个简单 Chart 的方法，就是从其他项目中复制并修改。例如，我们要简单地改写前面 MariaDB 的 Chart，令其使用本地的私有镜像仓库，就可以按照如下步骤进行。

- 下载 Chart：使用 helm fetch stable/mariadb 命令下载这一 Chart 的压缩包。
- 编辑 Chart。
- 利用 tar 解压之后，我们将目录重新命名为 mymariadb。
- 修改 templates 中的 deployment.yaml，简单地将其中的 image 字段硬编码为需要的镜像（当然不推荐这种用法，可以继续以变量的方式在 values 中进行设置）。
- 将 Chart.yaml 中的版本号修改为 0.1.1，name 为 mymariadb。
- 使用 helm package mymariadb 打包 Chart，会生成一个名为 mymariadb-0.1.1.tgz 的压缩包。
- 安装 Chart：通过 helm install mymariadb-0.1.1.tgz 命令即可将我们 "新" 生成的 Chart 安装到集群中。

10.12.15 搭建私有 Repository

在自建 Chart 之后自然需要搭建私有仓库。下面使用 Apache 搭建一个简单的 Chart 私有仓库，并将刚才新建的 mymariadb Chart 保存到私有仓库中，详情可参考 https://docs.helm.sh/ developing-charts/#chart-repo-guide。

Chart 仓库主要由前面提到的 Chart 压缩包和索引文件构成，通过 HTTP/HTTPS 对外提供服务。这里使用一个 Apache 应用来提供仓库服务。Apache 的设置如下。

- Apache 使用/var/web/repo 目录进行仓库的存储。
- 使用 http://127.0.0.1/repo 网址提供访问服务。

将前面生成的 mymariadb-0.1.1.tgz 文件复制到仓库的/var/web/repo 目录下。

接下来使用 helm repo index /var/web/repo --url http://127.0.0.1/repo 命令，Helm 将自动根据目录下的内容创建索引。在命令执行完毕后，可以看到在目录下多出一个 index.yaml 文件。

最后启动 Web Server。

为了能够使用这个私有仓库，需要将这个新的仓库地址加入 Helm 配置中：

```
$ helm repo add localhost http://127.0.0.1/repo
```

再次运行 helm search mysql 命令，会看到在 Chart 列表中多出 localhost/mymariadb 项目，也就是我们的新仓库中的 Chart。

现在可以通过 helm install localhost/mymariadb 命令安装私有仓库中的 Chart 了。

第 11 章
Trouble Shooting 指导

本章将对 Kubernetes 集群中常见问题的排查方法进行说明。

为了跟踪和发现在 Kubernetes 集群中运行的容器应用出现的问题，我们常用如下查错方法。

（1）查看 Kubernetes 对象的当前运行时信息，特别是与对象关联的 Event 事件。这些事件记录了相关主题、发生时间、最近发生时间、发生次数及事件原因等，对排查故障非常有价值。此外，通过查看对象的运行时数据，我们还可以发现参数错误、关联错误、状态异常等明显问题。由于在 Kubernetes 中多种对象相互关联，因此这一步可能会涉及多个相关对象的排查问题。

（2）对于服务、容器方面的问题，可能需要深入容器内部进行故障诊断，此时可以通过查看容器的运行日志来定位具体问题。

（3）对于某些复杂问题，例如 Pod 调度这种全局性的问题，可能需要结合集群中每个节点上的 Kubernetes 服务日志来排查。比如搜集 Master 上的 kube-apiserver、kube-schedule、kube-controler-manager 服务日志，以及各个 Node 上的 kubelet、kube-proxy 服务日志，通过综合判断各种信息，就能找到问题的成因并解决问题。

11.1 查看系统 Event

在 Kubernetes 集群中创建 Pod 后，我们可以通过 kubectl get pods 命令查看 Pod 列表，但通过该命令显示的信息有限。Kubernetes 提供了 kubectl describe pod 命令来查看一个 Pod 的详细信息，例如：

```
$ kubectl describe pod redis-master-bobr0
Name:                        Redis-master-bobr0
Namespace:                   default
Image(s):                    kubeguide/Redis-master
Node:                        k8s-node-1/192.168.18.3
Labels:                      name=Redis-master,role=master
Status:                      Running
Reason:
Message:
IP:                          172.17.0.58
Replication Controllers:     Redis-master (1/1 replicas created)
Containers:
  master:
```

```
    Image:          kubeguide/Redis-master
    Limits:
      cpu:          250m
      memory:       64Mi
    State:          Running
      Started:      Fri, 21 Aug 2015 14:45:37 +0800
    Ready:          True
    Restart Count:  0
  Conditions:
    Type          Status
    Ready         True
  Events:
    FirstSeen                          LastSeen                         Count   From
SubobjectPath          Reason          Message
    Fri, 21 Aug 2015 14:45:36 +0800    Fri, 21 Aug 2015 14:45:36 +0800  1
{kubelet k8s-node-1}   implicitly required container POD   pulled         Pod
container image "myregistry:5000/google_containers/pause:latest" already present on
machine
    Fri, 21 Aug 2015 14:45:37 +0800    Fri, 21 Aug 2015 14:45:37 +0800  1
{kubelet k8s-node-1}   implicitly required container POD   created        Created
with docker id a4aa97813908
    Fri, 21 Aug 2015 14:45:37 +0800    Fri, 21 Aug 2015 14:45:37 +0800  1
{kubelet k8s-node-1}   implicitly required container POD   started        Started
with docker id a4aa97813908
    Fri, 21 Aug 2015 14:45:37 +0800    Fri, 21 Aug 2015 14:45:37 +0800  1
{kubelet k8s-node-1}   spec.containers{master}                            created
Created with docker id 1e746245f768
    Fri, 21 Aug 2015 14:45:37 +0800    Fri, 21 Aug 2015 14:45:37 +0800  1
{kubelet k8s-node-1}   spec.containers{master}                            started
Started with docker id 1e746245f768
    Fri, 21 Aug 2015 14:45:37 +0800    Fri, 21 Aug 2015 14:45:37 +0800  1
{scheduler }                                              scheduled      Successfully assigned
Redis-master-bobr0 to k8s-node-1
```

通过 kubectl describe pod 命令，可以显示 Pod 创建时的配置定义、状态等信息，还可以显示与该 Pod 相关的最近的 Event 事件，事件信息对于查错非常有用。如果某个 Pod 一直处于 Pending 状态，我们就可以通过 kubectl describe 命令了解具体原因。例如，从 Event 事件中获知 Pod 失败的原因可能有以下几种。

- ◎ 没有可用的 Node 以供调度。
- ◎ 开启了资源配额管理，但在当前调度的目标节点上资源不足。

◎ 镜像下载失败。

通过 kubectl describe 命令，还可以查看其他 Kubernetes 对象，包括 Node、RC、Service、Namespace、Secrets 等，对每种对象都会显示相关的其他信息。

例如，查看一个服务的详细信息：

```
$ kubectl describe service redis-master
Name:                Redis-master
Namespace:           default
Labels:              name=Redis-master
Selector:            name=Redis-master
Type:                ClusterIP
IP:                  169.169.208.57
Port:                <unnamed>        6379/TCP
Endpoints:           172.17.0.58:6379
Session Affinity:    None
No events.
```

如果要查看的对象属于某个特定的 Namespace，就需要加上 --namespace=<namespace> 进行查询。例如：

```
$ kubectl get service kube-dns --namespace=kube-system
```

11.2 查看容器日志

在需要排查容器内部应用程序生成的日志时，我们可以使用 kubectl logs <pod_name> 命令：

```
$ kubectl logs redis-master-bobr0
[1] 21 Aug 06:45:37.781 * Redis 2.8.19 (00000000/0) 64 bit, stand alone mode, port 6379, pid 1 ready to start.
[1] 21 Aug 06:45:37.781 # Server started, Redis version 2.8.19
[1] 21 Aug 06:45:37.781 # WARNING overcommit_memory is set to 0! Background save may fail under low memory condition. To fix this issue add 'vm.overcommit_memory = 1' to /etc/sysctl.conf and then reboot or run the command 'sysctl vm.overcommit_memory=1' for this to take effect.
[1] 21 Aug 06:45:37.782 # WARNING you have Transparent Huge Pages (THP) support enabled in your kernel. This will create latency and memory usage issues with Redis. To fix this issue run the command 'echo never > /sys/kernel/mm/transparent_hugepage/enabled' as root, and add it to your /etc/ rc.local in order to retain the setting
```

```
after a reboot. Redis must be restarted after THP is disabled.
    [1] 21 Aug 06:45:37.782 # WARNING: The TCP backlog setting of 511 cannot be enforced
because /proc/sys/net/core/somaxconn is set to the lower value of 128.
```

如果在某个 Pod 中包含多个容器，就需要通过 -c 参数指定容器的名称来查看，例如：

```
kubectl logs <pod_name> -c <container_name>
```

其效果与在 Pod 的宿主机上运行 docker logs <container_id> 一样。

容器中应用程序生成的日志与容器的生命周期是一致的，所以在容器被销毁之后，容器内部的文件也会被丢弃，包括日志等。如果需要保留容器内应用程序生成的日志，则可以使用挂载的 Volume 将容器内应用程序生成的日志保存到宿主机，还可以通过一些工具如 Fluentd、Elasticsearch 等对日志进行采集。

11.3 查看 Kubernetes 服务日志

如果在 Linux 系统上安装 Kubernetes，并且使用 systemd 系统管理 Kubernetes 服务，那么 systemd 的 journal 系统会接管服务程序的输出日志。在这种环境中，可以通过使用 systemd status 或 journalctl 工具来查看系统服务的日志。

例如，使用 systemctl status 命令查看 kube-controller-manager 服务的日志：

```
# systemctl status kube-controller-manager -l
kube-controller-manager.service - Kubernetes Controller Manager
    Loaded: loaded (/usr/lib/systemd/system/kube-controller-manager.service; enabled)
    Active: active (running) since Fri 2015-08-21 18:36:29 CST; 5min ago
      Docs: https://github.com/GoogleCloudPlatform/kubernetes
  Main PID: 20339 (kube-controller)
    CGroup: /system.slice/kube-controller-manager.service
            └─20339 /usr/bin/kube-controller-manager --logtostderr=false --v=4 --master=http://kubernetes-master:8080 --log_dir=/var/log/kubernetes

Aug 21 18:36:29 kubernetes-master systemd[1]: Starting Kubernetes Controller Manager...
Aug 21 18:36:29 kubernetes-master systemd[1]: Started Kubernetes Controller Manager.
```

使用 journalctl 命令查看：

```
# journalctl -u kube-controller-manager
```

```
-- Logs begin at Mon 2015-08-17 16:43:22 CST, end at Fri 2015-08-21 18:36:29 CST. --
    Aug 17 16:44:14 kubernetes-master systemd[1]: Starting Kubernetes Controller Manager...
    Aug 17 16:44:14 kubernetes-master systemd[1]: Started Kubernetes Controller Manager.
```

如果不使用 systemd 系统接管 Kubernetes 服务的标准输出，则也可以通过日志相关的启动参数来指定日志的存放目录。

- --logtostderr=false：不输出到 stderr。
- --log-dir=/var/log/kubernetes：日志的存放目录。
- --alsologtostderr=false：将其设置为 true 时，表示将日志同时输出到文件和 stderr。
- --v=0：glog 的日志级别。
- --vmodule=gfs*=2,test*=4：glog 基于模块的详细日志级别。

在 --log_dir 设置的目录下可以查看各服务进程生成的日志文件，日志文件的数量和大小依赖于日志级别的设置。例如，kube-controller-manager 可能生成的几个日志文件如下：

- kube-controller-manager.ERROR；
- kube-controller-manager.INFO；
- kube-controller-manager.WARNING；
- kube-controller-manager.kubernetes-master.unknownuser.log.ERROR.20150930-173939.9847；
- kube-controller-manager.kubernetes-master.unknownuser.log.INFO.20150930-173939.9847；
- kube-controller-manager.kubernetes-master.unknownuser.log.WARNING.20150930-173939.9847。

在大多数情况下，我们从 WARNING 和 ERROR 级别的日志中就能找到问题的成因，但有时还需要排查 INFO 级别的日志甚至 DEBUG 级别的详细日志。此外，etcd 服务也属于 Kubernetes 集群的重要组成部分，所以不能忽略它的日志。

如果某个 Kubernetes 对象存在问题，则可以用这个对象的名字作为关键字搜索 Kubernetes 的日志来发现和解决问题。在大多数情况下，我们遇到的主要是与 Pod 对象相关的问题，比如无法创建 Pod、Pod 启动后就停止或者 Pod 副本无法增加，等等。此时，可以先确定 Pod 在哪个节点上，然后登录这个节点，从 kubelet 的日志中查询该 Pod 的完整日志，然后进行问题排查。对于与 Pod 扩容相关或者与 RC 相关的问题，则很可能在

kube-controller-manager 及 kube-scheduler 的日志中找出问题的关键点。

另外，kube-proxy 经常被我们忽视，因为即使它意外停止，Pod 的状态也是正常的，但会导致某些服务访问异常。这些错误通常与每个节点上的 kube-proxy 服务有着密切的关系。遇到这些问题时，首先要排查 kube-proxy 服务的日志，同时排查防火墙服务，要特别留意在防火墙中是否有人为添加的可疑规则。

11.4 常见问题

本节对 Kubernetes 系统中的一些常见问题及解决方法进行说明。

11.4.1 由于无法下载 pause 镜像导致 Pod 一直处于 Pending 状态

以 redis-master 为例，使用如下配置文件 redis-master-controller.yaml 创建 RC 和 Pod：

```
apiVersion: v1
kind: ReplicationController
metadata:
  name: redis-master
  labels:
    name: redis-master
spec:
  replicas: 1
  selector:
    name: redis-master
  template:
    metadata:
      labels:
        name: redis-master
    spec:
      containers:
      - name: master
        image: kubeguide/redis-master
        ports:
        - containerPort: 6379
```

执行 kubectl create -f redis-master-controller.yaml 成功，但在查看 Pod 时，发现其总是无法处于 Running 状态。通过 kubectl get pods 命令可以看到：

```
$ kubectl get pods
NAME                      READY    STATUS                                              RESTARTS   AGE
redis-master-6yy7o        0/1      Image: kubeguide/redis-master is ready, container is creating  0    5m
```

进一步使用 kubectl describe pod redis-master-6yy7o 命令查看该 Pod 的详细信息：

```
$ kubectl describe pod redis-master-6yy7o
Name:                    redis-master-6yy7o
Namespace:               default
Image(s):                kubeguide/redis-master
Node:                    127.0.0.1/127.0.0.1
Labels:                  name=redis-master
Status:                  Pending
Reason:
Message:
IP:
Replication Controllers:  redis-master (1/1 replicas created)
Containers:
  master:
    Image:               kubeguide/redis-master
    State:               Waiting
      Reason:            Image: kubeguide/redis-master is ready, container is creating
    Ready:               False
    Restart Count:       0
Conditions:
  Type          Status
  Ready         False
Events:
  FirstSeen                         LastSeen                          Count   From              SubobjectPath    Reason       Message
  Thu, 24 Sep 2015 19:19:25 +0800   Thu, 24 Sep 2015 19:25:58 +0800   3       {kubelet 127.0.0.1}               failedSync   Error syncing pod, skipping: image pull failed for k8s.gcr.io/pause:3.1, this may be because there are no credentials on this request. details: (API error (500): invalid registry endpoint https://gcr.io/v0/: unable to ping registry endpoint https://gcr.io/v0/v2 ping attempt failed with error: Get https://gcr.io/v2/: dial tcp 173.194.196.82:443: connection refused v1 ping attempt failed with error: Get https://gcr.io/v1/_ping: dial tcp 173.194.79.82:443: connection refused. If this private registry supports only HTTP or HTTPS with an unknown CA certificate, please add `--insecure-registry gcr.io` to the daemon's arguments. In the case of HTTPS, if you have access to the registry's CA certificate, no need
```

```
for the flag; simply place the CA certificate at /etc/docker/certs.d/gcr.io/ca.crt)
    Thu, 24 Sep 2015 19:19:25 +0800    Thu, 24 Sep 2015 19:25:58 +0800 3
{kubelet 127.0.0.1}    implicitly required container POD    failed Failed to pull
image "k8s.gcr.io/pause:3.1": image pull failed for k8s.gcr.io/pause:3.1, this may
be because there are no credentials on this request. details: (API error (500):
invalid registry endpoint https://gcr.io/v0/: unable to ping registry endpoint
https://gcr.io/v0/v2 ping attempt failed with error: Get https://gcr.io/v2/: dial
tcp 173.194.196.82:443: connection refused v1 ping attempt failed with error: Get
https://gcr.io/v1/_ping: dial tcp 173.194.79.82: 443: connection refused. If this
private registry supports only HTTP or HTTPS with an unknown CA certificate, please
add `--insecure-registry gcr.io` to the daemon's arguments. In the case of HTTPS,
if you have access to the registry's CA certificate, no need for the flag; simply
place the CA certificate at /etc/docker/certs.d/gcr.io/ca.crt
```

可以看到，该 Pod 为 Pending 状态。从 Message 部分显示的信息可以看出，其原因是 image pull failed for k8s.gcr.io/pause:3.1，说明系统在创建 Pod 时无法从 gcr.io 下载 pause 镜像，所以导致创建 Pod 失败。

解决方法如下。

（1）如果服务器可以访问 Internet，并且不希望使用 HTTPS 的安全机制来访问 gcr.io，则可以在 Docker Daemon 的启动参数中加上--insecure-registry gcr.io，来表示可以匿名下载。

（2）如果 Kubernetes 集群在内网环境中无法访问 gcr.io 网站，则可以先通过一台能够访问 gcr.io 的机器下载 pause 镜像，将 pause 镜像导出后，再导入内网的 Docker 私有镜像库，并在 kubelet 的启动参数中加上--pod_infra_container_image，配置为：

```
--pod_infra_container_image=<docker_registry_ip>:<port>/pause:3.1
```

之后重新创建 redis-master 即可正确启动 Pod。

注意，除了 pause 镜像，其他 Docker 镜像也可能存在无法下载的情况，与上述情况类似，很可能也是网络配置使得镜像无法下载，解决方法同上。

11.4.2 Pod 创建成功，但 RESTARTS 数量持续增加

创建一个 RC 之后，通过 kubectl get pods 命令查看 Pod，发现如下情况：

```
......
$ kubectl get pods
```

```
NAME            READY      STATUS       RESTARTS   AGE
zk-bg-ri3ru     0/1        Running      3          37s
......
$ kubectl get pods
NAME            READY      STATUS       RESTARTS   AGE
zk-bg-ri3ru     0/1        Running      5          1m
......
$ kubectl get pods
NAME            READY      STATUS       RESTARTS   AGE
zk-bg-ri3ru     0/1        ExitCode:0   6          1m
......
$ kubectl get pods
NAME            READY      STATUS       RESTARTS   AGE
zk-bg-ri3ru     0/1        Running      7          1m
```

可以看到 Pod 已经创建成功,但 Pod 一会儿是 Running 状态,一会儿是 ExitCode:0 状态,在 READY 列中始终无法变成 1/1,而且 RESTARTS(重启的数量)的数量不断增加。

这通常是因为容器的启动命令不能保持在前台运行。

本例中 Docker 镜像的启动命令为:

```
zkServer.sh start-background
```

在 Kubernetes 中根据 RC 定义创建 Pod,之后启动容器。在容器的启动命令执行完成时,认为该容器的运行已经结束,并且是成功结束(ExitCode=0)的。根据 Pod 的默认重启策略定义(RestartPolicy=Always),RC 将启动这个容器。

新的容器在执行启动命令后仍然会成功结束,之后 RC 会再次重启该容器,如此往复。其解决方法为将 Docker 镜像的启动命令设置为一个前台运行的命令,例如:

```
zkServer.sh start-foreground
```

11.4.3 通过服务名无法访问服务

在 Kubernetes 集群中应尽量使用服务名访问正在运行的微服务,但有时会访问失败。由于服务涉及服务名的 DNS 域名解析、kube-proxy 组件的负载分发、后端 Pod 列表的状态等,所以可通过以下几方面排查问题。

1. 查看 Service 的后端 Endpoint 是否正常

可以通过 kubectl get endpoints <service_name>命令查看某个服务的后端 Endpoint 列表，如果列表为空，则可能因为：

◎ Service 的 Label Selector 与 Pod 的 Label 不匹配；
◎ 后端 Pod 一直没有达到 Ready 状态（通过 kubectl get pods 进一步查看 Pod 的状态）；
◎ Service 的 targetPort 端口号与 Pod 的 containerPort 不一致等。

2. 查看 Service 的名称能否被正确解析为 ClusterIP 地址

可以通过在客户端容器中 ping <service_name>.<namespace>.svc 进行检查，如果能够得到 Service 的 ClusterIP 地址，则说明 DNS 服务能够正确解析 Service 的名称；如果不能得到 Service 的 ClusterIP 地址，则可能是因为 Kubernetes 集群的 DNS 服务工作异常。

3. 查看 kube-proxy 的转发规则是否正确

我们可以将 kube-proxy 服务设置为 IPVS 或 iptables 负载分发模式。

对于 IPVS 负载分发模式，可以通过 ipvsadm 工具查看 Node 上的 IPVS 规则，查看是否正确设置 Service ClusterIP 的相关规则。

对于 iptables 负载分发模式，可以通过查看 Node 上的 iptables 规则，查看是否正确设置 Service ClusterIP 的相关规则。

11.5 寻求帮助

如果通过系统日志和容器日志都无法找到问题的成因，则可以追踪源码进行分析，或者通过一些在线途径寻求帮助。下面列出了可给予相应帮助的常用网站或社区。

◎ Kubernetes 常见问题：https://kubernetes.io/docs/tasks/debug-application-cluster/troubleshooting/。
◎ Kubernetes 应用相关问题：https://kubernetes.io/docs/tasks/debug-application-cluster/debug-application/。
◎ Kubernetes 集群相关问题：https://kubernetes.io/docs/tasks/debug-application-cluster/debug-cluster/。

◎ Kubernetes 官方论坛：https://discuss.kubernetes.io/，可以查看 Kubernetes 的最新动态并参与讨论，如图 11.1 所示。
◎ Kubernetes GitHub 库问题列表：https://github.com/kubernetes/kubernetes/issues，可以在这里搜索曾经出现过的问题，也可以提问，如图 11.2 所示。
◎ StackOverflow 网站上关于 Kubernetes 的问题讨论：http://stackoverflow.com/questions/tagged/kubernetes，如图 11.3 所示。
◎ Kubernetes Slack 聊天群组：https://kubernetes.slack.com/，其中有许多频道，包括 #kubernetes-users、#kubernetes-novice、#kubernetes-dev 等，读者可以根据自己的兴趣加入不同的频道，与聊天室中的网友进行在线交流，如图 11.4 所示。还有针对不同国家的地区频道，例如中国区频道有#cn-users 和#cn-events。

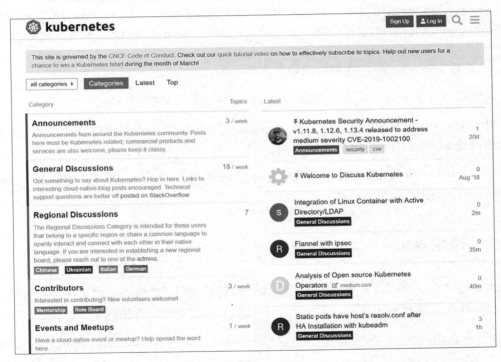

图 11.1　Kubernetes 官方论坛

第 11 章 Trouble Shooting 指导

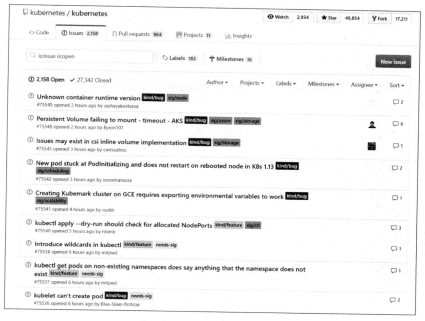

图 11.2　Kubernetes GitHub 库问题列表

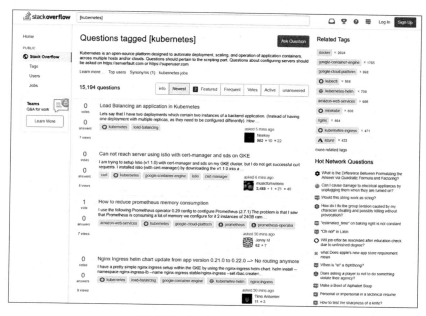

图 11.3　StackOverflow 网站

• 775 •

图 11.4　Kubernetes Slack 聊天群组

第 12 章
Kubernetes 开发中的新功能

本章对 Kubernetes 开发中的一些新功能进行介绍，包括对 Windows 容器的支持、对 GPU 的支持、Pod 的垂直扩缩容，并讲解 Kubernetes 的演进路线（Roadmap）和开发模式。

12.1 对 Windows 容器的支持

Kubernetes 从 1.5 版本开始，就引入了管理基于 Windows Server 2016 操作系统的 Windows 容器的功能。随着 Windows Server version 1709 的发布，Kubernetes 1.9 对 Windows 容器的支持提升为 Beta 版，Kubernetes 1.14 对 Windows 容器的支持提升为 GA 稳定版。目前 Windows Server 的最新版本是 2019，与 Kubernetes 的对接更加成熟。

目前 Windows Server 可以作为 Node 加入 Kubernetes 集群中，集群的 Master 仍需在 Linux 环境中运行。

在 Kubernetes 上改进了 Windows 容器的一些关键功能，包括：

◎ 对 Pod 网络模型的支持，可为一个 Pod 内的多个容器设置共享的网络命名空间（共享 kernel 模式）；
◎ 为每个 Pod 都设置了单个网络 Endpoint，以降低网络复杂性；
◎ 使用 Windows Server 的 Virtual Filtering Platform (VFP) Hyper-v Switch Extension 技术，实现了类似于 Linux iptables 的负载均衡器；
◎ Pod 支持容器运行时接口（CRI）和 Node 的统计信息；
◎ 支持使用 kubeadm 命令将 Windows Server 节点添加到集群中。

下面对 Windows Node 部署，以及 Kubernetes 支持的 Windows 容器特性和发展趋势进行讲解。

12.1.1 Windows Node 部署

需要在 Windows Server 上安装的 Kubernetes Node 相关组件包括 Docker、kubelet、kube-proxy 和 kubectl。

Docker 版本要求在 18.03 及以上，推荐使用最新版本。可参考官方文档 https://docs.docker.com/install/windows/docker-ee/#use-a-script-to-install-docker-ee 在 Windows Server 上安装 Docker。

从 Kubernetes 的版本发布页面（https://github.com/kubernetes/kubernetes/releases）下载 Windows Node 相关文件，例如 Kubernetes Windows Node 1.14 的下载页面为 https://github.com/kubernetes/kubernetes/blob/master/CHANGELOG-1.14.md#node-binaries，在该页面下载 kubernetes-node-windows-amd64.tar.gz 文件，将其解压缩后得到可执行文件 kubelet.exe、kube-proxy.exe、kubectl.exe 和 kubeadm.exe，如图 12.1 所示。

Node Binaries	
filename	sha512 hash
kubernetes-node-linux-amd64.tar.gz	75dc99919d1084d7d471a53ab60c743dc399145c99e83f37c6ba3c241b2c0b2ecc2c0d1b94690ff912e2a15b7c5595aa1d2
kubernetes-node-linux-arm.tar.gz	49013a4f01be8086fff332099d94903082688b9b295d2f34468462656da4709360025e9d84b069410c608977ef803079af6
kubernetes-node-linux-arm64.tar.gz	f8c0cb0c089cd1d1977c049002620b8cf748d193c1b76dd1d3aac01ff9273549c06a1e3dfe983dc40a95ee8b0719908e0cc
kubernetes-node-linux-ppc64le.tar.gz	48fc02c856a192388877189a43eb1cda531e548bb035f9dfe6a1e3c8d3bcbd0f8e14f29382da45702cb28a91126d13ede42
kubernetes-node-linux-s390x.tar.gz	d7c5f52cf602fd0c0d0f72d4cfe1ceaa4bad70a42f37f21c103f17c3448ceb2396c1bfa521eeeb9eef5f3173d84e4268704
kubernetes-node-windows-amd64.tar.gz	120afdebe844b06a7437bb9788c3e7ea4fc6352aa18cc6a00e70f44f54664f844429f138870bc15862579da632632dff2e7

图 12.1 Kubernetes Windows Node 二进制文件下载页面

下面介绍如何在 Windows Server 2019 上搭建 Kubernetes Node，详细的操作文档请参考 https://docs.microsoft.com/en-us/virtualization/windowscontainers/kubernetes/getting-started-kubernetes-windows。

1）制作 kubelet 运行容器所需的 pause 镜像。

下载 Microsoft 提供的 nanoserver 镜像（需要与 Windows Server 的版本相匹配）：

```
C:\> docker pull mcr.microsoft.com/windows/nanoserver:1809
```

将该镜像重命名为 microsoft/nanoserver:latest：

```
C:\> docker tag mcr.microsoft.com/windows/nanoserver:1809 microsoft/nanoserver:latest
```

编写 Dockerfile：

```
FROM microsoft/nanoserver
CMD cmd /c ping -t localhost
```

运行 docker build 命令制作 pause 镜像，并将镜像名设置为 kubeletwin/pause：

```
C:\> docker build -t kubeletwin/pause
```

2）将 Kubernetes Windows Node 的压缩包文件解压缩到 C:\k 目录下，得到 kubelet.exe、kube-proxy.exe 和 kubectl.exe 文件。

3）将连接 Kubernetes Master 所需的 kubeconfig 文件和 SSL 安全证书文件从 Linux Node 复制到 C:\k 目录下，将文件名改为 config。

可以使用 kubectl.exe 命令验证能否访问 Master：

```
[Environment]::SetEnvironmentVariable("KUBECONFIG", "C:\k\config", [EnvironmentVariableTarget]::User)
C:\k> kubectl config view
```

4）选择一个容器网络方案，目前支持的网络方案有如下 3 种。

（1）基于 3 层路由的网络方案。需要 Top of Rack（机柜上面安装的）交换机或路由器的支持，Windows 节点的容器 IP 池应由该交换机或路由器管理。在 Windows 节点上需要创建一个虚拟的 l2bridge 网桥，用于连通 Pod IP 与物理机之间的网络。ToR 交换机或路由器负责 Pod 的 IP 地址分配，并负责设置 Pod IP 与其他物理机联通的静态路由规则。如图 12.2 所示为基于 3 层路由的容器网络方案，这种网络方案与 Linux 的直接路由方案比较相似，但是对交换机或路由器的要求较高。

在图 12.2 中，物理机网络为 10.127.132.*，Linux Master 的 IP 地址为 10.127.132.128，Linux Node 的 IP 地址为 10.127.132.129，Windows Node 的 IP 地址为 10.127.132.213。Linux Node 上的容器网络 IP 范围为 10.10.187.0/26，Windows Node 上的容器网络 IP 范围为 10.10.187.64/26，要求 ToR 交换机或路由器管理 Windows 容器 IP 地址的分配并设置到其他主机的路由规则。

（2）基于 Open vSwitch 的网络方案。这种方案依赖 Open vSwitch，部署流程比较复杂。如图 12.3 所示为通用的网络部署架构。

第 12 章 Kubernetes 开发中的新功能

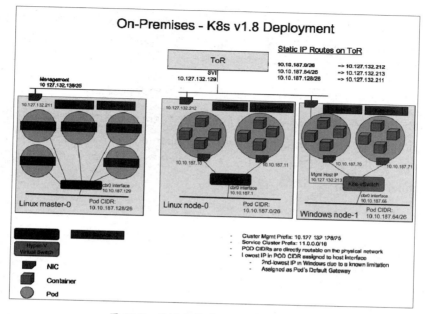

图 12.2 基于 3 层路由的容器网络方案

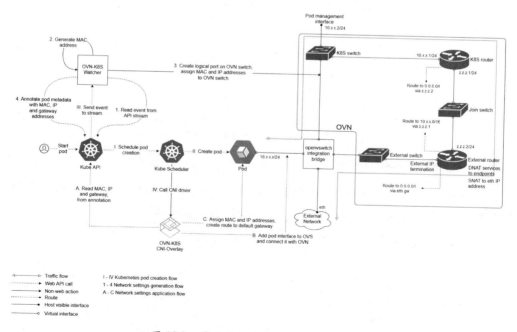

图 12.3 基于 Open vSwitch 的网络方案

(3）基于 Flannel 的网络方案。在 Linux 上为 Kubernetes 搭建 CNI 网络的 Flannel 系统时，要求 Linux 的 Kubernetes 环境也使用 Flannel 网络，能够实现 Windows 容器网络与 Linux 容器网络的互联互通。

5）下载 Windows Node 的部署脚本和配置文件并修改配置。

目前 Kubernetes 在 Windows Node 上的部署脚本和 CNI 网络插件可以从 GitHub 代码库（https://github.com/Microsoft/SDN/tree/master/Kubernetes）下载，将该代码库 windows 子目录的文件下载并保存到 C:\k 目录下，再从另外几个目录下载 CNI 网络插件相关的文件，例如 flannel、wincni 等，如图 12.4 所示。

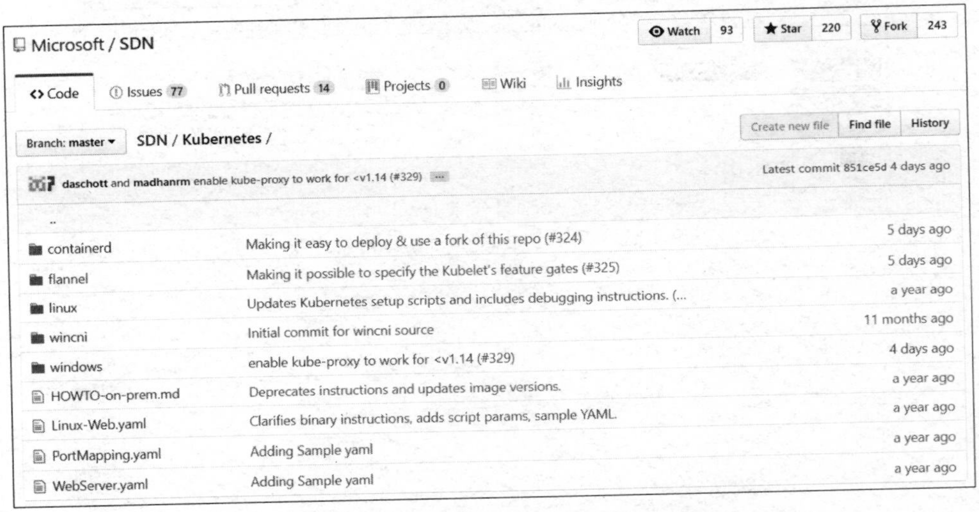

图 12.4　Windows Node 上的部署脚本下载页面

6）以 Flannel 为例配置 CNI 网络的关键信息。

下载 https://github.com/Microsoft/SDN/tree/master/Kubernetes/flannel 中的文件并将其保存到 C:\k 目录。还需要将 Flannel 的可执行文件 flanneld.exe 保存到 C:\flannel 目录下。

start-kubelet.ps1 脚本中的关键网络配置参数如下：

```
Param(
    [parameter(Mandatory = $false)] $clusterCIDR="10.244.0.0/16",
    [parameter(Mandatory = $false)] $KubeDnsServiceIP="169.169.0.100",
    [parameter(Mandatory = $false)] $serviceCIDR="169.169.0.0/16",
    [parameter(Mandatory = $false)] $KubeDnsSuffix="svc.cluster.local",
```

```
    [parameter(Mandatory = $false)] $InterfaceName="Ethernet",
    [parameter(Mandatory = $false)] $LogDir = "C:\k\logs",
    [ValidateSet("process", "hyperv")] $IsolationType="process",
    $NetworkName = "cbr0",
    [switch] $RegisterOnly
)
```

对于以下参数，可以在执行 start.ps1 时用命令行参数指定，也可以直接修改 start-kubelet.ps1 脚本进行设置。

- ◎ clusterCIDR：用于设置 Flannel 的容器 IP 池，需要与 kube-controller-manager 的 --cluster-cidr 保持一致。
- ◎ KubeDnsServiceIP：将其设置为 Kubernetes 集群 DNS 服务的 ClusterIP，例如 169.169.0.100。
- ◎ serviceCIDR：将其设置为 Kubernetes 集群 Service 的 ClusterIP 池，例如 169.169.0.0/16。
- ◎ InterfaceName：Windows 主机的网卡名，例如 Ethernet。
- ◎ LogDir：日志目录，例如 C:\k\logs。

在 Update-CNIConfig 函数内设置 Nameservers（DNS 服务器）的 IP 地址，例如 169.169.0.100：

```
function
Update-CNIConfig($podCIDR)
{
    $jsonSampleConfig = '{
  "cniVersion": "0.2.0",
  "name": "<NetworkMode>",
  "type": "flannel",
  "delegate": {
     "type": "win-bridge",
     "dns" : {
       "Nameservers" : [ "169.169.0.100" ],
       "Search": [ "svc.cluster.local" ]
     },
     "policies" : [
       {
         "Name" : "EndpointPolicy", "Value" : { "Type" : "OutBoundNAT", "ExceptionList": [ "<ClusterCIDR>", "<ServerCIDR>", "<MgmtSubnet>" ] }
       },
```

```
            {
                "Name" : "EndpointPolicy", "Value" : { "Type" : "ROUTE",
"DestinationPrefix": "<ServerCIDR>", "NeedEncap" : true }
            },
            {
                "Name" : "EndpointPolicy", "Value" : { "Type" : "ROUTE",
"DestinationPrefix": "<MgmtIP>/32", "NeedEncap" : true }
            }
        ]
    }
}'
```

7）运行 start.ps1 脚本，将 Windows Node 加入 Kubernetes 集群：

```
.\start.ps1 -ManagementIP "192.168.18.9" -ClusterCIDR "10.244.0.0/16"
-ServiceCIDR "169.169.0.0/16" -KubeDnsServiceIP "169.169.0.100"
```

将其中的-ManagementIP 参数设置为 Windows Node 的主机 IP 地址。

该脚本的启动过程如下。

（1）创建一个名为 External、类型为 L2Bridge 的 HNSNetwork 虚拟网卡，其管理 IP 为 Windows 主机 IP，如图 12.5 所示。

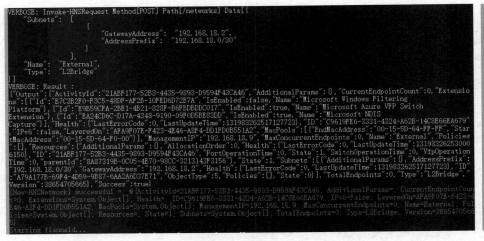

图 12.5　创建类型为 L2Bridge 的虚拟网卡信息

（2）启动 flanneld 程序，创建一个名为 cbr0、类型为 L2Bridge 的虚拟网卡（HNSNetwork），将其作为容器网络的网桥，之后 flanneld 正常退出，如图 12.6 所示。

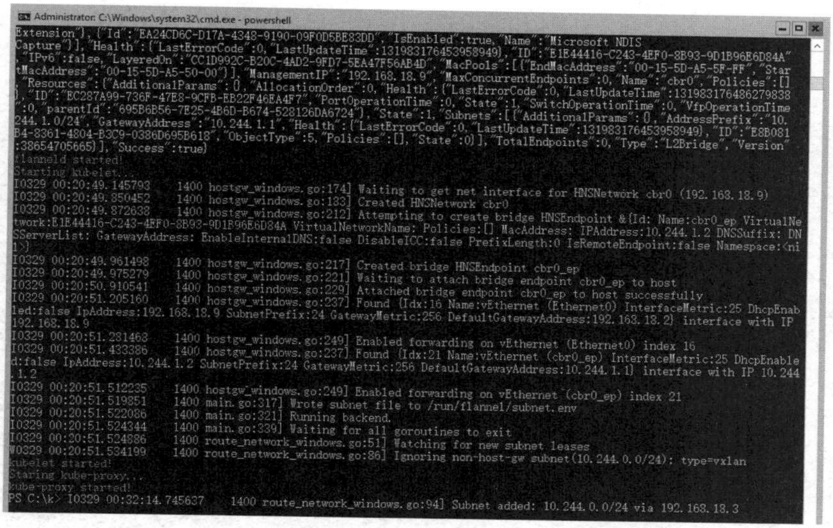

第 12 章 Kubernetes 开发中的新功能

图 12.6 创建名为 cbr0 的虚拟网卡

（3）在名为 cbr0 的虚拟网卡创建成功后，在一个新的 powershell 窗口启动 kubelet，如图 12.7 所示。

图 12.7 kubelet 的启动日志

（4）在一个新的 powershell 窗口启动 kube-proxy，如图 12.8 所示。

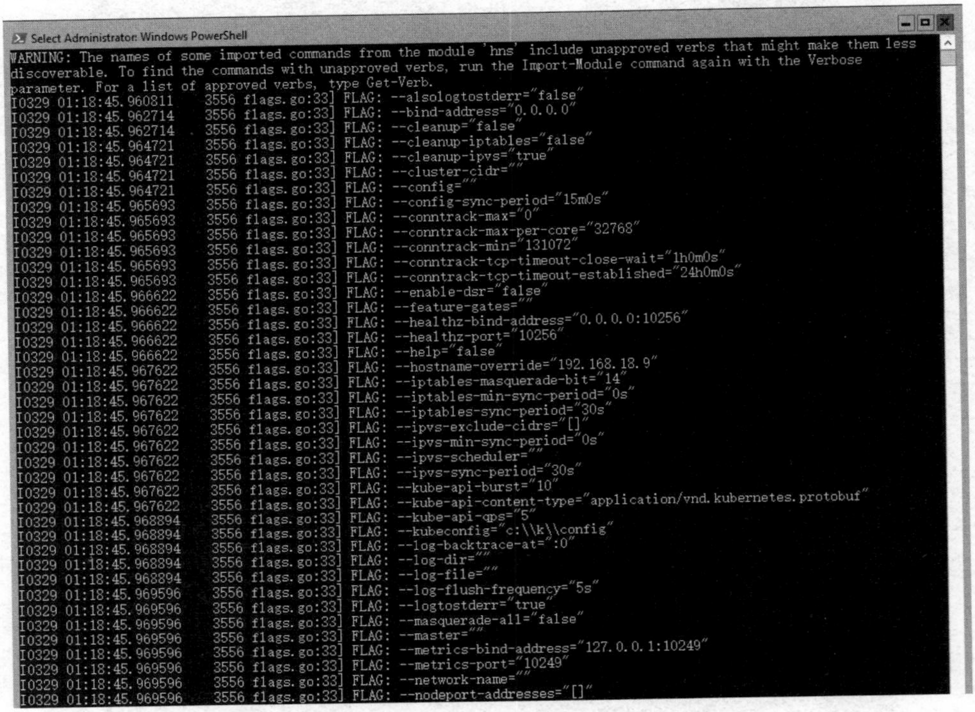

图 12.8 kube-proxy 的启动日志

（5）在脚本启动成功之后，可以在 Master 上查看新加入的 Windows Node：

```
# kubectl get nodes
NAME           STATUS   ROLES   AGE   VERSION
192.168.18.9   Ready    node    14s   v1.14.0
```

查看这个 Node 的 Label，可以看到其包含 "kubernetes.io/os=windows" 的 Label，与 Linux Node 有所区分（Linux Node 的标签为 "kubernetes.io/os=linux"）：

```
# kubectl get no --show-labels
NAME           STATUS   ROLES    AGE   VERSION   LABELS
192.168.18.9   Ready    <none>   53m   v1.14.0
beta.kubernetes.io/arch=amd64,beta.kubernetes.io/os=windows,kubernetes.io/arch=a
md64,kubernetes.io/hostname=192.168.18.9,kubernetes.io/os=windows
```

8）在 Windows Node 上部署容器应用。现在，我们可以像在 Linux Node 上部署容器应用一样,在 Windows Node 上部署 Windows 容器应用。下面是一个 Web 服务示例,可以从 https://github.com/Microsoft/SDN/blob/master/Kubernetes/flannel/l2bridge/manifests/simpleweb.yml 获得 YAML 配置文件,其中容器镜像使用的是 mcr.microsoft.com/windows/servercore:1809,并通过 powershell 命令行启动了一个 Web 服务。在 Deployment 的配置中需要设置 nodeSelector 为 "kubernetes.io/os: windows",以将 Windows 容器调度到 Windows Node 上:

```yaml
---
apiVersion: v1
kind: Service
metadata:
  name: win-webserver
  labels:
    app: win-webserver
spec:
  ports:
  - port: 80
    targetPort: 80
    nodePort: 40001
  selector:
    app: win-webserver
  type: NodePort
---
apiVersion: extensions/v1beta1
kind: Deployment
metadata:
  labels:
    app: win-webserver
  name: win-webserver
spec:
  replicas: 1
  template:
    metadata:
      labels:
        app: win-webserver
      name: win-webserver
    spec:
      containers:
      - name: windowswebserver
        image: mcr.microsoft.com/windows/servercore:1809
        command:
```

```yaml
        - powershell.exe
        - -command
        - "<#code used from https://gist.github.com/wagnerandrade/5424431#> ;
$$listener = New-Object System.Net.HttpListener ;
$$listener.Prefixes.Add('http://*:80/') ; $$listener.Start() ; $$callerCounts = @{} ;
Write-Host('Listening at http://*:80/') ; while ($$listener.IsListening)
{ ;$$context = $$listener.GetContext() ;$$requestUrl =
$$context.Request.Url ;$$clientIP =
$$context.Request.RemoteEndPoint.Address ;$$response =
$$context.Response ;Write-Host '' ;Write-Host('> {0}' -f $$requestUrl) ; ;$$count
= 1 ;$$k=$$callerCounts.Get_Item($$clientIP) ;if ($$k -ne $$null) { $$count +=
$$k } ;$$callerCounts.Set_Item($$clientIP, $$count) ;$$ip=(Get-NetAdapter |
Get-NetIpAddress); $$header='<html><body><H1>Windows Container Web
Server</H1>' ;$$callerCountsString='' ;$$callerCounts.Keys | %
{ $$callerCountsString+='<p>IP {0} callerCount {1} ' -f
$$ip[1].IPAddress,$$callerCounts.Item($$_) } ;$$footer='</body></html>' ;$$conte
nt='{0}{1}{2}' -f $$header,$$callerCountsString,$$footer ;Write-Output
$$content ;$$buffer =
[System.Text.Encoding]::UTF8.GetBytes($$content) ;$$response.ContentLength64 =
$$buffer.Length ;$$response.OutputStream.Write($$buffer, 0,
$$buffer.Length) ;$$response.Close() ;$$responseStatus =
$$response.StatusCode ;Write-Host('< {0}' -f $$responseStatus) } ; "
      nodeSelector:
        kubernetes.io/os: windows
```

用 kubectl 命令完成部署：

```
# kubectl apply -f simpleweb.yml
```

在 Pod 创建成功后，查看 Pod 的状态：

```
# kubectl get po -o wide
NAME                          READY   STATUS    RESTARTS   AGE
IP            NODE            NOMINATED NODE   READINESS GATES
win-webserver-848595695b-k8px9   1/1   Running   0          39s
10.244.2.20   192.168.18.9    <none>           <none>
```

查看 Service 的信息：

```
# kubectl get svc win-webserver
NAME            TYPE       CLUSTER-IP        EXTERNAL-IP   PORT(S)        AGE
win-webserver   NodePort   169.169.219.15    <none>        80:40001/TCP   74s
```

下面在 Linux 环境中访问 Windows 容器的服务。

使用容器 IP 访问 Windows 容器的服务：

```
# curl 10.244.2.20:80
<html><body><H1>Windows Container Web Server</H1><p>IP 10.244.2.20 callerCount 1 </body></html>
```

使用服务 IP 访问 Windows 容器的服务：

```
# curl 169.169.219.15:80
<html><body><H1>Windows Container Web Server</H1><p>IP 10.244.2.20 callerCount 2 </body></html>
```

对于类型为 NodePort 的服务，通过 Windows Server 的 IP 和 NodePort 也能访问 Windows 容器的服务：

```
# curl 192.168.18.9:40001
<html><body><H1>Windows Container Web Server</H1><p>IP 10.244.2.20 callerCount 1 <p>IP 10.244.2.20 callerCount 1 </body></html>
```

在 Windows 环境中使用容器 IP 或服务 IP 也能访问 Windows 容器的服务：

```
C:\> curl -UseBasicParsing 10.244.2.20:80


StatusCode        : 200
StatusDescription : OK
Content           : {60, 104, 116, 109...}
RawContent        : HTTP/1.1 200 OK
                    Content-Length: 127
                    Date: Fri, 29 Mar 2019 08:49:29 GMT
                    Server: Microsoft-HTTPAPI/2.0

                    <html><body><H1>Windows Container Web Server</H1><p>IP
10.244.2.20 callerCount 1 <p>IP 10.24...
   Headers        : {[Content-Length, 127], [Date, Fri, 29 Mar 2019 08:49:29 GMT],
[Server, Microsoft-HTTPAPI/2.0]}
   RawContentLength : 127


C:\> curl -UseBasicParsing 169.169.142.255:80


StatusCode        : 200
StatusDescription : OK
Content           : {60, 104, 116, 109...}
```

```
    RawContent        : HTTP/1.1 200 OK
                        Content-Length: 127
                        Date: Fri, 29 Mar 2019 08:49:39 GMT
                        Server: Microsoft-HTTPAPI/2.0

                        <html><body><H1>Windows Container Web Server</H1><p>IP
10.244.2.20 callerCount 2 <p>IP 10.24...
    Headers           : {[Content-Length, 127], [Date, Fri, 29 Mar 2019 08:49:39 GMT],
[Server, Microsoft-HTTPAPI/2.0]}
    RawContentLength  : 127
```

12.1.2 Windows 容器支持的 Kubernetes 特性和发展趋势

Kubernetes 在 1.14 版本中对 Windows 容器的支持达到 GA 稳定版，要求 Windows Server 的版本为 Windows Server 2019，支持的功能特性包括：

◎ 支持将 Secret 或 ConfigMap 挂载为环境变量或文件；
◎ Volume 存储卷支持 emptyDir 和 hostPath；
◎ 支持的持久化存储卷 PV 类型包括 FlexVolume（SMB+iSCSI 类型）、AzureFile 和 AzureDisk；
◎ 支持 livenessProbe 和 readinessProbe 健康检查机制；
◎ 支持容器的 postStart 和 preStop 命令；
◎ 支持 CRI 类型为 Dockershim；
◎ 支持的控制器包括 ReplicaSet、ReplicationController、Deployments、StatefulSets、DaemonSet、Job 和 CronJob；
◎ 支持 Service 的类型包括 NodePort、ClusterIP、LoadBalancer、ExternalName 和 Headless Service；
◎ 支持 CPU 和内存的资源限制；
◎ 支持 Resource Quotas；
◎ 支持基于任意指标数据的 HPA 机制；
◎ 支持的 CNI 网络插件包括 Azure-CNI、OVN-Kubernetes 和 Flannel（VxLAN 和 Host-Gateway 模式）；
◎ kubelet.exe 和 kube-proxy.exe 可以以 Windows 服务的形式运行。

已知的功能限制包括：

◎ 不支持容器以特权模式运行；
◎ 不支持 hostNetwork 网络模式；
◎ 不支持 CSI 插件；
◎ 不支持 NFS 类型的存储卷；
◎ 在挂载存储卷时无法设置 subpath；
◎ Windows 容器内的 OS 版本必须与宿主机 OS 的版本一致，否则容器无法启动。

Windows 容器在 Kubernetes 上计划完成的功能包括：

◎ 支持更多的 Overlay 网络方案，包括 Calico CNI 插件；
◎ 支持更多的 CRI 容器运行时；
◎ 支持通过 kubeadm 安装 Windows Node；
◎ 支持基于 Hyper-V 虚拟化技术在 1 个 Pod 中包含多个容器（目前在 1 个 Pod 中只能有 1 个容器）；
◎ 支持设置 terminationGracePeriodSeconds 实现对 Pod 的优雅删除；
◎ 支持设置 run_as_username 以自定义用户名运行容器内的程序。

可以预见，在不久的将来，Kubernetes 能完善 Windows 容器和 Linux 容器的混合管理，实现跨平台的容器云平台。

12.2 对 GPU 的支持

随着人工智能和机器学习的迅速发展，基于 GPU 的大数据运算越来越普及。在 Kubernetes 的发展规划中，GPU 资源有着非常重要的地位。用户应该能够为其工作任务请求 GPU 资源，就像请求 CPU 或内存一样，而 Kubernetes 将负责调度容器到具有 GPU 资源的节点上。

目前 Kubernetes 对 NVIDIA 和 AMD 两个厂商的 GPU 进行了实验性的支持。Kubernetes 对 NVIDIA GPU 的支持是从 1.6 版本开始的，对 AMD GPU 的支持是从 1.9 版本开始的，并且都在快速发展。

Kubernetes 从 1.8 版本开始，引入了 Device Plugin（设备插件）模型，为设备提供商提供了一种基于插件的、无须修改 kubelet 核心代码的外部设备启用方式，设备提供商只需在计算节点上以 DaemonSet 方式启动一个设备插件容器供 kubelet 调用，即可使用外部设备。目前支持的设备类型包括 GPU、高性能 NIC 卡、FPGA、InfiniBand 等，关于设备插件的说明详见官方文档 https://kubernetes.io/docs/concepts/extend-kubernetes/compute-storage-net/device-plugins。

下面对如何在 Kubernetes 中使用 GPU 资源进行说明。

12.2.1 环境准备

（1）在 Kubernetes 的 1.8 和 1.9 版本中，需要在每个工作节点上都为 kubelet 服务开启 --feature-gates="DevicePlugins=true" 特性开关。该特性开关从 Kubernetes 1.10 版本开始默认启用，无须手动设置。

（2）在每个工作节点上都安装 NVIDIA GPU 或 AMD GPU 驱动程序，如下所述。

使用 NVIDIA GPU 的系统要求包括：

◎ NVIDIA 驱动程序的版本为 361.93 及以上；
◎ nvidia-docker 的版本为 2.0 及以上；
◎ 配置 Docker 默认使用 NVIDIA 运行时；
◎ Kubernetes 的版本为 1.11 及以上。

Docker 使用 NVIDIA 运行时的配置示例（通常配置文件为 /etc/docker/daemon.json）如下：

```
{
    "default-runtime": "nvidia",
    "runtimes": {
        "nvidia": {
            "path": "/usr/bin/nvidia-container-runtime",
            "runtimeArgs": []
        }
    }
}
```

NVIDIA 设备驱动的部署 YAML 文件可以从 NVIDIA 的 GitHub 代码库 https://github.com/NVIDIA/k8s-device-plugin/blob/v1.11/nvidia-device-plugin.yml 获取：

```
apiVersion: extensions/v1beta1
kind: DaemonSet
metadata:
  name: nvidia-device-plugin-daemonset
```

```yaml
  namespace: kube-system
spec:
  updateStrategy:
    type: RollingUpdate
  template:
    metadata:
      annotations:
        scheduler.alpha.kubernetes.io/critical-pod: ""
      labels:
        name: nvidia-device-plugin-ds
    spec:
      tolerations:
      - key: CriticalAddonsOnly
        operator: Exists
      - key: nvidia.com/gpu
        operator: Exists
        effect: NoSchedule
      containers:
      - image: nvidia/k8s-device-plugin:1.11
        name: nvidia-device-plugin-ctr
        securityContext:
          allowPrivilegeEscalation: false
          capabilities:
            drop: ["ALL"]
        volumeMounts:
          - name: device-plugin
            mountPath: /var/lib/kubelet/device-plugins
      volumes:
        - name: device-plugin
          hostPath:
            path: /var/lib/kubelet/device-plugins
```

使用 AMD GPU 的系统要求包括：

◎ 服务器支持 ROCm（Radeon Open Computing platform）；
◎ ROCm kernel 驱动程序或 AMD GPU Linux 驱动程序为最新版本；
◎ Kubernetes 的版本为 1.10 及以上。

AMD 设备驱动的部署 YAML 文件可以从 NVIDIA 的 GitHub 代码库（https://github.com/RadeonOpenCompute/k8s-device-plugin/blob/master/k8s-ds-amdgpu-dp.yaml）获取：

```yaml
apiVersion: extensions/v1beta1
kind: DaemonSet
metadata:
  name: amdgpu-device-plugin-daemonset
  namespace: kube-system
spec:
  template:
    metadata:
      annotations:
        scheduler.alpha.kubernetes.io/critical-pod: ""
      labels:
        name: amdgpu-dp-ds
    spec:
      tolerations:
      - key: CriticalAddonsOnly
        operator: Exists
      containers:
      - image: rocm/k8s-device-plugin
        name: amdgpu-dp-cntr
        securityContext:
          allowPrivilegeEscalation: false
          capabilities:
            drop: ["ALL"]
        volumeMounts:
          - name: dp
            mountPath: /var/lib/kubelet/device-plugins
          - name: sys
            mountPath: /sys
      volumes:
        - name: dp
          hostPath:
            path: /var/lib/kubelet/device-plugins
        - name: sys
          hostPath:
            path: /sys
```

在完成上述配置后，容器应用就能使用 GPU 资源了。

12.2.2　在容器中使用 GPU 资源

GPU 资源在 Kubernetes 中的名称为 nvidia.com/gpu（NVIDIA 类型）或 amd.com/gpu（AMD 类型），可以对容器进行 GPU 资源请求的设置。

在下面的例子中为容器申请 1 个 GPU 资源：

```
apiVersion: v1
kind: Pod
metadata:
  name: cuda-vector-add
spec:
  restartPolicy: OnFailure
  containers:
   - name: cuda-vector-add
     image: "k8s.gcr.io/cuda-vector-add:v0.1"
     resources:
       limits:
         nvidia.com/gpu: 1 # requesting 1 GPU
```

目前对 GPU 资源的使用配置有如下限制：

◎ GPU 资源请求只能在 limits 字段进行设置，系统将默认设置 requests 字段的值等于 limits 字段的值，不支持只设置 requests 而不设置 limits；
◎ 在多个容器之间或者在多个 Pod 之间不能共享 GPU 资源；
◎ 每个容器只能请求整数个（1 个或多个）GPU 资源，不能请求 1 个 GPU 的部分资源。

如果在集群中运行着不同类型的 GPU，则 Kubernetes 支持通过使用 Node Label（节点标签）和 Node Selector（节点选择器）将 Pod 调度到合适的 GPU 所属的节点。

1. 为 Node 添加合适的 Label 标签

对于 NVIDIA 类型的 GPU，可以使用 kubectl label 命令为 Node 设置不同的标签：

```
# kubectl label nodes <node-with-k80> accelerator=nvidia-tesla-k80
# kubectl label nodes <node-with-p100> accelerator=nvidia-tesla-p100
```

对于 AMD 类型的 GPU，可以使用 AMD 开发的 Node Labeller 工具自动为 Node 设置合适的 Label。Node Labeller 以 DaemonSet 的方式部署，可以从 https://github.com/RadeonOpenCompute/k8s-device-plugin/blob/master/k8s-ds-amdgpu-labeller.yaml 下载 YAML

配置文件。在 Node Labeller 的启动参数中可以设置不同的 Label 以表示不同的 GPU 信息。目前支持的 Label 如下。

（1）Device ID，启动参数为 -device-id。

（2）VRAM Size，启动参数为 -vram。

（3）Number of SIMD，启动参数为 -simd-count。

（4）Number of Compute Unit，启动参数为 -cu-count。

（5）Firmware and Feature Versions，启动参数为 -firmware。

（6）GPU Family, in two letters acronym，启动参数为 -family，family 类型以两个字母缩写表示，完整的启动参数为 family.SI、family.CI 等。其中，SI 的全称为 Southern Islands；CI 的全称为 Sea Islands；KV 的全称为 Kaveri；VI 的全称为 Volcanic Islands；CZ 的全称为 Carrizo；AI 的全称为 Arctic Islands；RV 的全称为 Raven。

通过 Node Labeller 工具自动为 Node 设置的 Label 示例如下：

```
$ kubectl describe node cluster-node-23
Name:           cluster-node-23
Labels:         beta.amd.com/gpu.cu-count.64=1
                beta.amd.com/gpu.device-id.6860=1
                beta.amd.com/gpu.family.AI=1
                beta.amd.com/gpu.simd-count.256=1
                beta.amd.com/gpu.vram.16G=1
                beta.kubernetes.io/arch=amd64
                beta.kubernetes.io/os=linux
                kubernetes.io/hostname=cluster-node-23
......
```

2. 设置 Node Selector 指定调度 Pod 到目标 Node 上

以 NVIDIA GPU 为例：

```
apiVersion: v1
kind: Pod
metadata:
  name: cuda-vector-add
spec:
  restartPolicy: OnFailure
```

```
containers:
  - name: cuda-vector-add
    image: "k8s.gcr.io/cuda-vector-add:v0.1"
    resources:
      limits:
        nvidia.com/gpu: 1
nodeSelector:
  accelerator: nvidia-tesla-p100
```

上面的配置可确保将 Pod 调度到含有 "accelerator=nvidia-tesla-k80" Label 的节点运行。

12.2.3 发展趋势

发展趋势如下。

- GPU 和其他设备将像 CPU 那样成为 Kubernetes 系统的原生计算资源类型，以 Device Plugin 的方式供 kubelet 调用。
- 目前的 API 限制较多，Kubernetes 未来会有功能更丰富的 API，能支持以可扩展的形式进行 GPU 等硬件加速器资源的供给、调度和使用。
- Kubernetes 将能自动确保使用 GPU 的应用程序达到最佳性能。

12.3 Pod 的垂直扩缩容

Kubernetes 在 1.9 版本中加入了对 Pod 的垂直扩缩容（简称为 VPA）支持，这一功能使用 CRD 的方式为 Pod 定义垂直扩缩容的规则，根据 Pod 的运行行为来判断 Pod 的资源需求，从而更好地为 Pod 提供调度支持。

该项目的地址为 https://github.com/kubernetes/autoscaler，其中包含了三个不同的工具，分别是：

- 能够在 AWS、Azure 和 GCP 上提供集群节点扩缩容的 ClusterAutoScaler，已经进入 GA 阶段；
- 提供 Pod 垂直扩缩容的 Vertical Pod Autoscaler，目前处于 Beta 阶段；
- Addon Resizer，是 Vertical Pod Autoscaler 的简化版，能够根据节点数量修改 Deployment 资源请求，目前处于 Beta 阶段。

12.3.1 前提条件

垂直扩缩容目前的版本为 0.4,需要在 Kubernetes 1.11 以上版本中运行。该组件所需的运行指标由 Metrics Server 提供,因此在安装 Autoscaler 前要先启动 Metrics Server。

12.3.2 安装 Vertical Pod Autoscaler

首先使用 Git 获取 Autoscaler 的源码:

```
$ git clone https://github.com/kubernetes/autoscaler.git
```

下载结束之后,执行如下脚本,启动 VPA:

```
$ autoscaler/vertical-pod-autoscaler/hack/vpa-up.sh
customresourcedefinition.apiextensions.k8s.io/verticalpodautoscalers.autoscaling.k8s.io created
customresourcedefinition.apiextensions.k8s.io/verticalpodautoscalercheckpoints.autoscaling.k8s.io created
clusterrole.rbac.authorization.k8s.io/system:metrics-reader created
......
```

可以看到,在安装过程中生成了常见的 Deployment、Secret、Service 及 RBAC 内容,还生成了两个 CRD,接下来会用新生成的 CRD 设置 Pod 的垂直扩缩容。

12.3.3 为 Pod 设置垂直扩缩容

在下载的 Git 代码中包含一个子目录 example,可以使用其中的 redis.yaml 来尝试使用 VPA 功能。

查看其中的 redis.yaml 文件,可以看到 VPA 的定义:

```
$ cat autoscaler/vertical-pod-autoscaler/examples/redis.yaml
apiVersion: autoscaling.k8s.io/v1beta2
kind: VerticalPodAutoscaler
metadata:
  name: redis-vpa
spec:
  targetRef:
    apiVersion: "extensions/v1beta1"
    kind:       Deployment
```

```
      name:        redis-master
```

该定义非常简短：对名称为 redis-master 的 Deployment 进行自动垂直扩缩容。在 https://git.io/fhAeY 页面可以找到该对象的完整定义，除了包括 targetRef，还包括如下内容。

（1）UpdatePolicy：用于指定监控资源需求时的操作策略。

- ◎ UpdateMode：默认值为 Auto。
- ◎ Off：仅监控资源状况并提出建议，不进行自动修改。
- ◎ Initial：在创建 Pod 时为 Pod 指派资源。
- ◎ Recreate：在创建 Pod 时为 Pod 指派资源，并且在 Pod 的生命周期中可以通过删除、重建 Pod，将其资源数量更新为 Pod 申请的数量。
- ◎ Auto：目前相当于 Recreate。

（2）ResourcePolicy：用于指定资源计算的策略，如果这一字段被省略，则将会为在 targetRef 中指定的控制器生成的所有 Pod 的容器进行资源测算并根据 UpdatePolicy 的定义进行更新。

- ◎ ContainerName：容器名称，如果为 "*"，则对所有没有设置资源策略的容器都生效。
- ◎ Mode：为 Auto 时，表示为指定容器启用 VPA；为 Off 时，表示关闭指定容器的 VPA。
- ◎ MinAllowed：最小允许的资源值。
- ◎ MaxAllowed：最大允许的资源值。

通过 kubectl 将测试文件提交到集群上运行：

```
$ kubectl apply -f autoscaler/vertical-pod-autoscaler/examples/redis.yaml
verticalpodautoscaler.autoscaling.k8s.io/redis-vpa created
deployment.apps/redis-master created
```

在创建结束之后，VPA 会监控资源状况，大约 5 分钟后重新获取 VPA 对象的内容：

```
$ kubectl describe vpa redis-vpa
Name:         redis-vpa
Namespace:    kube-system
......
Spec:
  Target Ref:
    API Version:  extensions/v1beta1
```

```
          Kind:            Deployment
          Name:            redis-master
    Status:
      Conditions:
        Status:            True
        Type:              RecommendationProvided
      Recommendation:
        Container Recommendations:
          Container Name:  master
......
          Target:
            Cpu:           25m
            Memory:        262144k
......
```

可以看到，在 VPA 对象中已经有了新的推荐设置。接下来查看 Redis 的 Pod 资源请求：

```
$ kubectl describe po redis-master-679887b5c9-nb72t
......
      Requests:
        cpu:               25m
        memory:            262144k
......
```

不难看出，Pod 的资源状况和 Deployment 中的原始定义已经不同，和 VPA 中的推荐数量一致。

12.3.4 注意事项

注意事项如下。

- VPA 对 Pod 的更新会造成 Pod 的重新创建和调度。
- 对于不受控制器支配的 Pod，VPA 仅能在其创建时提供支持。
- VPA 的准入控制器是一个 Webhook，可能会和其他同类 Webhook 存在冲突，从而导致无法正确执行。
- VPA 无法和使用 CPU、内存指标的 HPA 共用。
- VPA 能够识别多数内存不足的问题，但并非全部。
- 尚未在大规模集群上测试 VPA 的性能。
- 如果多个 VPA 对象都匹配同一个 Pod，则会造成不可预知的后果。

◎ VPA 目前不会修改 limits 字段的内容。

12.4 Kubernetes 的演进路线和开发模式

Kubernetes 将每个版本的待开发功能由 SIG-Release 小组进行文档管理和发布，网址为 https://github.com/kubernetes/sig-release，可以跟踪每个大版本的功能列表，如图 12.9 所示。

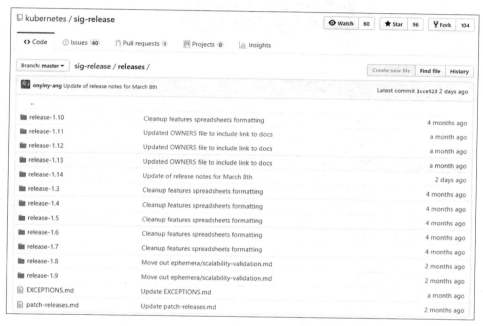

图 12.9 Kubernetes 各版本的 Release 详情页面

以 release-1.14 为例，从 release-1.14.md 文档中可以查看这个版本的详细信息，如图 12.10 所示。

版本发布的详细时间表如图 12.11 所示。

图 12.10　版本发布的信息页面

Timeline						
What	Who	Jan	Feb	Mar	WEEK	CI SIGNAL
Start of Release Cycle	Lead	Mon 7			week 1	master-blocking
Schedule finalized	Lead	Fri 11				
Team finalized	Lead	Fri 18			week 2	
Start Enhancements Tracking	Enhancements Lead	Tue 15				
1.14.0-alpha.1 released	Branch Manager	Tue 15				
Start Release Notes Draft	Release Notes Lead	Tue 22			week 3	
Begin Enhancements Freeze (EOD PST)	Enhancements Lead	Tue 29			week 4	master-blocking, master-upgrade
1.14.0-alpha.2 released	Branch Manager	Tue 29				
1.14.0-alpha.3 released	Branch Manager		Tue 12		week 6	
release-1.10 jobs removed	Test Infra Lead		Tue 12			
release-1.14 branch created	Branch Manager		Tue 19		week 7	

图 12.11　版本发布的时间表

第 12 章 Kubernetes 开发中的新功能

单击页面链接"Enhancements Tracking Sheet",可以查看该版本所包含的全部功能列表,按开发阶段分为 Alpha、Beta 和 Stable 三个类别,可以直观地看到各功能模块的实现阶段。每个功能都有 HTTP 链接,可以单击该链接跳转至 GitHub 中的相关 issue 页面,进一步查看该功能的详细信息。在 WIP Features by Release 表中还可以看到各功能特性在 Kubernetes 各版本中的开发过程,如图 12.12 所示。

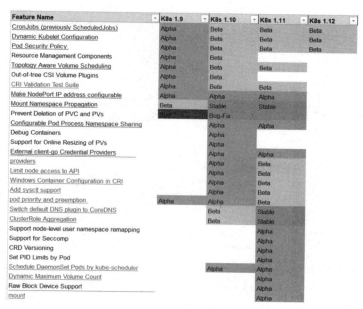

图 12.12 功能特性跟踪表格

每个版本中的每个 Feature 都由一个特别兴趣小组（Special Interest Group，SIG）负责开发和维护，各 SIG 小组的介绍可以在 https://github.com/kubernetes/community/blob/master/sig-list.md 找到，如图 12.13 所示。

SIGs and Working Groups

Most community activity is organized into Special Interest Groups (SIGs), time bounded Working Groups, and the community meeting.

SIGs follow these guidelines although each of these groups may operate a little differently depending on their needs and workflow.

Each group's material is in its subdirectory in this project.

When the need arises, a new SIG can be created

Master SIG List

Name	Label	Chairs	Contact	Meetings
API Machinery	api-machinery	* Daniel Smith, Google * David Eads, Red Hat	* Slack * Mailing List	* Regular SIG Meeting: Wednesdays at 11:00 PT (Pacific Time) (biweekly) * Kubebuilder and Controller Runtime Meeting: Wednesdays at 10:00 PT (Pacific Time) (monthly - second Wednesday every month)
Apps	apps	* Matt Farina, Samsung SDS * Adnan Abdulhussein, Bitnami * Kenneth Owens, Google	* Slack * Mailing List	* Regular SIG Meeting: Mondays at 9:00 PT (Pacific Time) (weekly)
Architecture	architecture	* Brian Grant, Google * Jaice Singer DuMars, Google * Matt Farina, Samsung SDS	* Slack * Mailing List	* Regular SIG Meeting: Thursdays at 19:00 UTC (biweekly)
Auth	auth	* Mike Danese, Google * Mo Khan, Red Hat * Tim Allclair, Google	* Slack * Mailing List	* Regular SIG Meeting: Wednesdays at 11:00 PT (Pacific Time) (biweekly)
Autoscaling	autoscaling	* Marcin Wielgus, Google	* Slack * Mailing List	* Regular SIG Meeting: Mondays at 14:00 UTC (biweekly/triweekly)

图 12.13　特别兴趣小组 SIG

目前已经成立的SIG小组有30个，涵盖了安全、自动扩缩容、大数据、AWS云、文档、网络、存储、调度、UI、Windows容器等方方面面，为完善Kubernetes的功能群策群力，共同开发。有兴趣、有能力的读者可以申请加入感兴趣的SIG小组，并可以通过Slack聊天频道与来自世界各地的开发组成员开展技术探讨和解决问题。同时，可以参加SIG小组的周例会，共同参与一个功能模块的开发工作。

反侵权盗版声明

电子工业出版社依法对本作品享有专有出版权。任何未经权利人书面许可，复制、销售或通过信息网络传播本作品的行为；歪曲、篡改、剽窃本作品的行为，均违反《中华人民共和国著作权法》，其行为人应承担相应的民事责任和行政责任，构成犯罪的，将被依法追究刑事责任。

为了维护市场秩序，保护权利人的合法权益，我社将依法查处和打击侵权盗版的单位和个人。欢迎社会各界人士积极举报侵权盗版行为，本社将奖励举报有功人员，并保证举报人的信息不被泄露。

举报电话：（010）88254396；（010）88258888
传　　真：（010）88254397
E-mail：dbqq@phei.com.cn
通信地址：北京市万寿路173信箱
　　　　　电子工业出版社总编办公室
邮　　编：100036